JACARANDA
SCIENCE QUEST 10
AUSTRALIAN CURRICULUM | FOURTH EDITION

T0362741

JACARANDA
SCIENCE QUEST 10

AUSTRALIAN CURRICULUM | FOURTH EDITION

GRAEME LOFTS

MERRIN J. EVERGREEN

CONTRIBUTING AUTHORS

Catherine Bellair | Jay Chew | Von Hayes | Adele Norton
Robert Stokes | Angela Stubbs | Ritu Tyagi

REVIEWED BY

Courtney Rubie, Wiradjuri woman

AUSTRALIAN
CURRICULUM
v9.0

A Wiley Brand

Fourth edition published 2023 by
John Wiley & Sons Australia, Ltd
Level 4, 600 Bourke Street, Melbourne, Vic 3000

First edition published 2011
Second edition published 2015
Third edition published 2018

Typeset in 10.5/13 pt TimesLT Std

ISBN: 978-1-394-15143-1

Front cover images: © STUART KINLOUGH / IKON
IMAGES / SCIENCE PHOTO LIBRARY

Illustrated by various artists, diacriTech and Wiley
Composition Services

Typeset in India by diacriTech

A catalogue record for this
book is available from the
National Library of Australia

NATIONAL LIBRARY OF AUSTRALIA

Printed in Singapore
M WEP324383 121124

The Publishers of this series acknowledge and pay their
respects to Aboriginal Peoples and Torres Strait Islander
Peoples as the traditional custodians of the land on which this
resource was produced.

This suite of resources may include references to (including
names, images, footage or voices of) people of Aboriginal
and/or Torres Strait Islander heritage who are deceased. These
images and references have been included to help Australian
students from all cultural backgrounds develop a better
understanding of Aboriginal and Torres Strait Islander Peoples'
history, culture and lived experience.

It is strongly recommended that teachers examine resources
on topics related to Aboriginal and/or Torres Strait Islander
Cultures and Peoples to assess their suitability for their own
specific class and school context. It is also recommended that
teachers know and follow the guidelines laid down by the
relevant educational authorities and local Elders or community
advisors regarding content about all First Nations Peoples.

All activities in this resource have been written with the safety
of both teacher and student in mind. Some, however, involve
physical activity or the use of equipment or tools. **All due care
should be taken when performing such activities**. To the
maximum extent permitted by law, the author and publisher
disclaim all responsibility and liability for any injury or loss
that may be sustained when completing activities described in
this resource.

The Publisher acknowledges ongoing discussions related to
gender-based population data. At the time of publishing, there
was insufficient data available to allow for the meaningful
analysis of trends and patterns to broaden our discussion of
demographics beyond male and female gender identification.

CONTENTS

■ SCIENCE INQUIRY

■ BIOLOGICAL SCIENCES

■ CHEMICAL SCIENCES

About this resource

NEW FOR

AUSTRALIAN CURRICULUM V9.0

JACARANDA
SCIENCE QUEST 10 AUSTRALIAN CURRICULUM
FOURTH EDITION

Developed by teachers for students

Tried, tested and trusted. Every lesson in the new *Jacaranda Science Quest* series has been carefully designed to support teachers and help students succeed by sparking their curiosity about the world around them.

Because both what and how students learn matter

Learning is personal

Whether students need a challenge or a helping hand, you'll find what you need to create engaging lessons.

Whether in class or at home, students can access carefully scaffolded and sequenced lessons to support in-depth Science Inquiry Skills development and step students through scientific inquiry with engaging interactive content and practical investigations. Automatically marked, differentiated question sets are all supported by detailed solutions — so students can get unstuck and progress!

Learning is effortful

Learning happens when students push themselves. With learnON, Australia's most powerful online learning platform, students can challenge themselves, build confidence and ultimately achieve success.

Learning is rewarding

Through real-time results data, students can track and monitor their own progress and easily identify areas of strength and weakness.

And for teachers, Learning Analytics provide valuable insights to support student growth and drive informed intervention strategies.

Learn online with Australia's most

- Trusted, curriculum-aligned content
- Engaging, rich multimedia
- All the teaching-support resources you need
- Deep insights into progress
- Immediate feedback for students
- Create custom assignments in just a few clicks.

Practical teaching advice and ideas for each lesson provided in teachON

Reading content and rich media including embedded videos and interactivities

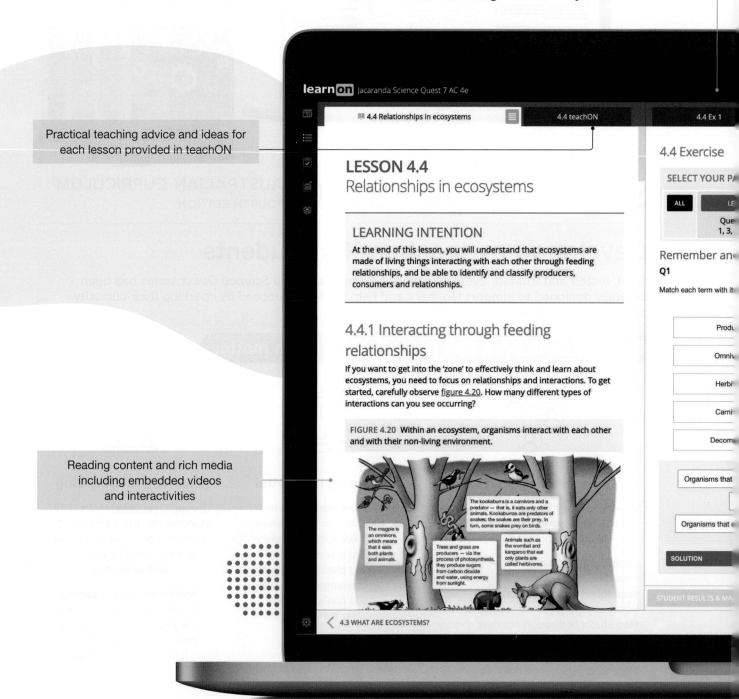

learnON Jacaranda Science Quest 7 AC 4e

4.4 Relationships in ecosystems   4.4 teachON   4.4 Ex 1

LESSON 4.4
Relationships in ecosystems

LEARNING INTENTION
At the end of this lesson, you will understand that ecosystems are made of living things interacting with each other through feeding relationships, and be able to identify and classify producers, consumers and relationships.

4.4.1 Interacting through feeding relationships
If you want to get into the 'zone' to effectively think and learn about ecosystems, you need to focus on relationships and interactions. To get started, carefully observe figure 4.20. How many different types of interactions can you see occurring?

FIGURE 4.20 Within an ecosystem, organisms interact with each other and with their non-living environment.

The kookaburra is a carnivore and a predator — that is, it eats only other animals. Kookaburras are predators of snakes; the snakes are their prey. In turn, some snakes prey on birds.

The magpie is an omnivore, which means that it eats both plants and animals.

Trees and grass are producers — via the process of photosynthesis, they produce sugars from carbon dioxide and water, using energy from sunlight.

Animals such as the wombat and kangaroo that eat only plants are called herbivores.

4.3 WHAT ARE ECOSYSTEMS?

4.4 Exercise

SELECT YOUR PA

ALL   LE

Que
1, 3,

Remember an
Q1

Match each term with it

Produ
Omniv
Herbi
Carni
Decom

Organisms that

Organisms that

SOLUTION

STUDENT RESULTS & MA

powerful learning tool, learnON

New! Quick Quiz questions
for skill acquisition

Differentiated question sets

Teacher and student views

Textbook questions

Fully worked solutions

eWorkbook

Practical investigation eLogbooks

Digital documents

Video eLessons

Interactivities

Extra teaching support resources

Interactive questions with immediate feedback

Get the most from your online resources

Online, these new editions are the complete package

Trusted Jacaranda theory, plus tools to support teaching and make learning more engaging, personalised and visible.

Interactive glossary terms help develop and support scientific literacy.

onResources link to targeted digital resources including video eLessons and weblinks.

Tables and images break down content, allowing students to understand complex concepts.

Brand new! Quick Quiz questions for skill acquisition in every lesson.

Three differentiated question sets, with immediate feedback in every lesson, enable students to challenge themselves at their own level.

Instant reports give students visibility into progress and performance.

Every question has immediate, corrective feedback to help students overcome misconceptions as they occur and get unstuck as they study independently — in class and at home.

Practical Investigation eLogbook

The **practical investigation eLogbook** ignites curiosity through science investigation work, with an extensive range of exciting and meaningful practical investigations. Aligned with the scientific method, students can develop rich science inquiry skills in conducting scientific investigations and communicating their findings, allowing them to truly think and act like scientists! The practical investigation eLogbook is supported with an unrivalled teacher and laboratory guide, which provides suggestions for differentiation and alteration, risk assessments, expected practical results and exemplary responses.

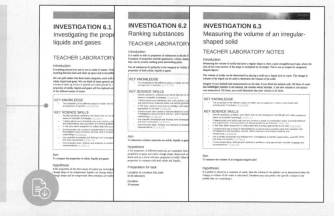

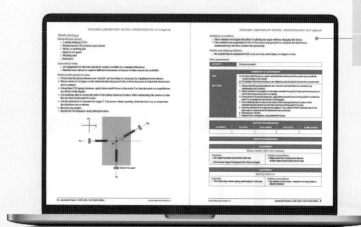

Enhanced practical investigation support includes practical investigation videos and an eLogbook with fully customisable practical investigations — including teacher advice and risk assessments.

eWorkbook

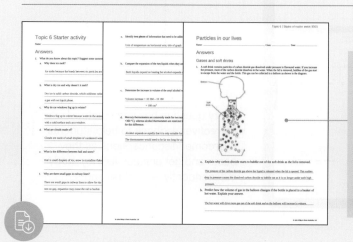

The **eWorkbook** is the perfect companion to the series, adding another layer of individualised learning opportunities for students, and catering for multiple entry and exit points in student learning. The eWorkbook also features fun and engaging activities for students of all abilities and offers a space for students to reflect on their own learning. The new eWorkbook and eWorkbook solutions are available as a downloadable PDF or a customisable Word document in learnON.

A wealth of teacher resources

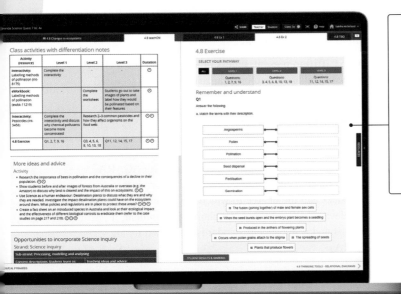

Enhanced teaching-support resources for every lesson, including:

- work programs and curriculum grids
- practical teaching advice
- three levels of differentiated teaching programs
- quarantined topic tests (with solutions)

Customise and assign

An inbuilt testmaker enables you to create custom assignments and tests from the complete bank of thousands of questions for immediate, spaced and mixed practice.

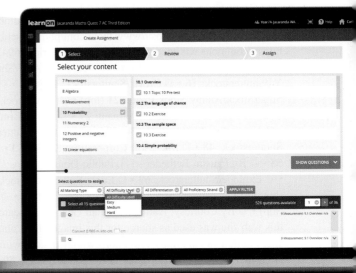

Reports and results

Data analytics and instant reports provide data-driven insights into progress and performance within each lesson and across the entire course.

Show students (and their parents or carers) their own assessment data in fine detail. You can filter their results to identify areas of strength and weakness.

Acknowledgements

The authors and publisher would like to thank the following copyright holders, organisations and individuals for their assistance and for permission to reproduce copyright material in this book.

Images

• © totojang1977/Shutterstock: **1** • © Dr. Gustave Stromberg/Bettmann/Getty Images: **2** • © 12_Tribes/Shutterstock: **5** • © By LSE library - https://www.flickr.com/photos/lselibrary/3833724834/in/set-72157623156680255/, No restrictions, https://commons.wikimedia.org/w/index.php?curid=9694262: **6** (top) • © Getty Images: **6** (bottom) • © Monkey Business Images/Shutterstock: **7, 162** (top) • © PRUSSIA ART/Shutterstock: **11** • © Plateresca/Shutterstock: **15** (bottom) • © Feng Yu/Shutterstock: **16** • © Crevis/Shutterstock: **18** (top) • © DTKUTOO/Shutterstock: **18** (bottom) • © RGB Ventures/SuperStock/Alamy Stock Photo: **21** (top) • © View Stock/View Stock/Getty Images: **21** (bottom) • © retro67/Shutterstock: **26** • © Pascale Warnant: **27** • © Derek Brumby/Shutterstock: **29** • © Matej Kastelic/Shutterstock: **32** (top) • © NatalieIme/Shutterstock: **34, 384** • © photong/Shutterstock: **35, 379** • © Toa55/Shutterstock: **47** • © Tupungato/Shutterstock: **48** • © Microsoft Corporation: **51, 52, 53** • © PixieMe/Shutterstock: **57** • © Vernier Software & Technology: **58** • © By Faustin Betbeder - [1], Public Domain, https://commons.wikimedia.org/w/index.php?curid=1490883: **66** • © Explode/Shutterstock: **69** • © LuckyBall/Shutterstock: **70** • © Ody_Stocker/Shutterstock: **75** (bottom), **107** (top), **107** (bottom) • © Andrew Syred/Science Source: **75** (top) • © Dr. R E Rowland: **75** (bottom) • © Preeda340/Shutterstock: **77** (top) • © By Courtesy: National Human Genome Research Institute - Talking Glossary of Genetics. The pdf version from this web site was used as source for this image file to obtain a better resolution than in the image embedded in the web site, Public Domain, https://commons.wikimedia.org/w/index.php?curid=1416628: **77** (bottom) • © EPA/TONI ALBIR/AAP Image: **80** (top) • © Kateryna Kon/Shutterstock: **80** (bottom), **215** • © Biophoto Associates/Science Source, DEPT. OF CLINICAL CYTOGENETICS, ADDENBROOKES HOSPITAL/Science Photo Library: **83** • © DEPT. OF CLINICAL CYTOGENETICS, ADDENBROOKES HOSPITAL/Science Photo Library, Public Domain: **84** • © By copied from http://www.pbs.org/wgbh/nova/photo51/images/befo-miescher.jpg, Public Domain, https://commons.wikimedia.org/w/index.php?curid=789048: **85** • © Science Photo Library/Alamy Stock Photo: **86** (left), **347** (left), **359** (top) • © World History Archive/Alamy Stock Photo: **86** (top), **179** • © SPL/Science Source: **86** (bottom) • © Soleil Nordic/Shutterstock: **87** (top), **96** • © GOLFX/iStock/Getty Images: **87** (bottom) • © Science History Images/Alamy Stock Photo: **87** (top), **117** (top) • © Wavebreakmedia/iStock/Getty Images, Vecton/Shutterstock: **88** • © Antonov Maxim/Shutterstock: **95** • © Achiichiii/Shutterstock: **104** (top) • © Designua/Shutterstock: **104** (bottom), **112, 443, 463** (top), **577** • © JOSE CALVO/Science Photo Library: **105** • © Emre Terim/Shutterstock: **106, 251** (top), **468** (top) • © danylyukk1/Shutterstock: **109, 467** • © Olando/Adobe Stock Photos: **105** • © Dr M.A. Ansary/Science Source: **115** • © wonderisland/Shutterstock: **116** • © Clker free Vecor images/Pixabay: **117** (middle) • © Prawny/Pixabay: **117** (middle) • © belander/Shutterstock: **117** (middle) • © barbaliss/Shutterstock: **117** (bottom) • © MSSA/Shutterstock: **117** (top) • © Sebastian Kaulitzki/Shutterstock: **118** (bottom), **118** (top) • © Zuzanae/Shutterstock: **119** (top), **121** • © kavzov/Shutterstock: **119** (bottom) • © VOLKER STEGER/Science Photo Library: **123** • Judith Kinnear: **124** • © Martin Shields/Alamy Stock Photo: **126** • © ellepigrafica/Shutterstock: **128** • © Ryan McVay/Stockbyte/Getty Images: **131** • © Vladimir Gjorgiev/Shutterstock, IKO-studio/Shutterstock: **133** (bottom) • © Dan Kosmayer/Shutterstock, Carl Stewart/Shutterstock: **133** (top) • © Champion studio/Shutterstock: **134** (bottom) • © Chad Baker/Jason Reed/Ryan McVay/Photodisc/Getty Images: **134** (top) • © By Hugo Iltis - Wellcome Library, London, CC BY 4.0, https://commons.wikimedia.org/w/index.php?curid=33070385: **137, 226** • © Voyagerix/Shutterstock, mentatdgt/Shutterstock, Grigvovan/Shutterstock: **141**

• © 'Queen Victoria & Royal Family' by M.W. Ridley. Illustration from Frank Leslie's illustrated newspaper, v. 44, no. 1137 (July 14 1877). - This image is available from the United States Library of Congress's Prints and Photographs divisionunder the digital ID pga.02489.This tag does not indicate the copyright status of the attached work. A normal copyright tag is still required. See Commons:Licensing for more information.. Licensed under Public Domain via Commons - https://wiki2.org/en/File:Queen_Victoria_%26_Royal_Family.jpg#/media/File:Queen_Victoria_%26_Royal_Family.jpg: **157** • © BlueRingMedia/Shutterstock: **158** • © Source: Adpated from Brezina et al (2013). Preimplantation Genetic Testing. Curr Obstet Gynecol Rep 2, 211–217 (2013): **159** • © S.Vidal/Shutterstock: **161** • © science photo/Shutterstock: **163** • © extender_01/Shutterstock: **170** (top) • © Dr. Brad Mogen, VISUALS UNLIMITED/Science Photo Library: **170** (bottom) • © Meletios Verras/Shutterstock: **174** (top) • © Bartosz Budrewicz/Shutterstock: **174** (bottom) • © E. Weil, C. Palmer: **177** • © CHAD HIPOLITO/Alamy Stock Photo, Chris Howes/Wild Places Photography/Alamy Stock Photo: **400** • © Emmanuel Buschiazzo: **178** (left) • © Broadbelt/Shutterstock: **178** (center) • © John A. Anderson/Shutterstock: **178** (top) • © of-fr/Shutterstock: **180** • © Michael Kuiper: **181** • © University of Illinois: **183** • © Brigida Soriano/Shutterstock: **203** • © Source: Weiss et al., The physiology and habitat of the last universal common ancestor Nature Microbiology, Springer Nature.: **207** • © By Hendrik Hollander - University of Amsterdam, Public Domain, https://commons.wikimedia.org/w/index.php?curid=2612040: **208** • © Wilm Ihlenfeld/Shutterstock: **211** • © davemhuntphotography/Shutterstock: **212** • © By Bufo_periglenes1.jpg: Charles H. Smith, vergrößert von Aglarechderivative work: Purpy Pupple (talk) - Bufo_periglenes1.jpg from U.S. Fish and Wildlife Service, Public Domain, https://commons.wikimedia.org/w/index.php?curid=40766836: **213** (top) • © Bruce Ellis/Shutterstock: **213** (bottom) • © Rich Carey/Shutterstock: **214** (top) • © Mares Lucian/Shutterstock: **214** (left) • © FiledIMAGE/Shutterstock: **214** (top) • © VectorMine/Adobe Stock Photos: **214** (bottom) • © Copyright Biosphoto/AUSCAPE All rights reserved/Auscape International Pty Ltd: **219** • © Erik Lam/Shutterstock: **220** (top) • © Prostock-studio/Shutterstock: **220** (bottom) • © Fancy Tapis/Shutterstock: **222** • © Source: Species Diversification, Bioninja: **223** (top) • © sunsetman/Shutterstock: **223** (bottom) • © Source: bioninja.com.au: **224** • © Flaxphotos/Shutterstock: **225** (top) • © Nasky/Shutterstock: **225** (bottom), **478** • © Potapov Alexander/Shutterstock: **227** • © Source:Bioninja: **228** • © Patsy A. Jacks/Shutterstock, Pexels/Pixabay, Perth Zoo, David Clode/Unsplash.com: **232** • © Kristian Bell/Shutterstock, hallam creations/Shutterstock, Binturong-tonoscarpe/Shutterstock, Filipe Frazao/Shutterstock: **233** (top) • © Vladimirkarp/Shutterstock, Igor Cheri/Shutterstock, photomaster/Shutterstock: **233** (middle) • © Magicleaf/Shutterstock, AnnstasAg/Shutterstock, ONYXprj/Shutterstock: **233** (bottom) • © Howard Sandler/Shutterstock: **235** (bottom) • © Arto Hakola/Shutterstock: **235** (top) • © Kristina Vackova/Shutterstock, frantisekhojdysz/Shutterstock: **236** (top) • © Julian W/Shutterstock, Karel Gallas/Shutterstock: **236** (middle) • © William Cushman/Shutterstock, Elena_sg80/Shutterstock: **236** (middle) • © Miroslav Hlavko/Shutterstock, Tim Zurowski/Shutterstock: **236** (bottom) • © Source: Encyclopædia Britannica, Inc: **238** • © Sabena Jane Blackbird/Alamy Stock Photo: **241** (top) • © Wiley art: **241** (top), **354** (bottom) • © Flak Kienas, 2009/Shutterstock: **241** (middle) • © Zadiraka Evgenii/Shutterstock: **241** (middle), **242** (bottom) • © Claude Huot/Shutterstock: **241** (middle) • © Serhej Calka/Getty Images: **241** (bottom) • © fotosub/Shutterstock: **242** (top) • © Breck P. Kent/Shutterstock: **242** (bottom) • © Lourens Smak/Alamy Stock Photo: **246** • © DreamBig/DreamBig/Shutterstock, Moehring/Shutterstock: **248** (top) • © Aldona Griskeviciene/Shutterstock: **248** (bottom) • © Roger Ressmeyer/Corbis/VCG/Getty Images: **251** (bottom) • © Vlad G/Shutterstock: **254** • © MilousSK/Shutterstock: **257** (top) • © Julien Tromeur/Shutterstock: **257** (bottom) • © LAGUNA DESIGN/Science Photo Library: **269** • © Ventin/Shutterstock: **270** • © Olga Popova/Shutterstock: **276** • © Source: Based on Isotones – Nuclides, 2020. Nuclear Power.: **278** • © John Wellings/Alamy Stock Photo: **279** • © images and videos/Shutterstock: **281** • © No on page credit required.: **282** (bottom) • © Olivier Le Queinec/Shutterstock: **282** (top) • © Source: Flatworld: **286** • © Source: Cartlidge, E. (2018). The battle behind the periodic table's latest additions. Nature, 558(7709), 175–176.: **287** • © Inna Bigun/Shutterstock, Science Photo Library/Alamy Stock Photo: **298** • © Belozersky/Shutterstock: **299** (top) • © gameover/Alamy Stock Photo: **299** (middle) • © Imfoto/Shutterstock: **299** (bottom) • © magnetix/Shutterstock: **300** (top), **316** • © pOrbital.com/Shutterstock: **300** (bottom) • © trgrowth/Shutterstock: **305** • © Vasilyev/Shutterstock: **310** (bottom) • © molekuul_be/Shutterstock: **310** (top) • © Michael Coddington/Shutterstock: **310** (bottom) • © Steve Cymro/Shutterstock: **313**, **325** (bottom), **365** (top) • © udaix/Shutterstock: **315** • © Source: Jkwchui, Creative Commons.: **317** (top) • © Steffen Foerster/Shutterstock: **317** (bottom) • © Peyker/Shutterstock: **318** • © Inna Bigun/Shutterstock, **463** (bottom) • © Turtle Rock Scientific/Science Source: **321** (bottom) • © Iakov Filimonov/Shutterstock: **336** (top) • © jarous/Shutterstock: **336** (bottom) • © DAVID M. PHILLIPS/Science Photo Library (SPL): **345** • © Dima Zel/Shutterstock: **346** • © MARTYN F. CHILLMAID/SCIENCE PHOTO LIBRARY/Alamy Stock Photo: **347** (left), **360** • © ANDREW LAMBERT PHOTOGRAPHY/SCIENCE PHOTO LIBRARY/Alamy Stock Photo: **347** (right), **368** • © Viktorija Reuta/Shutterstock: **354** (top) • © Paula Cobleigh/Shutterstock: **359** (bottom) • © Scharfsinn/Alamy Stock Photo: **361** • © Pierluigi.Palazzi/Shutterstock: **364** • © scanrail/Getty Images: **365** (bottom) • © sciencephotos/Alamy Stock Photo: **371** • © Ihor Matsiievskyi/Shutterstock: **374** • © s-ts/Shutterstock: **376** • © Lindsey Moore/Shutterstock: **381** • © Rovenko/Shutterstock: **385** (top) • © Mia Garrett/Shutterstock: **385** (bottom) • © BigBearCamera/Shutterstock: **386** • © Albert Russ/Shutterstock: **387** • © rktz/Shutterstock: **390** (top) • © Sansanorth/Shutterstock: **390** (bottom), **391** (bottom) • Adam J/Shutterstock: **392**

Text

• © Source: Jane Tiller & Paul Lacaze, Australians need more protection against genetic discrimination: health experts, 2021. The Conversation.: **183** • © Source: UK white goods Retrieved from: www.ukwhitegoods.co.uk/appliance-industry-news/226-washing-machine-news/1565-washclotheswiththinair: **521** • © Source: Associated Press: **560**

Every effort has been made to trace the ownership of copyright material. Information that will enable the publisher to rectify any error or omission in subsequent reprints will be welcome. In such cases, please contact the Permissions Section of John Wiley & Sons Australia, Ltd.

1 Investigating science

LESSON SEQUENCE

SCIENCE INQUIRY AND INVESTIGATIONS

Science inquiry is a central component of Science curriculum. Investigations, supported by a **Practical investigation eLogbook** and **teacher-led videos**, are included in this topic to provide opportunities to build Science inquiry skills through undertaking investigations and communicating findings.

LESSON
1.1 Overview

1.1.1 Introduction

Are you ethical? Does it matter? What influences your opinions, values and beliefs? How do your attitudes affect when, how and why you learn? How and why do you think the way that you do? Is it ever worth changing your mind? Why doesn't everyone think the same way as you do? Who are you and who are you yet to become?

These questions are all important to consider in science, not just when considering the content, but in the way you conduct investigations. The way you formulate investigation questions, design experiments, conduct investigations and form conclusions is often influenced by opinions, values and beliefs. It is important to ensure that these are considered to ensure accurate and valid data is obtained and findings are free of bias.

FIGURE 1.1 Einstein's theories were used to develop nuclear weapons — something he ethically opposed.

 Resources

Video eLesson Meet Professor Veena Sahajwalla (eles-1071)

Watch this video to learn about Professor Veena Sahajwalla, a leading expert in the field of recycling science, and founding director of the Centre for Sustainable Materials Research & Technology at UNSW. She is producing a new generation of green materials, products and resources made entirely, or primarily, from waste.

1.1.2 Think about science

1. Are there any examples of scientists altering their results in a scientific investigation?
2. What recent discoveries have allowed for more accurate results in investigations to be obtained?
3. Can unethical behaviour ever be justified?
4. How can simulations and models be used in investigations?
5. Is all secondary information on the internet accurate?
6. Who owns genetic material?
7. Is it your DNA or your environment that determines your beliefs and opinions?

1.1.3 Science inquiry

What makes good news?

We live in an age of information. In fact, you are continually being bombarded by it! How can you begin to make sense of all the information you receive? How can you better evaluate it? How can you incorporate this new information into what you already know to develop a better understanding of the world in which you live?

To effectively evaluate articles in the media and online, you need to be able to determine the authenticity and accuracy of the information presented. This includes being able to identify:
- what the facts are
- if the article is from a reliable journalist or media source
- how the information is presented
- if the author is an expert in the field or have they worked with experts in order to write the article?

Read the article headlines and opening paragraphs provided, and then answer the following questions.
1. For each article, consider the following.
    a. What do you think the article is about?
    b. What type of article do you think it is? Is it:

        i. sensational   ii. informative   iii. entertaining   iv. thought provoking?

    c. Use the internet to find further content from each article and find out more about the story by using search parameters such as the article headline and newspaper source.
    d. Analyse the language and style of writing used in the article. What kind of audience do you think this article was written for? Is the author presenting a biased view or providing statistics and graphs without including appropriate references?
    e. Do you think you need to be a scientist to understand what the author is writing about?
    f. Did the article headline grab your attention and make you want to read more? If not, how could it be improved?
    g. Research one of the events or issues mentioned and write your own article about it. Collate the class articles into a journal or newspaper.
2. The first of these articles was written more than ten years ago.
    a. What types of environmental and scientific problems did people face at the time?
    b. Are they similar or different to those we face today?
    c. Use the internet to find out more about the following issues mentioned in the articles:
        i. climate change
        ii. loss of biodiversity
        iii. nuclear power
        iv. destructive artificial intelligence
        v. conflict and war
        vi. poverty
        vii. Covid-19 pandemic
    d. How do you think people's opinions of the above issues have changed in the past ten years? Justify your answer.

Scientific consensus: Earth's climate is warming

It is unequivocal that the increase of CO_2, methane, and nitrous oxide in the atmosphere ... is the result of human activities ... Since systematic scientific assessments began in the 1970s, the influence of human activity on the warming of the climate system has evolved from theory to established fact.

NASA

Australia among the world's worst on biodiversity conservation

Australia is among the top seven countries worldwide responsible for 60% of the world's biodiversity loss between 1996 and 2008, according to a study published ... in the journal, *Nature*.

The Conversation.com

Fukushima accident

An earthquake and tsunami led to power loss in the Fukushima Daiichi nuclear plant. Without power, the cooling systems failed in three reactors, and their cores subsequently overheated. This led to a partial meltdown of the fuel rods, a fire in the storage reactor, explosions in the outer containment buildings and the release of radiation into the air and ocean.

Encyclopedia Britannica

Elon Musk: 'Mark my words — A.I. is far more dangerous than nukes'

Elon Musk has doubled down on his dire warnings about … artificial intelligence. … "And mark my words, AI is far more dangerous than nukes". … "It scares the hell out of me," … "It's capable of vastly more than almost anyone knows, and the rate of improvement is exponential."

CNBC.com

Russia gathering its forces to storm settlements near Sievierodonetsk, says Ukraine — as it happened

EU foreign policy chief, Josep Borrell, has said. 'We call on Russia to deblockade the ports… It is inconceivable, one cannot imagine that millions of tonnes of wheat remain blocked in Ukraine while in the rest of the world, people are suffering hunger,' Reuters reports… 'This is a real war crime…'

The Guardian.com

Over a quarter of a billion more people could fall into extreme levels of poverty and suffering this year

Over a quarter of a billion more people could crash into extreme levels of poverty in 2022 because of COVID-19, rising global inequality and the shock of food price rises supercharged by the war in Ukraine, reveals a new Oxfam brief today.

https://www.oxfam.org/en/press-releases

Australia's biosecurity emergency pandemic measures to end

As Australia moves towards living with COVID-19, the following emergency measures will also lapse:

- negative pre-departure tests for travellers entering Australia.
- restrictions on the entry of cruise vessels into and within Australian territory.
  International travellers … will still be required to provide proof of double vaccination against COVID-19.

https://www.health.gov.au/

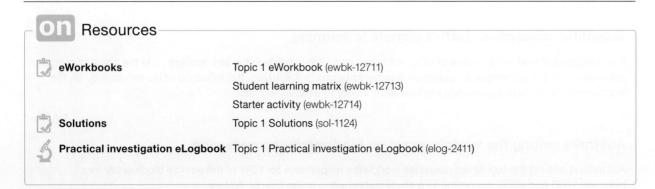

on Resources

eWorkbooks	Topic 1 eWorkbook (ewbk-12711)	
	Student learning matrix (ewbk-12713)	
	Starter activity (ewbk-12714)	
Solutions	Topic 1 Solutions (sol-1124)	
Practical investigation eLogbook	Topic 1 Practical investigation eLogbook (elog-2411)	

LESSON
1.2 Thinking flexibly

LEARNING INTENTION

At the end of this lesson you will be able to explain and provide examples of thinking (both flexibly and critically) in science, and describe the influence of attitudes and behaviours to the refinement of theories over time.

1.2.1 Thinking with an open mind

'One thing only I know, and that is that I know nothing.' This statement is often linked to a Greek philosopher called Socrates (470–399 BCE), who had a major impact on Western thinking and philosophy. This statement, however, also goes against what is commonly thought about science and scientists. Some consider that science will always have the answers and that scientists know all. Not only is such a belief untrue, it is also potentially dangerous. Thinking flexibly and with an open mind are better traits for a scientist to possess. The history of science and philosophy is littered with theories that at one time were considered to be answers, but were later discarded.

In science, it is important that you are not just able to think critically, but also able to think flexibly.

Being able to think critically involves being able to carefully consider information provided to you to help guide decision-making. Thinking flexibly involves being able to consider different points of view.

Science doesn't just involve conducting practical investigations and recording and analysing results. It is important to consider ethics, attitudes, opinions, values and beliefs. Being able to think flexibly and appreciate these different viewpoints is important because it allows you to think in a critical way and strengthen the conclusions you are able to draw.

Being able to think both critically and flexibly has allowed for some of the most exciting and amazing scientific discoveries.

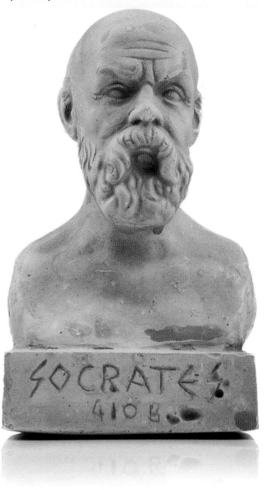

FIGURE 1.2 Socrates, a famous Greek philosopher

1.2.2 Making judgements

Opinions, values and attitudes involve making **judgements** about the desirability of something, whereas beliefs usually do not. Your **values** may involve making personal judgements and represent a deeper commitment than an attitude would. Values also act as standards in your decision making. **Opinions** can be expressed as a point of view that is based on known facts or available information. Although **beliefs** reflect what we think and know about the world, they do not have to be based on fact.

judgements opinions formed after considering available information

values a deep commitment to a particular issue and serve as standards for decision making

opinions personal views or judgements about something

beliefs feelings or mental acceptance that things are true or real

While you may see the world through the lenses in your eyes, your perceptions are filtered through your beliefs and assumptions.

Other lenses

Your family, cultural and social environments also play a part in how you perceive the world. Your attitudes, values and beliefs may be quite different due to the influence that these factors have on how you shape and organise your understanding of what happens around you. The time that you live in is also important. Imagine the effect this has had on scientists throughout different times in history.

Our attitudes can also be expressed by the distance that we place between ourselves and others. **Proximate rules** determine the physical distance (in zones) that is comfortable between people, depending on their relationships (see figure 1.3).

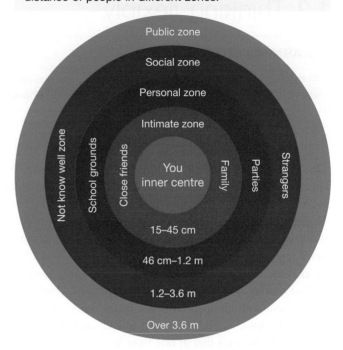

FIGURE 1.3 Proximate rules and the comfortable distance of people in different zones.

1.2.3 Why do we need to consider attitudes and opinions in science?

What is now considered science may also be described as a branch of philosophy. This branch is involved in trying to explain our observations from both inside and outside our bodies. There are many different ways to analyse the tree of knowledge that we call science. Three of these ways are:

- **inductionism** — suggests that scientific knowledge is proven knowledge and that large amounts of first-hand data, unbiased observations and a structured method can lead to theories that can become universal laws
- **falsification** — the philosopher Karl Popper (1902–1994) believed that no theory was ever proven beyond doubt. He believed that theories were just educated guesses and if they failed rigorous testing they should be thrown out.
- **paradigms** — or ways of thinking. Thomas Kuhn (1922–1996) saw science as being generated by basic theories or groups of ideas that are followed and defended by scientists. These paradigms are accepted even when data suggest that they may not be true. Only when the evidence against the theory becomes too great does the paradigm change, to be replaced by another, until it too is replaced.

FIGURE 1.4 Karl Popper

FIGURE 1.5 Thomas Kuhn

proximate rules rules that govern the physical distance that is comfortable between people

inductionism a theory stating that with enough evidence, scientific theories can become universal laws

falsification a credible hypothesis or theory should be able to be tested to be potentially disproved or contradicted by evidence

paradigms generally accepted perspectives, ideas or theories at a particular time

INVESTIGATION 1.1

What are my values and beliefs?

Aim

To reflect on and make decisions regarding a variety of claims

Method

1. On your own, score each of the following statements on a scale of 0 to 4 where 0 = strongly disagree and 4 = strongly agree.

0	1	2	3	4
Strongly disagree	Disagree	Neutral	Agree	Strongly agree

   a. Books are better than movies.

   b. Fiction is more interesting than non-fiction.

   c. Only wealthy students should get an education.

   d. Science classes should include science fiction stories.

   e. If something is too hard, it's not worth trying.

   f. Students who get below 50 per cent on a test do not deserve an education.

   g. At 15 years of age you have a sense of who you are.

   h. You are weak if you feel the need to belong.

   i. If you failed before, don't bother trying again.

   j. You can have ownership without possession.

2. For three of the statements in step 1, share your opinions by being involved in constructing a class 'opinionogram'.

   a. Divide the classroom into five zones, and assign a score of 0 to 4 to each zone.

   b. Each student now stands in the zone that indicates their score for the first statement.

   c. Discuss the reasons for your opinion with the students in your zone.

   d. Suggest questions that could be used to probe students in different opinion zones.

   e. With students in other zones, discuss their views and share with them the reasons for your opinion.

   f. Reflect on what you have heard from others. Decide if you want to change positions and, if so, change. Give a reason for why you are changing.

   g. Repeat steps (b)–(f) for two other statements.

Results

1. Outline your score for each of the given statements.

2. Construct graphs showing the opinion scales for each statement and comment on any observed patterns.

Discussion

1. a. Reflect on what you have learned about the opinions and perspectives of others.

   b. In your teams, discuss any insightful comments, ideas or opinions.

   c. Suggest questions that could be used to more closely probe reasons for your classmates' opinions. Share these probing questions with your class.

2. Suggest how you have demonstrated resilience, reflectiveness, responsibility and resourcefulness during this activity. Comment on things that you may change if you were to do the activity again.

Conclusion

Summarise your findings about opinions on different values and beliefs.

1.2.4 Communicating attitudes through paralanguage

Attitudes can be communicated both verbally and non-verbally. Rather than just using words, we also use our posture, use of space, gestures, facial expressions, and the tones, inflections, volume and pauses in our speech. The term **paralanguage** is used to describe this non-verbal communication, such as the way that you say something, rather than what you say.

The language of understanding

How can your use of language and non-verbal communication give the right impression about who you are? How do your attitudes affect when, how and why you learn? How can you make your learning and understanding more effective? Part of communication is using appropriate language and understanding what this language means, as shown in figure 1.6.

paralanguage non-verbal parts of communication; for example, the *way* something is said, rather than the words that are used

FIGURE 1.6 The language of understanding

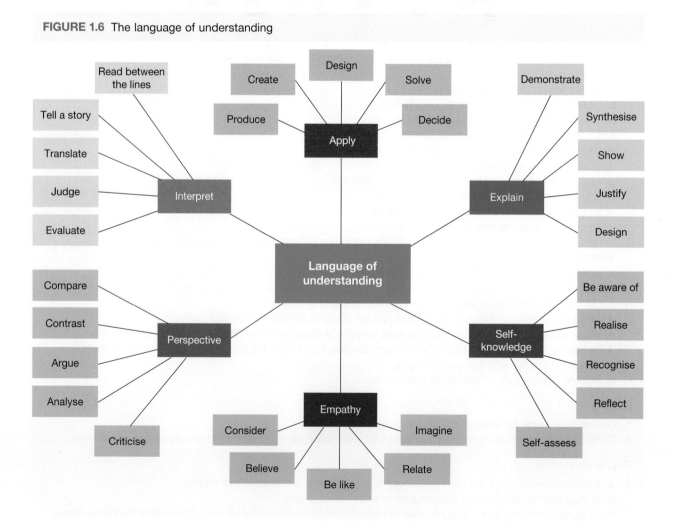

Thinking tools

You can organise and communicate your thoughts in many different ways. These thinking tools help you to improve your communication of ideas, organise thoughts, reflect attitudes and display ideas. Some examples include:

- mind maps
- priority grids
- SWOT analysis
- matrices
- cycle maps
- fishbone diagrams.

These thinking tools will be further explored in lessons through Science Quest 10.

 Resources

 eWorkbook Thinking tools and the language of learning (ewbk-12716)

1.2.5 Refinement of theories

What is a theory?

A **theory** is a well-supported explanation of a phenomena, based on investigations, research and observations. In science, theories are often tested using the scientific method.

It is important to note that theories are referred to as well-supported explanations, rather than proven explanations. Theories are formulated based on information available at the time, and provide an explanation of a phenomena.

Theories can change overnight, or take a very long time to change. Theories that were once popular and well accepted may be discarded when too much evidence builds up against them. They are replaced by a theory that better fits the observations.

As our ability to conduct investigations improves over time, our ability to interpret results and make observations is also changing. New technology has allowed us a greater understanding of science than ever before, thereby allowing us to adapt and adjust previously well-supported theories.

However, in five, 10 or 100 years time, new observations may replace the theories that are currently supported today.

> **theory** a well-supported explanation of a phenomenon based on facts that have been obtained through investigations, research and observations

---

SCIENCE AS A HUMAN ENDEAVOUR: Refining theories in astrophysics

Until recently, it was accepted that about 23 per cent of our universe was made up of stuff we can't even see. This invisible dark matter is said to lurk in the hearts of galaxies and keep the outermost stars from flying off into the void. It is thought to be responsible for the appearance of clusters of galaxies. But what if this isn't the case?

Newton's theories are again being questioned. A growing number of astrophysicists support a controversial new theory called Modified Newtonian Dynamics (MoND), which has led to some surprising predictions about the evolution of the universe. Previously, galaxies were thought to have formed from relatively dense pockets of matter with dark matter holding them together. The laws of the MoND theory suggest a different picture is possible. If correct, this theory could overthrow the established view of gravity and dark matter. These two areas underpin almost everything known about astronomy. MoND may also lead to a rethinking of Einstein's theory of relativity.

TABLE 1.1 The changing ideas of the universe

Year	Changing idea
1933	Fritz Zwicky coins the term 'dark matter' to describe unseen mass or 'gravitational glue' in galaxy clusters.
1978	Astronomers show that many galaxies are spinning too quickly to hold themselves together unless they are full of dark matter.
1983	Mordehai Milgrom publishes a modified gravity theory called MoND. It explains why galaxies don't fly apart without using dark matter, but remains at odds with Einstein's relativity.
1990s	Studies of galaxies and galaxy clusters show that their gravity bends light more strongly than is expected without dark matter. MoND researchers start devising improved theories to explain extra light bending.
1994	Jacob Bekenstein and Roger Sanders prove that any theory that resolves the light-bending issue and meshes MoND with relativity must involve at least three mathematical fields.
2000	New data on the cosmic microwave background reinforce the standard, dark matter picture of the universe.
2004	Jacob Bekenstein devises a version of MoND that is consistent with relativity.
2005	Constantinos Skordis and others show that relativistic MoND provides a good fit to the microwave background data.

 Resources

▶ **Video eLesson** Dark matter labs (eles-2688)

CASE STUDY: Newton (1643–1727) and Descartes (1596–1650)

Newton's theory of universal gravitation stated that everything was attracted to everything else. This would mean that the Sun's gravity would keep the Earth and other planets in orbit. Descartes, however, did not think that force could be transmitted through empty space and suggested that the Earth was in some kind of whirlpool that revolved around the Sun.

FIGURE 1.7 Newton and Descartes differed in their opinions on gravity and forces.

Another difference between these theories was their predictions about the shape of the Earth. Newton's theory suggested that the Earth would be flatter at the poles and fatter at the equator due to the effects of gravitational force. Descartes' theory suggested the opposite. In 1737, two expeditions left France to travel around the world and measure the curvature of the Earth to resolve the dispute. Upon their return, both expeditions provided measurements that supported Newton's prediction.

How do refinements of existing theories come about?

Theories are refined and adapted for a variety of reasons. This can be through:

- observations made in carefully planned laboratory-based or field-based experiments
- critical reinterpretation of previously accepted facts, producing a new framework
- new technologies that allow for changes to understanding and more depth of knowledge.

SCIENCE AS A HUMAN ENDEAVOUR: Einstein's impact

Albert Einstein's (1879–1955) contribution to modern physics is unique. Over a hundred years ago, when he was only 26 years old, he published a series of original theories that changed the way we see the universe. He published revolutionary ideas on the photoelectric effect, special relativity and Brownian motion.

In his study of Brownian motion, Einstein confirmed the existence of atoms. While other scientists were debating whether light was a particle or a wave, his theory of the photoelectric effect, which described the interaction of light and matter, suggested it was both.

His theory of special relativity examined the nature of space and time. The relativity theory is called 'special' because it doesn't include the effects of gravity. He showed how space and time could mix and match depending on your point of view. Special relativity stated that an atomic clock travelling at high speed in a jet plane ticks more slowly than a stationary clock. His theory also explained how an object could shrink in size and gain mass at the same time. It was this theory that led to the famous equation $E = mc^2$, which links energy and matter. This led to the realisation that huge amounts of energy are released in nuclear reactions. While this has provided some benefits, it has also led to detrimental applications such as the production and use of nuclear weapons.

FIGURE 1.8 A mushroom-shaped cloud is often associated with a nuclear explosion.

FIGURE 1.9 Einstein's 1939 letter to President Roosevelt

Albert Einstein
Old Grove Rd.
Nassau Point
Peconic, Long Island
August 2nd, 1939

F. D. Roosevelt,
President of the United States,
White House
Washington, D. C.

Sir:

Some recent work by E. Fermi and L. Szilard, which has been communicated to me in manuscript, leads me to expect that the element uranium may be turned into a new and important source of energy in the immediate future. Certain aspects of the situation which has arisen seem to call for watchfulness and, if necessary, quick action on the part of the Administration. I believe therefore that it is my duty to bring to your attention the following facts and recommendations.

In the course of the last four months it has been made probable — through the work of Joliot in France as well as Fermi and Szilard in America — that it may become possible to set up a nuclear chain reaction in a large mass of uranium, by which vast amounts of power and large quantities of new radium-like elements would be generated. Now it appears almost certain that this could be achieved in the immediate future.

This new phenomenon would also lead to the construction of bombs, and it is conceivable — though much less certain — that extremely powerful bombs of a new type may thus be constructed. A single bomb of this type, carried by boat and exploded in a port, might very well destroy the whole port together with some of the surrounding territory. However, such bombs might very well prove to be too heavy for transportation by air.

ACTIVITY: Einstein's letter to President Roosevelt

a. Find out what prompted Einstein to write the letter to President Roosevelt.
b. What were Einstein's thoughts on this application of theories that he had been involved in?
c. If you were in Einstein's situation, suggest how you would feel and what you would do. Present your thoughts in a letter that you would write to a close friend.

1.2.6 Shifting tides

What are the laws of nature? A physical law may be a hypothesis that has been confirmed by experiments so many times that it becomes universally accepted. Current research and advances in technology are increasingly leading some to question the constants or laws that have formed the basis for our science laws (including Einstein's theory of special relativity).

It is good to question what we think we know. Sometimes, the changes in technology and in our attitudes, values and beliefs can alter what we previously thought was a given. Questioning your assumptions can also lead you to deep insights.

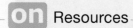 Resources

> **Video eLessons** Theoretical physicists of the twentieth century (eles-2687)
>
> Hesperides science (eles-1078)

1.2 Activities

| 1.2 Quick quiz | on | 1.2 Exercise |

Select your pathway

■ LEVEL 1	■ LEVEL 2	■ LEVEL 3
2, 3, 8	1, 4, 6, 9	5, 7, 10

These questions are even better in jacPLUS!
- Receive immediate feedback
- Access sample responses
- Track results and progress

Find all this and MORE in jacPLUS ▶

Remember and understand

1. Describe the difference between opinions and beliefs.
2. Fill in the word in the following sentence:
   _____ suggests that scientific knowledge is proven knowledge and that large amounts of first-hand data, unbiased observations and a structured method can lead to theories that can become universal laws.
3. **a.** State whether the following statements as true or false.

Statement	True or false?
i. Attitudes are a combination of feelings, beliefs and actions.	
ii. Beliefs involve making judgements about the desirability of something whereas opinions, values and attitudes do not.	
iii. Opinions can be expressed as a point of view based on fact whereas beliefs reflect what we think and know about the world, but don't need to be based on fact.	
iv. The history of science and philosophy is littered with theories that at one time were considered to be answers, but were later discarded.	
v. Laws in the new theory of Modified Newtonian Dynamics (MoND) could overthrow the established view of gravity and dark matter, question what we currently know about astronomy, and lead to the rethinking of Einstein's theory of relativity.	
vi. It can be useful for scientists to question their assumptions.	

   b. Justify your responses.
4. Match the language of learning categories with the appropriate question.

Language of learning	Questions
a. Apply	**A.** What does this mean?
b. Empathy	**B.** What are others aware of that I am missing?
c. Explain	**C.** What are my weaknesses? How do I best learn?
d. Interpret	**D.** How and where can I use this knowledge?
e. Perspective	**E.** Is it reasonable? Whose point of view is this?
f. Self-knowledge	**F.** Why is it so? How does it work?

Apply and analyse

5. A bias is a preference that may inhibit your impartial judgement.
   a. Give an example of how you are biased.
   b. Bias may be revealed by comments that are exaggerations, generalisations, imbalanced opinions stated as facts or emotionally charged words. Look through online articles and select two articles that show examples of bias. Bring these articles to school and discuss the bias with your class.
   c. Suggest why it is important to know your biases.
6. Research one of the following scientists and outline a theory that they have been involved in constructing: Charles Darwin, Marie Curie, Michael Faraday, Ernest Rutherford, Jean-Baptiste Lamarck, Francis Crick, Gregor Mendel, Albert Einstein.
7. Research and describe four examples of scientific theories that are no longer in favour.

Evaluate and create

8. Suggest ways in which you can use language positively when you are communicating with others.
9. a. Carefully examine the cartoon shown and then research Einstein's theory of relativity.
   b. On the basis of your findings, explain which ideas the cartoonist is trying to incorporate. Suggest how the cartoon could be improved.

10. Reflect on Einstein's quote 'Imagination is more important than knowledge'. In terms of science, what is your opinion on this statement? Justify your response.

Fully worked solutions and sample responses are available in your digital formats.

LESSON
1.3 Science and ethics

LEARNING INTENTION
At the end of this lesson you will be able to explain the influence of ethics in science and describe how this is influenced by goals, rights, needs and duties.

1.3.1 Difficult decisions

If you really wanted something, how far would you go to get it? What wouldn't you do?

If you wanted the lead in the school play, what would you do? Might you take up music lessons or buy the selecting teacher gifts? How about stealing a script so you can get that bit of extra practice in?

1.3.2 Goals, rights, needs and duties

Goals and rights

Liam wants to get a place in the school musical. This is Liam's **goal** — it is something he wants to achieve. However, Liam does not have a **right** to a place in the musical; although, as a student of the school, he does have the right to try for a place. A right is something we have if we can expect to be treated in a certain way, no matter what the consequences. A right is different from a need.

goal something that you want to achieve

right something you feel that you are entitled to

Needs and duties

A **need** is something we require. We all have the need to feel we are doing something worthwhile. If Liam gets a place in the band, he will have a **duty** towards his fellow band members. We often think of having a duty as being required to act in a certain way; for example, telling the truth. Liam may have several duties, such as learning the lyrics and attending rehearsal sessions.

Needs are often outlined using Maslow's hierarchy of needs, as shown in figure 1.11.

FIGURE 1.10 What are the goals, rights, needs and duties of Liam?

FIGURE 1.11 Maslow's hierarchy of needs

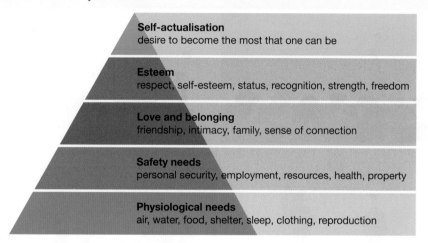

Self-actualisation
desire to become the most that one can be

Esteem
respect, self-esteem, status, recognition, strength, freedom

Love and belonging
friendship, intimacy, family, sense of connection

Safety needs
personal security, employment, resources, health, property

Physiological needs
air, water, food, shelter, sleep, clothing, reproduction

The most important needs are at the base of the hierarchy and are fundamental and primary needs.

Needs are something we require, and are different to wants, which is something we desire.

DISCUSSION

The COVID-19 pandemic brought to light the continual debate between wants and needs, and how opinions around these differ greatly between individuals.

The line between wants and needs often becomes more blurred for individuals in times of crisis, particularly in times of lockdown and economic uncertainty.

Discuss with those around you how and why individuals might view wants and needs differently in times of crisis, and describe four examples of how thoughts on wants and needs differed between individuals during the COVID-19 pandemic.

Duties versus goals

Duties often derive from goals and rights. For example, if you are accused of a crime and appear in court, you have a right to a lawyer, regardless of whether you are innocent or guilty. Your lawyer has a duty to try to get you acquitted — this is your lawyer's goal.

Some situations can become very complicated. For example, a dying man asks his doctor not to keep him alive any longer. Does the doctor have a duty to carry out the man's wishes because of the man's right to decide when and how to die? Or does the doctor have a duty to ignore the man's wishes because of the goal of preserving life?

need something that you require
duty moral obligation or responsibility

How are these related to science?

Scientists are also influenced by goals, rights, needs and duties. A goal of many scientists is to investigate the world around us and attempt to develop explanations of why and how it behaves as it does. Some scientists may also consider this to be their duty or the fulfilment of a need — or even their right to do so!

Science is often used to help us answer questions about how we can apply this knowledge. For example, if we want to know the effect of a particular diet, drug or some other factor on athletic performance, science can provide some answers.

The goals, rights and duties of scientific investigations become less clear when science is asked to provide us with answers about what we should do and how we should behave. **Ethics** are involved in shaping our ideas about what is right and wrong.

ethics the system of moral principles on the basis of which people, communities and nations make decisions about what is right or wrong

FIGURE 1.12 Should science delve into the mysteries of life? Who decides what will be researched and how discoveries will be used? Is science all about fame and fortune, or is it about seeking the truth? What is your image of science?

1.3.3 Ethics

Ethics involve your moral values. While some ethical values seem to be universal and widely accepted around the world, other ethical values vary — not only between countries, but also between different religions and communities. They may also vary within families, between different generations and throughout different times in history.

A particular scientific investigation or application of technology may be acceptable to one group of people, but highly offensive to another. Different belief systems might give rise to different ethical principles and practices. These may influence the types of scientific investigations performed and the ways in which they are conducted.

FIGURE 1.13 Ethics revolve around the idea of moral choice.

relation or from a
point of view.
Ethics [eth´iks] *n.*
moral choices to
value of human c
principles that o
for what is thous

Ethical values vary between countries, religions, communities and individuals — even between members of the same family. For example, capital punishment (the execution of a person for committing a crime), is considered by some to be right and by others to be wrong.

Science interacts with ethics in several ways, including:
- affecting the way in which science is conducted
- affecting the types of scientific research carried out
- in the conflict or match between scientific ideas and religious beliefs
- providing scientific community practices that act as a model for ethical behaviour.

The five key principles involved in ethics are:
- Integrity, truthfulness and transparency: honest reporting of any findings.
- Justice: equal access to benefits for various groups and all individuals are treated fairly and equitably.
- Beneficence: maximising benefits to an individual (do more good than harm).
- Non-maleficence: minimising harm caused to individuals.
- Respect and autonomy: valuing living things, both human and non-human, including customs, decision-making and freedom of choice. For living things that cannot make decisions, it ensures they are protected as required. This also includes ensuring that confidentiality is maintained.

It is important to consider ethics in conducting research and investigations. This includes in scientific investigations, medical research and in agriculture.

DISCUSSION

Discuss the following statements with your team.
- Scientists have a responsibility to consider the wider effects of their research.
- Individuals can influence the type of scientific research performed.
- The government controls what is done with scientific research.
- Companies should have total ownership of any research they financially support.

  Resources

    eWorkbooks Science and ethics (ewbk-12718)
                   Difficult decisions (ewbk-12720)

Scientific research and ethics

Scientific research is responsible for discoveries that have been of great value to humankind. A quick glance around us shows lots of products of science that increase our efficiency and improve our lifestyles. Scientific research is also responsible for discoveries that have had negative effects on individuals, communities, countries and our environment.

But when we talk about responsibility, is it science and the discoveries that are responsible, or is it the way in which the knowledge has been used? Who is responsible for how the knowledge is used? These issues are relevant to many examples of current scientific research.

FIGURE 1.14 An example of the movement of money in science

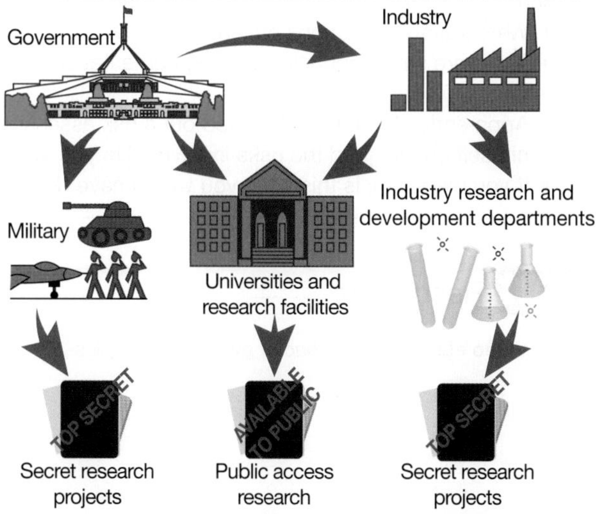

Government

Industry

Military

Universities and research facilities

Industry research and development departments

Secret research projects

Public access research

Secret research projects

1.3.4 Medical research

Medical research can be driven by need or greed. Sometimes it can provide important information, knowledge and understanding that can not only improve life, but also save it. Sometimes it can achieve this goal as well as make a lot of money for those involved in the research or its funding.

Taking risks

If acid inside your stomach eats into your stomach lining, an ulcer can result (figure 1.15). This very painful condition can also cause bleeding and can be difficult to treat. In some cases, surgery is required. It was thought that lifestyle factors, such as spicy food and stress, were key factors that triggered these painful ulcers.

One of the most well known examples of risk taking in research was seen in 2005, when Australian scientists Barry Marshall and Robin Warren received the Nobel Prize in Medicine for their research on stomach ulcers. They showed that the actual cause of many stomach ulcers was not lifestyle, but the presence of the bacteria *Helicobacter pylori* (figure 1.16). This revolutionary finding meant that ulcers could be treated with antibiotics.

Their discovery, however, was not recognised for a number of years. Their ideas faced strong opposition from the scientific community. Firm in his conviction that these bacteria were the real cause of ulcers, and that they could be easily cured by antibiotics, Marshall took a drastic step. He drank a container of *Helicobacter pylori* to infect himself! Fortunately for him (and us), although he experienced considerable discomfort, he was cured by antibiotics.

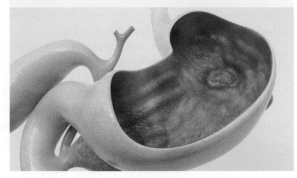

FIGURE 1.15 An illustration of a stomach ulcer

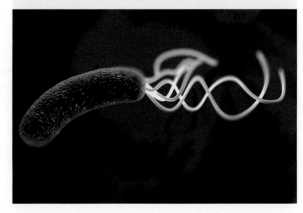

FIGURE 1.16 *Helicobacter pylori* bacteria in the human stomach cause stomach ulcers.

DISCUSSION

Discuss the following with those around you and share your ideas with the class.
- Were Marshall's actions ethical?
- There are strict regulations on experimentation on humans. Did this give him the right to infect himself?
- Was it his duty?
- Apparently Marshall had carried out a risk assessment and decided that the benefits of experimenting on himself outweighed the risks involved. Do you agree with his conclusion?
- If you were him, is this what you would have done?

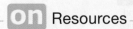 Resources

▶ **Video eLesson** *Heliobacter pylori* bacteria (eles-2691)

Drug trials

Lots of ethical issues are involved in drug trials and how these types of trials are conducted.

It is important in drug trials that all participants are aware of any risks involved, have full autonomy in the process, and are able to withdraw if they choose.

Often opinions differ in how drug trials should be conducted, such as those outlined in figure 1.17.

FIGURE 1.17 Different opinions of individuals on drug trials

A **The realist**
Drug trials are expensive and will add to the cost of the drugs, which is already high.

B **The humanist**
Testing takes time and we already know that these drugs have been effective. There are people dying who are in need of these drugs now.

C **The ethicist**
We have a responsibility to test these drugs to ensure that they are completely safe for all members of society. The most rigorous testing should always be carried out.

Sadly, there are many historical cases of drug trials that were not conducted ethically, such as the following.
- During the Tenofovir trials on HIV, individuals were given information in English, when they were French speakers, so were not informed of risks.
- In the Tuskegee syphilis experiments, some individuals were deliberately infected with syphilis and then given **placebos**, despite being informed they were given the treatment.
- During the TGN1412 trials in the United Kingdom, individuals were given anti-inflammatory drugs but were not properly informed of the risks involved.

> **placebo** a medicine or procedure that has no therapeutic effect, used as a control in testing new drugs

CASE STUDY: TGN1412

A new drug, TGN1412, was designed to treat leukaemia and certain autoimmune diseases such as rheumatoid arthritis. In rheumatoid arthritis, the body's immune system turns upon its own tissue and attacks it. The drug TGN1412 is a powerful antibody that works by binding to the immune system's T cells, causing them to activate and multiply rapidly.

TGN1412 made headlines in 2006 after its first trial on human subjects. It was given to six healthy young men in the United Kingdom and caused severe adverse reactions that required intensive care. One man's head swelled to three times its normal size, causing excruciating pain. The worst affected trial volunteer was 20-year-old Ryan Wilson, who was in a coma for three weeks after taking the drug.

Drug trial volunteers are mainly young people, and many are backpackers and students who are attracted to the payments made by pharmaceutical companies. Other controversies have arisen following drug trials in Nigeria and India, where it was unclear whether patients had given their informed consent.

Applications of medical research

Public institutions, such as universities, carry out medical research to increase our understanding and contribute to the development of possible solutions to current or potential future problems. Some of this research is linked to making money and some purely for the knowledge and understanding that it provides.

Medical research in private companies may also contribute to our knowledge, understanding and problem solving — their key goal, however, is to make a financial profit. The type of research being funded may be influenced more by its money-making potential than by its potential to reduce human suffering and improve quality of life.

1.3.5 Animal testing

Is it ethical to use animals in scientific research, such as that shown in figure 1.18? Animals are used in scientific research to test the effects of cosmetics, different surgical techniques, types of disease treatments, and to find out more about how their and our bodies function. During some of this research animals may experience pain, suffering and even death. There are many ethical issues related to the use of animals in scientific research, the types of animals used and whether the research itself is ethical.

1.3.6 Ethics in agriculture

With an increasing global human population comes the need for an increased food supply. Traditional plant breeding methods are being replaced with new technologies. One of these is the use of **genetic modification** (GM). This technology enables plants to be designed with features that increase crop yields and quality.

Some applications of genetic modification enable the development of crops that are resistant to herbicides (for example, canola), can make their own pesticides (for example, cotton) or contain added nutrients (for example, rice).

There is considerable debate about the use of genetic modification because it involves changing the plants at a molecular level: the actual DNA of the plant is modified. This technology can involve moving genes between different species, so the resulting plant is transgenic (contains DNA from different species).

genetic modification the technique of modifying the genome of an organism

FIGURE 1.18 Animal testing brings up many ethical considerations

FIGURE 1.19 Genetic modification is used to increase crop yields and quality.

DISCUSSION

Discuss the following questions in relation to genetically modified foods in agriculture.
- Is it right to interfere with nature?
- Does the addition of an animal gene to a plant make it suitable for vegetarians?
- Should GM foods show this status on their labels?
- Who should receive the profits?
- Who has ownership of the modified plants?

INVESTIGATION 1.2

Where do I stand on ethical issues in science?

Aim

To reflect on and make decisions regarding a variety of ethical issues in science

Method

1. On your own, score each of the following statements on a scale of 0 to 4, where 0 = strongly disagree and 4 = strongly agree.
   a. Immunisation of children should be compulsory.
   b. Genetic manipulation of food crops and animals should be illegal.
   c. IVF technology should be publicly funded.
   d. Nuclear reactors should be built in each Australian state and territory.
   e. Cosmetics should be tested on other animals prior to their availability to humans.
   f. The development of new drugs should be done by non-profit organisations rather than those that may make a profit.
   g. If an effective but expensive drug is available to cure a life-threatening disease, it should be available to everyone, not just those who can afford it.
   h. Genetically modified food should be clearly labelled as such.
   i. Close relatives of humans, such as monkeys and chimpanzees, should not be used as animals in scientific research that tests the effectiveness of treatments against various diseases.
   j. Scientists should be allowed to experiment on themselves.
2. For at least three of the statements, share your opinions by being involved in constructing a class 'opinionogram'.
   a. Divide the classroom into five zones, and assign a score of 0 to 4 to each zone.
   b. Each student should stand in the zone that indicates their score for the first statement.
   c. Have a member of the class record the number of students at each point of the scale.
   d. Discuss the reasons for your opinion with the students in your zone.
   e. Suggest questions that could be used to probe students in different opinion zones.
   f. Share reasons for your opinion with students in other zones and listen to their reasons for their stance.
   g. Reflect on what you have heard from others. Decide if you want to change positions and, if so, change. Give a reason for why you are changing.
   h. Have a member of the class record the number of students at each point of the scale.
   i. Repeat steps (b)–(h) for the other two statements.
   j. Reflect on what you have learned about the opinions and perspectives of others.
   k. Suggest questions that could be used to more closely probe reasons for your classmates' opinions. Share these probing questions with your class.

Results

1. Outline your score for each of the given statements.
2. Construct graphs showing the opinion scales for each statement and comment on any observed patterns.
3. Construct a PMI chart for each statement based on opinions and statements made by others in the class.

Discussion

1. Research two of the issues above. Construct a table with reasons for and against. Compare and discuss your table with others.
2. Select one of the statements (ensure it is different from the statements debated in step 2 of the Method) and organise a class debate.

Conclusion

Summarise your findings about opinions on ethical issues in science.

1.3 Activities

1.3 Quick quiz **on**	1.3 Exercise

These questions are even better in jacPLUS!
- Receive immediate feedback
- Access sample responses
- Track results and progress

Find all this and MORE in jacPLUS ▶

Select your pathway

■ LEVEL 1	■ LEVEL 2	■ LEVEL 3
1, 2, 6	3, 5, 8	4, 7, 9

Remember and understand

1. **a.** State whether the following statements as true or false.

Statement	True or false?
i. A right is something we have if we can expect to be treated in a certain way, no matter what the consequences.	
ii. Scientists can be influenced by goals, rights, needs and duties.	
iii. The goals, rights and duties of scientific investigations become clearer when science is asked to provide us with answers about what we should do and how we should behave.	
iv. Ethics are often involved in shaping our ideas about what is right and wrong.	
v. The type of research being funded by some private companies may be influenced more by its money-making potential than by its potential to reduce human suffering and improve quality of life.	
vi. Nobel prize recipient Barry Marshall drank a container of *Helicobacter pylori* to infect himself to support his conviction that the actual cause of many stomach ulcers was bacterial rather than lifestyle.	

   b. Justify your responses.
2. Outline the five main principles of ethics.
3. Describe some ethical issues related to animal testing.
4. Outline some of the arguments against using genetically modified crops.

Apply and analyse

5. **a.** Laura is a member of the pre-musical performance squad. Liam would like to be a member of the squad. Think about this situation and the goals, rights, needs and duties that Laura and Liam each have, and then copy and complete the following table.

   TABLE Identifying and specifying goals, rights, needs and duties

Person	Goals	Rights	Needs	Duties
Laura				
Liam				
Teacher in charge of casting				
Audience for the musical				
Rest of the cast				

   b. How people behave in any situation is largely determined by how they perceive the relative importance of their goals, rights, needs and duties.
      i. Describe how Liam may behave if he perceives that his goals and needs are of greater importance than those of others.
      ii. Contrast this with the behaviour you may expect if he perceives his goals as being less important than those of others.
      iii. How do you think Laura and Liam should behave towards each other?

6. Analyse each of the levels in Maslow's hierarchy of needs. How does this apply to your everyday life?
7. Drug trials were conducted in 2020 to test a vaccine against COVID-19. Provide recommendations on how integrity, justice, beneficence, non-maleficence and autonomy can be assured.

Evaluate and create

8. Comment on the following statement: 'Scientific discoveries should belong to everyone'. Outline your opinions of this and write three paragraphs justifying your response.
9. We have had to face some very complex and difficult issues because of recent scientific and technological advances. Examples of issues being faced in Australia include:
   - compulsory immunisation of children
   - genetic manipulation of food crops and animals to optimise such things as their resistance to pests and their growth rate
   - irradiating food to maximise its shelf life
   - public funding of IVF technology
   - reducing irrigation to improve water quality of rivers
   - building a new nuclear reactor in Australia.

   Research one of the preceding issues and create a report summarising ethical considerations around this.

Fully worked solutions and sample responses are available in your digital formats.

LESSON
1.4 Designing and conducting your own investigation

LEARNING INTENTION

At the end of this lesson you will be able to design and conduct investigations with attention to variables, reports and scientific processes.

1.4.1 The scientific method

As a science student you are required to undertake scientific investigations. These investigations will not only help you understand scientific concepts, they can be a lot of fun! Scientists around the world all follow what is known as the **scientific method**. This allows scientists to examine each other's work and build on the scientific knowledge gained. An important aspect of science is being able to reproduce someone else's experiment. The more evidence a scientist has about a theory, the more accepted the theory will be.

The scientific method is summarised in figure 1.20.

The skills you will develop in conducting scientific investigations include the following:
- questioning and predicting
- planning and conducting
- processing, modelling and analysing
- evaluating
- communicating scientifically.

1.4.2 Begin with a plan

Whenever you take a trip away from home, you need to plan ahead and have some idea of where you are going. You need to know how you are going to get there, what you need to pack and have some idea of what you are going to do when you get there.

It's the same with an experimental investigation. Planning ahead increases your chances of success. It's easier if you can break an investigation down into steps.

scientific method a systematic approach to planning, undertaking research, and analysing data and observations

FIGURE 1.20 The scientific method

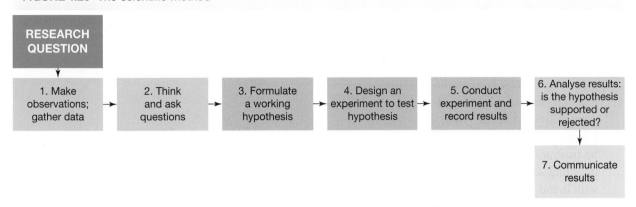

Designing your own investigation requires a great appreciation of your learning, and requires you to be creative, organised and be able to persevere and adapt when things don't work the way you expect.

Part of being a scientist is through the action of refinement — there are no failures in science, but, instead, lessons that allow you to adjust your hypothesis and try something new. That is all part of the excitement of scientific investigation.

FIGURE 1.21 What sorts of questions do you ask yourself to decide whether you should take on new learning?

The four Rs

In order to plan an investigation, it helps to consider the four Rs: resilience, reflectiveness, responsibility and resourcefulness.

- **Resilience** is about believing in yourself and having the ability to tolerate sometimes feeling a little uncomfortable. As learning is an emotional business, your ability to tolerate emotions is important. Learning is not always fast and smooth; there can be frustrating flat spots, exhilarating highs and upsetting setbacks. Resilience helps you to stick with it and recover from any disappointments. It is important in learning to help you tolerate your emotional seesaw. The components of resilience are shown in figure 1.22.
- **Reflectiveness** is being self-aware and mindful of what could be and what has been. It involves being open-minded and sometimes standing back and looking at the big picture; asking yourself if your own assumptions are getting in the way of the truth.
- **Responsibility** is being able to manage yourself and your learning. It's about monitoring your progress and thinking about other options and different perspectives.
- **Resourcefulness** is knowing what tools you have and when to use them. It's about taking responsible risks and using a range of appropriate learning tools and strategies.

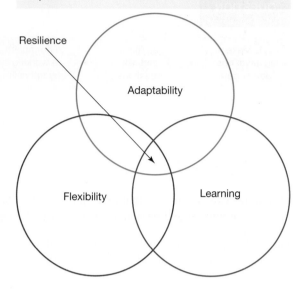

FIGURE 1.22 Resilience involves many different aspects

1.4.3 Keeping records

A **logbook** is an essential part of a long scientific investigation. It provides you with a complete record of your investigation, from the time you begin to search for a topic. Your logbook will make the task of writing your report very much easier.

A logbook is just like a diary. Make an entry whenever you spend time on your investigation. Each entry should be clearly dated. It's likely that the first entry will be a mind map or list of possible topics. Other entries might include:

- notes on background research conducted in the library. Include all the details you will need for the bibliography of your report (see section 1.4.11)
- a record of the people that you asked for advice (including your teacher), and their suggestions
- diagrams of equipment, and other evidence that you have planned your experiments carefully
- all of your 'raw' results, in table form where appropriate
- an outline of any problems encountered and how you solved them
- first drafts of your reports, including your thoughts about your conclusions

An online logbook

An exercise book can be used as a logbook, but there are several advantages in maintaining your logbook online in the form of a **blog** or in a program such as OneNote. If you choose to use a blog to record your investigation, there are many sites that will allow you to set up a free blog. Your teacher might be able to provide some suggestions.

resilience the ability to tolerate feeling a little uncomfortable sometimes

reflectiveness being self-aware, open-minded and sometimes standing back and looking at the big picture

responsibility being able to manage yourself and your learning

resourcefulness taking responsible risks and using a range of appropriate learning tools and strategies

logbook a complete record of an investigation from the time a search for a topic is started

blog a personal website or web page where an individual can upload documents, diagrams, photos and short videos

Once you set up a blog, every entry you make will be dated automatically. You can upload documents, diagrams, photos and short videos. You can also add links to other sites and invite friends, family and teachers to post comments about your progress.

You should take some precautions if you decide to use a blog as a logbook.

- Limit your posts to those related to your science investigation. Don't use your logbook blog for social networking.
- Do not include your address or phone number.
- If your blog is on the internet (rather than a school intranet):
  - do not post any photos of yourself in school uniform or any other clothing that will identify where you go to school
  - do not include your full name, address, phone number or the name of your school in the blog; use only your first name or a nickname
  - use privacy settings or use a password to ensure that only trusted school friends, family and your teacher have access to the blog.

FIGURE 1.23 A blog used as a logbook for a student research investigation

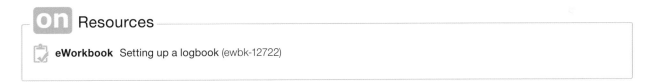

Resources

eWorkbook Setting up a logbook (ewbk-12722)

1.4.4 Finding a topic

Your investigation is much more likely to be of high quality if you choose a topic that you will enjoy working on. These steps might help you choose a good topic.

1. Think about your interests and hobbies. They might give you some ideas about investigation topics.
2. Make a list of your ideas.
3. Brainstorm ideas with a partner or in a small group. You might find that exchanging ideas with others is very helpful.
4. Find out what other students have investigated in the past. Although you will not want to cover exactly the same topics, investigations performed by others might help you to think of other ideas.
5. Do a quick search in the library or at home for books or newspaper articles about topics that interest you. Search the internet. You might also find articles of interest in magazines or journals. You could use a table (such as table 1.2) to organise your ideas.

FIGURE 1.24 When designing a research question it helps to think about your interests, think about your hobbies and brainstorm your ideas with others.

TABLE 1.2 A record of topic area and relevant resources

Topic area	Name of book, magazine, website, etc.	Chapter or article	Topic ideas

From observations to ideas

Many ideas for scientific investigations start with a simple observation. Some well-known investigations and inventions from the past started that way.

Even discoveries that were made by accident (such as the discovery of penicillin by Fleming) would not have been made without observation skills.

Other important 'ingredients' in these discoveries are curiosity and the ability to ask questions and form ideas that can be tested by experiment and further observation.

DISCUSSION

Write a list of things you are curious about (you may look around the room to help you) — this may be something you observe around you or even something related to one of your hobbies. Discuss this with the class and see if you can come up with a list of ideas for an investigation.

1.4.5 Formulating a question

Once you have decided on your topic, you need to determine exactly what you want to investigate. It is better to start with a simple, very specific question than a complicated or broad question.

For example, the topic 'basketball' is very broad. Many simple questions could be asked about basketball that you could use as a basis of an investigation.
- What size basketball bounces the highest?
- What angle should you throw the ball from the free throw line to shoot a goal?
- What surface of basketball court allows you to run the fastest?

Your question needs to be realistic and testable. In defining the question, you need to consider whether:

- you can obtain the background information that you need
- the equipment that you need is available
- the investigation can be completed in the time you have available
- the question is safe to investigate.

FIGURE 1.25 What questions might you ask about the topic of basketball?

In science, it is important that you continually revise and refine your investigation question. You may find that your question is too broad and doesn't provide clear observations or findings that allow you to draw conclusions. You may find that your question cannot be answered through investigation or observations that are available to you. You might even find that when designing your investigation, you do not have the time and resources to answer your investigation question. It is important that as you formulate questions, you keep track of the refinement and revision of this question. It is often helpful to record a note to yourself about WHY you altered your question of investigation.

1.4.6 Creating an aim and hypothesis using variables

What are variables?

In every investigation, there are different variables. Variables are observations or measurements that can change during an experiment. You should only change one variable at a time in an experiment.

When determining your **aim** and **hypothesis**, you need to be able to identify your independent and dependent variable.

> The **independent variable** is the one that is deliberately manipulated by an investigator during an experiment (what is being tested).
>
> The **dependent variable** is the one that is measured or observed by the investigator during an experiment. It is a variable that may change when the independent variable is changed.

For example, if you were performing an experiment to find out which brand of fertiliser was best for growing a particular plant, the independent variable would be the brand of fertiliser. The dependent variable would be the heights of the plants after a chosen number of days.

A third type of variable, known as **controlled variables**, are also important in designing investigations. These will be explored in section 1.4.7.

Aim

Your investigation should have a clear and realistic aim. Your aim should be very specific and related to your variables. The aim of an investigation is its purpose, or the reason for doing it.

> When writing an aim, you should always link together the dependent and independent variables. Some examples of aims are as follows.
> - To determine how the size of wheels affects the speed of a toy car
> - To compare the effect of different fertilisers on the growth of pea plants
> - To find out whether different coloured lights affect the growth of algae in an aquarium
> - To determine which metal is the most reactive in hydrochloric acid

aim a statement outlining the purpose of an investigation

hypothesis a suggested, testable explanation for observations or experimental results; it acts as a prediction for the investigation

independent variable the variable that you deliberately change during an experiment to observe its effect on another variable

dependent variable the variable that is being affected by the independent variable; that is, the variable you are measuring.

controlled variables the conditions that must be kept constant throughout an experiment

Hypothesis

A hypothesis is a statement that is a prediction for your investigation. Your hypothesis should relate to your aim and should be **testable** and **falsifiable** with an experimental investigation.

The results of your investigation will either support (agree with) or not support (disagree with) the hypothesis. A hypothesis cannot be proven correct, but rather have evidence that further supports the statement made.

There are many ways to formulate a hypothesis. It is important that it is a statement and not a question.

One way to write a hypothesis is through the IF...THEN... format. Some examples of hypotheses (based on the earlier aims) are:
- IF the wheels of a toy car are increased in size, THEN the speed of the car will increase
- IF the fertiliser with the most nutrients is used, THEN the pea plants will grow at a faster rate
- IF different lights are shone on algae in an aquarium, THEN algae under the red light will have the highest growth rate
- IF different metals are placed in hydrochloric acid, THEN group I metals will be the most reactive and produce the most hydrogen gas

1.4.7 Designing an experiment

In order to complete a successful investigation, you need to make sure that your experiments are well designed. Once you've decided exactly what you are going to investigate, you need to be aware of:
- which variables need to be controlled and which variables can be changed
- whether a control is necessary
- what observations and measurements you will make and what equipment you will need to make them
- the importance of repeating experiments (replication) to make your results more reliable
- how you will record and **analyse** your data.

A poorly designed investigation is likely to produce a conclusion that is not valid.

Controlling variables

When you are testing the effect of an independent variable on a dependent variable, all other variables should be kept constant. Such variables are called controlled variables. For example, in the fertiliser experiment, the type of plant, amount of water provided to each plant, soil type, amount of light, temperature and pot size are all controlled variables. The process of controlling variables is also known as **fair testing**.

The need for a control

Some experiments require a **control**. A control is needed in the fertiliser experiment to ensure that the result is due to the fertilisers and not something else. The control in this experiment would be a pot of plants to which no fertiliser was added. All other variables would be the same as for the other three pots.

testable able to be supported or proven false through the use of observations and investigation

falsifiable can be proven false

analyse examine methodically and in detail to answer a question or solve a problem

fair testing a test that changes only one variable and controls all other variables when attempting to answer a scientific question

control an experimental set-up in which the independent variable is not applied

FIGURE 1.26 It is important to include a control group that does not have the independent variable applied.

Valid experiments

A **valid** experiment gathers data on what it actually set out to measure. If your aim was to find out whether watering plants with sea water affects their growth rate, comparing the number of radish seeds that germinate after one week when watered with tap water or sea water would not be a valid method because it does not actually measure growth rate. It tests the effect of sea water on seed germination.

Reliable experiments

Replication is the repeating of an experiment to make sure you have collected **reliable data**. In the case of the fertiliser experiment, a more reliable result could be obtained by setting up multiple pots for each brand of or having multiple seedlings in each pot. An average result for each brand or the control could then be calculated.

A reliable experiment provides consistent results when repeated, even if it is repeated on different days and under slightly different conditions; for example, in a different room or with a different researcher collecting the data. Replication increases the reliability of an experiment. This can involve simply doing the same experiment a few times, or having different groups repeat the same experiment and pooling the data gathered by each group when writing the report.

 **Resources**

| | eWorkbook | Variables and controls (ewbk-12724) |
| | Video eLesson | A good experiment? (eles-2630) |

1.4.8 Getting approval

You should now be ready to write a plan for your investigation. You should not commence any experiments until your plan has been approved by your science teacher. You plan should include:

FIGURE 1.27 Write out a plan for your investigation.

1. **Title**
   The likely title — you may decide to change it before your work is completed. Usually, your title is your research question. The title should be in the form of a question; for example, How does watering grass seeds with a detergent solution affect their growth?

2. **The problem**
   A statement of the question that you intend to answer. Include a hypothesis. A hypothesis is an educated guess about the outcome of your experiments. It is usually based on observations and is always abled to be tested by further observations or measurements.

3. **Outline of your experiment**
   Outline how you intend to go about answering the question. This should briefly outline the experiments that you intend to conduct.

4. **Equipment**
   List here any equipment that you think will be needed for your experiments.

5. **Resources**
   List here the sources of information that you have already used and those that you intend to use. This list should include library resources, organisations and people.

valid sound or true experiment that can be supported by other scientific investigations

replication repeating of an experiment to make sure you have collected reliable data

reliable data consistent data that is achieved when an investigation is replicated

Safety

When getting approval from your teacher, it is important to show that you have considered safety (and ethics, which was covered earlier in the topic).

Some general safety precautions that will help to ensure you and others are not harmed include:

- Wearing protective clothing. This might include laboratory coat, safety glasses and gloves.
- Being aware of the position of safety equipment, such as fire blanket, fire extinguisher, safety shower and eye wash.
- Reading labels carefully to confirm contents and concentration of chemicals or pathogenic agents.
- Correct disposal of equipment and chemicals, including damaged equipment (i.e. broken glassware) and cleaning and packing up of equipment.
- Conducting an investigation as outlined in your approved plan. Don't vary your plan without approval from your teacher.

FIGURE 1.28 Safety equipment is vital in various experiments.

Often, hazards are addressed through a **risk assessment**, which allows for the identification of hazardous chemicals and equipment, the risks involved and what procedures need to be followed to work safely with these. You may be asked to create a risk assessment by your teacher as part of the submission of your plan.

1.4.9 Conducting investigations and gathering data

Once your plan has been approved by your teacher, you may begin your experiments.

Details of how you conducted your experiments should be recorded in your logbook. All observations and measurements should be recorded. Use tables where possible to record your data.

Where appropriate, measurements should be repeated and an average value determined. All measurements — not just the averages — should be recorded in your logbook.

Photographs should be taken if appropriate.

You might need to change your experiments if you get results you don't expect. Any major changes should be checked with your teacher.

FIGURE 1.29 All observations and measurements should be recorded.

Precision and accuracy

As you plan and carry out your investigation you need to ensure that the data you collect is **precise** and **accurate**. Choosing the most appropriate instruments to make your measurements is important.

Precise measurements

Precision is the degree to which repeated measurements produce the same result. It tells us how close a series of measurements are to each other, as can be seen in figure 1.30. If there is a large variation in the results, the precision is low. If the results are all very similar, and only vary by a very small amount, then the precision is high.

risk assessment a procedure that identifies the potential hazards of an experiment and gives protective measures to minimise the risk

precise multiple measurements of the same investigation being close to each other

accurate an experimental measurement that is close to a known value

The degree of precision of the measurements taken in an experiment depends on the instruments that have been used. If you want to measure the length of your classroom, you could use a trundle wheel with marks every 10 cm, or you could use a tape measure marked in millimetres. The tape measure would provide the most precise measurement. Similarly, to measure 100 mL of water, you could use a measuring cylinder that is graduated in millilitres or you could use a measuring cup that is marked every 100 mL. The measuring cylinder would provide a more precise measurement than the cup. A set of scales that measures mass to two decimal places is more precise than one that measures mass to one decimal place.

Accurate results

Accuracy is different to precision. Accuracy refers to how close an experimental measurement is to a known value. Sometimes results that are not precise can still be accurate, if the average of them is close to the actual value. This can be seen in figure 1.30. A small measuring cylinder can provide a reasonably precise measurement of a volume of water but, if it is not read at eye level, the measurement may not be accurate. A set of bathroom scales might display a reading with two decimal places but, if you use it on carpeted floor, it may not provide an accurate measurement of your mass if it is designed to be used on a hard floor. To ensure that your results are accurate, you should use measuring instruments correctly, and in some instances, it may be necessary to **calibrate** the instruments. To calibrate a set of scales, for example, you could place an object that has a mass of exactly 100.00 g on the scale and adjust the scale until it reads exactly 100.00 g.

If an archer is precise, their arrows hit close to one another. If an archer is accurate, their arrows hit close to the target.

FIGURE 1.30 Comparing accuracy and precision

Not accurate
not precise

Accurate but
not precise

Not accurate but
precise

Accurate and
precise

Choosing equipment for precision

Choosing the correct piece of equipment is critical to ensure that your results are precise. Your bathroom scales and the electronic scales in a science laboratory both measure mass, but the laboratory scales are more precise. Your school might have different sets of scales that measure to one or two decimal places. Scales that measure to two decimal places are more precise. High precision scales are needed for some of the senior chemistry experiments.

For measuring instruments with a scale, such as thermometers, rulers and measuring cylinders, the graduations (lines) on the scale give an indication of the precision of the instrument. Generally, an instrument with smaller gradations is more precise.

calibrate to check or adjust a measuring instrument to ensure accurate measurements

SAMPLE PROBLEM 1: Precision

Which of the rulers in the diagram below is the more precise?

Ruler A

Ruler B

THINK

Look at the number of divisions on each ruler between each marked measurement.
On ruler A, between 100 and 200, there are 50 divisions.
On ruler B, between 2 cm and 4 cm, there are 20 divisions.

WRITE

As ruler A has more graduations over the same space as ruler B, ruler A is more precise.

Measuring volumes of liquids

When liquids are placed in a vessel, the surface of the liquid is often curved (as shown in figure 1.31). This curved surface of a column of liquid is called a **meniscus**. Measurements must be recorded from the bottom of this meniscus.

FIGURE 1.31 You always should measuring liquids from the base of the meniscus

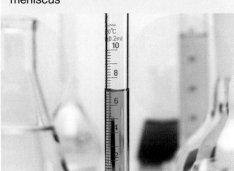

SAMPLE PROBLEM 2: Measuring readings of a meniscus

What is the measurement of this liquid in a measuring cylinder?

THINK

1. The liquid level should be read from the bottom of the meniscus, not where it touches the glass. Imagine a line drawn across from the bottom of the meniscus to the glass.

2. Look for the scale marking below the liquid level and above the liquid level. These are 55 mL and 60 mL. To calculate the volume between the two scale markings, subtract the smaller reading from the larger reading.

3. There are five divisions between these two scale markings. To determine the size of each small scale marking, divide the volume calculated in step 2 by 5.

WRITE

$60 - 55 = 5$ mL

$5 \div 5 = 1$
Each scale division is 1 mL.

meniscus the curve seen at the top of a liquid in response to its container

4. To read the measurement of the liquid level, count up from the lower scale marking, 55, to the liquid level; this is two scale divisions.

$55 + 2 = 57$ mL

Ensuring equipment is accurate

Measurements can be very precise, but incorrect. Every so often current affairs TV programs bring attention to service stations that overcharge customers for petrol by having faulty petrol pumps that give inaccurate readings of the amount of petrol delivered by the pump. For each litre of petrol pumped, the machine might give a reading of 1.1 L and the customer is charged accordingly. The machine is quite precise, but not accurate.

CASE STUDY: Calibrating a pH meter

Some measuring instruments require calibration to ensure that they provide accurate measurements. The calibration might be part of the manufacturing process, or it may need to be carried out by the user regularly. A pH meter is a device that needs to be calibrated regularly. pH is a measure of the acidity of a substance. You can measure pH with a universal indicator.

For a more precise reading a pH meter can be used. It is a device that is placed in the solution and it gives a reading of the pH to one or two decimal places. Over time it can lose its calibration and give inaccurate readings. A reading of 6.25 might be displayed when the solution actually has a pH of 5.38. To calibrate the pH meter you place it in solutions of known pH and adjust the device until it reads the correct values for these solutions. You can then use the meter to measure the pH of a solution with an unknown concentration.

FIGURE 1.32 A pH meter needs to be calibrated regularly to ensure it gives accurate readings.

1.4.10 Graphing variables

Many different types of data can be collected in scientific experiments. Data is often presented in tables or as graphs.

Tables

Tables can be used to record data to help separate and organise your information. All tables should:
- have a heading
- display the data clearly, with the independent variable in the first column and the dependent variable in later columns
- include units in the column headings and not with every data point
- be designed to be easy to read.

TABLE 1.3 The effect of different brands of fertilisers on the height of seedlings.

Fertiliser	Height of seedling (cm)				
	Day 2	Day 4	Day 6	Day 7	Day 8
Brand X	2	3	5	6	9
Brand Y	3	5	7	9	11
Brand Z	1	2	3	5	7
Control	0	0.6	1.8	2.5	4

Graphs

Graphs can help you see patterns and trends in your data. Once your data is recorded in a table, you need to work out the best graph to choose. This is often affected by the type of data you have (is your data qualitative or quantitative).

If you use a graph to show your results, you would normally graph the independent variable (the one you changed) on the *x*-axis, and the dependent variable (the one you measured) on the *y*-axis. When the dependent variable changes with time, you can graph time on the *x*-axis and the dependent variable on the *y*-axis. For example, in the fertiliser experiment, two types of graphs could be used, a line graph or a column graph (bar chart).

FIGURE 1.33 Some examples of graphs used in a fertiliser experiment

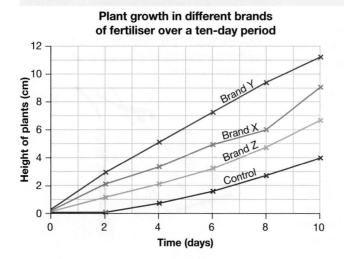

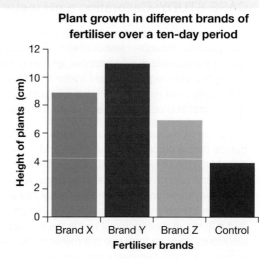

Different types of graphs are appropriate in different circumstances.

- *Scatterplots* require both sets of data to be numerical. Each dot represents one observation. A scatterplot can easily show trends between data sets, and correlations can be seen. A line of best fit may be added to show the overall trend in the data.
- *Line graphs:* These are scatterplots with the dots joined. The dots are usually joined using a straight line, but sometimes the line is curved. They are used for continuous data. Always used if the data is recorded over time.
- *Bar/column* graphs can be used when one piece of data is qualitative and the other is quantitative. The bars are separated from each other. The horizontal axis has no scale because it simply shows categories. The vertical axis has a scale showing the units of measurement.
- *Histograms* are a special kind of bar graph that show continuous categories, and are often used when examining frequency. The bars are not separated.
- *Pie charts* and *divided bar charts* are used to show frequencies or portions of a whole. This includes percentages or fractions.

FIGURE 1.34 Features of a line graph

3. Setting up and labelling the axes

Graphs represent a relationship between two variables.

Usually the independent variable is plotted on the horizontal *x*-axis and the dependent variable on the vertical *y*-axis.

After deciding on the variable for each axis, you must clearly label each axis with the variable and its units. The units are written in brackets after the name of the variable.

2. Title

Tell the reader what the graph is about! The title describes the results of the investigation or the relationship between variables.

1. Grid

Graphs should always be drawn on grid paper so values are accurately placed. Drawing freehand on lined or plain paper is not accurate enough for most graphs.

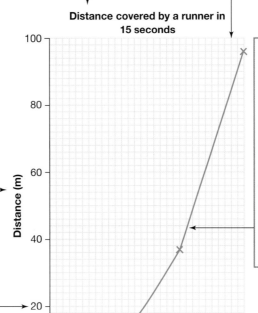

Distance covered by a runner in 15 seconds

6. Drawing the line

A line is then drawn through the points.

A line that follows the general direction of the points is called a 'line of best fit' because it best fits the data. It should be on or as close to as many points as possible.

Some points follow the shape of a curve, rather than a straight line. A curved line that touches all the points can then be used.

4. Setting up the scales

Each axis should be marked into units that cover the entire range of the measurement. For example, if the distance ranges from 0 m to 96 m, then 0 m and 100 m could be the lowest and highest values on the vertical scale. The distance between the top and bottom values is then broken up into equal divisions and marked. The horizontal axis must also have its own range of values and uniform scale (which does not have to be the same scale as the vertical axis). The most important points about the scales are:
- they must show the entire range of measurements
- they must be uniform; that is, show equal divisions for equal increases in value.

5. Putting in the values

A point is made for each pair of values from the data table (the meeting point of two imaginary lines from each axis). The points should be clearly visible. Only include a point for (0, 0) if you have the data for this point.

TABLE 1.6 Data table

Distance (m)	Time (s)
0	4
8	5
37	10
96	15

SAMPLE PROBLEM 3: Choosing types of graphs

Identify the type of graph that would be most appropriate to display the following data:

a. Data from Melbourne Zoo showing how the mass of a baby elephant has increased over time

b. The mass of each elephant at Melbourne Zoo

c. The proportion of visitors using various modes of transport to travel to Melbourne Zoo.

THINK

a. The mass of one elephant is a number that changes over time, so it is quantitative data. Mass can take any numerical value, so it is continuous data.

b. We compare the mass of different elephants by showing the name of each elephant and its mass at a set point in time. The name of each elephant is qualitative, and the mass of each elephant is quantitative (continuous).

c. The proportion of visitors using various modes of transport shows fractions or percentages of a whole.

WRITE

Mass is continuous data, so a line graph would be the best choice.

As we have both qualitative and quantitative data, a bar or column graph would be the best choice.

As the data is showing the proportion of people using different modes of transport, the best choice would be a pie chart or divided bar chart.

---

on Resources

eWorkbook   Organising and evaluating results (ewbk-12726)

Interactivity   Sector graphs or pie graphs (int-4061)

1.4.11 Writing your report

You can begin writing your report as soon as you have planned your investigation, but it cannot be completed until your observations are complete. Your report should be typed or neatly written on A4 paper and presented in a folder. It should begin with a table of contents, and the pages should be numbered. Your report should include the following headings (unless they are inappropriate for your investigation).

Scientific report structure

Abstract

The abstract provides the reader with a brief summary of your whole investigation. Even though this appears at the beginning of your report, it is best not to write it until after you have completed the rest of your report.

Introduction

Present all relevant background information. Include a statement of the problem that you are investigating, saying why it is relevant or important. You could also explain why you became interested in the topic.

Aim

State the purpose of your investigation; that is, what you are trying to find out. Include the hypothesis.

Materials and methods

Describe in detail how you did your experiments. Begin with a list or description of equipment that you used. You could also include photographs of your equipment if appropriate. The method description must be detailed

---

38 Jacaranda Science Quest 10 Australian Curriculum Fourth Edition

enough to allow somebody else to repeat your experiments. It should also convince the reader that your investigation is well controlled. Labelled diagrams can be used to make your description clear. Using a step-by-step outline makes your method easier to follow.

Results

Observations and measurements (often referred to as data) are presented here. Data should, wherever possible, be presented in table form for ease of reading. Graphs can be used to help you and the reader interpret data. Each table and graph should have a title. Make sure you use the most appropriate type of graph for your data.

Your results should allow for trends and patterns to be seen, and for any outliers (unusual results) to be observed.

When showing your results, you should consider the mean (average), median (middle values) and range to help best show your data, and which type of data will best reflect your findings.

Discussion

Discuss your results here. Begin with a statement of what your results indicate about the answer to your question. Explain how your results might be useful. Any weaknesses in your design or difficulties in measuring could be outlined here. Explain how you could have improved your experiments. What further experiments are suggested by your results?

You should also outline how error has affected your results. Errors may include:
- **Systematic error** — errors that affect accuracy (how close a measurement is to the actual value). This is often due to equipment error; for example, a 30 cm ruler is actually only 29 cm, so all measurements are wrong using this ruler.
- **Random error** — chance errors that may occur that affect how close results are to each other (precision). These are often mitigated by repeating investigations and taking an average. These are normal errors that occur; for example, slight differences in reaction time when recording the time taken for a reaction to occur. Parallax errors are random errors caused by reading scales from a different angle or position.

As well as the above, you should also address any uncertainties in the data that you obtain, and factors that may influence the accuracy and precision of your data and findings.

In your discussion, you may also link your information to similar findings or other scientific reports. In using secondary data, you should ensure that this is valid, and from a source that is reliable, such as in peer-reviewed scientific journals.

You may use this secondary scientific evidence alongside your primary data that you gathered to help support your findings. Remember, there may be multiple explanations for your data, so secondary sources can assist you in formulating conclusions.

Conclusion

This is a brief statement of what you found out. It is a good idea to read your aim again before you write your conclusion. Your conclusion should also state whether your hypothesis was supported. You should not be disappointed if it is not supported. In fact, some scientists deliberately set out to reject hypotheses!

Bibliography

Make a list of books, other printed or audio-visual material and websites to which you have referred. The list should include enough detail to allow the source of information to be easily found by the reader. Arrange the sources in alphabetical order.

The way a resource is listed depends on whether it is a book, magazine (or journal) or website. For each resource, list the following information in the order shown:
- author(s), if known (book, magazine or website)
- title of book or article, or name of website
- volume number or issue (magazine)
- URL (website) and the date you accessed the web page
- publisher (book or magazine), if not in title
- place of publication, if given (book)
- year of publication (book, magazine or website)
- chapter or pages used (book).

systematic error an error (often due to equipment) that affects the accuracy of results

random error an error that affects the precision of results

Some examples of different sources are listed below:

- Taylor, N, Stubbs A., Stokes, R. (2020) *Jacaranda Chemistry 2 VCE Units 3 & 4.* 2nd edition. Milton: John Wiley & Sons.
- Gregg, J, (2014), 'How Smart are Dolphins?' *Focus Science and Technology*, Issue 264, February 2014, BBC, pages 52–57.
- Australian Marine Wildlife Research & Rescue Organisation, http://www.amwrro.org.au, 2014.

Acknowledgements

List the people and organisations who gave you help or advice. You should state how each person or organisation assisted you.

on **Resources**

📋 **eWorkbook**   Components of a practical report (ewbk-12728)

1.4 Activities

learn on

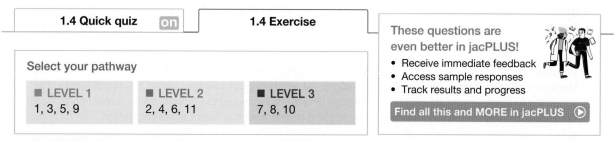

1.4 Quick quiz **on**	1.4 Exercise

These questions are even better in jacPLUS!
- Receive immediate feedback
- Access sample responses
- Track results and progress

Find all this and MORE in jacPLUS ▶

Select your pathway

■ LEVEL 1	■ LEVEL 2	■ LEVEL 3
1, 3, 5, 9	2, 4, 6, 11	7, 8, 10

Remember and understand

1. Construct a flowchart to show the steps that you need to take before beginning your experiments.
2. What is the advantage of repeating an experiment several times?
3. Describe the difference between an independent and a dependent variable. Provide an example of each.
4. Outline an example of the use of a control in an experiment.
5. In which section of your report do you describe possible improvements to your experiments?

Apply and analyse

6. For each problem described, identify the independent and dependent variable and three other variables that would need to be controlled.
   a. Josie wanted to find out whether adding salt to a pot of water causes it to boil faster.
   b. Charlotte would like to investigate if plants watered with pure water grow faster than those watered with lemonade.
   c. Jayden is testing the hypothesis that wearing a swimming cap makes you swim faster.
   d. Shinji is testing the myth that classical music played to a baby in the womb results in them having a higher IQ.
   e. Nikita has heard that most people shrink slightly (in height) throughout the day and stretch out at night. She would like to know whether this is true.
7. Discuss the advantages and disadvantages of using a blog as a logbook for your investigation.
8. Why is it better to write the abstract of a scientific report last, even though it appears at the beginning?
9. Describe why it is important to write a risk assessment before you conduct an investigation.

Evaluate and create

10. You wish to conduct a practical investigation to determine how fast 100 mL water takes to boil in different sized containers.
    a. Write a research question for this topic.
    b. Write an aim and hypothesis for this investigation.
    c. Summarise the materials you would use in this investigation.
    d. Write a method summarising how you will explore this topic.

11. A student explored the growth of a plant when watered with different solutions and obtained the following results. Show this information in a graph and analyse the data obtained.

TABLE The effect of watering plants with different types of solutions on plant growth

Watering solution	Growth of plant (cm)		
	Day 1	Day 2	Day 3
Water	2.1	5.4	7.1
Lemonade	3.2	6.7	9.4
Soda water	2.5	5.9	8.2

Fully worked solutions and sample responses are available in your digital formats.

LESSON
1.5 SkillBuilder — Controlled, dependent and independent variables

LEARNING INTENTION

At the end of this lesson you will be able to identify independent, dependent and controlled variables.

onlineonly

Why do we need to manage variables in an investigation?

In an investigation, it is important to only change one of the variables at a time and then observe what that change brings about in the other variables.

Go online to access:
- **Tell me:** an overview of the skill and its application in science
- **Show me:** a video and a step-by-step process to explain the skill
- **Let me do it:** an interactivity, question set and SkillBuilder activity for you to practice and consolidate your understanding of the skill.

on Resources

LESSON
1.6 SkillBuilder — Writing an aim and forming a hypothesis

LEARNING INTENTION

At the end of this lesson you will be able to write aims and hypotheses.

Why do we need to write aims and form hypotheses?

In science, we conduct investigations to draw conclusions and gather data and results. Every investigation requires an aim — a short statement of what we are trying to achieve. Alongside an aim, the ability to formulate predictions is important in science. This is performed through the use of a hypothesis. Being able to write aims and hypotheses are vital skills for any scientist.

If this → → then this

Go online to access:
- **Tell me:** an overview of the skill and its application in science
- **Show me:** a video and a step-by-step process to explain the skill
- **Let me do it:** an interactivity, question set and SkillBuilder activity for you to practice and consolidate your understanding of the skill.

on Resources

eWorkbook	SkillBuilder — Writing an aim and forming a hypothesis	(ewbk-12732)
Video eLesson	Writing an aim and forming a hypothesis	(eles-4155)
Interactivity	Writing an aim and forming a hypothesis	(int-8089)

LESSON
1.7 SkillBuilder — Measuring and reading scales

LEARNING INTENTION

At the end of this lesson you will be able to read and record measurements accurately.

Why do we need to measure and read scales?

When conducting experiments, it is critical that measurements and data are recorded accurately. Whether measuring volume or temperature, or interpreting alternate scales, it is important that they are recorded accurately.

In science, a scale or set of numbered markings generally accompanies each measuring device. For example, your ruler measures length, and its scale has markings enabling you to measure with an accuracy of 0.1 cm. When reading a scale, it is important to determine what each of the markings on the scale represents.

A small measuring cylinder can provide a reasonably precise measurement of a volume of water but, if it is not read at eye level, the measurement may not be accurate. Measurements should always be made with your eye in line with the reading you are taking. When scales are read from a different angle, the reading is not accurate. This type of reading error is called parallax error.

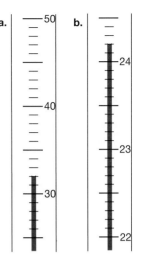

Go online to access:
- **Tell me:** an overview of the skill and its application in science
- **Show me:** a video and a step-by-step process to explain the skill
- **Let me do it:** an interactivity, question set and SkillBuilder activity for you to practice and consolidate your understanding of the skill.

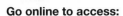 Resources

eWorkbook	SkillBuilder — Measuring and reading scales (ewbk-12734)	
Video eLesson	Measuring and reading scales (eles-4153)	
Interactivity	Reading scales (int-0201)	

LESSON
1.8 SkillBuilder — Creating a simple column or bar graph

LEARNING INTENTION

At the end of this lesson you will be able to construct simple column or bar graphs.

online only

What is a column or bar graph?

Column graphs show information or data in columns. In a bar graph, the bars are drawn horizontally and in column graphs, they are drawn vertically. They can be hand drawn or constructed using computer spreadsheets.

Go online to access:
- **Tell me:** an overview of the skill and its application in science
- **Show me:** a video and a step-by-step process to explain the skill
- **Let me do it:** an interactivity, question set and SkillBuilder activity for you to practice and consolidate your understanding of the skill.

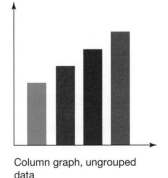

Column graph, ungrouped data

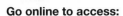 Resources

eWorkbook	SkillBuilder — Creating a simple column or bar graph (ewbk-12736)	
Video eLesson	Creating a simple column or bar graph (eles-1639)	
Interactivity	Creating a simple column or bar graph (int-3135)	

LESSON
1.9 SkillBuilder — Drawing a line graph

LEARNING INTENTION

At the end of this lesson you will be able to construct line graphs.

on line only

What is a line graph?

A line graph displays information as a series of points on a graph that are joined to form a line. Line graphs are very useful to show change over time. They can show a single set of data, or they can show multiple sets, which enables us to compare similarities and differences between two sets of data at a glance.

Go online to access:
- **Tell me:** an overview of the skill and its application in science
- **Show me:** a video and a step-by-step process to explain the skill
- **Let me do it:** an interactivity, question set and SkillBuilder activity for you to practice and consolidate your understanding of the skill.

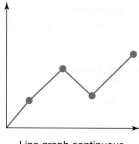

Line graph continuous, ungrouped data

on Resources

eWorkbook	SkillBuilder — Drawing a line graph (ewbk-12738)	
Video eLesson	Drawing a line graph (eles-1635)	
Interactivity	Drawing a line graph (int-3131)	

LESSON
1.10 Sample scientific investigation

LEARNING INTENTION

At the end of this lesson you will be able to design your own investigation using a sample investigation as an example.

1.10.1 Investigating muddy water

Sean, a Year 10 student, conducted an experimental investigation to compare the turbidity (cloudiness) of water in the following three locations:
- a creek near his school
- a creek near his home
- a river near his home.

His search for information in the library revealed that the cloudiness was caused by particles of soil (and sometimes pollution) suspended in the water. Sean chose his topic because he was interested in the environment. He felt that clean water was the right of all living things. His research and background knowledge led him to form the hypothesis that 'the clearest water will be in the river'.

Sean took water samples from each of the three locations on four days. He found a method of measuring turbidity from a library book. It involved adding a chemical called potash alum to a sample of water in a jar. The potash alum makes the particles of suspended soil clump together and fall to the bottom of the jar. A layer of mud is formed. The height of the mud at the bottom is then measured.

A summary of Sean's method, including a list of materials and equipment required, is shown. You will notice that Sean used a fourth sample. It was needed as a control and contained distilled water. This was to ensure that there was nothing in the pure water to cause a layer at the bottom of the jar when the potash alum was added. His results are in table 1.4.

TABLE 1.4 Results table measuring the levels of mud in water samples from three different areas

Water sample	Height of mud (mm)															
	Day 1				Day 2				Day 3				Day 4			
	Test			Average	Test			Average	Test			Average	Test			Average
	1	2	3		1	2	3		1	2	3		1	2	3	
1. Home creek	3.5	4.0	5.0	4.2	5.0	4.5	5.0	4.8	4.5	5.0	4.5	4.3	5.0	4.5	4.0	4.5
2. School creek	2.5	2.0	2.0	2.2	3.0	2.5	2.5	2.7	2.0	2.5	2.5	2.3	2.0	2.0	2.5	2.2
3. Barnes River	1.0	0.5	0.0	0.5	2.0	1.0	1.5	1.5	0.5	1.0	0.5	0.7	0.5	0.5	0.5	0.5
4. Distilled water	0.0	0.0	0.0	0.0	0.0	0.0	0.0	0.0	0.0	0.0	0.0	0.0	0.0	0.0	0.0	0.0

Sean's Investigation

Materials

- 4 large jars or bottles with lids for collecting water samples (capacity of about 1 L each)
- 4 identical jam jars with lids, labelled 1, 2, 3 and 4
- metal teaspoon (not plastic, in case it breaks)
- potash alum (potassium aluminium sulfate)
- 4 water samples from different locations
- ruler with 1-millimetre graduations
- 100 mL measuring cylinder
- permanent marker

Method

1. Water samples (about 1 litre each) were collected from a specific part of the creeks and river on the same day.
2. Each of three clean jars was filled to the same level with the water samples — a labelled jar for each location. A fourth labelled jar was filled to the same level with distilled water.
3. One level teaspoon of potash alum was added to each jar. Lids were put on the jars and the jars were shaken.
4. The jars were left for 30 minutes to allow the particles to settle.
5. The height of the layer of mud on the bottom of each jar was measured and recorded.
6. The jars were emptied and washed and the experiment was repeated three more times.
7. Water samples were collected from the same locations on three other days over a ten-day period and the entire experiment was repeated three more times.

1.10.2 Analysing the data

Sometimes it is necessary to refine the raw data (the data initially collected), presenting them in a different way. Sean was planning to use his average measurements to make a column graph. He decided to simplify his table so that it was easier to construct the column graph. The simplified table (shown in table 1.5) and column graph (shown in figure 1.35) make it easier for others to read the results, and easier for Sean to see patterns and draw conclusions.

TABLE 1.5 Average heights of mud in water from three different areas

Water sample	Average height of mud (mm)			
	Day 1	Day 2	Day 3	Day 4
1. Home creek	4.2	4.8	4.3	4.5
2. School creek	2.2	2.7	2.3	2.2
3. Barnes River	0.5	1.5	0.7	0.5

FIGURE 1.35 Sean's graph makes it easier to see patterns and draw conclusions.

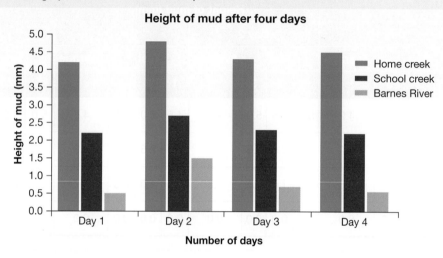

1.10.3 Being critical

Sean was pleased with his results and was able to draw conclusions. In the discussion section of his report, he suggested that further studies be done. The turbidity was affected by weather conditions and the sampling needed to be done over a longer period, and in different weather conditions. Sean had recorded weather details on each day that he sampled water and was able to explain the very high mud level in the river on day 2. It is almost always possible to suggest improvements to your experiments.

1.10.4 Drawing conclusions

Sean's hypothesis, that the clearest water would be in the river, was supported. The conclusions he drew were as follows:

1. The home creek has the muddiest water, with sample values ranging from heights of 4.2 to 4.8 mm of mud per 200 mL of water. The school creek has moderate amounts of mud compared to the other two samples. Sample values ranged from 2.2 to 2.7 mm of mud per 200 mL of water. The river water is the clearest, with sample values of 0.5 to 1.5 mm of mud per 200 mL of water.
2. Weather conditions can alter the amount of mud in water bodies by either adding run-off from drains or stirring up the water. This was particularly noticeable in the samples taken from the river site on day 2, which followed a period of rain.

FIGURE 1.36 Chemical waste running into a river. How might you test for such materials in a water sample from this site?

on Resources

eWorkbooks   Scientific investigation examples (ewbk-12740)

Drawing conclusions (ewbk-12742)

1.10 Activities

learn on

| 1.10 Quick quiz **on** | 1.10 Exercise |

Select your pathway

| ■ LEVEL 1 | ■ LEVEL 2 | ■ LEVEL 3 |
| 1, 3 | 2, 5 | 4, 6 |

These questions are even better in jacPLUS!
- Receive immediate feedback
- Access sample responses
- Track results and progress

Find all this and MORE in jacPLUS ⊙

Remember and understand

1. For Sean's experiment, identify the following.
   a. The independent variable   b. The dependent variable   c. The variables he controlled

Apply and analyse

2. Explain why a sample of distilled water was included in Sean's experiment.
3. Explain why Sean repeated the experiment three times on four separate days.
4. Explain why Sean used a column graph rather than another type of graph to present his results.

Evaluate and create

5. Suggest how Sean could improve the reliability and accuracy of his experiment.
6. In your opinion, is Sean's conclusion valid? Justify your answer.

Fully worked solutions and sample responses are available in your digital formats.

LESSON
1.11 Using secondary sources to draw conclusions

LEARNING INTENTION

At the end of this lesson you will be able to explain the importance of secondary evidence and use evidence that is valid and assists in drawing conclusions.

1.11.1 Ensuring validity in secondary sources

Part of science is being able investigate not just primary sources that you obtain in direct observation, but also secondary sources. It is important that any secondary evidence you use is valid and provides appropriate evidence to assist in drawing conclusions.

Secondary evidence that can assist you should have:
- a basis in facts derived from studies with high validity and minimal bias
- statistical evidence to support conclusions
- a clear distinction between correlation and causation — two variables may often have some correlation (they both increase, for example), but have no causation (one variable does not cause the change in value in the other)
- data from investigations that have a reproducible and reliable method (for example, using a large sample size and various control groups)
- peer-reviewed research formed from scientific ideas.

Using valid secondary evidence allows for the development of evidence-based arguments and for better conclusions to be drawn.

1.11.2 Is the evidence reliable?

When exploring secondary evidence, it is important to ensure that it is reliable.

Much of the information available to the public often lacks reliability and can confuse individuals about the key scientific ideas.

Many argue that we are currently in an age of information overload. We are constantly being bombarded with information from a variety of sources, many of these associated with the media. Some of the information that you are exposed to may not be accurate or the whole story. The information may be **biased** in the selection, emphasis, word choice and context used. It is important that when interpreting information, you are aware of these possible biases. You also need to be aware of your own biases!

In making sense of this new information, you need to focus on the ways in which you build your knowledge. How you — as a 'learner' at the centre of your learning — use your senses (for example, sight, hearing, smell, touch and taste) to *perceive* your world, and emotion, reason and language to *interpret* what you sense.

FIGURE 1.37 Do you know the full story looking at a headline?

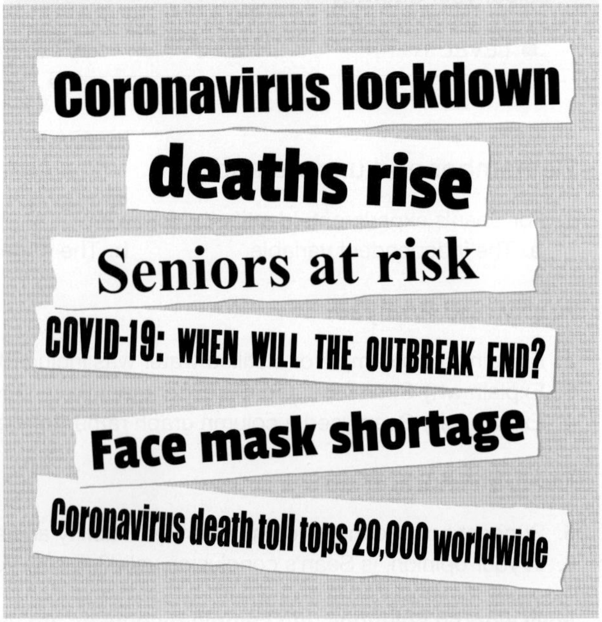

Coronavirus lockdown deaths rise

Seniors at risk

COVID-19: WHEN WILL THE OUTBREAK END?

Face mask shortage

Coronavirus death toll tops 20,000 worldwide

---

biased inclined towards a preference or prejudice

Considering all types of science-related media is important, to help critically analyse the validity of information and to evaluate conclusions.

Where can we get reliable information?

Government websites and those from educational institutions and established organisations are usually reliable. When using other websites, it is a good idea to look for dates and research the authors to see if the information is verifiable and current. You should try to use multiple sites to verify the data, and be cautious of sites that can be altered by the public (such as Wikipedia or social media). Google Scholar can be a good starting point to help find reliable sources of information.

DISCUSSION

Are websites that can be edited by the general public always unreliable? Discuss this and present your ideas to the class.

It can be helpful to look at the domain name of a website (the letters towards the end of the URL). Remember that a website address ending in '.au' means that it is registered in Australia, but some non-governmental Australian websites are registered in the United States and will not include the '.au'.

Peer-reviewed journals in science are also reliable sources of evidence, as the articles have been reviewed for quality of research, experimental reproducibility, accuracy and validity, and adhere to the required standards of a journal, ensuring articles are not biased.

It is also important to ensure that sites or authors of content being researched do not have any conflict of interest. For example, if an article is talking about an incredible new medication and data on this, but the author or organisation are funded by the drug manufacturer, a potential conflict of interest exists.

peer-reviewed evaluated scientifically, academically, or professionally worked by others working in the same field

TABLE 1.6 Examples of generic top-level domain names

Domain name contains	Source
.gov.au	Australian federal, state, territory and local government entities; cannot be edited by the general public
.gov	United States government entities; cannot be edited by the general public
.edu.au	Limited to Australian educational institutions including schools and universities; cannot be edited by the general public
.edu	Largely limited to American higher educational institutions (universities and colleges; not schools); cannot be edited by the general public
.org	Intended for use by not-for-profit organisations; some web pages can be edited by the public (for example, Wikipedia.org)
.com.au	Intended for commercial use, but now used within Australia across a range of sectors including businesses, not-for-profits, schools, individual people
.com	Intended for commercial use, but now used within the United States across a range of sectors including businesses, not-for-profits, schools, individual people
.net	Intended for networking technology organisations such as internet service providers, but now used across a range of sectors including businesses, not-for-profits, schools, individual people

Resources

eWorkbooks Evaluating media reports (ewbk-12744)

Summarising secondary sources (ewbk-12746)

1.11 Activities

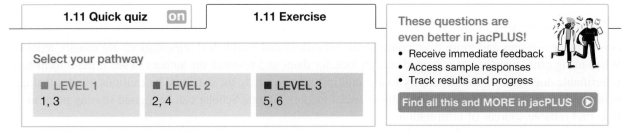

| 1.11 Quick quiz on | 1.11 Exercise |

Select your pathway

■ LEVEL 1	■ LEVEL 2	■ LEVEL 3
1, 3	2, 4	5, 6

These questions are even better in jacPLUS!
- Receive immediate feedback
- Access sample responses
- Track results and progress

Find all this and MORE in jacPLUS ▶

Remember and understand

1. Explain the difference between a secondary and a primary source.
2. Outline three factors that are seen in reliable sources of secondary evidence.

Apply and analyse

3. Describe how you may be able to use editable websites such as Wikipedia to find reliable sources of information.
4. Explain why you might also use secondary evidence in a scientific report if you have already gathered primary evidence.
5. Explain the link between bias and conflict of interest.

Evaluate and create

6. A student has used the following sources of information for their secondary evidence to support findings they made in an investigation:
   - three peer-reviewed journals
   - an article by the Department of Health on the Australian government website
   - a Twitter post by a well-known immunologist
   - two articles from the *Herald Sun* website (heraldsun.com.au) or similar.

   Evaluate each of the sources of secondary evidence, summarising assumptions that can be made about their reliability and validity.

Fully worked solutions and sample responses are available in your digital formats.

LESSON
1.12 Using spreadsheets

LEARNING INTENTION

At the end of this lesson you will be able to use a spreadsheet to record, graph and analyse data.

1.12.1 The advantages of spreadsheets

A spreadsheet is a computer program that can be used to organise data into columns and rows.

Once the data are entered, mathematical calculations, such as adding, multiplying and averaging, can be carried out easily using the spreadsheet functions.

Spreadsheets have many advantages over handwritten or word-processed results. For example, with spreadsheets you can:
- make calculations quickly and accurately
- change data or fix mistakes without redoing the whole spreadsheet
- use the spreadsheet's charting function to present your results in graphic form.

1.12.2 Elements of a spreadsheet

Although there are a number of spreadsheet programs available, they all have the same basic features and layout, as shown in example 1. The data shown are from a student research project about the different factors on the growth of bean plants.

1.12.3 Entering data into cells

You can enter different types of data into a cell:
- a number or value
- a label; that is, text (for titles and headings)
- a formula (an instruction to make a calculation).

Decide in which cell you want to insert the data (the active cell). Type the data in the cell and press 'Enter'. To edit or change the data, simply highlight the cell and type in the new data — it will replace the old data when you press 'Enter'.

ELEMENTS OF A SPREADSHEET: Example 1

- At the top of the spreadsheet are the toolbar and formula bar.
- A *row* is identified by a number; for example, 'row 1' or 'row 2'.
- A *column* is identified by a letter; for example, 'column A' or 'column B'.
- A *cell* is identified by its column and row address. For example, 'cell G3' refers to the cell formed by the intersection of column G with row 3. In this example, cell G3 is the active cell (shown by its heavy border). The active cell address and its contents (once data are entered) are shown to the left of the formula bar.
- A *range* is a block of cells. For example, 'range C3:F4' includes all the cells in columns C through to F and rows 3 through to 4.

FIGURE 1.38 Key components in an Excel spreadsheet

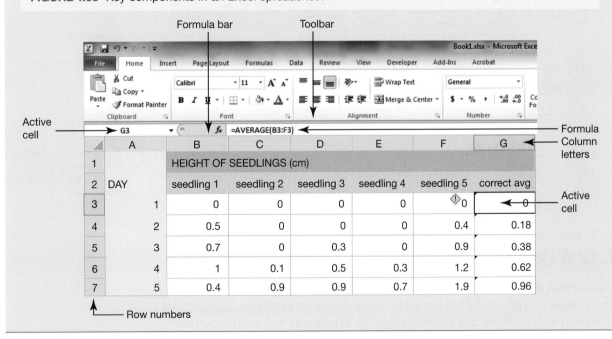

1.12.4 Creating formulae

To create a formula, you need to start with a special character or symbol to indicate that you are keying in a formula rather than a label or value. This is usually one of the symbols =, @ or +, depending on the spreadsheet program. For example, a formula to add the contents of cell B1 to cell C1 would take one of the following forms: =B1+C1 or @B1+C1 or +B1+C1.

Once you have entered the formula in a cell, the result of the calculation, rather than the formula, will be shown. The formula can be seen in the status bar when the cell is active (see example 2). If you subsequently needed to change the values in B1 or C1, the spreadsheet will automatically use the formula to recalculate and show the new result.

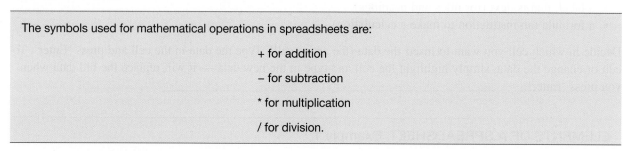

The symbols used for mathematical operations in spreadsheets are:

+ for addition

– for subtraction

* for multiplication

/ for division.

CREATING FORMULAE: Example 2

The spreadsheet in example 1 has been further developed. Formulae have now been entered to average the heights of the seedlings.

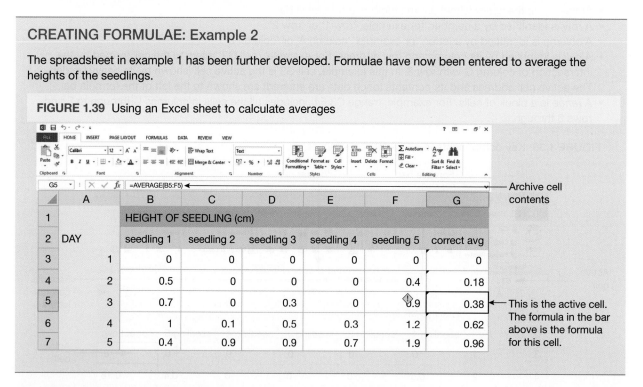

FIGURE 1.39 Using an Excel sheet to calculate averages

1.12.5 Using functions

Some common types of calculations are built into the spreadsheet, so that you don't always need to type out the full formulae. These are called **functions**. All functions have two parts: the name and a value (called the **argument**) that the function will operate on. The value is normally placed in parentheses, (), and can be written as a set of numbers or as a range (a block of cells). For example, a function to calculate the average of the amounts entered in cells B1, B2, B3 and B4 would be written: =AVERAGE(B1:B4).

functions common type of calculation built into spreadsheets

argument value that a function in a spreadsheet will operate on

TABLE 1.7 Common spreadsheet functions

Name	Application	Example	Result
AVERAGE	calculates the average of the argument values	=AVERAGE(1,2,3,4)	2.5
COUNT	counts the number of values in the argument	=COUNT(A3:A6)	4
MAX	returns the largest value in the argument	=MAX(1,9,5)	9
MIN	returns the smallest value in the argument	=MIN(1,9,5)	1
MODE	returns the most common value in the argument	=MODE(1,1,5,5,1)	1
MEDIAN	returns the median value of the argument values	=MEDIAN(1, 2, 3, 5, 6)	3
ROUND	rounds the argument to the number of decimal places specified	=ROUND(12.25,1)	12.3
SUM	calculates the sum of the values in the argument	=SUM(1,9,5)	15
MEDIAN	returns the median (middle) value of the argument values	=MEDIAN(1,2,3,4,5)	3

1.12.6 Copying cells

Spreadsheets have a command that allows you to copy a formula or value from one cell to another cell (or into a range of cells). This is usually found in the *Edit* menu in the Home tab (*Fill Down* or *Fill Right*).

The way a formula is copied depends on whether the cell references use:
- **relative referencing**, which you use when you want the cell address in the formula to change according to the relative location of the cell that you have copied it to. Example 2 in section 1.12.4 uses relative referencing. The formula AVERAGE(B5:F5) in the active cell G5 was copied downwards, so that there was no need to type the formulae in the rest of the column. The formula in the next cell (G6) is therefore AVERAGE(B6:F6) and so on.
- **absolute referencing**, which you use when you want a cell address in the formula to be constant, no matter where it is copied to. Absolute referencing is denoted by the symbol $ placed in the cell address. For example, B3 (see example 3).

> **relative referencing** used in a spreadsheet when the cell address in the formula is changed
>
> **absolute referencing** used in a spreadsheet when a cell address in the formula remains constant, no matter where it is copied to

COPYING CELLS: Example 3

FIGURE 1.40 Using absolute referencing in a spreadsheet

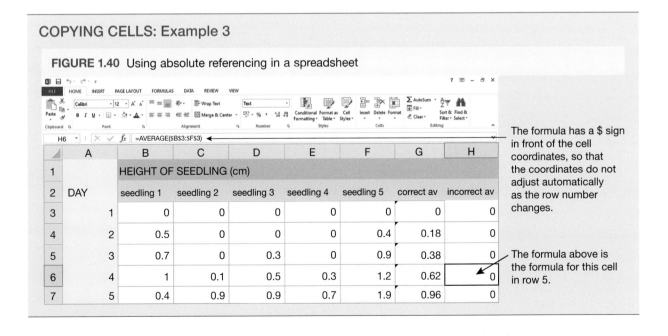

The formula has a $ sign in front of the cell coordinates, so that the coordinates do not adjust automatically as the row number changes.

The formula above is the formula for this cell in row 5.

1.12.7 Formatting cells

Investigate your spreadsheet program (most come with a tutorial) to learn how to use other useful features such as:

- adding and deleting rows or columns (useful if you have forgotten to include some calculations in your planning or decide you don't need some items)
- changing column widths (to show the full cell contents when the data are longer than the default column width) and changing row heights so that you can use larger font sizes for titles and headings
- inserting horizontal or vertical lines to improve the presentation of your spreadsheet
- changing cell formats to control how the data are to be displayed, such as using different fonts and character styles (underlining, bold, italic).

You can also format numeric values in a variety of ways. For example, the *Fixed* or *Number* format will display values to the number of specified decimal places. The *Percent* format will display values as a percentage, to the number of specified decimal places.

Once you have keyed in your data and included any necessary calculations, print out your spreadsheet and save it to a disk so that you can store the document and use it later.

1.12.8 Spreadsheet graphics

The three main types of graphs — pie, bar and line graphs — can usually be produced by a spreadsheet. It means that you can easily display your results graphically, but you still need to decide which is the most appropriate type of graph for your data.

The first step in producing a spreadsheet graph is to select the block of the cells that contains the data to be graphed. Use the spreadsheet's charting function, which usually brings up a window where you can indicate the type of graph, and add title and label details. When you are satisfied with the result, you can display and print out your graph.

 Resources

 eWorkbook Calculating using a spreadsheet (ewbk-12748)

1.12 Activities

learn on

1.12 Quick quiz   on

1.12 Exercise

These questions are even better in jacPLUS!
- Receive immediate feedback
- Access sample responses
- Track results and progress

Find all this and MORE in jacPLUS

Select your pathway

■ LEVEL 1
1

■ LEVEL 2
2, 3

■ LEVEL 3
4

Remember and understand

1. Look at the section of a spreadsheet provided and answer the following questions.

G5	fx	=AVERAGE(B5:F5)					
	A	B	C	D	E	F	G
1		HEIGHT OF SEEDLING (cm)					
2	DAY	seedling 1	seedling 2	seedling 3	seedling 4	seedling 5	correct avg
3	1	0	0	0	0	0	0
4	2	0.5	0	0	0	0.4	0.18
5	3	0.7	0	0.3	0	0.9	0.38

   a. What does cell G3 contain?
   b. Does cell E2 contain a value or a label?
   c. If the formula in cell G4 is AVERAGE(B4:F4), what would the formula be in cells G5 and G6?

2. The following table 'Time taken to evaporate different substances' shows the results of an experiment that tested the amount of time taken for eucalyptus oils and other substances (0.1 mL of each) to evaporate at a constant temperature. The experiment was performed twice.
   a. Enter the data into a spreadsheet.
   b. Use the spreadsheet function to calculate the average time that each substance took to evaporate.

TABLE Time taken to evaporate different substances

Substance	Time (s)	
	Trial 1	Trial 2
Methylated spirits	4.17	1.85
Turpentine	63.48	43.02
Water	54.42	57.05
Oil from *E. rossi*	195.92	191.23
Oil from *E. nortonii*	103.99	105.39

Apply and analyse

3. The following table shows the distance travelled by Jesse at 3-second intervals during a 100-metre sprint. The data were recorded during the sprint by attaching a paper tape to Jesse's waist. As he ran, the tape was pulled through a timer that printed a dot every 3 seconds.

TABLE Distance and speed traveled in 3-second intervals

Time (s)	Distance travelled in time interval (m)	Average speed for time interval (m/s)
0	0	
3	35	
6	25	
9	15	
12	15	
15	10	

a. Enter the data into a spreadsheet. Calculate the average speed travelled in each 3-second interval by applying a formula to the first cell in the column, and then copying it down. Remember that average speed can be calculated by dividing the distance travelled by the time taken:

$$\text{Speed} = \frac{\text{distance}}{\text{time}}$$

b. Calculate Jesse's average speed over the total time.

Evaluate and create

4. The following data were collected by two car servicing centres in Canberra, at the request of a student. The table shows the level of carbon monoxide and carbon dioxide emissions (as a percentage of total emissions) from cars of various ages.

TABLE Carbon monoxide and carbon dioxide emissions of cars by year of manufacture

Year car manufactured	Carbon monoxide (%)	Carbon dioxide (%)
1977	3.17	11.8
1983	2.48	13.6
1985	3.70	11.4
1987	1.60	13.1
1989	1.08	10.2
1996	0.19	15.2

a. Enter the data into a spreadsheet and create a graph to display these results.
b. Create formulae to work out the average carbon monoxide and carbon dioxide emissions for:
   i. cars manufactured up to 1985
   ii. cars manufactured from 1987 onwards.
c. Car manufacturers were required to install catalytic converters in cars made after 1986. Catalytic converters cut down carbon monoxide emissions by converting some of the carbon monoxide to carbon dioxide. What can you conclude from these data about the success of catalytic converters?

Fully worked solutions and sample responses are available in your digital formats.

LESSON
1.13 SkillBuilder — Using a spreadsheet

online only

How do you use a spreadsheet to record, analyse and graph your results?

Spreadsheets, through programs such as Excel, provide very powerful ways to identify trends and patterns in your data. They allow for data to be recorded in cells to format tables, and allow for the quick analysis of data and creation of different types of graphs.

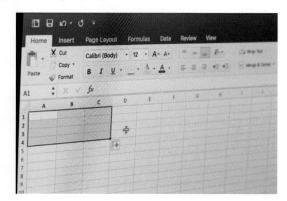

Go online to access:
- **Tell me:** an overview of the skill and its application in science
- **Show me:** a video and a step-by-step process to explain the skill
- **Let me do it:** an interactivity, question set and SkillBuilder activity for you to practice and consolidate your understanding of the skill.

on Resources

eWorkbook	SkillBuilder — Using a spreadsheet (ewbk-12750)
Video eLesson	Using a spreadsheet (eles-4230)
Interactivity	Using a spreadsheet (int-8160)

LESSON
1.14 Using data loggers and databases

1.14.1 The data logger

A data logger is a device that stores a large number of pieces of information (data) sent to it by sensors attached to it.

The data logger can transfer this data to another device, such as a graphing calculator or, more commonly, a computer, which can use data logger software or a spreadsheet program to manipulate the data (see lesson 1.12). Usually the computer or calculator graphs the collected data, and we can use these graphs to see patterns and trends easily.

When can a data logger be used?

Data loggers are particularly useful whenever an experiment requires several successive measurements. Sometimes, these measurements will take place over several hours or days — such as when measuring the way air pressure varies with the weather. Sometimes, many measurements must be taken over a short time interval — such as when measuring changes in air pressure as sound waves pass by. Data loggers are very flexible and can help scientists gather and analyse data for these types of experiments, as well as many others.

CASE STUDY: Using data loggers in endothermic and exothermic reactions

In an experiment, we investigate temperature changes in chemical processes. In addition to the laboratory equipment required for this experiment, including safety glasses, we will need a data logger with a temperature sensor attached to it. The data logger will need to be attached to a computer on which the data logger software has been installed.

Citric acid and baking soda

For this part of the experiment, we will need baking soda, citric acid, a beaker, a foam cup, other necessary laboratory equipment such as safety glasses, as well as a data logger and temperature sensor. We will use 30 mL of citric acid and 10 g of baking soda. These items are shown in figure 1.41. We set the run time to 200 seconds and the data collection rate to once per second. We insert the temperature sensor into the acid, press a button on the data logger to start data collection and then add the baking soda to the acid. The data logger collects the data, which the computer software automatically graphs after completion, as shown in figure 1.42.

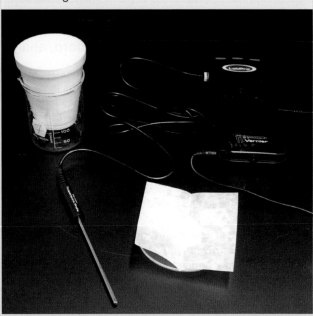

FIGURE 1.41 The equipment required for citric acid and baking soda

FIGURE 1.42 Graphed data from a data logger in the investigation

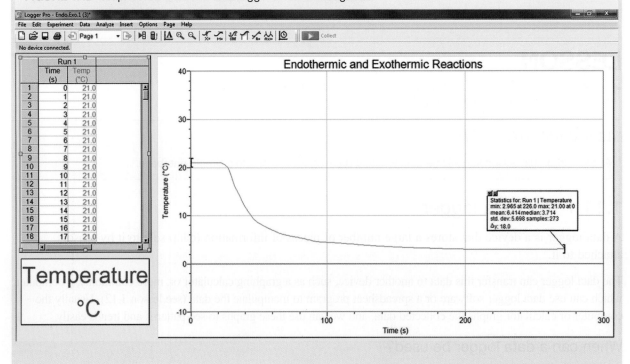

INVESTIGATION 1.3

Exothermic and endothermic processes

Aim

To use a data logger to measure change in temperature and compare this to using a thermometer

Materials

- safety glasses
- 10 g baking soda (dissolved in 100 mL water)
- 30 mL citric acid (10 g dissolved in 100 mL water)
- thermometer
- 4 large test tubes
- test tube rack
- balance
- stirring rod
- temperature probe
- data logger
- laptop
- bench mat
- 10 mL measuring cylinder
- magnesium ribbon
- sand paper
- 0.5 M hydrochloric acid

Method

Part 1: Magnesium in hydrochloric acid

1. Pour 10 mL of 0.5 M hydrochloric acid into a test tube in a test-tube rack. Place a thermometer in the test tube and allow it to come to a constant temperature. Record the temperature of the solution.
2. Clean a 10 cm piece of magnesium ribbon using the sandpaper until it is shiny on both sides. Coil the magnesium ribbon and place it into the test tube of hydrochloric acid.
3. Observe the temperature of the solution as the magnesium reacts with the hydrochloric acid. Record the final temperature of this solution.
4. Repeat the above process using a data logger and temperature problem instead of the thermometer.

Part 2: Citric acid and baking soda

5. Pour 10 mL of citric acid solution into a test tube in a test tube rack.
6. Place a thermometer in the test tube and allow it to come to a constant temperature. Record the temperature of the solution.
7. Use a balance to weigh 3 g of baking soda; add it to the citric acid solution in the test tube and stir gently.
8. Observe the temperature of the solution as the baking soda dissolves in the citric acid solution. Record the final temperature of this solution.
9. Repeat the above process using a data logger and temperature probe instead of the thermometer.

Results

Record your results for both your investigation with the thermometer and with the data logger.

Discussion

1. How did the results from the thermometer and results for the data logger compare?
2. What variables did you need to keep constant?
3. Outline any errors that may have occurred in your investigation.
4. Describe three suggestions for improvement.
5. Which method was more accurate? How do you know?

Conclusion

Summarise your findings about using a data logger to measure change in temperature and compare this with a thermometer.

1.14.2 Using databases

Databases are simply information or data arranged in one or more tables. We use databases every day, for example, when we look up information in the index of a book.

An electronic database is a powerful computer application and is an important tool for a business, an organisation or a scientist. A database's design is crucial to its usefulness, so a database must be designed with ease of searching uppermost in mind. Many programs can be used to create electronic databases, including Microsoft Access.

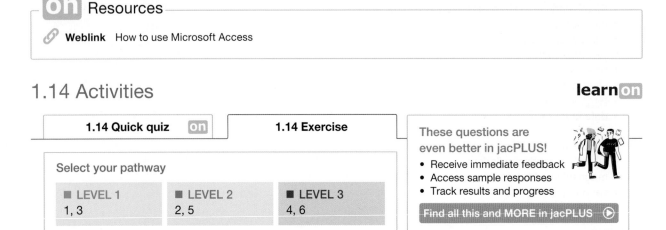

1.14 Activities

learn on

1.14 Quick quiz on   1.14 Exercise

Select your pathway

| ■ LEVEL 1 | ■ LEVEL 2 | ■ LEVEL 3 |
| 1, 3 | 2, 5 | 4, 6 |

These questions are even better in jacPLUS!
- Receive immediate feedback
- Access sample responses
- Track results and progress

Find all this and MORE in jacPLUS ⊙

Remember and understand

1. Describe the purpose of a data logger.
2. List three examples where you could use a database in everyday life.
3. Outline two disadvantages of using a data logger.

Apply and analyse

4. Look at the graph shown in the figure 1.42 of the collected data produced by the computer for *citric acid and baking soda*.
   a. What was the temperature of the acid at the start of the experiment?
   b. What was the lowest temperature that the solution of citric acid and baking soda reached? How long after first adding the baking soda did this occur?
   c. Is dissolving baking soda in citric acid an exothermic or endothermic process? How do you know?
5. Sensors are the devices that take the measurements that the data logger collects. Outline scientific investigations that could use data collected by sensors that measure:
   a. electric current
   b. acidity of solutions
   c. concentration of carbon dioxide in the air
   d. total dissolved solids (salt content)
   e. light intensity.

Evaluate and create

6. a. List the advantages of using a data logger over taking the measurements manually.
   b. Design an experiment in which using a data logger provides an advantage over manual data collection.

Fully worked solutions and sample responses are available in your digital formats.

LESSON
1.15 Review

Access your topic review eWorkbooks

on Resources

■ Topic review Level 1	■ Topic review Level 2	■ Topic review Level 3
ewbk-12760	ewbk-12762	ewbk-12764

1.15.1 Summary

Thinking flexibly

- It is important to think both critically (use information to guide decision-making) and flexibly (consider different viewpoints) in science.
- Some ways to analyse science include: inductionism (proven knowledge through first-hand data), falsification (theories are educated guesses that cannot be proven beyond reasonable doubt) and through paradigms (ways of thinking).
- Both verbal and non-verbal parts of communication are important to convey meaning and attitudes.
- Paralanguage is used to describe non-verbal parts of communication. This communication is about the way something is said, including pitch, speed, inflections, facial expressions and gestures.
- A theory is a well-supported explanation of a phenomena, based on investigations, research and observations. In science, theories are often tested using the scientific method.

Science and ethics

- In science, ethical considerations need to be balanced with goals, rights, needs and duties.
- Ethics involve your moral values and interact with science in a variety of ways.
- There are five main principles of ethics: integrity, justice, beneficence, non-maleficence and respect/ autonomy.

Designing and conducting your own investigation

- You should follow the scientific method when conducting an investigation.
- After you select a topic, you should turn this into a research question, which is specific and able to be investigated.
- Variables are factors that an investigator can control, change or measure.
- An independent variable is manipulated by the investigator (e.g. the type of metal examined).
- A dependent variable is measured by the investigator and is influenced by the independent variable.
- A controlled variable is one that is kept the same in an investigation across all trials — there are usually numerous controlled variables in an experiment
- An aim is a one- to two-sentence outline of the purpose of the investigation, linking the dependent and independent variables.

- A hypothesis is a tentative, testable and falsifiable statement for an observed phenomenon and acts as a prediction for the investigation.
- Part of scientific method or process is designing an experiment, conducting the experiment and analysing the results.
- You should ensure that the data you collect is precise and accurate.
- Graphs and tables are very important and display your results to see trends and patterns.

Sample scientific investigation

- In the sample investigation, an investigation is being conducted on the turbidity of water in three locations.
- The data provided is able to be analysed to create a graph, allowing patterns to be more easily seen.
- The data can be used to support or not support the hypothesis.
- Conclusions can be made about the turbidity of water in the three locations using the data.

Using secondary sources to draw conclusions

- It is vital to ensure that secondary data used is both reliable and valid.
- Secondary data can help support the findings made in investigations and the primary data obtained.
- When using secondary data, it should not be biased and should have no conflict of interest.

Using spreadsheets

- Spreadsheets and databases can help organise your information.
- Spreadsheets contain many functions that can be used to extract information from data or perform calculations automatically.

Using data loggers and databses

- Data loggers allow you to get more accurate information over time.
- Some data loggers are also able to plot the data they record to provide accurate graphs automatically.
- Databases are information or data arranged in one or more tables. Spreadsheets can be used to draw conclusions about the data in a database.

1.15.2 Key terms

absolute referencing   used in a spreadsheet when a cell address in the formula remains constant, no matter where it is copied to

accurate   an experimental measurement that is close to a known value

aim   a statement outlining the purpose of an investigation

analyse   examine methodically and in detail to answer a question or solve a problem

argument   value that a function in a spreadsheet will operate on

beliefs   feelings or mental acceptance that things are true or real

biased   inclined towards a preference or prejudice

blog   a personal website or web page where an individual can upload documents, diagrams, photos and short videos

calibrate   to check or adjust a measuring instrument to ensure accurate measurements

control   an experimental set-up in which the independent variable is not applied

controlled variables   the conditions that must be kept constant throughout an experiment

dependent variable   the variable that is being affected by the independent variable; that is, the variable you are measuring.

duty   moral obligation or responsibility

ethics   the system of moral principles on the basis of which people, communities and nations make decisions about what is right or wrong

fair testing   a test that changes only one variable and controls all other variables when attempting to answer a scientific question

falsifiable   can be proven false

falsification   a credible hypothesis or theory should be able to be tested to be potentially disproved or contradicted by evidence

functions   common type of calculation built into spreadsheets

genetic modification   the technique of modifying the genome of an organism

goal   something that you want to achieve

human error   mistakes made by the person performing the investigation

hypothesis a suggested, testable explanation for observations or experimental results; it acts as a prediction for the investigation

independent variable the variable that you deliberately change during an experiment to observe its effect on another variable

inductionism a theory stating that with enough evidence, scientific theories can become universal laws

judgements opinions formed after considering available information

logbook a complete record of an investigation from the time a search for a topic is started

meniscus the curve seen at the top of a liquid in response to its container

need something that you require

opinions personal views or judgements about something

paradigms generally accepted perspectives, ideas or theories at a particular time

paralanguage non-verbal parts of communication; for example, the *way* something is said, rather than the words that are used

peer-reviewed evaluated scientifically, academically, or professionally worked by others working in the same field

placebo a medicine or procedure that has no therapeutic effect, used as a control in testing new drugs

precise multiple measurements of the same investigation being close to each other

proximate rules rules that govern the physical distance that is comfortable between people

random error an error that affects the precision of results

reflectiveness being self-aware, open-minded and sometimes standing back and looking at the big picture

relative referencing used in a spreadsheet when the cell address in the formula is changed

reliable data consistent data that is achieved when an investigation is replicated

replication repeating of an experiment to make sure you have collected reliable data

resilience the ability to tolerate feeling a little uncomfortable sometimes

resourcefulness taking responsible risks and using a range of appropriate learning tools and strategies

responsibility being able to manage yourself and your learning

right something you feel that you are entitled to

risk assessment a procedure that identifies the potential hazards of an experiment and gives protective measures to minimise the risk

scientific method a systematic approach to planning, undertaking research, and analysing data and observations

systematic error an error (often due to equipment) that affects the accuracy of results

testable able to be supported or proven false through the use of observations and investigation

theory a well-supported explanation of a phenomenon based on facts that have been obtained through investigations, research and observations

valid sound or true experiment that can be supported by other scientific investigations

values a deep commitment to a particular issue and serve as standards for decision making

---

 Resources

 eWorkbooks
Study checklist (ewbk-12752)

Literacy builder (ewbk-12753)

Crossword (ewbk-12755)

Word search (ewbk-12757)

Reflection (ewbk-12759)

 Solutions
Topic 1 Solutions (sol-1124)

 Practical investigation eLogbook
Topic 1 Practical investigation eLogbook (elog-2411)

 Digital document
Key terms glossary (doc-40480)

1.15 Activities

1.15 Review questions

Select your pathway

■ LEVEL 1
1, 2, 6, 13

■ LEVEL 2
3, 4, 8, 9, 14

■ LEVEL 3
5, 7, 10, 11, 12

These questions are even better in jacPLUS!
• Receive immediate feedback
• Access sample responses
• Track results and progress

Find all this and MORE in jacPLUS ▶

Remember and understand

1. Match the words in the following list with their meanings.

Words	Meanings
a. Conclusion	A. Concerns that deal with what is morally right or wrong
b. Abstract	B. The variable that is deliberately changed in an experiment
c. Discussion	C. The part of a journal article where a brief overview of the article is given
d. Results	D. A list of steps to follow in an experiment
e. Hypothesis	E. The answer to the aim or the problem
f. Ethical considerations	F. A list of equipment needed for the experiment
g. Independent variable	G. The variable that is measured in an experiment
h. Dependent variable	H. States what was seen or measured during an experiment. May be presented in the form of a table or graph.
i. Method	I. A sensible guess to answer a problem
j. Apparatus	J. The part of a report where problems with the experiment and suggestions for improvements are discussed

2. Describe the difference between an aim and a hypothesis in an investigation.
3. Explain and provide five examples of paralanguage.
4. List some of the factors affecting the decision about whether money is spent on finding a cure for a particular disease.

Apply and analyse

5. Should farmers be allowed to plant the type of crop they believe produces the best yield, irrespective of whether others object to the manner in which the crop was bred? Justify your response.
6. Should the labelling of genetically modified foods be compulsory? Explain your response.
7. Carefully observe the flowchart provided.

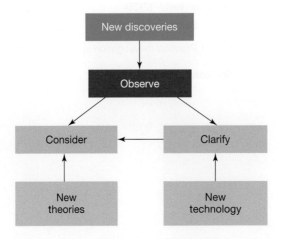

a. Do you think that this figure effectively summarises how we mix new information to rethink ideas and improve our scientific knowledge?

b. Can you think of an example of how our scientific knowledge has developed in a way similar to that suggested by the model?

c. Suggest how the model could be improved.

8. Research examples of drugs or tests that are known to encourage people to tell the truth.

a. Find out how they work.

b. Do you believe that you have the right not to be forced to tell the truth? Explain why.

c. Do you think that lie detector tests and truth serums or treatments should be used to force people to tell the truth or test whether they are telling the truth? Justify your response.

Evaluate and create

9. Miranda wants to test the following hypothesis: Hot soapy water washes out tomato sauce stains better than cold soapy water.

TABLE The effect of water temperature on stains during washing

Water temperature (°C)	Observations
20	Dark stain left after washing
40	Faint stain left after washing
60	No stain left after washing
80	No stain left after washing

a. List the equipment she will need.

b. Identify the independent and dependent variables in this investigation.

c. List five variables that will need to be controlled.

d. Outline a method that could be used to test the hypothesis.

e. Miranda's results are shown in the table Write a conclusion based on Miranda's results.

10. Gemina and Habib want to investigate whether the type of surface affects how high a ball bounces. Habib thought the ball would probably bounce the highest off a concrete floor. They dropped tennis balls from different heights onto a concrete floor, a wooden floor and carpet. Their results are shown in the following table.

TABLE The average height of a ball bouncing on different surfaces at different distances

Distance ball dropped (cm)	Average height of bounce (cm)		
	Concrete	Wood	Carpet
25	22	14	8
50	46	34	18
75	70	50	26
100	94	66	34
125	X	85	Z
150	128	94	48
175	129	Y	50
200	130	100	51

a. Write a hypothesis for this experiment.

b. Construct a line graph of Gemina and Habib's results.

c. Use your graph to estimate the values X, Y and Z.

d. Identify two variables that had to be kept constant in this experiment.

e. Identify two trends in the results.

f. Do the results support the hypothesis you wrote?

g. Predict how high the tennis ball would bounce off each floor if it was dropped from a height of 225 cm.

11. Charles Darwin's theory of evolution sat unpublished for over ten years before it was published as *On the Origin of Species*.

   a. Research and outline why it took him such a long time to make his ideas public.
   b. Many caricatures of Charles Darwin have appeared over the years. Suggest what the creator of the cartoon is suggesting. Do you consider this accurate in terms of Darwin's theory? Explain your answer.
   c. Outline the key ways in which the theory of evolution differed from the accepted theological view of the time.
   d. Identify the other scientist who is responsible for proposing the theory of evolution. Suggest why he is not as well known as Charles Darwin.
   e. Identify at least five other people involved in the development and acceptance of the theory of evolution and state their key contribution.
   f. If a scientist were to propose a new theory about creation that significantly differed from the currently accepted view, suggest how this might be received by the scientific community and the general public.
   g. Suggest a possible alternative to the theory of evolution. Provide reasons to support your theory.

12. Use the issues mind map provided to help you identify various perspectives on one of the following issues.
   • Watering of gardens should be illegal.
   • Cars should be driven only when there are at least four occupants.
   • Scientists should be allowed to research whatever they want.
   • If a vaccine for the dangerous variant of H5N1 is synthesised, it should be given only to children under 10 years of age.

13. Scientific ideas and theories can change over time. Does this mean that those previously accepted were totally wrong? Discuss and justify your response.

14. Carefully read through each of the following statements. For each statement, decide whether you agree or disagree and then justify your response.
   • Scientific discoveries and understanding often rely on developments in technology and technological advances.
   • Financial backing from governments or commercial organisations determines the types of scientific research and development that are carried out.
   • Scientific understanding, models and theories are changed over time through a process of review by the scientific community.
   • The focus of scientific research can be influenced by the current values and needs of society.
   • Advances in science and technologies can have a significant effect on people's lives.
   • Scientific knowledge should be used to evaluate whether you should accept claims, explanations or predictions.

Fully worked solutions and sample responses are available in your digital formats.

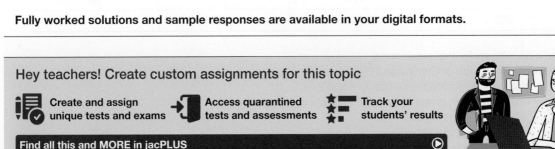

Online Resources

 Resources

Below is a full list of **rich resources** available online for this topic. These resources are designed to bring ideas to life, to promote deep and lasting learning and to support the different learning needs of each individual.

1.1 Overview

eWorkbooks
- Topic 1 eWorkbook (ewbk-12711)
- Student learning matrix (ewbk-12713)
- Starter activity (ewbk-12714)

Solutions
- Topic 1 Solutions (sol-1124)

Practical investigation eLogbook
- Topic 1 Practical investigation eLogbook (elog-2411)

Video eLesson
- Meet Professor Veena Sahajwalla (eles-1071)

1.2 Thinking flexibly

eWorkbook
- Thinking tools and the language of learning (ewbk-12716)

Practical investigation eLogbook
- Investigation 1.1: What are my beliefs and values? (elog-0739)

Video eLessons
- Dark matter labs (eles-2688)
- Theoretical physicists of the twentieth century (eles-2687)
- Hesperides science (eles-1078)

1.3 Science and ethics

eWorkbooks
- Science and ethics (ewbk-12718)
- Difficult decisions (ewbk-12720)

Practical investigation eLogbook
- Investigation 1.2: Where do I stand on ethical issues in science? (elog-0741)

Video eLesson
- *Heliobacter pylori* bacteria (eles-2691)

1.4 Designing and conducting your own investigation

eWorkbooks
- Setting up a logbook (ewbk-12722)
- Variables and controls (ewbk-12724)
- Organising and evaluating results (ewbk-12726)
- Components of a practical report (ewbk-12728)

Video eLesson
- A good experiment? (eles-2630)

Interactivity
- Sector graphs or pie graphs (int-4061)

1.5 SkillBuilder — Controlled, dependent and independent variables

eWorkbook
- SkillBuilder — Controlled, dependent and independent variables (ewbk-12730)

Video eLesson
- Controlled, dependent and independent variables (eles-4156)

Interactivity
- Controlled, dependent and independent variables (int-8090)

1.6 SkillBuilder — Writing an aim and forming a hypothesis

eWorkbook
- SkillBuilder — Writing an aim and forming a hypothesis (ewbk-12732)

Video eLesson
- Writing an aim and forming a hypothesis (eles-4155)

Interactivity
- Writing an aim and forming a hypothesis (int-8089)

1.7 SkillBuilder — Measuring and reading scales

eWorkbook
- SkillBuilder — Measuring and reading scales (ewbk-12734)

Video eLesson
- Measuring and reading scales (eles-4153)

Interactivity
- Reading scales (int-0201)

1.8 SkillBuilder — Creating a simple column or bar graph

eWorkbook
- SkillBuilder — Creating a simple column or bar graph (ewbk-12736)

Video eLesson
- Creating a simple column or bar graph (eles-1639)

Interactivity
- Creating a simple column or bar graph (int-3135)

1.9 SkillBuilder — Drawing a line graph

eWorkbook
- SkillBuilder — Drawing a line graph (ewbk-12738)

Video eLesson
- Drawing a line graph (eles-1635)

Interactivity
- Drawing a line graph (int-3131)

1.10 Sample scientific investigation

eWorkbooks
- Scientific investigation examples (ewbk-12740)
- Drawing conclusions (ewbk-12742)

1.11 Using secondary sources to draw conclusions

eWorkbooks
- Evaluating media reports (ewbk-12744)
- Summarising secondary sources (ewbk-12746)

1.12 Using spreadsheets

eWorkbook
- Calculating using a spreadsheet (ewbk-12748)

1.13 SkillBuilder — Using a spreadsheet

eWorkbook
- SkillBuilder —Using a spreadsheet (ewbk-12750)

Video eLesson
- Using a spreadsheet (eles-4230)

Interactivity
- Using a spreadsheet (int-8160)

1.14 Using data loggers and databases

Practical investigation eLogbook
- Investigation 1.3: Exothermic and endothermic processes (elog-0743)

Weblinks
- How to use Microsoft Access

Teacher-led video
- Investigation 1.3: Exothermic and endothermic processes (tlvd-10814)

1.15 Review

eWorkbooks
- Topic review Level 1 (ewbk-12760)
- Topic review Level 2 (ewbk-12762)
- Topic review Level 3 (ewbk-12764)
- Study checklist (ewbk-12752)
- Literacy builder (ewbk-12753)
- Crossword (ewbk-12755)
- Word search (ewbk-12757)
- Reflection (ewbk-12759)

Digital document
- Key terms glossary (doc-40480)

To access these online resources, log on to **www.jacplus.com.au**

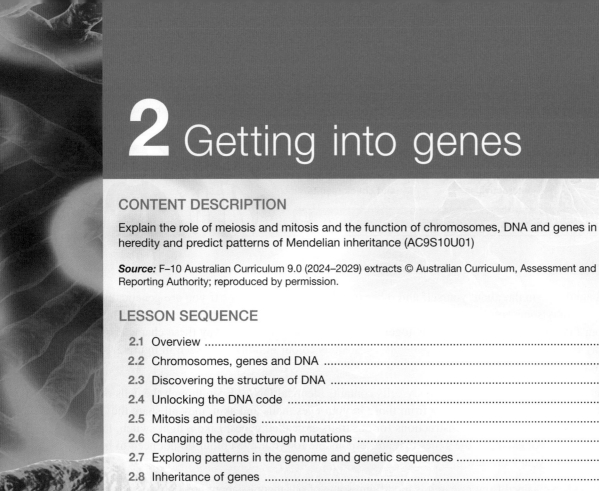

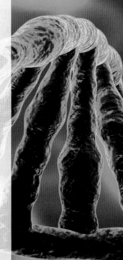

2 Getting into genes

CONTENT DESCRIPTION

Explain the role of meiosis and mitosis and the function of chromosomes, DNA and genes in heredity and predict patterns of Mendelian inheritance (AC9S10U01)

Source: F–10 Australian Curriculum 9.0 (2024–2029) extracts © Australian Curriculum, Assessment and Reporting Authority; reproduced by permission.

LESSON SEQUENCE

SCIENCE INQUIRY AND INVESTIGATIONS

Science inquiry is a central component of Science curriculum. Investigations, supported by a **Practical investigation eLogbook** and **teacher-led videos**, are included in this topic to provide opportunities to build Science inquiry skills through undertaking investigations and communicating findings

LESSON
2.1 Overview

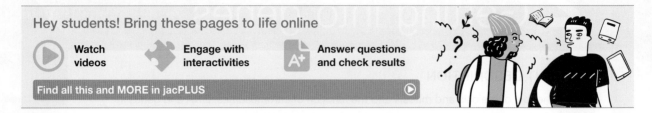
2.1.1 Introduction

Have you noticed anything similar about yourself and other members of your family? If you are genetically related, you share not only some similar characteristics, but also similar DNA (deoxyribonucleic acid). This molecule links your family generations genetically together. Patterns also exist in the way these characteristics are inherited. Many traits are inherited between generations, such as colour blindness (as seen in figure 2.1), hair colour and even the shape of your hairline.

Did you know that the nucleus of each of your body cells contains identical DNA? Even though the cells of your stomach look and function very differently from those in your eyes, nails and skin, they all share the same DNA. Sections of DNA that provide instructions for our traits are called genes. Like a light switch, these can be switched on or off. For example, the genes that are responsible for your eye colour do not need to be switched on in the cells of your stomach!

Increasingly, the knowledge gained from research is modifying some of our theories about how DNA is inherited. For example, research is suggesting that some of an organism's experiences and lifestyle choices can lead to a change in which genes are switched on or off — and that this change can be inherited by their next generation.

Our ability to control and analyse DNA is rapidly advancing — not just for medical treatments, but in creating genetically modified organisms. Technologies are also being developed that can edit an organism's genome!

In this topic, you will gain and understand of the relationship between your DNA and your genes, and how these are inherited across generations.

FIGURE 2.1 Can you read each of the numbers? This test is known as the Ishihara test, which examines colour blindness, a genetically inherited trait that is more common in males.

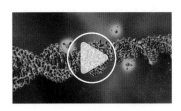
2.1.2 Think about genes

1. Do you have a Darwin's point?
2. Why don't hairs grow on stomach linings?
3. What do monks, peas, mathematics and genes have in common?
4. What do Xs and Ys have to do with sex?
5. What do telomeres have to do with aging?
6. How can you get bacteria to produce human insulin?
7. Why doesn't the number of chromosomes double with each generation?
8. Do all my family have the same blood type?
9. Are mutations always bad news?
10. What's this about 'jumping genes'?
11. Designer babies — should we or shouldn't we?
12. What does CRISPR have to do with editing your genome?
13. Who owns information about your DNA?

2.1.3 Science inquiry

elog-2426

INVESTIGATION 2.1

Do you fit into your genes or do they fit into you?

Aim

To observe the inheritance of traits across generations and make predictions

Materials

- pen
- paper
- collection of photographs of members of your family (or a group member's family)

PART A: Focusing on eye colour

Method

In your team, carefully observe the figure provided showing the Davis and Swift families and share your observations.

Results

Construct a table and record any observations you made.

Discussion

1. If the following couples had another child, suggest what their eye colour may be and give a reason for your suggestion.
   a. Ken and Margaret Davis
   b. Kevin and Gwenda Swift
   c. Geoff and Linda Davis
2. Suggest a reason Geoff (with brown eyes) and Linda (blue eyes) had brown-eyed and blue-eyed children, whereas Ken (with brown eyes) and Margaret (blue eyes) had only children with brown eyes.

3. Martin Swift's fiancée, Justine, has blue eyes, but both of her parents have brown eyes. If Justine and Martin have children together, what colour eyes do you think may be possible? Discuss reasons for your response.

PART B: Focusing on family photos

Method

1. In pairs, select one team member to bring to school a collection of photographs of as many individuals from their family as they can.
2. Carefully observe each photograph of the family members, looking for similarities.

Results

Construct a table with the family features that you have observed and indicate which family members have these features.

Discussion

1. Can you identify any patterns or make any interesting suggestions about these observations? If so, discuss and record these.
2. Discuss which characteristics may be passed from parent to child and which may not.
3. Make a summary of your discussion to share with other teams. Add any other interesting points from these discussions to your team summary.

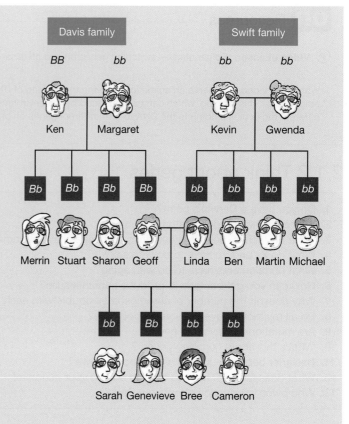

PART C: Focusing on your thoughts

Method

1. Conduct some research on each of the following topics:
   a. Because June and Frank have five sons, the chance of their next child being a daughter is increased.
   b. People who have committed very violent crimes should be sterilised because their children will also be violent.
   c. Parents should be allowed access to technologies that enable them to select the gender and specific characteristics of their children.
   d. Technologies that alter the gametes (sperm and ova) should be illegal.
2. Discuss each of the issues with your group and answer the discussion questions provided.

Results

Write a summary of your findings (both your own and your classmates).

Discussion

For each of the scenarios listed, state whether you agree, disagree or don't know. Justify your responses, providing clear reasoning for your answer.

Conclusion (for Parts A, B and C)

Summarise your findings from the three activities, outlining your understanding of patterns of inheritance and issues and debates around genetics.

Extension activity

As a class, conduct a class debate on the different issues in Part C. Place some students on the 'for' side and some on the 'against' side, with each group outlining reasons and arguments for the debate.

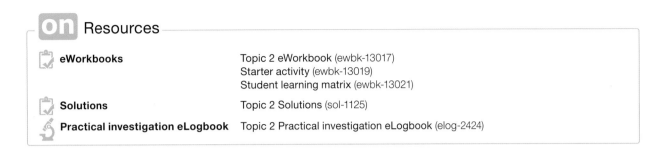
LESSON
2.2 Chromosomes, genes and DNA

LEARNING INTENTION

At the end of this lesson you will be able to describe the relationship between DNA, genes, chromosomes, nuclei and cells and distinguish between different types of cells and chromosomes.

2.2.1 Cells

We are living organisms. The key distinguishing feature of all living things is that they are cellular — comprised of basic biological units known as **cells**.

The human body is made up of trillions of cells, allowing us to have amazing complexity. Some organisms are much simpler. Organisms such as amoeba, for example, are only made up of a single cell.

The cells that make up humans and other animals differ in their structure and function. Some different types of cells include:

- skin cells
- nerve cells
- stem cells
- fat cells
- blood cells
- muscle cells.

With the exception of sex cells and red blood cells, all of the cells within an organism contain identical DNA in their nuclei.

> **cell** the smallest unit of life and the building blocks of living things

FIGURE 2.2 Cells in the body differ in structure and function

(a) Star-shaped (e.g. motor neuron cells)

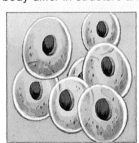

(b) Spherical (e.g. egg cells)

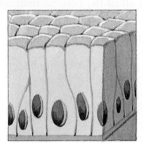

(c) Columnar (e.g. gut cells)

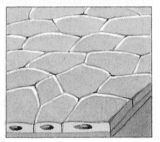

(d) Flat (e.g. skin cells)

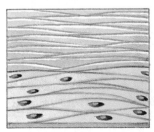

(e) Elongated (e.g. human smooth muscle cells)

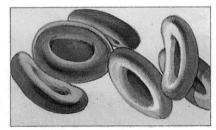

(f) Disc-shaped (e.g. human red blood cells)

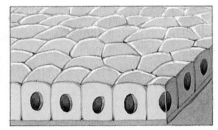

(g) Cuboidal (e.g. human kidney cells)

2.2.2 DNA

Where did you get those pointed ears, big nose or long toes from? Features or traits that are inherited are passed from one generation to the next in the form of a genetic code. This code is written in a molecule called **deoxyribonucleic acid (DNA)** and is located within the **nucleus** of your cells.

int-5870

FIGURE 2.3 DNA is found in the nucleus of cells. Prior to cell division, it coils up into structures that, when stained, can be seen with a microscope. These structures are called chromosomes.

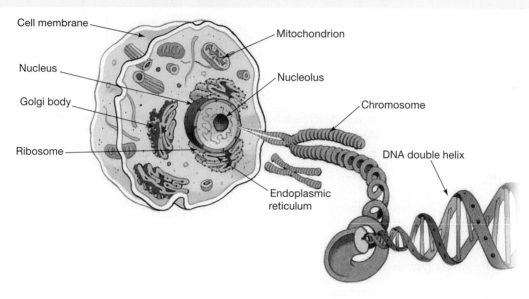

FIGURE 2.4 The hierarchical relationship between cells, nuclei, chromosomes, genes and DNA

Cell → Nucleus → Chromosome → Gene → DNA

Most of your body cells contain all of the genetic instructions needed to make another you. Your DNA, however, is more than just a genetic blueprint of instructions; it is also a unique 'ID tag' and an ancient 'book' that holds secrets both from your ancestral past and for your possible futures.

The structure of DNA will be further explored in lesson 2.3.

2.2.3 Genes

Each genetic instruction that codes for a particular trait (for example, shape of the ear lobe, blood group or eye colour) is called a **gene**. Genes contain the instructions for making a particular polypeptide (protein). Genes are made up of specific segments of DNA and are organised into larger structures called chromosomes, which are located within the nucleus of the cell.

FIGURE 2.5 Linked genes and unlinked genes

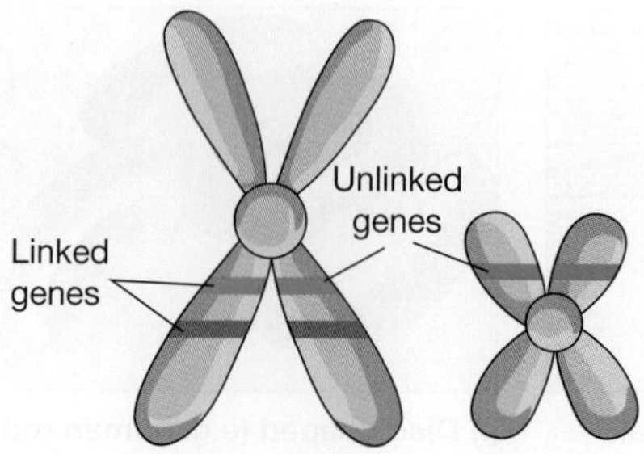

Linked genes

Unlinked genes

deoxyribonucleic acid (DNA) a substance found in all living things that contains its genetic information

nucleus roundish structure inside a cell that contains DNA and acts as the control centre for the cell

gene segment of a DNA molecule with a coded set of instructions in its base sequence for a specific protein product; when expressed, may determine the characteristics of an organism

The location of a gene on a specific chromosome is called its **locus**. Genes that are located on the same chromosome are described as being linked.

2.2.4 Chromosomes

Prior to cell division your DNA replicates itself, and this long double-stranded molecule (2 to 3 metres) bunches itself up into 46 little packages called **chromosomes**. They are called chromosomes (*chromo* = 'coloured' + *some* = 'body') because scientists often stain them with various dyes so that they are easier to see.

Chromosomes are only visible when a cell is about to divide or is in the process of dividing. When your cells are not dividing, chromosomes are not visible because the coils are unwound and the DNA is spread throughout the nucleus.

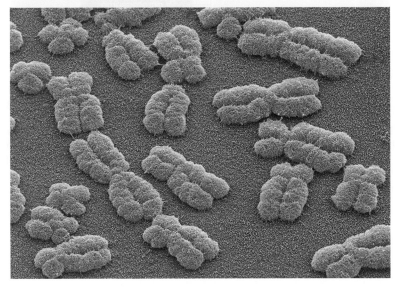

FIGURE 2.6 Scanning electron micrograph showing double-stranded chromosomes

locus position occupied by a gene on a chromosome

chromosomes tiny thread-like structures inside the nucleus of a cell that contain the DNA that carries genetic information

centromere a section of a chromosome that links sister chromatids

sister chromatids identical chromatids on a replicated chromosome

telomere a cap of DNA on the tip of a chromosome that enables DNA to be replicated safely without losing valuable information

kinetochore a region on a chromosome associated with cell division

chromatid one identical half of a replicated chromosome

Chromosomes have various structures that are present in both replicated and non-replicated forms. The **centromere** is a region in which **sister chromatids** and the short and long arms of chromosomes are attached. The **telomere** is the cap on the end of each chromosome. The **kinetochore** is a region associated with the centromere that is important in cell division. When chromosomes are replicated and appear in the 'X' shape, the two parts of the chromosome are each referred to as a **chromatid** or, more specifically, sister chromatids.

ewbk-13022

int-8123

FIGURE 2.7 The structure of a chromosome. The chromosome may be in an unreplicated or replicated form.

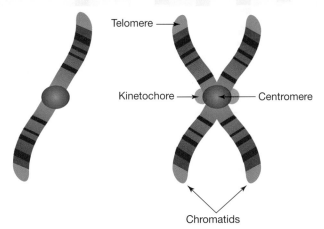

Telomere

Kinetochore — Centromere

Chromatids

FIGURE 2.8 Fluorescent-dyed chromosomes showing stained centromere

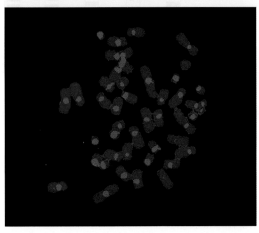

Chromosomes — more than one type

Chromosomes can be divided into two main types: **autosomes** and **sex chromosomes**.

X chromosomes and Y chromosomes are two types of sex chromosomes.

FIGURE 2.9 Exploring the types of chromosomes

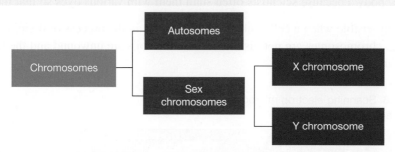

Autosomes

Of the 46 chromosomes in your **somatic cells**, 44 are present in both males and females and can be matched into 22 pairs on the basis of their relative size, position of centromere and stained banding patterns. These are called autosomes. They are numbered from 1 to 22 on the basis of their size: chromosome 1 being the largest of the autosomes and chromosome 22 the smallest (see figure 2.10).

The members of each matching pair of chromosomes are described as being **homologous** (these matching pairs can be seen in figure 2.12). Those that don't match are called non-homologous. For example, two number 21 chromosomes would be referred to as homologous, but a number 21 chromosome and a number 11 chromosome would be non-homologous.

autosomes non-sex chromosomes

sex chromosomes chromosomes that determine the sex of an organism

somatic cells cells of the body that are not sex cells

homologous chromosomes with matching centromeres, gene locations, sizes and banding patterns

FIGURE 2.10 Human chromosomes in order of size with banding patterns and centromere

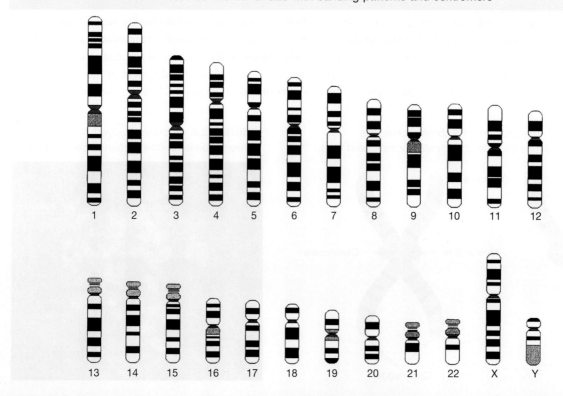

Sex chromosomes

The two remaining chromosomes are the sex chromosomes. In humans, these differ between males and females. Females possess a pair of X chromosomes (XX) and males have an X chromosome and a Y chromosome (XY). The sex chromosomes are important in determining an individual's biological sex (whether they are male or a female), as shown in figure 2.11.

FIGURE 2.11 The combination of sex chromosomes from the sperm and egg determines the sex of an individual.

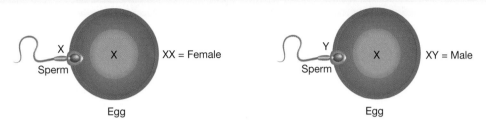

2.2.5 Chromosome patterns — karyotypes

Differences between the chromosome pair size, shape and banding can be used to distinguish them from each other. Scientists use these differences to construct a **karyotype**, as seen in figure 2.12. Cells about to divide are treated and stained, mounted on slides for viewing, and photographed.

These photographs are cut up and rearranged into pictures that show the chromosomes in matching pairs in order of size from largest to smallest. Karyotyping can reveal a variety of chromosomal disorders such as Down syndrome and Turner syndrome.

The sex of an individual can also be determined using karyotyping. In humans, females possess two similarly sized X sex chromosomes. In males, however, their sex chromosomes are not matching — they possess an X chromosome and a smaller Y chromosome. In figure 2.12, the individual is male.

karyotype an image that orders chromosomes based on their size

FIGURE 2.12 A human male karyotype

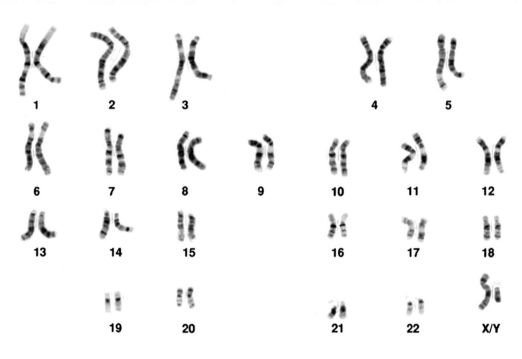

EXTENSION: Too many or too few?

Sometimes a genetic mistake or mutation can occur that results in more or less of a particular type of chromosome. Down syndrome is an example of a **trisomy** mutation in which there are three number 21 chromosomes instead of two. Turner syndrome is an example of a **monosomy** mutation that results in only one sex chromosome (XO).

TABLE 2.1 Some examples of chromosome changes and approximate incidence rates

Chromosome change	Resulting syndrome	Approximate incidence rate
Addition: whole chromosome		
Extra number 21 (47, +21)	Down syndrome	1/700 live births
Extra number 18 (47, +18)	Edwards syndrome	1/3000 live births
Extra number 13 (47, +13)	Patau syndrome	1/5000 live births
Extra sex chromosome (47, XXY)	Klinefelter syndrome	1/1000 male births
Extra Y chromosome (47, XYY)	N/A	1/1000 male births
Deletion: whole chromosome		
Missing sex chromosome (46, XO)	Turner syndrome	1/5000 female births
Deletion: part chromosome		
Missing part of number 4	Wolf–Hirschhorn syndrome	1/50 000 live births
Missing part of number 5	Cri-du-chat syndrome	1/10 000 live births

on Resources

📋 **eWorkbook**   Karyotype (ewbk-13026)

🔗 **Weblinks**   Karyotype - case studies
                  Virtual karyotype

2.2.6 Types of cells and cell division

All cells come from pre-existing cells. Cell division is a vital process in which replicated chromosomes are divided to form new cells.

Mitosis and meiosis are two types of cell division (figure 2.13) involved in cell production. The two types of cells produced are **gametes** (or sex cells) and somatic cells (body cells).

FIGURE 2.13 Examining cell division and the types of cells produced

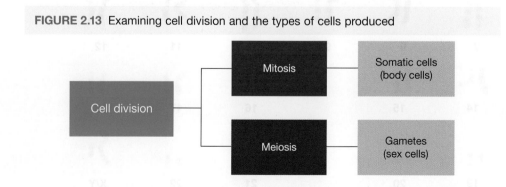

trisomy a condition where there are three copies of a particular chromosome instead of two

monosomy a condition where there is only one copy of a particular chromosome instead of two

gametes reproductive or sex cells such as sperm or ova

Somatic cells

Mitosis is the type of cell division used to produce new cells that your body requires for replacement, growth and repair. The cells produced are identical to each other and to the original cell, with the same number of chromosomes. Mitosis will be further examined in section 2.5.3.

Cells of your body that are not your sex cells are often referred to as body cells or somatic cells. With the exception of your red blood cells (which lose their nucleus when mature so they can carry more oxygen), all of your somatic cells contain chromosomes in pairs within their nucleus. Humans have 23 pairs.

Sex cells

Meiosis is the type of cell division used to produce sex cells (or gametes). **Ova** (singular ovum) and **sperm** are two types of gametes, seen in females and males respectively.

Meiosis results in the chromosome number being halved, so instead of pairs of chromosomes in each resulting cell, only one chromosome comes from each pair. Cells with one of every chromosome are referred to as **haploid**.

Genetic variation can increase the survival of a species. Meiosis contributes to this. The cells that are produced by this type of cell division are different from each other and from the original 'parent' cell. The random mixing of gametes produced by meiosis also contributes to variation within the species. Meiosis will be further examined in section 2.5.4.

Mixing gametes together

The genetic information that you received from your mother was packaged into 23 chromosomes in the nucleus of her egg cell (ovum). Your father's genetic information was packaged into 23 chromosomes in the nucleus of the sperm that fertilised your mother's egg cell.

When these gametes fused together at **fertilisation**, the resulting **zygote** contained 23 pairs of chromosomes (one pair from each parent) — a total of 46 chromosomes.

FIGURE 2.14 Meiosis leads to the formation of gametes, which come together during fertilisation.

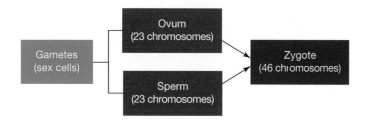

mitosis cell division process that results in new genetically identical cells with the same number of chromosomes as the original cell

meiosis cell division process that results in new cells with half the number of chromosomes of the original cell

ova female reproductive cells or eggs

sperm male reproductive cell

haploid the possession of one copy of each chromosome in a cell

fertilisation penetration of the ovum by a sperm

zygote a cell formed by the fusion of male and female reproductive cells

SCIENCE AS A HUMAN ENDEAVOUR: Has the secret of age reversal been discovered?

In the 1970s, Tasmanian-born scientist Dr Elizabeth Blackburn made a discovery that contributed to our understanding of how cells age and die. She showed how the presence of a cap of DNA called a telomere on the tip of the chromosome enabled DNA to be replicated safely without losing valuable information in the rest of the chromosome. Each time the cell divides, however, these telomeres shorten. When the telomeres drop below a certain length, the cell stops dividing and dies. This is a normal part of ageing.

Blackburn and her colleagues later discovered an enzyme, **telomerase**, that was involved in maintaining and repairing telomeres. In 2009, Blackburn and her colleagues were awarded the Nobel Prize in Physiology and Medicine for their work on how chromosomes are protected by telomeres and the enzyme telomerase.

FIGURE 2.15 Dr Elizabeth Blackburn

FIGURE 2.16 The telomeres shorten during mitosis, the cell division that produces somatic cells.

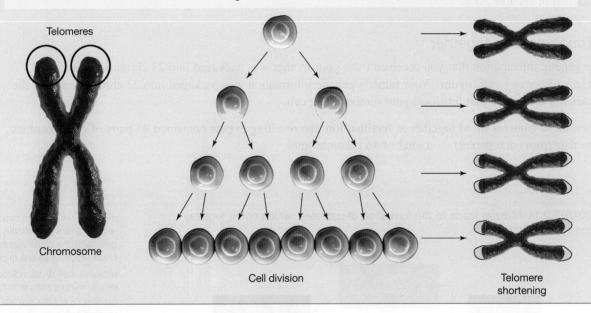

Telomeres

Chromosome

Cell division

Telomere shortening

DISCUSSION

Other scientists are now involved in finding out more about the exciting possibilities that our understanding of this telomeres process may open up. In 2010, for example, Mariela Jaskelioff and her colleagues in the USA genetically engineered mice with short telomeres and inactive telomerase to see what would happen when they turned this enzyme back on. Their results showed that after four weeks, new brain cells were developing and tissue in several organs had regenerated — and the mice were living longer. If this happens in mice, what might future research suggest for humans?

telomerase an enzyme involved in maintaining and repairing a telomere

INVESTIGATION 2.2

Extracting DNA

Aim

To extract DNA from ground wheatgerm

Materials

- 1 teaspoon of finely ground wheatgerm
- 14 mL of isopropyl alcohol (or equivalent)
- 1 mL of liquid detergent
- 20 mL of hot tap water (50–60 °C)
- test tube
- measuring cylinders
- rubber stopper
- test tube rack
- Pasteur pipette and bulb
- glass stirring rod

Method

1. Construct a table for the results, allowing room for observations in the form of a diagram.
2. Add the wheatgerm and hot water to a test tube. Twist the stopper in and shake for 3 minutes.
3. Add 1 mL of detergent and mix gently with the glass rod for about 5 minutes. Do not create foam. (If you do create foam, suck it out with the Pasteur pipette.)
4. Tilt the tube at an angle and slowly pour in the alcohol so that it sits at the bottom.
5. Collect the DNA with the glass rod. Feel it with your fingers and make your final observations.

Results

Note and sketch your observations in a table:
a. immediately after adding the alcohol
b. at 3- and 15-minute intervals
c. after you have collected and removed the DNA.

Discussion

1. What colour did you expect DNA to be? Why do you think it was the colour that you observed?
2. How could you confirm that it really was DNA? Relate your response to scientific knowledge (remember to reference your secondary information).
3. Why do you think detergent needed to be added?
4. Suggest improvements to the experimental design.

Conclusion

Summarise your findings for this investigation.

 Resources

 eWorkbook Genes and chromosomes (ewbk-13024)

2.2 Activities

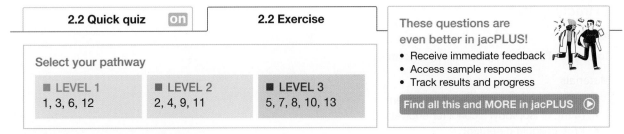

| 2.2 Quick quiz **on** | 2.2 Exercise |

Select your pathway

| ■ LEVEL 1 | ■ LEVEL 2 | ■ LEVEL 3 |
| 1, 3, 6, 12 | 2, 4, 9, 11 | 5, 7, 8, 10, 13 |

These questions are even better in jacPLUS!
- Receive immediate feedback
- Access sample responses
- Track results and progress

Find all this and MORE in jacPLUS ⏵

Remember and understand

1. **a.** State whether the following statements are true or false.

Statements	True or false?
i. Human female gametes are called ova and male gametes are called sperm.	
ii. Mitosis is the type of cell division that is used to produce sex cells.	
iii. All chromosomes are the same size.	
iv. Non-homologous chromosomes have the same relative size, position of centromere and stained banding patterns.	
v. Human males possess a pair of X chromosomes (XX), whereas human females possess an X and a Y chromosome (XY).	
vi. Sometimes a genetic mistake or mutation can occur that results in more or less of a particular type of chromosome.	
vii. Differences between the chromosome pair size, shape and banding can be used to distinguish them from each other and to construct a karyotype.	
viii. Telomeres (caps of DNA on the tip of the chromosome), which enable DNA to be replicated safely without losing valuable information, shorten each time the cell divides.	

   b. Rewrite any false statement to make it true.

2. Match the term to its description in the following table.

Term	Description
a. Autosomes	**A.** A molecule that contains genetic information and is located inside the nucleus of a eukaryotic cell
b. Chromosome	**B.** A piece of genetic information that contains the instructions for making a particular polypeptide
c. DNA	**C.** Another name for body cells that are not sex cells
d. Fertilisation	**D.** Another name for sex cells
e. Gametes	**E.** The fusion of the sperm and the ovum
f. Gene	**F.** The location of DNA within your cells
g. Linked	**G.** The name given to non-sex chromosomes
h. Locus	**H.** Genes located on the same chromosome
i. Nucleus	**I.** The location of a gene on a chromosome
j. Somatic cells	**J.** A structure containing genes, located in the nucleus of eukaryotic cells and only visible when the cell is about to divide or is in the process of dividing

3. Are chromosomes always visible in a cell? Explain.
4. Explain how a karyotype can determine the following.
   a. The gender of an individual
   b. Chromosomal abnormalities

5. Distinguish between the following pairs of terms.
   a. Ovum and sperm
   b. X chromosome and Y chromosome
   c. Sex chromosomes and autosomes
   d. Somatic cells and sex cells
   e. Mitosis and meiosis
   f. Homologous and non-homologous
   g. Gene and DNA
6. Refer back to and carefully and observe figure 2.10, which shows some features of human chromosomes, and answer the questions.
   a. Is chromosome 6 larger or smaller than chromosome 2?
   b. Which is smaller, the X chromosome or the Y chromosome?

Apply and analyse

7. **SIS** Suggest why chromosomes are stained with dyes. What observations would you expect under a microscope if a dye wasn't used?
8. The following questions refer to figures A and B.

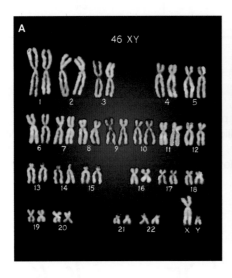 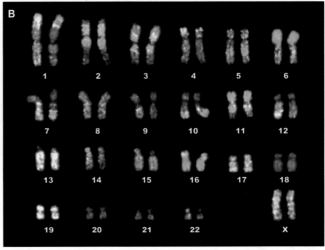

a. Carefully observe figures A and B and suggest features that are useful in pairing the chromosomes.
b. On the basis of information in the karyotype, suggest the sex of A and B. Justify your responses.
c. Suggest why karyotyping can be carried out only on cells that are about to divide.

9. **SIS** Observe the figures provided that show the chromosomes belonging to four different types of organisms.

Kangaroo (6 pairs)   Human (23 pairs)   Domestic fowl (18 pairs)   Fruit fly (4 pairs)

a. Suggest whether the chromosomes are from somatic cells or sex cells. Justify your response.
b. Suggest which organisms possess chromosomes:
   i. most like humans
   ii. least like humans.
   Justify your responses.
c. Do any observations in part **b** surprise you? Why?

Evaluate and create

10. **SIS** Each species has a particular number of chromosomes. The following table shows some examples of the number of chromosomes in the body cells of some organisms.

TABLE Number of chromosomes in body cells (non-sex cells) of some living things

Species of living thing	Number of chromosomes in each body cell	Species of living thing	Number of chromosomes in each body cell
Chimpanzee	48	Tomato	24
Euglena (unicellular)	90	Cabbage	18
Fruit fly	8	Frog	26
Human	46	Housefly	12
Koala	16	Pig	40
Onion	16	Platypus	52
Shrimp	254	Rice	24
Sugarcane	80	Sheep	54

  a. Construct a column graph using the data in the table provided.
  b. Identify the species with the following.
    i. Highest total number of chromosomes
    ii. Lowest total number of chromosomes
  c. Carefully observe your graph, looking for any patterns. Discuss possible reasons for these.
  d. Do you think that the number of chromosomes reflects the intelligence of an organism? Provide reasons for your response.
  e. Suggest the number of chromosomes in the sex cells of the following.
    i. Housefly
    ii. Sheep

11. Carefully observe the karyotypes A and B.

  a. Suggest the gender of the individual in A and in B. Justify your responses.
  b. Use your observations and table 2.1 to suggest the following.
    i. The type of chromosome change shown in each figure
    ii. The name of the resulting genetic disorders
  c. One of these disorders is also sometimes described as trisomy 21. Suggest a reason for this description.
12. If each cell nucleus has about a metre of DNA, how does it all fit in? Draw a diagram to show this.

13. **SIS** **a.** Research and report on the following aspects of Elizabeth Blackburn's work and explain the following.
    i. Her contribution to our understanding of DNA
    ii. Her stance on stem cell science that resulted in her losing her position on the President's Council on Bioethics
  b. i. What is bioethics?
    ii. What is the Presidential Commission for the Study of Bioethical Issues, and what does it have to do with science? How does it differ from the President's Council on Bioethics?
    iii. Find out more about the types of issues that have been considered by the Commission. Write four clear bullet points summarising these issues, and outline your personal opinions on them.

Fully worked solutions and sample responses are available in your digital formats.

LESSON
2.3 Discovering the structure of DNA

LEARNING INTENTION

At the end of this lesson you will be able to outline the key events in the discovery and understanding of the structure of DNA.

2.3.1 Learning about DNA

As discussed in lesson 2.2, DNA is the abbreviation for deoxyribonucleic acid. As the name suggests, it is a type of nucleic acid. It was not until 1869 that people were formally introduced to DNA, when it was discovered by Friedrich Miescher. Working in a laboratory located within a castle in Germany, Miescher — a young Swiss postgraduate student — isolated DNA from the nuclei of white blood cells taken from pus on bandages. Miescher named the compound he isolated 'nuclein'.

FIGURE 2.17
Friedrich Miescher

2.3.2 The nucleotide

In 1929, over 50 years after its discovery, Phoebus Levene showed that DNA was made up of repeating units called **nucleotides**.

Each of these nucleotides consisted of:
- a sugar (**deoxyribose** in DNA)
- a phosphate group
- a **nitrogenous base** (adenine (A), thymine (T), guanine (G) or cytosine (C)).

nucleotides compounds (DNA building blocks) containing a sugar part (deoxyribose or ribose), a phosphate part and a nitrogen-containing base that varies

deoxyribose the sugar in the nucleotides that make up DNA

nitrogenous base a component of nucleotides that may be one of adenine, thymine, guanine, cytosine or uracil

FIGURE 2.18 a. The structure of a nucleotide **b.** A nucleotide containing the nitrogenous base guanine

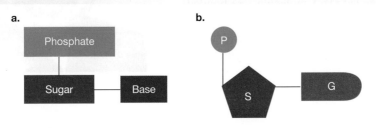

Phoebus Levene (see figure 2.19) suggested that the nucleotides could be joined to form chains. Although his theory was correct in terms of the chain formation, it was incorrect in other aspects of its structure. His tetranucleotide model, which proposed that DNA had equal amounts of all four nitrogenous bases, contributed to scientists of the time favouring proteins, rather than DNA, as the carrier of genetic information.

The experiments of Alfred Hershey and Martha Chase in 1953 supported those of Oswald Avery in 1943 (figure 2.20), suggesting that DNA was the molecule through which genetic information was carried between generations, not proteins.

FIGURE 2.19 Phoebus Levene

FIGURE 2.20 Oswald Avery

2.3.3 A with T and G with C

In 1950, Erwin Chargaff (figure 2.21) contributed to our understanding of the structure of DNA by his careful and thorough analysis of the four types of nucleotides and their ratios in DNA. His research led to the concept of base pairing.

This concept states that in DNA:
- every adenine (A) binds to a thymine (T)
- every cytosine (C) binds to a guanine (G).

This is now known as **Chargaff's rule**.

Therefore, any strand of DNA should have equal numbers of A and T, and equal numbers of C and G.

The different bases in the double stranded DNA molecules are bonded using a special bond known as a hydrogen bond. While C and G bond together with three hydrogen bonds, T and A bond together with two hydrogen bonds. The base-pairing rule and the presence of hydrogen bonds can be seen in figure 2.22.

FIGURE 2.21 Erwin Chargaff

Chargaff's rule a rule that states the pairing of adenine with thymine and cytosine with guanine

FIGURE 2.22 Observing DNA base pairing within a strand of DNA. Think of base pairing like puzzle pieces matching together.

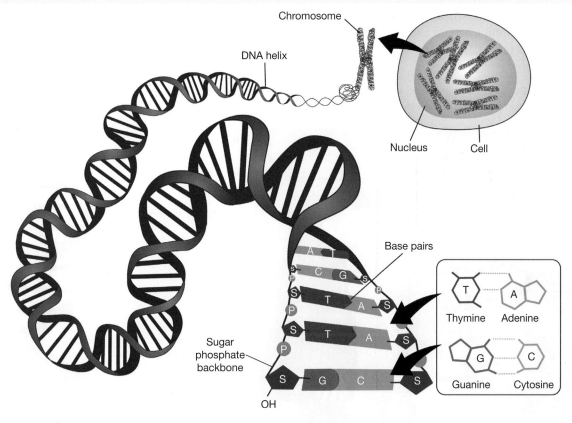

Chromosome

DNA helix

Nucleus

Cell

Base pairs

Sugar phosphate backbone

OH

T · · · · · A
Thymine   Adenine

G · · · · · C
Guanine   Cytosine

2.3.4 Understanding the structure of DNA

The next piece of the puzzle of solving the structure of DNA was contributed by Rosalind Franklin. Rosalind Franklin and Maurice Wilkins had decided to crystallise DNA so that they could make an X-ray pattern of it (as seen in figure 2.23). They specialised in making images using the properties of X-rays of biological molecules so the images could be analysed to find out information about their three-dimensional structures. Franklin's X-ray picture of a DNA molecule provided important clues about the shape of the molecule.

FIGURE 2.23 Rosalind Franklin's X-ray picture provided important clues about the shape of the DNA molecule.

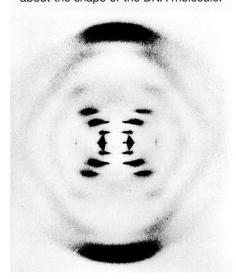

FIGURE 2.24 Rosalind Franklin provided a key clue to solve the structure of DNA.

2.3.5 Discovering the double helix

James Watson and Francis Crick were building a DNA model to try to solve its structure. They were shown Franklin's image of DNA, which strongly suggested that DNA had a helical shape. They used this information, as well as that from Chargaff and other researchers (such as their US colleague Linus Pauling), to successfully solve the structure of DNA. At last the structure was identified!

eles-4213

FIGURE 2.25 a. Watson (left) and Crick (right) with their model of part of a DNA molecule in 1953
b. A schematic structure of a DNA double helix

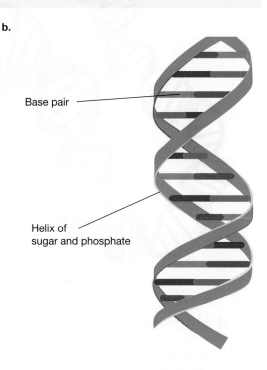

Base pair

Helix of
sugar and phosphate

DISCUSSION

Rosalind Franklin and Maurice Wilkins were critical in discovering the structure of DNA; however, most of the credit is given to Watson and Crick. If Maurice Wilkins and Rosalind Franklin had a more harmonious working relationship, do you think Franklin would have been involved in writing the scientific paper where the structure of DNA was first described, and that she would have been given the same credit for discovering the structure of DNA as Watson and Crick?

INVESTIGATION 2.3

Constructing a model of DNA

Aim

To construct a DNA model in order to represent the structure of DNA

Materials

Various materials to assist you to create a model — some suggestions include pipe cleaners, straws, paper, icypole sticks, paperclips, toothpicks or playdough

Method

1. In groups of two, construct a model of DNA using items you have available. In your model, you need to ensure that you:
   • clearly show the four different types of nitrogenous bases
   • have at least eight base pairs in total, ensuring you follow Chargaff's rule
   • are able to show the double helix structure of DNA.
2. Label all parts of your DNA model. You may choose to use a legend or key to assist you.

Results

Draw a clearly labelled diagram of your DNA model.

Discussion

1. Evaluate your model.
   a. Which aspects of the structure of DNA does your model show accurately?
   b. In what ways is your model different from an actual DNA molecule?
2. a. How many of each of the following nitrogenous bases were in your model?
      i. Adenine
      ii. Guanine
      iii. Cytosine
      iv. Thymine
   b. How did you represent each of these?
3. Bonds that connect the two strands of DNA are known as hydrogen bonds. What did you use to represent these hydrogen bonds?
4. DNA when stretched out is incredibly long. However, it can condense in such a way that the total length is very small. This can occur by wrapping it around special proteins known as histones (visualise this as being like wrapping toilet paper tightly around a toilet roll). If you had to adjust your model to show this, how would you do this? Summarise the process you would use.

Conclusion

Summarise your investigation, outlining how your model showed the structure of DNA.

on Resources

eWorkbook	DNA (ewbk-13030)	
Video eLessons	Rosalind Franklin and Watson and Crick (eles-1782)	
	DNA structure (eles-4211)	
Interactivity	Complementary DNA (int-0133)	
Weblink	The history of DNA timeline	

2.3 Activities

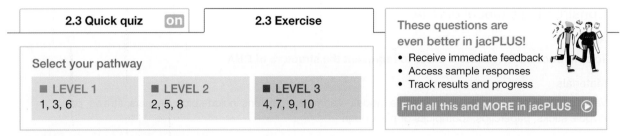

2.3 Quick quiz **on**   **2.3 Exercise**

Select your pathway

■ LEVEL 1	■ LEVEL 2	■ LEVEL 3
1, 3, 6	2, 5, 8	4, 7, 9, 10

These questions are even better in jacPLUS!
- Receive immediate feedback
- Access sample responses
- Track results and progress

Find all this and MORE in jacPLUS ▶

Remember and understand

1. **a.** State whether the following statements are true or false.

Statements	True or false?
i. DNA are the subunits of nucleotides.	
ii. DNA was first isolated from the nuclei of white blood cells taken from pus on bandages.	
iii. Miescher named the compound he extracted from cell nuclei 'nuclein'.	
iv. Prior to DNA being proposed as the molecule that passes genetic information between generations, it was thought to be protein molecules.	
v. According to Chargaff's rule, adenine binds to guanine and cytosine binds to thymine.	
vi. A nucleotide is made up of a phosphate group, a sugar, and a nitrogenous base.	
vii. All nucleotides that make up DNA contain the same type of nitrogenous base.	
viii. DNA has the structure of a double helix.	

   b. Rewrite any false statement to make it true.

2. Match the scientist to their scientific contribution in the following table.

Scientist	Scientific contribution
a. Erwin Chargaff	**A.** The scientist who first isolated (discovered) DNA.
b. Francis Crick	**B.** The scientist who showed that DNA was made up of repeating units called nucletotides joined to form chains.
c. Friedrich Miescher	**C.** The scientist who, together with Alfred Hershey and Oswald Avery, suggested that, rather than protein, DNA was the molecule through which genetic information was carried between generations.
d. Martha Chase	**D.** The scientist who, together with James Watson, suggested DNA has the structure of a double helix.
e. Phoebus Levene	**E.** The scientist whose research led to the concept of base pairing, which states that in DNA every adenine (A) binds to a thymine (T), and every cytosine (C) binds to a guanine (G).
f. Rosalind Franklin	**F.** The scientist whose X-ray diffraction picture provided important clues about the shape of the DNA molecule.

3. Use a diagram to show how the nucleotides that make up DNA are organised.
4. Outline what the research of Hershey, Chase and Avery suggested.

Apply and analyse

5. **SIS** Chargaff's rule was an important discovery in gaining understanding about the structure of DNA.
    a. Describe what is meant by Chargaff's rule.
    b. An examination determined that a specific fragment of DNA was 23 per cent adenine. What percentage would you expect the following to be?
        i. Thymine   **ii.** Guanine   **iii.** Cytosine
6. Describe the contributions of Rosalind Franklin, Maurice Wilkins, James Watson and Francis Crick to the discovery of the structure of DNA.
7. **SIS** Find out more about DNA and how knowledge about its structure is being used in research and other applications. Write five clear examples of this.

Evaluate and create

8. **SIS** Assume Watson and Crick used the scientific method to allow them to discover the structure of DNA. Write an aim and hypothesis they may have used to assist in their discovery.
9. **SIS** Investigate more about the history of how we have obtained our genetic knowledge. Present your findings as a timeline.
10. Investigate, provide examples and evaluate the effect that our increased knowledge about the structure and function of DNA has had on the following.
    a. Our species
    b. Other species
    c. Our planet

Fully worked solutions and sample responses are available in your digital formats.

LESSON
2.4 Unlocking the DNA code

LEARNING INTENTION

At the end of this lesson you will be able to explain how the code in DNA is transcribed into mRNA and then translated into proteins.

2.4.1 The universal genetic code

Did you know that all living things share the same genetic letters? This universal genetic language provides strong evidence that all life on Earth evolved from one ancient cell line.

All DNA is made up of nucleotides and the same four nitrogenous bases: adenine, guanine, cytosine and thymine. The order these bases appear is what makes each organism unique.

FIGURE 2.26 The hierarchical relationship between the cell, nucleus, DNA and nucleotides

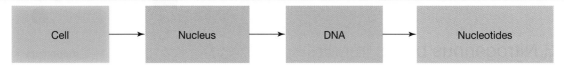

Cell → Nucleus → DNA → Nucleotides

Like other eukaryotic organisms, DNA is located within the nucleus and mitochondria of your cells. These differences will be considered later. In this topic, we will focus mostly on nuclear DNA.

DNA provides the blueprint or instructions for **proteins** to be created, with different sections of DNA (genes) coding for particular proteins.

> **proteins** molecules, such as enzymes, haemoglobin and antibodies made up of amino acids

2.4.2 Stepping down the DNA ladder

Nucleotides are the **monomers** of nucleic acids. So, like other **nucleic acids**, DNA molecules are **polymers** of nucleotides.

Each nucleotide is made up of three parts: a sugar part, a phosphate part and a nitrogenous base (refer to figure 2.27).

While the sugar (in this case, deoxyribose) and phosphate are always the same for each nucleotide in DNA, the nitrogenous base may vary. The four possible bases in DNA are adenine (A), thymine (T), cytosine (C) and guanine (G).

FIGURE 2.27 Reviewing the parts that make up a nucleotide

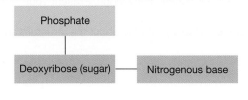

FIGURE 2.28 The different nitrogenous bases in DNA

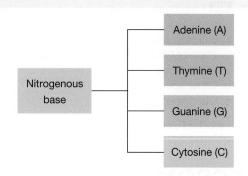

The nucleotides are joined together in a chain. The sugar and phosphate parts make up the outside frame and the nitrogenous bases are joined to the sugar parts (refer to figure 2.29). In DNA, two strands, joined by hydrogen bonds, come together to form a double-stranded DNA molecule (refer to figure 2.30).

FIGURE 2.29 Nitrogenous bases are attached to the sugar part of the nucleotide.

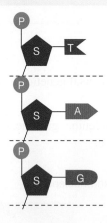

FIGURE 2.30 DNA forming a double stranded molecule

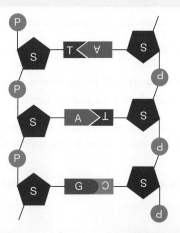

2.4.3 Nitrogenous bases in pairs

A DNA molecule is made up of two chains of nucleotides. Hydrogen bonds join them at their complementary (or matching) nitrogenous base pairs. Adenine binds to thymine and cytosine to guanine. This matching of the nitrogenous bases is often referred to as the base-pairing rule or Chargaff's rule.

monomers molecules that are the building blocks of larger molecules known as polymers

nucleic acids molecules composed of building blocks called nucleotides, which are linked together in a chain

polymers molecules made of repeating subunits of monomers joined together in long chains

For example, a segment of DNA that has one strand with the code GATTACA would have a complementary strand of CTAATGT. The bases in its double-stranded view would be as follows:

GATTACA

CTAATGT

DNA molecules have the appearance of a double helix or spiral ladder. Using the spiral ladder metaphor, DNA could be considered as having a sugar–phosphate backbone or frame, and rungs or steps that are made up of **complementary base pairs** of nitrogenous bases joined together by hydrogen bonds.

complementary base pairs in DNA, specific base pairs will form between the nitrogenous bases adenine (A) and thymine (T) and between the bases cytosine (C) and guanine (G)

triplet a sequence of three nucleotides in DNA that can code for an amino acid

amino acid an organic compound that forms the building blocks of proteins

2.4.4 Unlocking DNA codes

The sequence of nucleotides in DNA is often described in terms of the nitrogenous bases that they contain. These are often read in groups of three — this group of three is known as a **triplet** (as seen in figure 2.31, with the triplet TAG).

The DNA code is read three bases at a time. Although some of these DNA triplets code for a start (for example, TAC) or stop (for example, ATT, ATC or ACT) instruction, most triplets code for a particular **amino acid**. The triplet GAT, for example, codes for the amino acid aspartic acid. The process of making proteins will be further explored in section 2.4.6.

The sequence of these triplets in DNA contains the genetic information to assemble your body's proteins from amino acids. This includes all of your hormones, enzymes, antibodies and many other proteins that are essential for your survival. If one of these triplets (or its bases) are incorrect or missing, this may result in a protein not being coded for or produced — which could result in death.

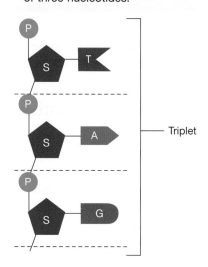

FIGURE 2.31 A triplet is made of three nucleotides.

FIGURE 2.32 The links between DNA and proteins

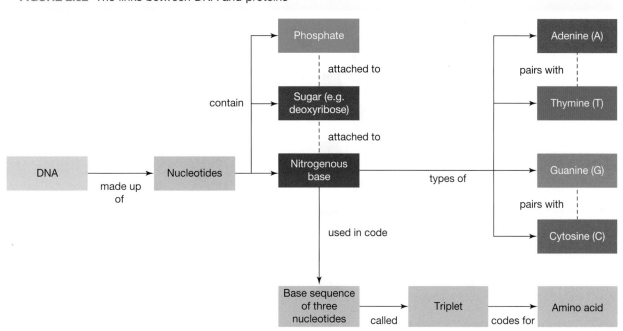

2.4.5 Introducing RNA

Like DNA, **ribonucleic acid (RNA)** is a type of nucleic acid and is made up of nucleotides. Its nucleotides, however, are different from those of DNA.

Unlike DNA, RNA:
- contains the sugar **ribose** (instead of deoxyribose)
- contains the nitrogenous base uracil (instead of thymine)
- is shorter and single-stranded, so it can fit through a nuclear pore.

These differences are highlighted in figure 2.33.

RNA comes in three mains forms: **messenger RNA (mRNA)**, **transfer RNA (tRNA)** and **ribosomal RNA (rRNA)**. All are involved the process of protein synthesis.

2.4.6 Protein synthesis: reading the code

The instructions for making proteins are coded for in the sequence of nitrogenous bases in DNA. DNA, however, cannot get out of the nucleus because it is too big to fit through the nuclear pores. So, DNA makes mRNA by a process called **transcription**. The RNA then moves to a **ribosome** in the cytoplasm where the genetic message is translated into a protein in a process known as **translation**.

ribonucleic acid (RNA) a type of nucleic acid that contains ribose sugar

ribose the sugar found in nucleotides of RNA

messenger RNA (mRNA) single-stranded RNA transcribed from a DNA template that then carries the genetic to a ribosome to be translated into a protein

transfer RNA (tRNA) molecules located in the cytosol that transport specific amino acids to complementary mRNA codons in the ribosome

ribosomal RNA (rRNA) a special type of RNA that forms the structure of ribosomes

transcription the process by which the genetic message in DNA is copied into a mRNA molecule

ribosome organelle found in the cells of all organisms in which translation occurs

translation the process in which amino acids are joined in a ribosome to form a protein

FIGURE 2.33 Comparing the differences between **a.** DNA and **b.** RNA

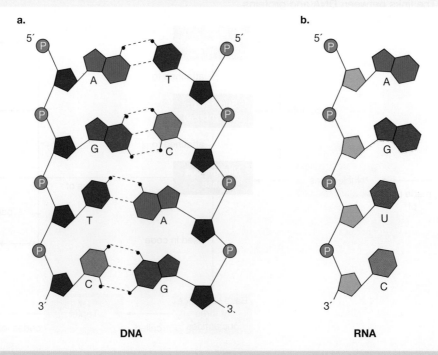

DNA   RNA

FIGURE 2.34 The process of reading DNA to produce a protein

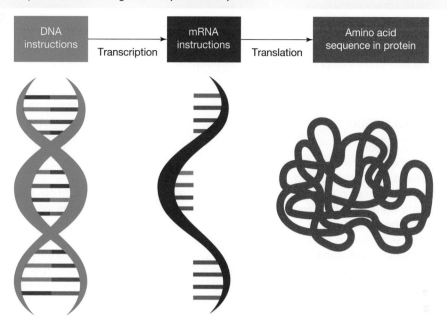

ACTIVITY: Metaphors and DNA

With increased knowledge and understanding, previous metaphors used to describe DNA are increasingly appearing to be inadequate in describing its complexities. The double helix, for example, describes its shape but not its function.

a. Find out more about two of the following metaphors and suggest reasons each is becoming less useful.
  - Double helix
  - Computer code of life
  - Chemical building block
  - Alphabet of life
  - Blueprint
  - Book of life
b. In six words or fewer, suggest a metaphor that could be used to communicate what DNA is all about — especially to those who do not have a background in biology. Provide reasons to support the use of your metaphor.

2.4.7 Transcription

The first step in making a protein involves the unzipping of the gene's DNA. When the relevant part of the DNA strand is exposed (the template strand), a special copy of the sequence is produced in the form of messenger RNA (mRNA) (figure 2.36). The process of making this complementary mRNA copy of the DNA message is called transcription.

FIGURE 2.35 A section of the DNA unzips so that the mRNA copy can be made.

FIGURE 2.36 Using a DNA template to make an mRNA transcript

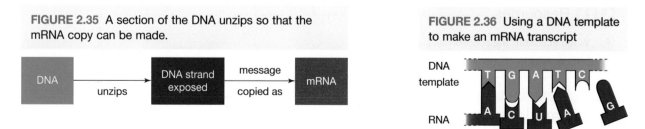

During transcription, an RNA molecule is formed with bases complementary to the DNA's base sequence.

When transcription finishes, modifications are made to the DNA before it leaves the nucleus. A special cap and tail are added to either side of the mRNA. Sections of DNA known as introns are removed (non-coding sequences), leaving sections called exons (which have the information required to form a protein in translation).

As its name suggests, messenger RNA (mRNA) passes through the pores of the nuclear membrane into the cytoplasm to take its genetic copy of the protein instruction message to ribosomes. These may be free floating in the cytosol or attached to the rough endoplasmic reticulum.

The steps of transcription

In the process of transcription:
1. DNA is unzipped (the two strands are separated).
2. One strand is used as a template strand.
3. RNA nucleotides that are complementary to the DNA template are added to form a mRNA strand using a special enzyme known as RNA polymerase. (The complementary mRNA codon for the start triplet TAC in DNA, for example, would be AUG.)
4. The newly synthesised mRNA is released from the template.
   For example, a DNA template with the sequence TACGACG would be transcribed into mRNA as AUGCUGC. *Remember, there is no T in RNA — it is instead replaced with a U.*

FIGURE 2.37 Summarising the process of transcription

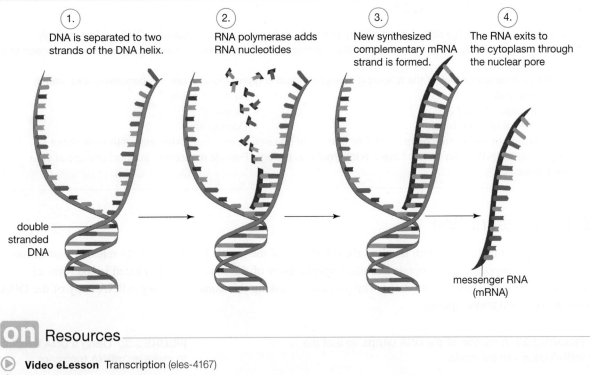

1. DNA is separated to two strands of the DNA helix.
2. RNA polymerase adds RNA nucleotides
3. New synthesized complementary mRNA strand is formed.
4. The RNA exits to the cytoplasm through the nuclear pore

double stranded DNA

messenger RNA (mRNA)

on Resources

▶ **Video eLesson** Transcription (eles-4167)

🧩 **Interactivity** Transcription (int-8125)

2.4.8 Translation

Ipsa scientia potestas est. Unless you speak Latin, you will need some help to translate this sentence! Once it is translated, you can then do something with it. This is similar to the meaning of the sentence: Knowledge itself is power.

Once the mRNA has reached the ribosome, its message needs to be translated into a protein. The mRNA is read in groups of three. This group of three is known as a **codon**. Each codon provides the instructions to add a specific amino acid, as shown in figure 2.38.

The ribosome and another type of molecule called transfer RNA (tRNA) are involved in this process. tRNA already located in the surrounding cytosol collects and transfers the appropriate amino acid to its matching code on the mRNA. These amino acids are joined by peptide bonds to make a protein (or polypeptide), as seen in figure 2.38.

The steps of translation

In the process of translation:

1. mRNA enters the ribosome and is read one codon at a time (always starting with AUG).
2. A tRNA molecule brings the matching amino acid for the codon to the ribosome. The amino acid is determined by the codon chart, as seen in figure 2.38.
3. The appropriate amino acid is added to form a protein (or polypeptide).
4. The next codon is read, and another amino acid is added.
5. Amino acids continue to be added, joining together as a protein.
6. When a STOP codon is reached, translation stops and the new protein is released.

FIGURE 2.38 Amino acid codon chart. AUG is the start codon and is usually first in an mRNA sequence.

First letter	Second letter: U	Second letter: C	Second letter: A	Second letter: G	Third letter
U	UUU Phe	UCU Ser	UAU Tyr	UGU Cys	U
U	UUC Phe	UCC Ser	UAC Tyr	UGC Cys	C
U	UUA Leu	UCA Ser	UAA Stop	UGA Stop	A
U	UUG Leu	UCG Ser	UAG Stop	UGG Trp	G
C	CUU Leu	CCU Pro	CAU His	CGU Arg	U
C	CUC Leu	CUC Pro	CAC His	CGC Arg	C
C	CUA Leu	CCA Pro	CAA Gln	CGA Arg	A
C	CUG Leu	CCG Pro	CAG Gln	CGG Arg	G
A	AUU Ile	ACU Thr	AAU Asn	AGU Ser	U
A	AUC Ile	ACC Thr	AAC Asn	AGC Ser	C
A	AUA Ile	ACA Thr	AAA Lys	AGA Arg	A
A	AUG Met	ACG Thr	AAG Lys	AGG Arg	G
G	GUU Val	GCU Ala	GAU Asp	GGU Gly	U
G	GUC Val	GCC Ala	GAC Asp	GGC Gly	C
G	GUA Val	GCA Ala	GAA Glu	GGA Gly	A
G	GUG Val	GCG Ala	GAG Glu	GGG Gly	G

So, for example, if a mRNA code (shown in figure 2.39) is:

AUG – UAC – GGU

The amino acids in this protein would be (in order)

Met – Tyr – Gly

codon sequence of three bases in mRNA that codes for a particular amino acid

FIGURE 2.39 mRNA codons code for particular amino acids.

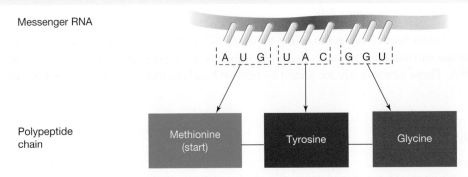

As seen in figure 2.38, there are three stop codons that provide an end point for protein synthesis. These are UAA, UAG and UGA. So for the following mRNA code, you will first need to find the start amino acid (Met), then determine the amino acids for the codons but do not go past the stop codon.

mRNA   UGAUC-AUG-AUC-UCG-UAA-GAU-AUC

Amino acid   Met - Ile - Ser - Stop

Each tRNA molecule (as shown in figure 2.40) has three bases called an anti-codon. This anti-codon is complementary to the codon on the mRNA (for example, if the codon is CAU, the anti-codon is GUA). This is what enables the tRNA to ensure the correct amino acid is being delivered.

FIGURE 2.40 The process of translation

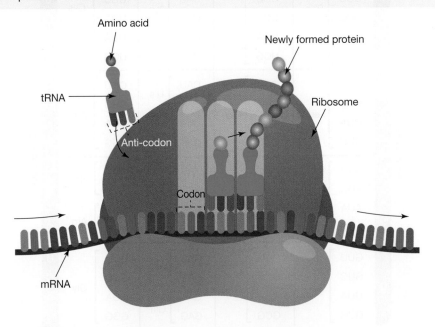

 Resources

- ▶ **Video eLesson**   Translation (eles-4168)
- ✦ **Interactivity**   Translation (int-8126)

2.4.9 Precious proteins

ewbk-13032

int-8127

FIGURE 2.41 Summarising the process of protein synthesis

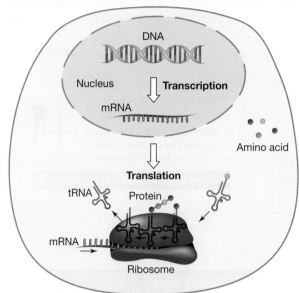

Why are proteins so important?

Proteins form parts of cells, regulate many cell activities and even help defend against disease.

- Your heart muscle tissue contains special proteins that can contract, enabling blood containing haemoglobin and hormones to be pumped through your body. Haemoglobin is a protein that carries oxygen necessary for cellular respiration.
- Many hormones are proteins. Insulin, glucagon and adrenaline, for example, are hormones that influence activities of your cells.
- Enzymes are also made up of protein and can be involved in regulating metabolic activities such as those in chemical digestion and respiration.
- Antibodies are examples of proteins that play a key role in your immune system in its defence against disease.

Plants also rely on proteins for their survival. Their growth and many other essential activities are regulated by hormones (such as auxins) and enzymes.

The production of proteins or polypeptides through transcription and translation is a carefully regulated process (figure 2.41). The formation of proteins through this process is vital for the survival of all organisms — from single-celled bacteria to complex organisms such as humans.

ACTIVITY: Rhymes of protein synthesis

Rhymes such as the following help us remember new information. Read or sing it, spelling out the triplets and codons with your fingers. Create your own rhyme about protein synthesis.

DNA is in my genes	**DNA bases read times three**	**ATT, ACT, ATC**
Tells me how to make proteins	Always starting with TAC	Stop making proteins for memRNA
Got my genes from Mum and Dad	mRNA codon would be AUG	codons for this would be
Mixed them up and made me glad	DNA triplets tell the story of me	UAA, UGA, UAG
DNA is in my genes	DNA bases read times three	Always in a group of three
Tells me how to make proteins	Always starting with TAC	Stop making proteins for me

---

2.4.10 Switched on or off?

Different genes are responsible for different characteristics, such as the colour of flower petals, the markings on a snail shell, or a person's blood group or eye colour. Every body cell in an organism has the same set of genes, called a genome, but not all genes are active. When genes are 'active' they are able to produce a protein and when genes are 'inactive' they do not produce the protein product coded for by the gene.

Some have to be switched on to act and some have to be switched off at different stages in the life of a cell. This is why hairs do not grow on the stomach lining and cheek cells do not grow on toenails.

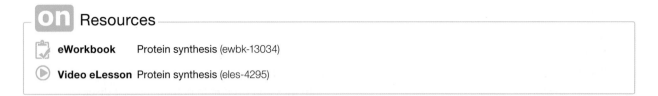
2.4 Activities

learn on

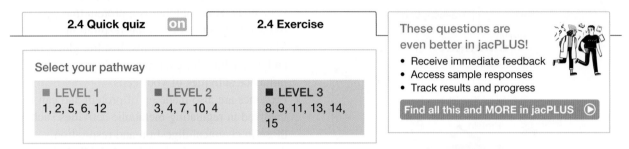

2.4 Quick quiz	**on**	2.4 Exercise

Select your pathway

■ LEVEL 1	■ LEVEL 2	■ LEVEL 3
1, 2, 5, 6, 12	3, 4, 7, 10, 4	8, 9, 11, 13, 14, 15

These questions are even better in jacPLUS!
- Receive immediate feedback
- Access sample responses
- Track results and progress

Find all this and MORE in jacPLUS ▶

Remember and understand

1. **a.** State whether the following statements are true or false.

Statements	True or false?
i. The instructions for making proteins are coded for in the sequence of nitrogenous bases in DNA.	
ii. The triplet code in mRNA is referred to as an anti-codon.	
iii. The process of making a complementary mRNA copy of the DNA message is called translation.	
iv. mRNA passes through the pores of the nuclear membrane into the cytoplasm to take its genetic copy of the protein instruction message to ribosomes, which may be free floating in the cytosol or attached to the rough endoplasmic reticulum.	
v. tRNA transfers the appropriate amino acid to its matching code on the mRNA, and then these amino acids are joined together by peptide bonds to make a protein.	

   b. Rewrite any false statement to make it true.
2. Identify the correct term for each of the following abbreviations:
   a. mRNA   **b.** tRNA   **c.** DNA
3. Match the terms to their description in the following table.

Terms	Description
a. A, T, G, C	**A.** Building blocks that make up DNA
b. A, U, G, C	**B.** Complementary base that pairs with thymine in DNA
c. Adenine	**C.** Complementary base to adenine in RNA
d. Nucleotide	**D.** First step in protein synthesis that results in the production of mRNA
e. Ribosome	**E.** Four possible types of nitrogenous bases in DNA
f. Phosphate	**F.** Four possible types of nitrogenous bases in RNA
g. Transcription	**G.** Second step in protein synthesis that results in production of a protein
h. Translation	**H.** Site of protein synthesis
i. Triplet	**I.** The sequence of three nucleotides in DNA that code for a particular amino acid
j. Uracil	**J.** The part of a nucleotide identical in DNA and RNA

4. What is meant by the base-pairing rule? Use a diagram in your response.
5. Explain the importance of protein synthesis.

Apply and analyse

6. A section of the DNA code, showing triplet, codon and amino acids, is shown in the following table.

TABLE DNA triplets and corresponding mRNA codons and amino acids

DNA triplet	mRNA codon	Amino acid
AAT	UUA	Leucine (leu)
ACG	UGC	Cysteine (cys)
TAC	AUG	Start/methionine (met)
ATT	UAA	Stop
CGG	GCC	Alanine (ala)
CAT	GUA	Valine (val)
ATG	UAC	Tyrosine (tyr)
CCA	GGU	Glycine (gly)

A template strand of DNA is found to be:

TAC CAT CGG CCA ATG ACG CGG CGG ATT

Use the provided amino acid table to suggest the following.
   a. The corresponding mRNA strand
   b. Amino acids sequence for the protein formed
   c. The anti-codons that match each codon
7. Part of a protein was found to have the following sequence of amino acids:

met — val — ala — gln — lys — trp

Identify four possible mRNA sequences that could lead to the production of this protein.
8. **SIS** A section of the same gene in three separate species is shown. (*Note:* these are from the template strands used in transcription.)

Human:	AAA	GCG	GCA
Chimpanzee:	AAG	GCC	GCT
Gorilla:	AAA	ACG	TCA

   a. How many differences appear in the DNA sequence between both the chimpanzee and gorilla from humans?
   b. Transcribe and translate each of these DNA strands.
   c. How many differences appear in the amino acid sequence between both the chimpanzee and gorilla from humans?
   d. Is this the same as your findings in part **a**?
9. All cells of a particular living thing, such as a spider, have the same sets of genetic instructions, but not all of that organism's cells have the same structure and function. Suggest what causes this and why cell specialisation is so important. Discuss your response.
10. Create a table showing the differences between transcription and translation. Ensure you show the location, inputs and outputs of each process.

Evaluate and create

11. **SIS** Scientists have discovered a gene switch that has restored youthful vigour to aging, failing brains in rats. Results from investigations suggest an 'on switch' for genes involved in learning. Injection of an enzyme flips the switch on and improves the learning and memory performance of older rats. Find out more about this type of research or other research that involves switching on genes. Discuss your findings.

12. Construct a Venn diagram or matrix table to summarise the similarities and differences between the following.
    a. DNA and RNA
    b. Codons and triplets

13. Construct flowcharts, diagrams or concept maps to show connections or links between the following terms.
    a. Nitrogenous base, sugar, phosphate, deoxyribose, ribose, DNA, RNA, uracil, thymine, guanine, cytosine, adenine
    b. DNA, mRNA, transcription, translation, amino acids, protein

14. **SIS** One of the longest genes to transcribe is the DMD gene, which codes for a protein known as dystrophin. The gene itself makes up around 0.08 per cent of the human genome. Data for this gene is shown in the provided table.

TABLE Data on the DMD gene in humans

Gene	2.3 million bases (2.3 megabases)
Primary transcript	2.1 million bases (2.1 megabases)
Mature transcript	14 000 bases
Protein	3685 amino acids

a. Why is the length of the primary transcript smaller than the length of the gene?
b. Why is the length of the mature mRNA transcript shorter than the primary mRNA transcript? Draw a clearly labelled diagram to show this.
c. The gene for dystrophin takes around 16 hours to be transcribed. If this same rate of transcription was used to transcribe the CFTR gene (which is approximately 190 000 bases long), what length of time would this be expected to take?

15. **SIS** Muscular dystrophy is a disease in which the DMD gene contains a mutation, affecting the production of the dystrophin protein. The two forms of this disease are Duchenne (DMD) and Becker (BMD). Examples of the different causes of these muscular dystrophies are shown in the table provided.

TABLE The number of exons and amino acids in different genes

	Exons	Amino acids in dystophin protein
Normal	79	3685
DMD	79	940
BMD	70	3200

a. Draw a column graph showing the number of exons and the amino acids for each scenario.
b. DMD is more severe than BMD, with individuals wheelchair bound much earlier. Those with DMD also have a much lower life expectancy. Using the results shown and your knowledge of the genetic code, suggest a reason this might be the case.

Fully worked solutions and sample responses are available in your digital formats.

LESSON
2.5 Mitosis and meiosis

LEARNING INTENTION

At the end of this lesson you will be able to differentiate between the two types of cell division, mitosis, and meiosis, and be able to explain the importance of the role of meiosis in ensuring variation.

All cells arise from pre-existing cells. That's pretty amazing when you really think about it! This means that all organisms living today originated from cells from the past. The cells you are made up of come from an unbroken line of cells. Where, when and who did your original cell come from?

2.5.1 Cell division in eukaryotes

Scientists are still grappling with many questions about the origin of life. Maybe you will be the one to shed new light on some possible answers in the future? What we do know, however, is there are two key types of cell division.

- Mitosis is the type of cell division involved in growth, development and repair of tissues. Some eukaryotic organisms also use mitosis for **asexual reproduction**.
- Organisms that use **sexual reproduction** to reproduce use another type of cell division in their reproductive process called meiosis.

asexual reproduction
reproduction that does not involve fusion of sex cells (gametes)

sexual reproduction
reproduction that involves the joining together of male and female gametes

FIGURE 2.42 Mitosis and meiosis are two types of cell division.

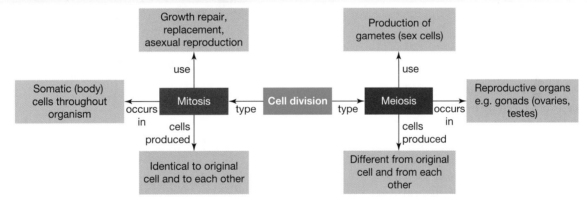

2.5.2 Nucleus, chromosomes and DNA

As outlined in lesson 2.2, all eukaryotic cells have a nucleus, which contains genetic information with instructions that are necessary to keep the cell (and organism) alive. This information is contained in structures called chromosomes, which are made up of a chemical called deoxyribonucleic acid (DNA) wrapped around protein called histones, which allows for DNA to be tightly packed.

FIGURE 2.43 DNA is contained in the chromosomes, which are located in the nuclei of cells.

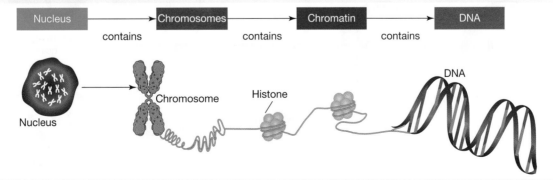

Counting chromosomes

Within the somatic cells (or body cells) of an organism is usually a particular number of chromosomes that is characteristic for that species.

> In humans, the total number of chromosomes in a somatic cell is 46. These chromosomes appear as 23 pairs in each cell. The term describing chromosomes in pairs is **diploid**, because of the two sets of chromosomes.
>
> Our gametes (or sex cells), however, contain only one set of chromosomes. They are referred to as being haploid.

You may see the symbol *n* used to identify the haploid number. The diploid number would be identified as 2*n*.

In humans and many other organisms, our diploid cells are produced through mitosis and our haploid cells are produced through meiosis.

diploid the possession of two copies of each chromosome in a cell

clones genetically identical copies

2.5.3 Mitosis

What happens when skin wears away and damaged tissues need repairing? How do seedlings grow into giant trees? How did you get to be so big?

Throughout the life of multicellular organisms, mitosis is the type of cell division that is used for:

- growth
- replacement
- repair
- asexual reproduction.

Mitosis involves division of the nucleus. The cells produced by mitosis are genetically identical to each other and to the original cell. They have the same number of chromosomes and DNA instructions. Because they have identical genetic information, they are described as being **clones** of each other.

FIGURE 2.44 Eukaryotic unicellular organisms such as **a.** *Amoeba* and **b.** *Euglena* divide by binary fission involving mitosis. Unlike meiosis, mitosis produces identical cells.

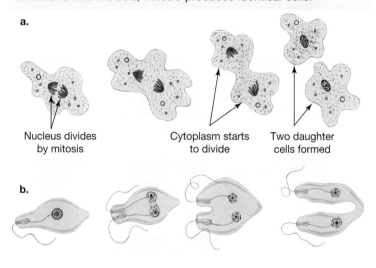

a.

Nucleus divides by mitosis · Cytoplasm starts to divide · Two daughter cells formed

b.

FIGURE 2.45 Mitosis results in the production of identical daughter cells.

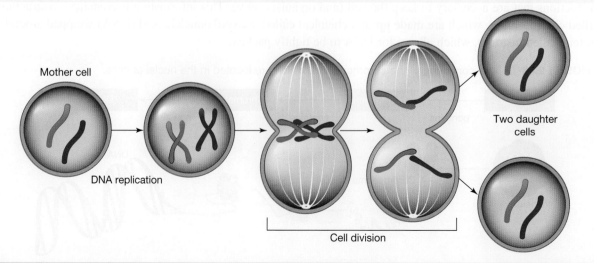

Mother cell

DNA replication

Two daughter cells

Cell division

Mitosis is an important part of the cell cycle. The cell cycle includes the following stages:
- *Interphase:* DNA replication occurs, alongside the growth of the cell, including replication of other organelles. During this stage, individual chromosomes cannot be seen under the microscope.
- *Mitosis:* The nucleus divides, and then each chromosome is split by the centromere. Mitosis is made up of:
  - Prophase: Chromatin condenses and chromosomes become visible. The nucleus disappears ('plump').
  - Metaphase: Chromosomes line up along the middle of the cell at the equatorial plate ('middle').
  - Anaphase: Identical chromatids separate and each move to an opposite side of the cell ('away').
  - Telophase: Nuclear membranes reform around the chromatids ('two').
- *Cytokinesis:* The cytoplasm splits, leading to the formation of two identical daughter cells. This involves the cell membrane pinching inwards so that a new membrane is formed, dividing the cell in two. In plate cells, a cell wall is also formed using a cell plate.

FIGURE 2.46 The stages of the cell cycle, including interphase, when DNA is replicated, and mitosis, when the nucleus divides

ewbk-13036

int-8128

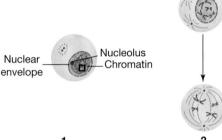

Nuclear envelope — Nucleolus, Chromatin

| **1** **Interphase** Chromosomes replicate | **2** **Prophase** Nuclear membrane disappears and chromosomes become visible | **3** **Metaphase** Chromosomes align at the equatorial plate | **4** **Anaphase** Chromatids separate | **5** **Telophase** Nuclei reform | **6** **Cytokinesis** Two diploid daughter cells are formed |

elog-2432

INVESTIGATION 2.4

Observing cell division

Aim

To observe the stages of mitosis under the microscope

Materials

- Light microscope
- Prepared slides of cells undergoing mitosis

Method

1. Collect slides of cells under mitosis.
2. Place a slide on the light microscope, starting on the lowest magnification.
3. Focus the cells and try to find an area where lots of different stages of mitosis are visible (like the image shown). The best area to look at is usually a region of high growth, such as the growing tip of a plant.
4. Sketch your observations in your results. *Note:* if your equipment allows, you may take a photo of your results (through the microscope lens) instead of sketching, and attach this to your report.
5. Increase the magnification and focus the cells. Sketch your observations. Label different mitotic stages you can see.
6. Increase the magnification one more time and focus the cells and sketch your observations (this higher magnification can be harder to focus, so continue to use the fine focus knob to assist).

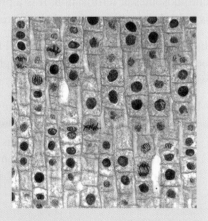

Results

Sketch and label the observed cells at each of the magnifications.
When sketching:
- List the magnification and include a title.
- Use a pencil to draw and a ruler for labelling.
- Do not shade or colour your image.

Remember, you do not have to draw every single cell; you may just choose a section of the image to draw.

Discussion

1. What stage were most of the cells in? Why do you think this is?
2. Compare your results to another member of your class. Did you receive similar results?
3. What proportion of your cells were in the following stages?
   a. Interphase
   b. Prophase
   c. Metaphase
   d. Anaphase
   e. Telophase

Conclusion

Summarise the findings for this investigation.

 Resources

 eWorkbook   Mitosis (ewbk-13038)

 Video eLessons Amoeba (eles-2694)
                 Euglena (eles-2695)
                 Stages of mitosis (eles-4215)

 Interactivities Mitosis (int-3027)
                 The stages of mitosis (int-3028)

2.5.4 Meiosis

Why do gametes only have one set of chromosomes? If they didn't, each time the egg and sperm nuclei combined during fertilisation, the number of chromosomes in the next generation of cells would double! (For example, if each gamete had 46 chromosomes, the resulting cell after fertilisation would have 92 chromosomes.)

Meiosis is the kind of nuclear division that prevents the doubling of chromosomes at fertilisation. It is a process in which the chromosome number is halved. It is often referred to as a reduction division.

In humans, this means the parent cell that is to undergo meiosis would initially be diploid ($2n$) and the resulting daughter cells or gametes produced by meiosis would be haploid (n).

Meiosis is made up of two main stages, as outlined in figure 2.47:
- Meiosis I (the first division), in which homologous chromosomes are separated. (A homologous pair is a matching set of chromosomes; for example, the two copies of chromosome 1 are homologous.)
- Meiosis II (the second division), in which sister chromatids are separated (in a similar way to mitosis).

FIGURE 2.47 The two stages of meiosis

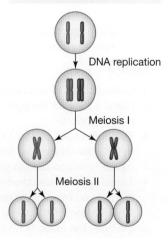

Meiosis I

At the start of meiosis I, interphase occurs, and DNA is replicated.

Following DNA replication, the following stages occur in meiosis I.
- *Prophase I:* Chromosomes condense and become visible. The nuclear membrane disappears. Chromosomes find their homologous pair. (A special process known as crossing over can also occur in which chromosomes may swap sections of DNA.)
- *Metaphase I:* Chromosomes line up in their homologous pairs at the equatorial plate.
- *Anaphase I:* The homologous pairs separate to opposite sides of the cell.
- *Telophase I:* The nuclei in the two new cells reforms.
- *Cytokinesis:* The cells split into two cells. These cells are now haploid (half the chromosome number).

FIGURE 2.48 The first division in meiosis I

| **Prophase I** | **Metaphase I** | **Anaphase I** | **Telophase I** | **Cytokinesis** |
| Chromosomes condense and crossing-over occurs | Chromosomes align at the equatorial plate | Chromosomes separate | Nuclei reform | Cells divide, forming two daughter cells |

Meiosis II

After meiosis I (shown in figure 2.48), the next stage is meiosis II (figure 2.49). This is similar to mitosis, in which chromosomes separate at the centromere and split into two chromatids.

It is important to note that homologous pairs are no longer present (because each member of each homologous pair is in separate cells).

Interphase does not occur in meiosis II because chromosomes are already in their replicated form.

The following stages occur in meiosis II.
- *Prophase II:* Chromatin condenses and chromosomes become visible. The nucleus disappears.
- *Metaphase II:* Chromosomes line up along the middle of the cell at the equatorial plate.
- *Anaphase II:* Identical chromatids separate and each move to an opposite side of the cell.
- *Telophase II:* Nuclear membranes reform around the chromatids.
- *Cytokinesis:* The cytoplasm splits, leading to four daughter cells being formed.

FIGURE 2.49 The second division in meiosis II

| **Prophase II** | **Metaphase II** | **Anaphase II** | **Telophase II** | **Cytokinesis** |
| Start of 2nd division and two daughter cells present | Chromosomes line up along equatorial plate | Chromatids separate | 4 nuclei reform | Cells divide forming four haploid daughter cells |

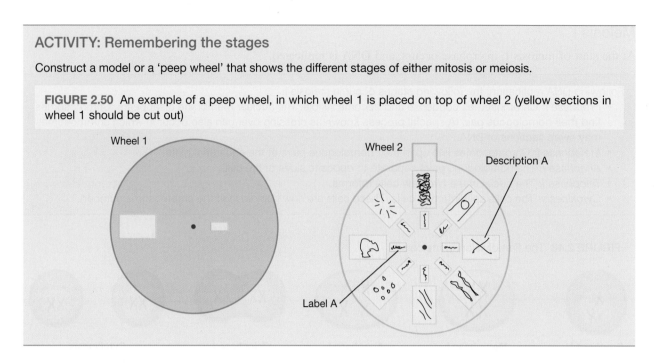
2.5.5 Meiosis mix-up and variation

Each parent produces gametes by the process of meiosis. Within each gamete are chromosomes from each parent. Chromosomes carried in the sperm and inherited from the father are referred to as **paternal chromosomes**, and chromosomes from the ovum and inherited from the mother are referred to as **maternal chromosomes**.

Meiosis provides sexually reproducing organisms with a source of **variation**. Variation within a species can provide some individuals with an increased chance of surviving over others. Depending on the environment and selection pressures at a particular time, different variations may be advantageous. High variation among individuals will increase the chance that some will survive to reproduce. This improves the chances of the species surviving.

paternal chromosomes chromosomes carried in the sperm

maternal chromosomes chromosomes from the ovum

variation differences between cells or organisms

This variation occurs through meiosis due to:

- *Independent assortment:* The way in which chromosomes line up during metaphase I of meiosis is random. Imagine three pairs of chromosomes — half of these came from your father and half came from your mother. They may line up differently, as seen in figure 2.51.

FIGURE 2.51 The different ways that three pairs of homologous chromosomes can line up in meiosis

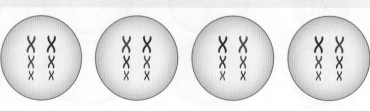

Given that humans have 23 pairs of chromosomes, there are around 8 388 608 (2^{23}) possible ways to divide these chromosomes into each type of gamete.

- *Crossing over:* Homologous chromosomes exchange genetic material during meiosis I. This results in a section of one chromosome swapping its genetic information with another. For example, genes that were once on a paternal chromosome can be transferred or crossed over onto a maternal chromosome and vice versa.

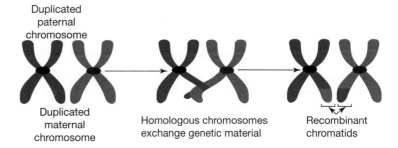

FIGURE 2.52 Crossing over results in the swapping of genetic material between homologous chromosomes.

Duplicated paternal chromosome

Duplicated maternal chromosome

Homologous chromosomes exchange genetic material

Recombinant chromatids

2.5.6 Fertilisation

In humans, fertilisation occurs when a haploid gamete from each parent fuses together to form a diploid zygote. But which sperm will fertilise the ovum? The identity of the lucky sperm that will contribute its genetic information to the next generation depends largely on chance.

Depending on which sperm fertilises the egg, many genetic combinations are possible. This is another source of genetic variation that can give sexually reproducing organisms an increased chance of survival.

The zygote contains 23 paternal chromosomes from its father and 23 maternal chromosomes from its mother. Each pair of chromosomes will consist of a chromosome from each parent. The zygote divides rapidly by mitosis to form an embryo that will also use this type of cell division to develop and grow. Each time this process occurs, cells with this complete new set of chromosomes are produced.

You are a product of both meiosis and mitosis.

FIGURE 2.53 Fertilisation and the link between mitosis and meiosis

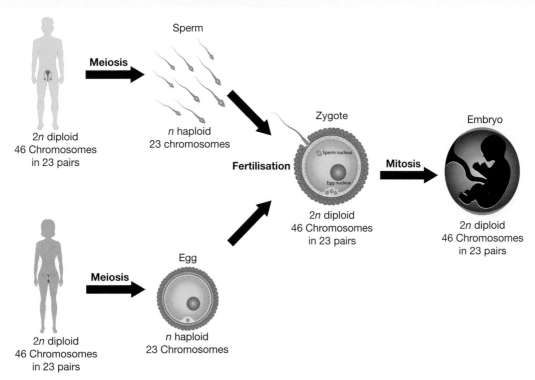

Sperm

Meiosis

2n diploid
46 Chromosomes
in 23 pairs

n haploid
23 chromosomes

Fertilisation

Zygote

Sperm nucleus

Egg nucleus

2n diploid
46 Chromosomes
in 23 pairs

Mitosis

Embryo

2n diploid
46 Chromosomes
in 23 pairs

Egg

Meiosis

2n diploid
46 Chromosomes
in 23 pairs

n haploid
23 Chromosomes

CASE STUDY: Boy or girl?

When a friend or family member is expecting a baby, one of the first questions people wonder or ask is whether it will be a boy or a girl. Probability suggests a 50 per cent chance either way.

Human somatic cells contain 22 pairs of autosomes and a pair of sex chromosomes. The sex chromosome difference is abbreviated, so that females are described as being XX and males as being XY.

As a result of meiosis, gametes will contain only one sex chromosome. Human females (XX) can only produce gametes that contain an X chromosome. Human males (XY), however, will produce half of their gametes with an X chromosome and the other half with a Y chromosome. So, if a gamete containing a Y chromosome fuses with the ovum (which contains an X chromosome), the resulting zygote will be male (XY). Likewise, if the ovum is fertilised by an X-carrying gamete, then a female (XX) will result. This is shown in figure 2.54.

FIGURE 2.54 Is the mother or father the key determiner of the gender of the child?

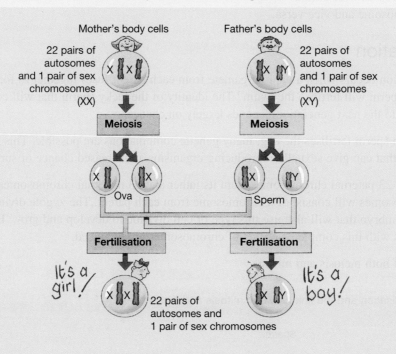

CASE STUDY: Twins — or more!

Sometimes in the very early stages of division following fertilisation, clusters of a few cells develop into two separate individuals. If this happens, identical twins result because each cluster has the same genetic make-up as the other.

Usually, only one ovum is released at a time. However, if several are released, twins can result from fertilisation by different sperm. In this case, the babies are not identical because they have different genetic make-ups.

FIGURE 2.55 Identical or fraternal twins — one sperm or more?

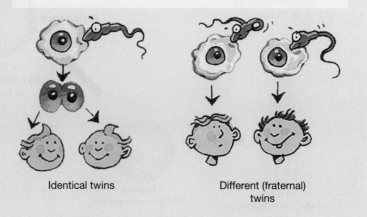

Identical twins   Different (fraternal) twins

EXTENSION: Gender-determining factors in other species

The gender-determining factors of other animals can be quite different from those of humans. In birds, for example, it is the female that has different sex chromosomes, Z and W, and the male has two Z chromosomes. In some reptiles, gender is determined by the temperature at which the egg is kept rather than chromosomes. The temperature of the sand in which some crocodiles and turtles bury their eggs can determine whether the offspring will be male or female. The gender of brush turkey chicks is also determined partly by temperature.

elog-2434

INVESTIGATION 2.5

What's the chance?

Aim

To simulate the chance of a male or female being conceived at fertilisation

Materials

- 20-cent coin

Method

1. After reading the instructions and before you carry out the experiment, write a clear hypothesis, predicting the number of times you will toss heads and the number of times you will toss tails.

TABLE A comparison of the times heads or tails were recorded when a coin was tossed 50 times

	Number of heads	Percentage of heads	Number of tails	Percentage of tails
Individual tosses				
Combined class result				

2. Toss a coin 50 times and record your results.

Results

1. Count the number of heads and tails and record the data in a table like the one shown.
2. Calculate the percentage chance of obtaining heads and the percentage chance of obtaining tails.
3. Combine the results of the whole class and calculate the percentage chance of obtaining heads and tails.
4. Draw a graph of your results.

Discussion

1. Analyse your data.
   a. Was your prediction supported or not?
   b. Were the percentage results obtained for 50 tosses the same as or different from the total class results? Suggest reasons for the similarities or differences.
2. If you tossed a coin 1000 times, would you obtain similar results?
3. What is the chance of obtaining heads each time you toss the coin?
4. If heads represented a sperm carrying an X chromosome and tails represented a sperm carrying a Y chromosome, suggest how this activity could link to the chances of a male or female baby being conceived.
5. Suggest a strength, a limitation and an improvement for this investigation.

Conclusion

Summarise the findings for this investigation.

2.5 Activities

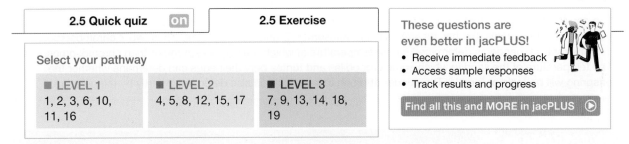

| 2.5 Quick quiz **on** | 2.5 Exercise |

Select your pathway

■ LEVEL 1	■ LEVEL 2	■ LEVEL 3
1, 2, 3, 6, 10, 11, 16	4, 5, 8, 12, 15, 17	7, 9, 13, 14, 18, 19

These questions are even better in jacPLUS!
- Receive immediate feedback
- Access sample responses
- Track results and progress

Find all this and MORE in jacPLUS ▶

Remember and understand

1. The following diagram shows the process of mitosis and meiosis.

   Fill in the blanks to complete the differences between mitosis and meiosis.

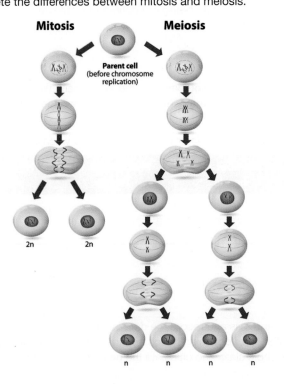

TABLE Comparison of meiosis and mitosis

Feature	Mitosis	Meiosis
Occurrence	During _____ and _____	During formation of _____ and _____
Places of occurrence	Occurs in all _____ cells	Occurs in _____ cells
Number of divisions	_____	_____
Number of daughter cells	_____ cells	_____ cells
Number of chromosomes	_____	_____
Type of reproduction	_____	_____
Genetic information	Daughter cells are genetically _____ to the parent cells.	Daughter cells are genetically _____ to the parent cells.
Occurrence of recombination or crossing over	_____	_____

2. **a.** State whether the following statements are true or false.

Statements	True or false?
i. All cells come from pre-existing cells.	
ii. Cells produced by mitosis are genetically identical to each other and to the parent cell.	
iii. Mitosis involves division of the cytoplasm whereas cytokinesis involved division of the nucleus.	
iv. Meiosis prevents the doubling of chromosomes at fertilisation because it produces cells with half the chromosome number of the original cell.	
v. Variation within a species can provide some individuals with an increased chance of surviving over others.	
vi. Little to no variation among individuals within a species improves the chances of the species surviving.	
vii. One way in which meiosis increases variation in sexually reproducing organisms is in terms of the number of combinations in which chromosomes can be divided up into the gametes.	
viii. Crossing over of genetic information between each pair of chromosomes in meiosis is a source of variation in a species.	

   b. Rewrite any false statement to make it true.

3. Match the term with its definition in the following table.

Term	Definition
a. Diploid	**A.** Cells with four sets of chromosomes
b. Fertilisation	**B.** Cells with one set of chromosomes (e.g. as in human gametes)
c. Gamete	**C.** Cells with two sets of chromosomes (e.g. as in human somatic cells)
d. Haploid	**D.** Reduces the number of chromosomes in the daughter cells by half that of the parent cell
e. Meiosis	**E.** Sex cell
f. Mitosis	**F.** The fusion of gametes
g. Tetraploid	**G.** The result of two gametes fusing together
h. Zygote	**H.** Type of cell division important for growth, repair and replacement

4. List two differences between sexual and asexual reproduction.
5. State the names of the two main types of cell division. Explain why each is used.
6. **a.** List three functions of mitosis.
   b. Describe the features of the cells produced by mitosis.
7. Copy and complete the following table.

TABLE Comparison of meiosis and mitosis

Type of cell division	Why use it?	Where does it occur?	Features of cells produced
Mitosis			
Meiosis			

8. With the use of a diagram, explain how the sex of a human baby is determined.

Apply and analyse

9. Distinguish between the following pairs of terms.
   a. Cytokinesis and mitosis
   b. Prophase I and Prophase II
   c. Diploid and haploid
   d. Gamete and zygote
   e. Fertilisation and meiosis
   f. Somatic cells and gametes

▶

10. If a woman has already given birth to three boys, what are the chances of her next child being a girl? Justify your response.
11. A few genetic traits, such as hairiness in ears, are due to genes carried on the Y chromosome. Would males and females have the same chance of having the trait? Justify your response.
12. **SIS** The following steps show the method for an onion root tip mitosis experiment.
    1) Use a scalpel to cut 2 to 3 onion roots.
    2) Place the roots onto a glass slide.
    3) Pour 4 to 5 drops of 1M hydrochloric acid onto the roots.
    4) Place the slide with the roots onto a hotplate (60 °C for 2 minutes).
    5) Add 2 to 3 drops of Toluidine blue onto the root tips.
    6) Place a coverslip over the roots.
    7) Examine the slide under the microscope with low magnification, followed by a higher magnification.
    Explain why the root tips are stained with Toluidine blue.
13. **SIS** Using the following table, suggest the possible effect of increasing global temperatures on turtles, crocodiles and lizards.

TABLE Temperature control of sex in some reptiles

Reptile	Cold 20–27 °C	Warm 28–29 °C	Hot > 30 °C
Turtle	Male	Male or female	Female
Crocodile	Female	Male	Female
Lizard	Female	Male or female	Male

Evaluate and create

14. The kind of job a man does can affect whether he produces more or less Y sperm or any sperm at all. Chemicals and hormones washed into waterways or used in producing food can affect fertility. Research an example of an environmental impact on fertility and report your findings. Make sure you quote the sources of your information.
15. In many cultures throughout history, a woman has been blamed for not producing sons and has been divorced. From a biological point of view, could this be justified? Explain your answer.
16. Copy and complete the following Venn diagram, choosing from the following terms:
    Word bank: somatic, only, body, gonads, gametes, anywhere, different, identical, chromosomes, cell division, eukaryotes.

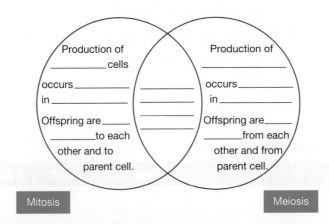

17. The Y chromosomes of human males are shorter than the X chromosomes. Would the same number of genes be carried by both chromosomes? Justify your response with clear examples.
18. **SIS** **a.** How many sets of chromosomes do you think an organism would have if it was identified as *4n* and tetraploid?
    b. Outline an investigation you may conduct to determine if an organism is tetraploid.
19. **a.** Create a model to allow you to describe each stage of meiosis.
    b. Explain why interphase only occurs before meiosis I and not meiosis II.

Fully worked solutions and sample responses are available in your digital formats.

LESSON
2.6 Changing the code through mutations

LEARNING INTENTION

At the end of this lesson you will be able to describe the relationship between DNA, chromosomes and mutations, and outline the factors that contribute to causing mutations.

2.6.1 What are mutations?

Errors or changes in DNA, genes or chromosomes can have a variety of consequences. These genetic mistakes are called **mutations**.

An example of a DNA mutation is polydactyly, which sees those affected develop extra fingers and toes, as seen in figure 2.56.

on Resources

▶ **Video eLesson**   Polydactyl cat (eles-2698)

FIGURE 2.56 Polydactyly is often caused by a DNA mutation.

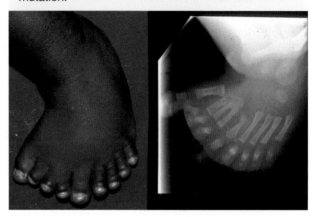

2.6.2 DNA replication

FIGURE 2.57 DNA replication is semi-conservative — a combination of original (blue) and new (purple) strands.

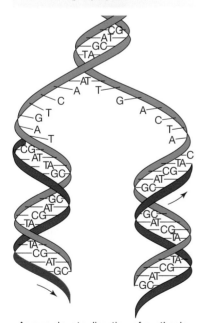

Arrows denote direction of synthesis.

DNA is very stable and can be replicated into exact copies of itself. This process is called **DNA replication** and enables genetic material to be passed on unchanged from one generation to the next. DNA replication begins with the 'unzipping' of the paired strands. A new complementary strand is made for each original DNA strand. This results in the formation of two new double-stranded DNA molecules, each containing one new DNA strand and one original DNA strand. This process is vital in cell division, occurring during interphase both at the start of mitosis and meiosis.

The process of DNA replication has a number of checkpoints to test for any mistakes that may be made, so that they can be corrected or destroyed. Sometimes, however, the mistakes get through this screening process. When this happens, we say that a mutation has occurred.

mutations changes to DNA sequence, at the gene or chromosomal level

DNA replication process that results in DNA making a precise copy of itself

2.6.3 Mutagenic agents

Mutations can happen by chance or have a particular cause.
- When the cause of the mutation cannot be identified it is called a **spontaneous mutation**.
- When the cause can be identified it is referred to as an **induced mutation**.

A factor that triggers mutations in cells is called a **mutagen** or mutagenic agent.

Examples of mutagenic agents include:
- radiation, such as ultraviolet radiation, nuclear radiation and X-rays
- chemical substances, such as formalin, asbestos, tobacco and benzene (which used to be common in pesticides)
- infectious agents, such as human papillomavirus (HPV).

As a result of the thinning of the ozone layer in the atmosphere, we are exposed to increasing amounts of UVB radiation that can damage (or mutate) our DNA. This can lead to the development of skin cancers. Protective clothing and sunscreens can help reduce our exposure to this dangerous, potentially mutagenic environmental radiation.

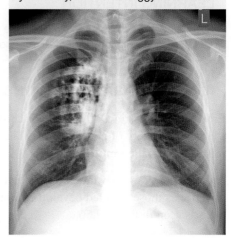

FIGURE 2.58 Mutations may result in cancerous growths (tumours) within your body, shown as foggy areas here.

 Resources

> ▶ **Video eLesson**   DNA and Hiroshima (eles-1781)

2.6.4 Errors in the code

Changes in the genetic code due to mutations may result in a particular protein not being made or a faulty version being produced.

In some cases, the production of an essential enzyme may be impaired, disrupting chemical reactions and resulting in the deficiencies or accumulation of other substances. This may cause the death of the cell and, eventually, the organism.

Point mutations

Occasionally, errors can occur during DNA replication as DNA is being copied. This means that the instructions carried by the code are not followed exactly.

This may be the result of:
- an incorrect pairing of bases
- the substitution of a different nucleotide
- the deletion or insertion of a nucleotide.

The mutation, where a single nucleotide is affected, is known as a **point mutation**. Such a point mutation can change the genetic message by coding for a different amino acid, leading to the production of a different or non-functional protein through the process of protein synthesis (as explored in lesson 2.4). The deletion or insertion mutations are often more serious because they can cause a frameshift, altering the groups of three in which the sequence is usually read. This can have severe consequences. Just like changing letters in a word can change its meaning (as seen in table 2.2), changes in the DNA sequence can change the meaning of the genetic code.

spontaneous mutation a mutation of DNA that cannot be explained or identified

induced mutation a mutation of DNA in which the cause can be identified

mutagen agent or factor that can induce or increase the rate of mutations

point mutation a mutation at one particular point in the DNA sequence, such as a substitution or single base deletion or insertion

FIGURE 2.59 Different types of mutations

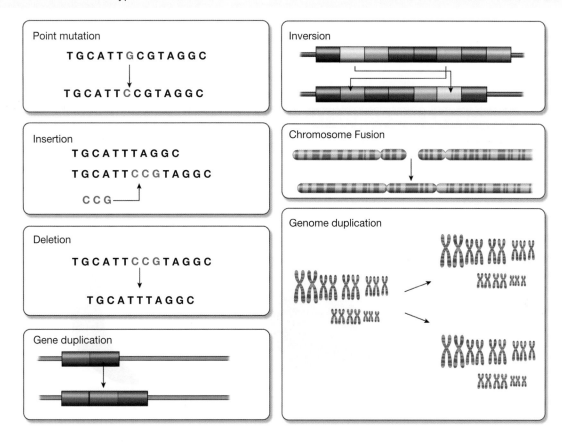

TABLE 2.2 Analogy of how mutations lead to a change in meaning from an initial message. The initial message of 'post' is changed with the insertion, deletion, inversion or substitution of letters (shown in bold).

Original	Insertion	Deletion	Inversion	Substitution
post	post**er**	pot	po**ts**	p**e**st

Types of point mutations

Substitution mutations

One of the most common point mutations is that of a substitution mutation, in which one base is replaced with another.

Substitution mutations can be one of three types:
- Missense mutations: where one amino acid is swapped for another (as seen in sickle-cell anaemia, outlined in the case study).
- Silent mutations: no change occurs in the amino acid.
- Nonsense: where an amino acid is changed to a STOP codon (causing a shortened protein).

CASE STUDY: Sickle-cell anaemia

Sickle-cell anaemia is a disease that is usually associated with a mutation in the gene that codes for one of the polypeptides that make up haemoglobin in red blood cells. In this mutation, an adenine base is substituted by a thymine base. The result is misshapen red blood cells that can clump together and block blood vessels.

TABLE 2.3 Exploring mutations in sickle-cell anaemia

	Normal red blood cell	Sickle-cell red blood cell
Coding DNA sequence (complementary)	CTG ACT CCT G**A**G	CTG ACT CCT G**T**G
Template DNA sequence	GAC TGA GGA C**T**C	GAC TGA GGA C**A**C
Complementary RNA sequence	CUG ACU CCU G**A**G	CUG ACU CCU G**U**G
Amino acid sequence	leu — thr — pro — **glu**	leu — thr — pro — **val**
Phenotype of red blood cell	Normal doughnut-shaped blood cell 	Sickle-shaped blood cell 

Frameshift mutations

Another type of mutation that may result from additions or deletions is a frameshift mutation. In this case, every amino acid from that point is altered, leading to a completely different amino acid code.

Imagine the following sentence:

THE FAT CAT SAT

The insertion of one 'letter' completely changes the groups of three letters, so the sentence now reads:

THT EFA TCA TSA T

These mutations change the amino acid sequence drastically, because they alter all the codons after the mutation. Examples of these mutations can be seen in table 2.4.

TABLE 2.4 Examples of frameshift mutations

	Normal	Insertion	Deletion
Coding DNA sequence (complementary)	CTG ACT CCT	CT**A** GAC TCC T	CGA CTC CT
Template DNA sequence	GAC TGA GGA	GAT CTG AGG A	GCT GAG GA
Complementary RNA sequence	CUG ACU CCU	CUA GAC UCC U	CGA CUC CU
Amino acid sequence	leu — thr — pro	leu — asp— ser	arg — leu

Chromosomal mutations

Point mutations relate to changes in the DNA sequence in genes; however, mutations can also involve chromosomes. These changes may involve the addition or deletion of entire chromosomes, or the deletion, addition or mixing of genetic information from segments of chromosomes. Some examples of disorders that result from chromosome mutations are shown in table 2.5.

TABLE 2.5 Examples of human chromosome abnormalities (mutations). The risk of these mutations increases with maternal age.

Chromosome abnormality	Resulting disorder	Incidence (per live births)
Extra chromosome number 21	Down syndrome	1 in 700
Missing sex chromosome (XO)	Turner syndrome	1 in 5000
Extra sex chromosome (XXY)	Klinefelter syndrome	1 in 1000

FIGURE 2.60 Two examples of karyotypes with chromosomal abnormalities

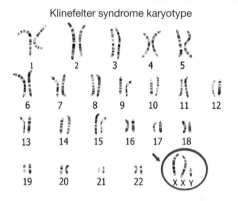

2.6.5 Mutants unite!

Not all mutations are harmful. Some mutations can increase the survival chances of individuals within a population, and hence the survival of their species. Other mutations (such as that seen in figure 2.61) are neither harmful or beneficial to individuals who possess them.

FIGURE 2.61 Heterochromaia (different eye colours) often occurs in somatic cells and is not inherited.

Spray resistance

Pesticides kill the majority of insects sprayed. Some insects within the population, however, may survive because they possess slight variations or mutations in their genes that give them resistance to the pesticide. The mutated gene in the surviving insects is passed on to their offspring, who gain that resistance too. While the insects without the resistance die out, those with resistance increase in numbers.

Good for you, but not for me

When we look at natural selection as a mechanism for evolution in topic 3, we see how mutations can be a very important source of new genetic material. While such mutations can be beneficial for the survival of the species under threat, they are not necessarily beneficial to humans. The resistance of bacteria to antibiotics, for example, has resulted in selection for antibiotic-resistant bacteria. This means we are unable to use these antibiotics to treat diseases caused by these resistant bacteria, because the drugs are no longer effective.

Malaria and sickle-cell mutation

Malaria is a disease that is very common in many parts of Africa, Asia and South America. It is caused by a parasite that is transmitted by a species of mosquito. When the mosquito bites an individual, the parasite then grows in red blood cells of its human host. This disease is one of the main global causes of human disease-related deaths.

The mutation that results in sickle-cell anaemia can increase your resistance to malaria. If you are a carrier for this trait, the parasite cannot grow as effectively in your red blood cells, which means you are less likely to die from malaria than people in the population without the **allele**.

2.6.6 Not all mutations are inherited

Only mutations that have occurred in the germline cells such as the sex cells or gametes (sperm and ova) are inherited. In sexually reproducing organisms, mutations that occur in somatic cells are not passed on to the next generation.

allele alternate form of a gene for a particular characteristic

on	Resources	
eWorkbook	Mutations (ewbk-13044)
Video eLesson	Types of mutations (eles-4214)
Weblink	Scientists warn against vitamins

2.6 Activities

learn on

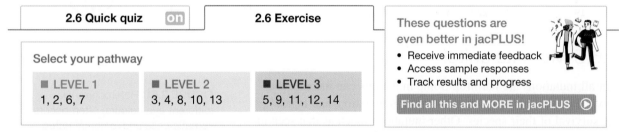

2.6 Quick quiz on	2.6 Exercise

Select your pathway

■ LEVEL 1	■ LEVEL 2	■ LEVEL 3
1, 2, 6, 7	3, 4, 8, 10, 13	5, 9, 11, 12, 14

These questions are even better in jacPLUS!
- Receive immediate feedback
- Access sample responses
- Track results and progress

Find all this and MORE in jacPLUS ▶

Remember and understand

1. a. State whether the following statements are true or false.

Statements	True or false?
i. Mutations in both germline and somatic cells are passed on to the next generation.	
ii. Errors or changes in DNA, genes or chromosomes are called mutations.	
iii. Radiation (for example, ultraviolet radiation, nuclear radiation and X-rays) is not considered to be a mutagenic agent.	
iv. Changes in the genetic code due to mutations may result in a particular protein not being made or a faulty version being produced.	
v. If you are a carrier for sickle-cell anaemia, the malaria parasite cannot grow as effectively in your red blood cells, which means you are less likely to die from malaria than people in the population without the allele.	
vi. All mutations are harmful.	

b. Rewrite any false statements to make them true.

2. **a.** Name the process by which DNA makes copies of itself.
   b. Explain why the model used to describe the process identified in part **a** is called semi-conservative. Include a diagram in your response.
   c. Explain why it is important for DNA replication to produce exact copies of the original DNA.

3. **a.** Describe what is meant by the term *mutagenic agent*. Provide an example.
   b. Distinguish between the terms *spontaneous mutation* and *induced mutation*.
   c. Identify two disorders associated with chromosome mutations.
   d. Describe two examples of mutations that can increase chances of survival.

4. Outline the relationship between sickle-cell anaemia and mutated DNA.

Apply and analyse

5. **a.** Explain the difference between a missense, nonsense and silent mutation.
   b. An initial DNA template strand is as follows: AAA GCG TAC.
      If the second adenine (A) is changed to a guanine (G), what type of mutation has occurred (refer back to the genetic code on figure 2.38).
   c. How does this type of mutation differ from a frameshift mutation? Which would be more likely have significant effects?

6. Are mutations always detrimental? Provide an example to justify your response.

7. **SIS** Suggest why radiographers wear special protective clothing and use remote controls for taking X-rays.

8. **SIS** The karyotype shows an individual with Turner syndrome.
   a. Describe how this karyotype differs from that of a normal human karyotype.
   b. Suggest how this mutation may have resulted.

9. The karyotype in figure 2.60 shows an individual with Down syndrome.
   a. Describe how this karyotype differs from that of a normal human karyotype.
   b. Suggest how this mutation may have resulted.
   Search online for Down syndrome research. Use your own knowledge and information found to answer the following questions.
   c. Suggest why the DSCR1 gene is of importance.
   d. On which chromosome is the DSCR1 gene located?
   e. Outline the advantage suggested by the research of possessing an extra copy of the DSCR1 gene.

Turner syndrome karyotype

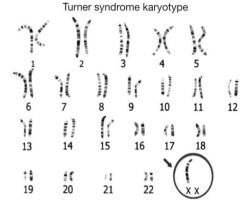

10. **SIS** Research and report on one of the following topics.
    • Antibiotic resistance and bacteria
    • Pesticide resistance and insects
    • Sickle cell anaemia, malaria and heterozygote advantage
    • Antioxidant vitamins, brightly coloured vegetables and mutations
    • Breast cancer and the BRCA gene

Evaluate and create

11. Over the years, considerable debate has occurred over the use of taking vitamin supplements. James Watson, one of the scientists who proposed the double helical structure of DNA, is of the opinion that taking high doses of some vitamins can interfere with cancer treatment.
    a. Access the **Scientists warn against vitamins** weblink and read the article.
    b. Summarise the key points.
    c. Use the internet to find evidence for or against James Watson's claims.
    d. Outline and justify your own opinion.

12. **SIS** Some people claim that eating brightly coloured fruits and vegetables that are high in antioxidants can reduce the occurrence of mutations within your body.
    a. Research this claim and summarise your findings.
    b. Formulate a relevant hypothesis that could be investigated scientifically, including identification of independent, dependent and controlled variables.
    c. To enable collection of reliable data to test your hypothesis, design an investigation that also addesses any safety or ethical issues.
    d. Describe results from your investigation that would support your hypothesis.

13. **SIS** The following graph shows the frequency of children being born with Down syndrome based on maternal age.
   Carefully examine the graph. DS is an abbreviation for Down syndrome.
   a. State the axis label of the
      i. *x*-axis
      ii. *y*-axis.
   b. Suggest how you could improve the axis labels.
   c. Suggest a title for the graph.
   d. What do you think the fractions in the graph represent?
   e. Formulate a hypothesis that would be relevant to the graph.
   f. Describe the pattern or trend.
   g. Suggest an interpretation of the data.

14. **SIS** Carefully examine the provided graph on the rate of sex-linked mutations in differing levels of radiation.
   a. State the axis label (including the units) of the
      i. *x*-axis
      ii. *y*-axis.
   b. In the graph, identify the
      i. independent variable
      ii. dependent variable.
   c. Suggest a title for the graph.
   d. Explain what is meant by a sex-linked mutation.
   e. Use the graph to estimate the rate of sex-linked mutations at 2000 Roentgen units of radiation.
   f. Suggest a reason for the use of error bars on the graph.
   g. Formulate a hypothesis that would be relevant to the graph.
   h. Describe the pattern or trend. Incorporate the axis labels in your description.
   i. Suggest an interpretation of the data.
   j Propose a question that could be used to focus further relevant research.

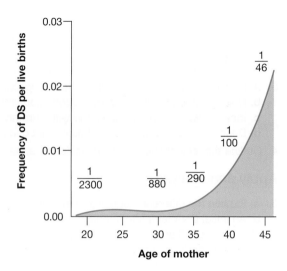

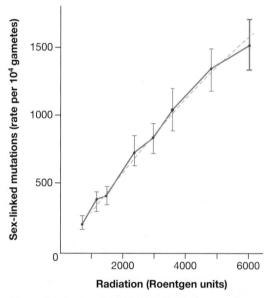

Fully worked solutions and sample responses are available in your digital formats.

LESSON
2.7 Exploring patterns in the genome and genetic sequences

LEARNING INTENTION

At the end of this lesson you will be able provide examples of how advances in technology have enabled us to sequence the human genome and explore patterns in our genetic sequences.

2.7.1 Who do you think you are?

You are incredibly special. Much of who and what you are is determined by genes. Genes determine many of the traits and characteristics that make you, you.

As a result of rapid advances in technology, the time and cost of the sequencing of your genes has dramatically reduced. As a consequence, huge amounts of genetic information is being generated.

Will you be one of the new generation of scientists who combine genetics and **bioinformatics** to make sense of how variations in our genetic sequences can affect not just what we look like, but also our health?

> **bioinformatics** the science of analysing biological data through computers, particularly around genomics and molecular genetics

2.7.2 DNA sequencing

Determining the sequence of nucleotides along a section of DNA used to be a long and tedious process. With the development of automated DNA sequencers, it is now faster and less costly. The process involves the addition of nucleotides that have been tagged with fluorescent dyes. A different coloured dye is used to tag each type of nucleotide (containing A, T, C or G). Some of the tagged nucleotides will attach to the DNA sample fragments. Light signals from these tagged nucleotides are detected. Computers then analyse this data and construct the DNA sequence of the sample.

The automation and use of computer analysis in DNA sequencing is an example of bioinformatics, because it involves the use of computer technology to manage and analyse biological data. The technology related to DNA sequencing is still developing and soon (if not already), millions of sequences may be able to be determined at the same time. What new scientific discoveries will result?

Similar to DNA sequencing, the genes themselves can be sequenced. Gene sequencing identifies the order of nucleotides along a gene.

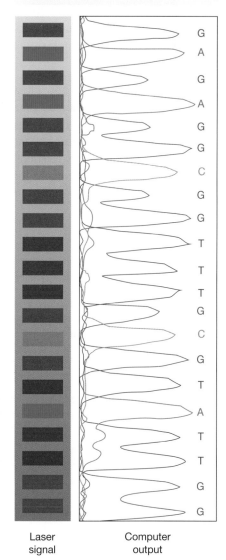

FIGURE 2.62 DNA sequencers identify the base sequence of sections of a DNA fragment.

Laser signal   Computer output

FIGURE 2.63 A DNA sequencing machine. These machines are capable of processing huge numbers of DNA sequences. One machine can sequence 400 to 600 million bases over a ten-hour run.

2.7.3 What is a genome?

A **genome** is the complete complement of genetic material in a cell or organism. Although there may be small variations within the members of a species, each species has a unique genome. The study of genomes is called **genomics**.

The genome size is often described in terms of the total number of base pairs (or bp). Due to the size of genomes, sometimes the genome size is described in mega base (Mb) pairs, with one mega base pair equal to a million base pairs.

genome the complete set of genes present in a cell or organism
genomics the study of genomes

2.7.4 The Human Genome Project (HGP)

Completed in 2003, the Human Genome Project (HGP) was an international investigation to identify, sequence and study the genetic instructions within humans. The key findings from the HGP were that the human genome has a size of around three billion base pairs (or 3000 Mb) and contains around 20 000 to 25 000 genes. It was also found that all humans share about 99.9 per cent of their DNA. The complexity of this can be seen in figure 2.64.

As a result of the HGP, we now know the number, location and sequence of human genes. This new knowledge may lead to developments in the diagnosis and treatment of genetic diseases, and in our construction of evolutionary relationships between organisms. But is there also cause for concern? Who will have access to the information and how will it be regulated?

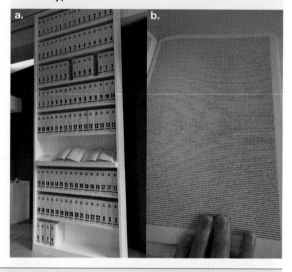

FIGURE 2.64 a. The print-out of the HGP in the Wellcome Collection's 'Medicine Now' exhibition contains 116 volumes. **b.** One page from the chromosome 6 volume. Note the very small typeface.

DISCUSSION

Now that the human genome has been sequenced, it poses many questions, such as those outlined in figure 2.65.

Discuss your ideas around each of these questions.

What are some other considerations that should be made around benefits and possible issues around the sequencing of the genome?

FIGURE 2.65 Different issues are related to sequencing the genome.

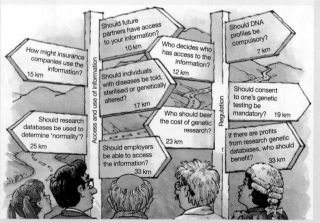

Resources

 eWorkbook Blood samples (ewbk-12770)

 Weblink Culture and science

2.7.5 We've all got genomes!

Sequencing a genome can be costly, so not all organisms have been sequenced. Often the decision may be based on a commercial or research focus. After the sequence of the bases within the genome has been completed, computers are used to analyse it to identify the genes it contains. Although the bases within the sequences are universal (the same across all organisms), both the genome size and the number of genes it contains vary between organisms.

The study of various genomes has revealed that the same genes that cause a fly to be a fly are also used to make a human a human. Parts of our genome are virtually interchangeable with those of our close primate 'cousins' — with human and chimpanzee genomes differing less than 2 per cent. Rather than revealing the source of our diversity and uniqueness, knowledge of our genome has brought us closer to that of other life on Earth.

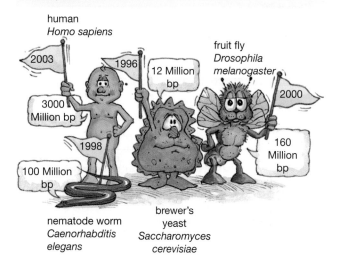

FIGURE 2.66 Comparing the genome sizes (and the year they were sequenced) of different species. What trends and patterns can you see, where bp stands for base pairs?

human
Homo sapiens
2003
3000 Million bp
1998
100 Million bp

nematode worm
Caenorhabditis elegans

1996
12 Million bp

brewer's yeast
Saccharomyces cerevisiae

fruit fly
Drosophila melanogaster
2000
160 Million bp

DISCUSSION

Since mice and humans diverged from a common ancestor millions of years ago, most of the DNA that codes for functional genes has remained similar, whereas the non-coding DNA has mutated and is now extremely different. What might the purpose of this non-coding DNA be? Why do you think the functional genes remained similar?

genome maps maps that describe the order and spacing of genes on each chromosome

linkage analysis use of markers to scan the genome and map genes on chromosomes

2.7.6 What are genome maps?

Genome maps describe the order of genes and the spacing between them on each chromosome.

A team of scientists at the Walter and Eliza Hall Institute in Melbourne are using statistical models and fast computers to identify possible locations of particular genes within genomes. Information from families in the investigation is collected so that pedigrees can be constructed. They then use markers to scan the genome and perform a **linkage analysis** in their attempt to map the gene.

The locus for the cystic fibrosis gene is on chromosome 7. This chromosome is shown in figure 2.67. Polydactyly, cystic fibrosis and one form of colour blindness are linked genes — they are located on the same chromosome.

The team analyse the pedigree, trait and genotyping information using probability models that measure the significance of the linkage. Linkage analysis has already proved successful in mapping the genes for Huntington's disease, muscular dystrophy, and the breast cancer genes *BRCA1* and *BRCA2*.

FIGURE 2.67 Some of the genes located on human chromosome 7

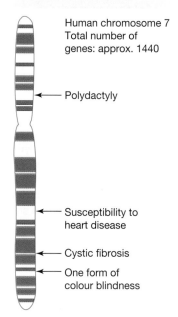

Human chromosome 7
Total number of genes: approx. 1440

Polydactyly

Susceptibility to heart disease

Cystic fibrosis

One form of colour blindness

SCIENCE AS A HUMAN ENDEAVOUR: Fast computers, statistical genetics and markers

Statistical methods have been used to establish gene linkage and estimate recombination fractions (due to crossing over in meiosis) since the 1930s. British scientists Julia Bell and John Haldane were the first to establish linkage between haemophilia and colour blindness with X-linked genes in 1937, and Jan Mohr found linkage between blood group types on an autosome in 1954. figure 2.68 shows genomes maps for human chromosomes 19 and 20.

FIGURE 2.68 Genome map of chromosome 19 and 20. One gene that can be seen is BCAS1, which codes for breast carcinoma-amplified sequence 1. This gene is often amplified in individuals with breast cancer.

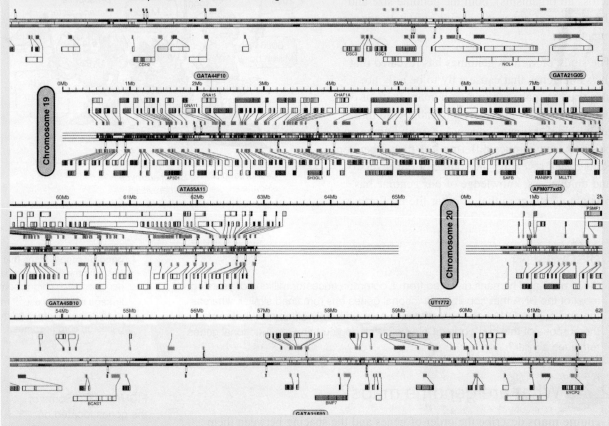

Molecular markers — to map

It was not until around 1980 that DNA sequence differences were used as molecular markers. The combination of these new markers with the use of **restriction fragment length polymorphisms (RFLPs)**, new multi-locus mapping methods, suitable algorithms, and the affordability and availability of fast computers revolutionised human genetic mapping.

Polymerase chain reaction — to amplify

In the late 1980s, the **polymerase chain reaction (PCR)** technique was beginning to revolutionise **molecular genetics**. This technique enabled amplification of tiny amounts of DNA, increasing the amount and hence the depth to which it could be studied, including those associated with inherited diseases (figure 2.69). This technique is used to diagnose COVID-19, even if only a small amount of the virus is present (which can't be detected by a rapid test).

restriction fragment length polymorphisms (RFLPs) variations in the lengths of DNA fragments in individuals with different alleles of a gene

polymerase chain reaction (PCR) a process which amplifies small amounts of DNA

molecular genetics study of genetics at a molecular level

FIGURE 2.69 During PCR, DNA is doubled after every cycle

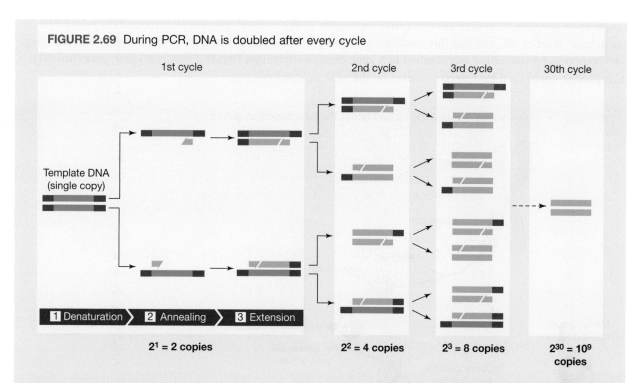

1st cycle   2nd cycle   3rd cycle   30th cycle

Template DNA (single copy)

| 1 Denaturation | 2 Annealing | 3 Extension |

$2^1 = 2$ copies $2^2 = 4$ copies $2^3 = 8$ copies $2^{30} = 10^9$ copies

SNPs — more markers

More recent research has focused on **single nucleotide polymorphisms (SNPs)**. SNPs (pronounced 'snips') are variations in your genome in which a single nucleotide has been substituted. More than 10 million SNPs have already been mapped in the human genome.

With advances in technology, identification of SNPs within specific genes is becoming more accessible and affordable. This knowledge can be used to guide lifestyle and medical treatment choices. Who do you think should have access to this genetic information? How might others use it?

FIGURE 2.70 SNPs are single nucleotide substitutions.

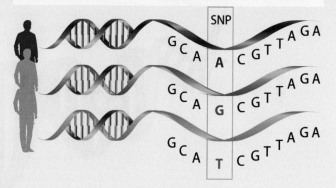

SNP

G C A A C G T T A G A

G C A G C G T T A G A

G C A T C G T T A G A

From HCG to International HapMap Project

The International HapMap Project aimed to describe patterns between human genetic variation, and health and disease. The HapMap is an abbreviation for 'haplotype map'. A haplotype is the inherited block of several SNPs (single nucleotide polymorphisms) together on a chromosome. The HapMap describes haplotypes, their locations in the genome and their frequency in different populations. A variety of other projects and research groups are now working with this aim. What will they discover and what impact will it have on your future?

2.7.7 Epigenetics

While the HGP and its technologies provided us with information about the sequence of DNA, it is only part of the story.

To understand more about its function, we may also need to know more about the DNA of our ancestors. Maybe environmental triggers can switch on or off particular genes? If some of these involve lifestyle triggers, could the events our ancestors experienced affect us? **Epigenetics** suggests that this may be the case.

single nucleotide polymorphisms (SNPs) genetic differences between individuals that can result from single base changes in their DNA sequences

epigenetics the study of the effect of the environment on the expression of genes

Epigenetics suggests that some environmental exposures or experiences may chemically modify your DNA by 'switching' it on or off, and that this modification can be inherited. For example, your great-grandmother may have experienced something that resulted in a gene being switched on (when it was previously switched off). This change could then have been inherited by the next generation.

FIGURE 2.71 How various epigenetics factors affect the expression of genes

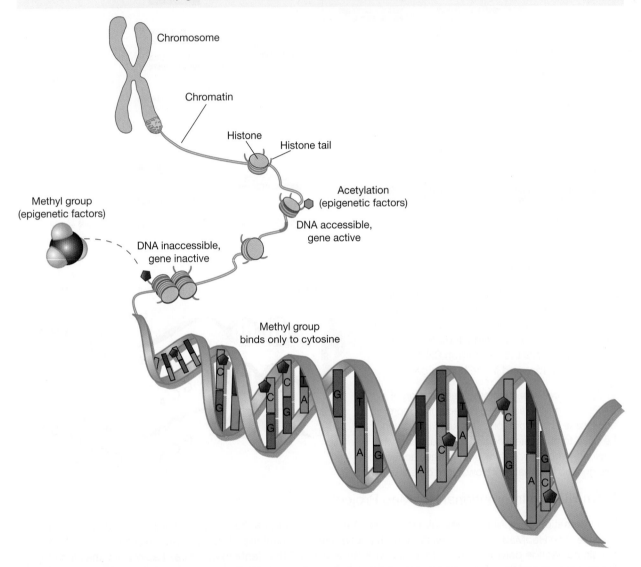

DISCUSSION

Imagine that the results of a personal genome scan suggested that you have a 25 per cent chance of developing a particular disease.
1. If you were told that environmental factors (such as diet and exercise) were more important than possessing the gene for this disease, how would this affect your future lifestyle?
2. Share your decision and justification with others.

Weblinks Epigenetic transformation — you are what your grandparents ate
The Human Genome Project

2.7 Activities

learn on

2.7 Quick quiz on	2.7 Exercise

Select your pathway

■ LEVEL 1 1, 5, 6	■ LEVEL 2 2, 4, 7, 8	■ LEVEL 3 3, 9, 10

These questions are even better in jacPLUS!
- Receive immediate feedback
- Access sample responses
- Track results and progress

Find all this and MORE in jacPLUS ▶

Remember and understand

1. **a.** State whether the following statements are true or false.

Statements	True or false?
i. A genome is the complete complement of genetic material (in a haploid set of chromosomes) in a cell or organism.	
ii. Variations in genetic sequences can affect not just what we look like but also our health.	
iii. As a result of rapid advances in technology, the time and cost of the sequencing of genes has dramatically increased.	
iv. Because each species has a unique genome, all humans have the same genome.	

b. Rewrite any false statement to make it true.

2. Match the term with its description in the following table.

Term	Description
a. Chromosome	**A.** A segment of double-stranded DNA that contains information that codes for the production of a particular protein
b. Epigenetics	**B.** More than 10 million of these variations in which a single nucleotide has been substituted have been mapped in the human genome
c. Gene	**C.** The complete complement of genetic material in a cell or organism
d. Genome	**D.** The structure on which genes are located
e. Genome maps	**E.** The study of genomes
f. Genomics	**F.** Theory that some environmental exposures or experiences may chemically modify your DNA by 'switching' it on or off, and that this modification can be inherited
g. PCR	**G.** These describe the order of genes and the spacing between them on each chromosome
h. SNPs	**H.** The technique used to amplify the amount of DNA

3. Describe the relationship between the following terms.
   a. Gene, genome, genomics, genome map
   b. Gene, gene sequencing, nucleotides, nitrogenous bases, DNA

Apply and analyse

4. **SIS** **a.** Describe three reasons the discoveries made in the Human Genome Project were important.
   b. Did the sequencing of the human genome answer our questions about why humans are unique? Explain.
   c. Now that the human genome has been mapped, suggest three questions that could be asked.
5. **SIS** **a.** Explain the difference between genetics and epigenetics.
   b. Provide two examples of epigenetic changes in DNA.
   c. Can epigenetic changes be inherited? Explain your response.
6. If you have a high risk of dying from a genetic disorder in 15 years, who do you think should know about it? Justify your response.
7. **SIS** Different genetic instructions within and between species are due to different nucleotide sequences in their genes. The following table shows part of the sequences of different genes from various organisms.

TABLE Comparing a section of DNA in three different organisms

Type of organism	Section of gene sequence
Duck	TAG GGG TTG CAA TTC AGC ATA GGG ATC
Human	TTG TGG TTG CTT TTC ACC ATT GGG TTC
Bacteria	AAT GAA TGT AAC AGG GTT GAA TTA AAA

   a. Suggest how they are similar.
   b. Suggest how they are different.
   c. Suggest a reason for the similarities and differences.

Evaluate and create

8. Personal genome scans can provide a lot of information about your genetic disposition for particular diseases and disorders. They do not, however, always guarantee that you will develop the disease.
   a. Find out more about the relationship between genotype, phenotype and environmental factors.
   b. Find out how these factors relate to the use, accuracy and effectiveness of personal genome scans.
   c. Based on your research, suggest regulations around the use of these scans.
9. Guidelines have been developed for companies in the US that supply 'custom DNA' or DNA sequences to order. These guidelines have been introduced to make it harder for bioterrorists to build dangerous viruses as potential bioweapons. However, concern exists that, because these rules are voluntary and most custom DNA is made outside the US, they may have limited value.
   a. Find out more about custom DNA and bioterrorism or bioweapons and DNA, and explain how these relate to gene sequencing.
   b. Research the current guidelines or regulations for producing and selling 'custom DNA' and describe your findings.
10. **SIS** **a.** Three of the most common types of genetic ancestry testing for genealogy are Y chromosome testing, mitochondrial DNA testing and single nucleotide polymorphism testing (SNPs).
    i. What are SNPs and why might they be used in ancestry DNA tests?
    ii. Find out one reason you would use Y chromosome testing rather than mitochondrial DNA testing.
    b. Ancestry DNA, My Heritage DNA, 23 and Me and FamilyTree DNA are four companies that offer DNA testing kits for the purpose of exploring ethnicity and family matches.
    i. Research one of these companies and the DNA testing that they offer.
    ii. If you were to be DNA tested, which company would you choose? Justify your choice.
    iii. Create a PMI around the use of DNA testing kits.
    iv. What is your opinion on the use of genetic genealogy?

Fully worked solutions and sample responses are available in your digital formats.

LESSON
2.8 Inheritance of genes

LEARNING INTENTION

At the end of this lesson you will be able to describe patterns of inheritance of simple dominant/recessive characteristics through generations of a family.

2.8.1 From one generation to the next

Have you ever browsed through the family photo album and looked at family members at different ages? Did any look like you? Which features do you share with them? Do certain characteristics seem to appear and disappear from one generation to the next? How could this happen?

The passing on of characteristics from one generation to the next is called **inheritance**. The study of inheritance involves a branch of science called **genetics**.

These characteristics or features are examples of your **phenotype**.

Your phenotype is determined by both your **genotype** and your environment. Your genotype is determined by genetic information in the chromosomes that you received in the gametes of your parents.

FIGURE 2.72 Traits are passed on between generations.

2.8.2 It's not all about your genes!

Environmental factors contribute to characteristics that make up your phenotype. Your weight, for example, although influenced by genetic factors, is also influenced by what you eat and how active you are. Exposure to and use of chemicals in your environment (such as pollution, hair dyes, tanning lotions and make-up), stress, intensity of sunlight and temperature ranges are other examples of environmental factors that can contribute to your phenotype.

FIGURE 2.73 Phenotypes are influenced by genotype and environment.

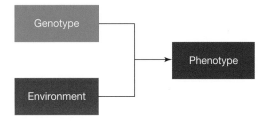

inheritance genetic transmission of characteristics from parents to offspring

genetics study of inheritance

phenotype characteristics or traits expressed by an organism

genotype genetic instructions (contained in DNA) inherited from parents at a particular gene locus

INVESTIGATION 2.6

How does the environment affect phenotype?

Aim

To determine how the environment affects the phenotypes in plants

Materials

- 10 seedlings grown from cuttings of the same plant
- potting mix in two small pots

Method

1. Write a clear hypothesis for this investigation.
2. Plant five of the seedlings in pot A and five in pot B.
3. Place pot A in a dark cupboard and pot B near a window.
4. Leave the plants undisturbed for two weeks. Water both pots when necessary. Ensure you use the same amount of water for both plants. After two weeks compare the plants in both pots.

Results

Copy and complete the following table.

TABLE Observations of seedlings in a pot in darkness (pot A) and a pot near a window (pot B)

	Pot A	Pot B
Number of seedlings that are still alive		
Colour of leaves		
Average height of seedlings		
Average number of leaves per seedling		

Discussion

1. Explain how you calculated the average number of leaves and the average height of the seedlings.
2. Answer the following for this experiment.
   a. What is the independent variable?
   b. What is the dependent variable?
   c. Which environmental factors were controlled?
3. Why is it important to use seedlings grown from cuttings of the same plant for this experiment?
4. Why were five seedlings planted in each pot?
5. Construct graphs of your data.
6. Comment on observed patterns in your data.
7. Explain why this experiment demonstrates that environmental factors play a part in determining the phenotype of an organism.

Conclusion

Summarise the findings of this investigation.

INVESTIGATION 2.7

Genetics database

Aim

To increase awareness of a variety of inherited traits

a. Widow's peak
b. Straight hairline

a. Unattached earlobes b. Attached earlobes

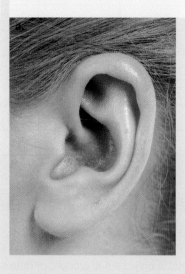

When you clasp your hands, is your right or left thumb on top?

Identify the number. If you cannot see it, you may be colour blind.

Do you have a cleft chin?

Method

Explore the phenotypes for the different traits of ten classmates. You may need to refer to the pictures provided to work out what each characteristic means.

TABLE Genetics database observations

Name of student				
Widow's peak?				
Right thumb over left when clasping hands?				
Cleft chin?				
Right-handed?				
Ear lobes attached?				
Freckles?				
Gap between front teeth?				
Hair naturally straight?				
Colour blind?				

Results

Copy and complete the table shown. Enter data for ten students in the table (add some more columns to allow for this).

Discussion

1. The database you have created contains only a small amount of data so you do not need to set up a query to search for particular data to save time. (It would probably take you more time to set up the query than it would take to look through the data manually!) Can you think of examples of databases that contain so much information that it would take days to search the data manually?
2. Does your school keep a computerised database of student details? What type of information is kept in the database?

Conclusion

Summarise the findings for this investigation.

2.8.3 A product of chance

The similarities and differences in how you look compared to your relatives are partly due to chance.

When fertilisation takes place, the zygote receives a pair of each set of chromosomes, the maternal and paternal chromosomes. Located within these chromosomes are the genes for particular characteristics.

These alternative forms or expressions of a gene are called alleles. You have a particular combination of alleles in your genotype (one allele is inherited from each parent). For example, in figure 2.74, the gene for eye colour is shown. Two variations of this are shown: blue or brown eyes. The different alleles of eye colour lead to these different traits and are represented using letters.

dominant a trait (phenotype) that requires only one allele to be present for its expression in a heterozygote

recessive a trait (phenotype) that will only be expressed in the absence of the allele for the dominant trait

2.8.4 Recessive and dominant traits

The alleles and in turn the genotype an individual has determines the trait expressed. These traits can be **dominant** or **recessive**.

- A trait is dominant if at least one allele for the trait is present. If alleles for both the dominant and recessive traits are present for a gene, the dominant trait will be expressed (as it 'masks' the recessive trait).
- A trait is recessive if two copies of the allele for the trait are required for the trait to be expressed. If the allele for both a dominant and recessive trait is present, the recessive trait is hidden.

Each of these alleles is often represented using a letter. In the family generations diagram shown in figure 2.74, the expression of the gene for eye colour is shown. The allele for brown eyes is denoted as a capital letter B, because brown is the dominant trait. The allele for blue eyes has been denoted by a lower-case letter b, because this trait is recessive to the brown eye trait.

Let's look at this example in the Davis and Swift families in figure 2.74.

FIGURE 2.74 The inheritance of eye colour in a family

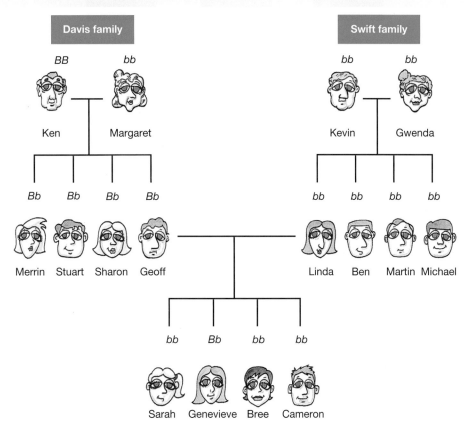

A common area of confusion and misconception around dominant and recessive comes into play when referring to alleles. Many resources incorrectly refer to a dominant or recessive allele.

In terms of biology, the expression of the genotype as a particular phenotype (the 'trait') is accepted as what is dominant or recessive, rather than the allele itself.

2.8.5 Mix and match

As previously outlined, the combination of the alleles that you have for a particular gene is called your genotype.

If your alleles for that gene are the same (for example, BB or bb), then you are described as **homozygous** (or pure breeding) and if they are different (for example, Bb), then you are **heterozygous** (or hybrid) for that trait.

homozygous a genotype in which the two alleles are identical
heterozygous a genotype in which the two alleles are different

A genotype of *BB* can be described as **homozygous dominant**, while a genotype of *bb* can be described as **homozygous recessive**.

In **complete dominance**, you can determine which trait is dominant by looking at the heterozygote. The trait that the heterozygote expresses is the dominant trait.

FIGURE 2.75 Different types of genotypes

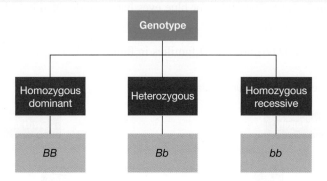

2.8.6 Are you a carrier?

If a person is heterozygous for a trait and possesses alleles for both the recessive and the dominant trait, the dominant trait will be expressed in the phenotype. The recessive phenotype will only be observed when the allele for the dominant trait is not present.

The term **carrier** is also used to describe someone who is heterozygous for a particular trait. People may not be aware they are the carrier of an allele for a recessive trait (such as blue eyes and red hair) because the trait does not show in their phenotype. They may, however, have children who show the recessive trait.

Can you suggest how two brown-eyed parents (dominant trait) could have a child who has blue eyes (recessive trait)?

homozygous dominant a genotype where both alleles for the dominant trait are present

homozygous recessive a genotype where both alleles for the recessive trait are present

complete dominance a type of inheritance where traits are either dominant or recessive

carrier an individual heterozygous for a characteristic who does not display the recessive trait

FIGURE 2.76 Alleles on chromosomes inherited from each of your parents contribute to your genotype.

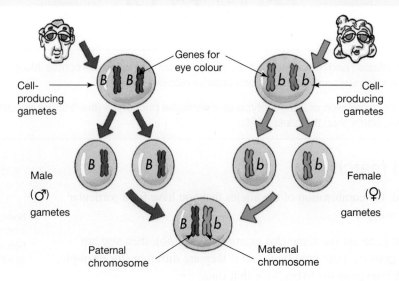

SCIENCE AS A HUMAN ENDEAVOUR: Mendel's memos

Gregor Mendel (1822–1884), an Austrian monk, carried out experiments on pea plants in a monastery garden for 17 years. His work was unknown for about 35 years. When it was discovered in 1900, he became known as the 'father of genetics'. From his experiments, Mendel was able to explain patterns of inheritance of certain characteristics. Why did Mendel use pea plants and not cabbages? Pea plants are easily grown in large numbers and have easily identifiable characteristics that have either/or alternatives. Mendel could control their breeding by taking pollen from a particular pea plant and putting it on the stigma of another. Pea plants can also be self-pollinated.

Mendel crossed a pure-breeding tall plant with a pure-breeding short plant. A plant is classed as pure breeding for a characteristic if it has not shown the alternative characteristic for many generations.

Mendel showed the factor for shortness had not disappeared because when he crossed the tall offspring (called the F_1 generation) with each other, about a quarter of those offspring (called the F_2 generation) were short. He called shortness a recessive factor because it was hidden or masked in the F_1 generation. We now know that Mendel's 'factors' are genes. The alternative forms of the factors are alleles.

Mendel investigated one trait at a time. He bred plants for single characteristics such as height. He worked out that if many pure-breeding tall and short plants were crossed and then the first generation (F_1 generation) was also crossed, the ratio of tall to short plants would be about 3:1. He repeated these experiments many times using the other characteristics of the pea plants and came up with similar ratios. This is called the **monohybrid ratio**.

FIGURE 2.77 Gregor Mendel

monohybrid ratio the 3:1 ratio of a particular characteristic for offspring produced by heterozygous parents, controlled by autosomal complete dominant inheritance

FIGURE 2.78 Pea plants showing the characteristics Mendel used in his experiments

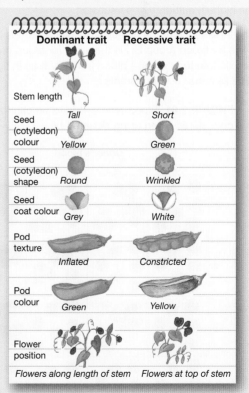

Dominant trait	Recessive trait
Stem length	
Tall	Short
Seed (cotyledon) colour — Yellow	Green
Seed (cotyledon) shape — Round	Wrinkled
Seed coat colour — Grey	White
Pod texture — Inflated	Constricted
Pod colour — Green	Yellow
Flower position — Flowers along length of stem	Flowers at top of stem

FIGURE 2.79 Mendel's experiments were well designed, and his record-keeping was meticulous.

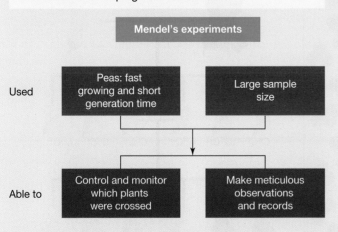

Mendel's experiments

Used
- Peas: fast growing and short generation time
- Large sample size

Able to
- Control and monitor which plants were crossed
- Make meticulous observations and records

EXTENSION: Degrees of dominance

In complete dominance, the expression of one trait is dominant over the other. This results in both the homozygous dominant and heterozygous genotypes being expressed as the same phenotype. In two other types of inheritance, neither allele is dominant over the other.

- In codominance, the heterozygote has the characteristics of both parents. An example of this type of inheritance is seen in the human blood groups.
- In incomplete dominance, the heterozygotes show a phenotype that is intermediate between the phenotypes of the homozygotes. An example of this type of inheritance is seen in the flower colour of snapdragons (seen in figure 2.82).

The phenotype of the heterozygote can indicate the type of inheritance.

You may find variations in the definitions of the terms *recessive*, *dominant*, *codominance* and *incomplete dominance* in various resources. New technologies and new knowledge can modify how we see, understand and communicate our knowledge. This eventually results in the creation, modification or replacement of terminology and theories that are used by a majority or enforced by those with the highest authority or persuasion.

FIGURE 2.80 Different types of dominance

Heterozygotes and types of inheritance

Complete dominance
R = red flower
r = white flower

Codominance
I^A = blood group A
I^B = blood group B

Incomplete dominance
C^R = red flower
C^W = white flower

Rr → Red flower

$I^A I^B$ → Blood group AB

$C^R C^W$ → Pink flower

The inheritance of the human ABO blood groups is by codominance. I^A is the allele for blood group A; I^B is the allele for blood group B; i is the allele for blood group O. The type of blood group you have determines who you can donate to or receive blood from (as seen in figure 2.81). Which blood type are you? Are you the same blood type as either of your parents?

eles-4224

FIGURE 2.81 Blood type needs to be carefully considered when using donated blood.

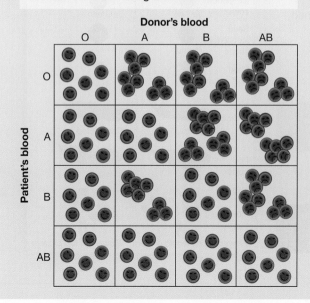

FIGURE 2.82 An example of the inheritance of a trait showing incomplete dominance

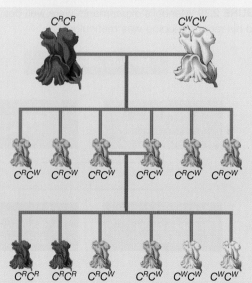

Both codominance and incomplete dominance can be considered examples of partial dominance. The common feature of these types of inheritance is that the heterozygote will show or express a phenotype that is different from the phenotype of an individual with either homozygous genotype.

Partial dominance can result in offspring that express a phenotype not observed in either parent.

ACTIVITY: Learning the key terms

On your own, in pairs or in teams, create a rhyme, song or poem that effectively uses as many of the key terms in this lesson as possible. An example is provided. Add movements or actions for each line and share it with your class.

Alleles are alternative forms of genes
Sometimes showing, sometimes behind the scenes
Genotypes are made up of two of them
Homozygotes have two the same
Heterozygotes have one of each kind
From each parent, alleles you will find

2.8 Activities

learn on

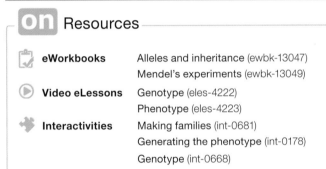

2.8 Quick quiz on	2.8 Exercise

Select your pathway

■ LEVEL 1	■ LEVEL 2	■ LEVEL 3
1, 2, 4, 7	3, 5, 8, 10	6, 9, 11, 12

These questions are even better in jacPLUS!
- Receive immediate feedback
- Access sample responses
- Track results and progress

Find all this and MORE in jacPLUS ▶

Remember and understand

1. Match the term with its description in the following table.

Term	Description
a. Fertilisation	**A.** Characteristics determined by genotype and influenced by environment
b. Gametes	**B.** Combination of alleles for a particular trait
c. Genetics	**C.** Sex cells
d. Genotype	**D.** The branch of science that studies inheritance
e. Inheritance	**E.** The fusion of male and female gametes
f. Phenotype	**F.** The passing of characteristics from one generation to the next

▶

2. a. State whether the following statements are true or false.

Statements	True or false?
i. When fertilisation takes place, the zygote receives either two maternal chromosomes or two paternal chromosomes in each pair of chromosomes.	
ii. A capital letter is used for the allele of the dominant trait (for example, *B*) and a lower-case letter is used for the allele of the recessive trait (for example, *b*).	
iii. The recessive trait can only be expressed if the allele for the dominant trait is not present.	
iv. The phenotype describes the combination of alleles present whereas the genotype describes the expression of the trait (for example, brown or blue eyes).	
v. People are not usually aware of being a carrier because it does not show in their phenotype, and they may have children who show the trait.	
vi. In complete dominance, both the homozygous dominant and heterozygous genotypes express the same phenotype.	

b. Rewrite any false statements to make them true.

3. For each of the following provide a definition and example.
  a. Allele
  b. Gene
  c. Heterozygous genotype
  d. Homozygous dominant genotype
  e. Homozygous recessive genotype
4. A widow's peak is dominant to a straight hairline. Identify the hairline of an individual with the genotype:
  a. *HH* **b.** *Hh* **c.** *hh*
5. State how many alleles are on a homologous pair of chromosomes for a particular trait. Provide a reason for your response.
6. Distinguish between the following pairs of terms.
  a. Inheritance and genetics
  b. Genotype and phenotype
  c. Dominant trait and recessive trait
  d. Pure breeding and hybrid

Apply and analyse

7. Mendel obtained a ratio of 3 tall to 1 short plants in the offspring when he crossed pure-breeding tall and short plants. Convert this monohybrid ratio of 3:1 into the following.
  a. A fraction
  b. A percentage
8. A purebred pink flower was bred with a purebred white flower. All the offspring were heterozygous with pink flowers.
  a. Which trait is dominant?
  b. Which trait is recessive?
  c. Write the possible genotypes for plants with pink flowers.
  d. Write the possible genotypes for plants with white flowers.
9. Refer to Investigation 2.6.
  a. Propose an experiment that you can run to investigate the effect of another environmental factor on the phenotypes in plants.
  b. Suggest a way to increase the reliability of the experiment.

Evaluate and create

10. Construct a Venn diagram with the headings 'Determined by genetics' and 'Determined by environmental factors', and then place the following terms in the most appropriate category: eye colour, tattoo, skin colour, cleft chin, freckles, colour blind, hair colour, scar, widow's peak.

11. Investigate two of the following genetic conditions: Huntington's disease, Tay–Sachs, cystic fibrosis, fragile X syndrome or PKU.
    a. Briefly describe each disease.
    b. Is the disease dominant or recessive?
    c. Write the genotype for an individual with the disease.
    d. Based on your findings, do you believe that individuals should find out if they have the allele for the diseases you researched? Justify your response.

12. **SIS** With reference to information provided regarding Mendel's experiments, answer the following questions.
    a. Suggest why the ability to self-pollinate and cross-pollinate was an advantage for the pea plants in Mendel's experiments.
    b. Suggest four strengths in the design of Mendel's experiments.
    c. Suggest the phenotype of a pea plant that showed the following.
       i. Dominant traits for seed colour, shape and coat colour
       ii. Recessive traits for stem length and flower position
       iii. A dominant trait for pod texture, but a recessive trait for pod colour
    d. Construct a simple table to include the phenotype and genotype for the trait of each plant. Use an appropriate letter to match each of the characteristics Mendel studied.

Fully worked solutions and sample responses are available in your digital formats

LESSON
2.9 Punnett squares and predicting inheritance

LEARNING INTENTION

At the end of this lesson you will be able to use Punnett squares to predict characteristics of offspring.

2.9.1 Mixing your genes?

Selecting a mate can be one of the most crucial decisions in your life. This selection process involves both conscious and unconscious choices. Next time you look at that special person, take a *really* good look. One day you might be mixing your genes together! What might the result be?

FIGURE 2.83 The genes of parents affect the genes of their offspring.

This is a question that not only faces individuals selecting a mate but those who choose to reproduce through assisted technologies. The combination of genes greatly determines the traits of an offspring.

According to **Mendelian inheritance**, genes inherited from either parent segregate into gametes at an equal frequency.

The chances of having offspring that show a particular trait depends on the trait's type of inheritance; that is, whether it is inherited by complete dominance, codominance, incomplete dominance or **sex-linked inheritance**.

Mendelian inheritance an inheritance pattern in which a gene inherited from either parent segregates into gametes at an equal frequency

sex-linked inheritance an inherited trait coded for by genes located on sex chromosomes

2.9.2 Punnett squares: Predicting possibilities

Reginald Punnett (1875–1967) was a geneticist who supported Mendel's ideas. He repeated Mendel's experiments with peas and also did his own genetic experiments on poultry. Punnett is responsible for designing a special type of diagram, which is named after him. A **Punnett square** is a diagram that is used to predict the outcome of a genetic cross for each offspring.

A Punnett square shows which alleles for a particular trait are present in the gametes of each parent. It then shows possible ways in which these can be combined. The alleles in each of the parent's genotypes for that trait are put in the outside squares and then combined to show the possible genotypes of the offspring.

FIGURE 2.84 In a Punnett square, alleles from each parent's genotype are used to determine the possible genotypes of the offspring.

Punnett square for $Bb \times Bb$

B = allele for brown eyes

b = allele for blue eyes

	Father	
Possible gametes	B	b
B	BB	Bb
b	Bb	bb

(Mother — rows)

Offspring probabilities

Genotype: $\frac{1}{4}$ BB: $\frac{1}{2}$ Bb: $\frac{1}{4}$ bb

Phenotype: $\frac{3}{4}$ brown eyes: $\frac{1}{4}$ blue eyes

Reviewing alleles

Depending on the type of inheritance, the alleles that may be inherited and shown on a Punnett square can be represented in slightly different ways. This was introduced earlier in section 2.8.4. Representing alleles currently is vital for use in Punnett squares. Some examples are shown in table 2.6.

When you are selecting letters to use to represent your alleles, ensure you use letters in which the capital and lower case is easily distinguished (for example, B and b or A and a). If this is not possible, place a line above the allele for the recessive trait to make them easier to distinguish (for example, C and c or S and s). If alleles are on sex chromosomes, the sex chromosome (X or Y) should also be shown.

TABLE 2.6 Representing alleles

Type of inheritance	How to write the alleles	How to write the genotypes
Complete dominance (e.g. dominant/recessive) on an autosome	Use a capital letter for the allele of the dominant trait (e.g. B) and a lower-case version of the same letter for the allele for the recessive trait (e.g. b).	BB, Bb or bb
Incomplete dominance or codominance on autosomes	Different letters are used to represent each allele, and these are written as a superscript on a shared capital letter (e.g. C^R and C^W or I^A and I^B).	$I^A I^A$, $I^A I^B$ or $I^B I^B$
Sex-linked (e.g. X-linked inheritance)	If the allele is located on the sex chromosome X, the capital letter X is used with a superscript (e.g. X^B and X^b).	$X^B X^B$, $X^B X^b$, $X^b X^b$, $X^B Y$ or $X^b Y$

The examples in section 2.9.3 show examples of Punnett squares with complete dominance. Figures 2.91 and 2.93 in the extension boxes show examples of Punnett squares involving codominance and sex linkage.

Punnett square a diagram used to predict the outcome of a genetic cross

2.9.3 Creating Punnett squares

Remember Linda Swift and Geoff Davis from lesson 2.8? The inheritance of eye colour was shown in their family. Brown eyes were shown to be dominant to blue eyes. figure 2.85 shows that Geoff has brown eyes (*Bb*), Linda has blue eyes (*bb*) and their children have either brown or blue eyes.

But how were the alleles from each parent inherited? The mix of alleles that Linda and Geoff contributed to their children's genetic make-up shown in figure 2.86 can be simplified by using a Punnett square.

This example can also be seen in figure 2.87. The Punnett square created tells us that the chance of producing a child with the combination *Bb* (heterozygous) is two out of four or 50 per cent, and the chance of producing the combination *bb* (homozygous recessive) is also two out of four or 50 per cent.

Hence, each of Linda and Geoff's children have a 50 per cent chance of having blue eyes and a 50 per cent chance of having brown eyes (you can also write this as a simplified ratio — 1 brown:1 blue eyes).

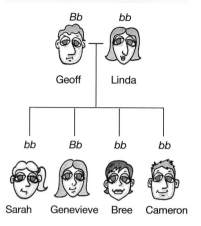

FIGURE 2.85 The inheritance of blue eyes is recessive to the inheritance of brown eyes.

FIGURE 2.86 The chromosomes showing the different alleles of eye colours (*Note:* the entire chromosome has been coloured to show the allele. In reality, the gene for eye colour would only be a very small section of the chromosome.)

	Linda	Geoff	Genevieve	Cameron	Sarah	Bree
Combination of alleles	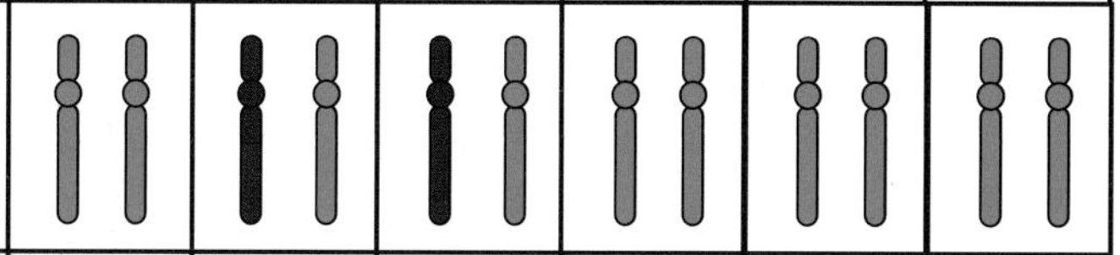					
Genotype	*bb*	*Bb*	*Bb*	*bb*	*bb*	*bb*
Phenotype	Blue eyes	Brown eyes	Brown eyes	Blue eyes	Blue eyes	Blue eyes

FIGURE 2.87 Punnett squares show us the chance of offspring inheriting particular combinations.

Punnett square *Bb* × *Bb*

B = allele for brown eyes

b = allele for blue eyes

		Linda	
	Possible gametes	*b*	*b*
Geoff	*B*	*Bb*	*bb*
	b	*Bb*	*bb*

Offspring probabilities

Genotype: $\frac{1}{2}$ *Bb*: $\frac{1}{2}$ *bb*

Phenotype: $\frac{1}{2}$ brown eyes: $\frac{1}{2}$ blue eyes

To create a Punnett square:
1. Write down the cross that is occurring, showing the genotypes of the two parents.
2. Create a grid that is 3 x 3 squares.
3. Show gametes that are passed on from one parent in the first column (leave the first square blank). So, for example, in figure 2.87, Linda, who is *bb*, will only be able to pass on *b* in their gametes.
4. Show gametes that are passed on from the other parent in the first row (leaving the first square blank). So, in the example in figure 2.87, Geoff, who is *Bb*, will have *B* in half their gametes and *b* in the other half.
5. Fill in the boxes by taking looking at the gamete from the mother and gamete from the father, and combining these to make the genotype (if this leads to a heterozygote, the allele for the dominant trait should be listed first).
6. Use the Punnett square to determine probabilities. These may be expressed as fractions, percentages or ratios.

It is important to note that the chance of inheritance calculated for one child is not dependent on the inheritance of another. As with a coin toss, each is an independent event.

Punnett square between two heterozygous individuals

Let's look at another example of completing a Punnett square, outlined in figure 2.88.

Two parents, Amal and Marco, decided to have a child. They wish to determine the chance that their child will have unattached earlobes, which is dominant over attached earlobes. Both Amal and Marcoare heterozygous (*Ee*) for this gene trait.

The ratio of genotypes is 1 *EE* : 2 *Ee* : 1 *ee* (or a 25 per cent chance of *EE*, a 50 per cent chance of *Ee* and a 25 per cent chance of *ee*).

The ratio of phenotypes is 3 unattached earlobes:1 attached earlobes (or a 75 per cent chance of unattached, and a 25 per cent chance of attached).

FIGURE 2.88 A Punnett square for Amal and Marco.

Punnett square for *Ee* × *Ee*

E = allele for unattached earlobes

e = allele for attached earlobes

Amal

	Possible gametes	*E*	*e*
Marco	*E*	*EE*	*Ee*
	e	*Ee*	*ee*

CASE STUDY: Genetic inheritance and cystic fibrosis

About 1 in 2500 people suffer from an genetic disorder called cystic fibrosis (CF). The CF gene is located on chromosome number 7. Cystic fibrosis is a recessive trait. In the CF allele, one amino acid in a chain of 1480 amino acids is not produced, causing a faulty protein to be synthesised. This results in the production of large amounts of thick mucus by cells lining the lungs and in the pancreas, where digestive juices are secreted. The mucus interferes with the working of the respiratory and digestive systems. Infection readily occurs and sufferers tend to have a shortened life span.

Checking to see if you are a CF carrier

Since the identification of the defective allele in 1989, the DNA of parents-to-be can be analysed to find out if they carry the allele for cystic fibrosis. Around 1 in 25 Australians are thought to be carriers of this allele. This means they could have a child with cystic fibrosis. For example, if both parents are carriers, the chance is 1 in 4 that they will have a child with cystic fibrosis.

FIGURE 2.89 The likelihood of a child suffering from cystic fibrosis

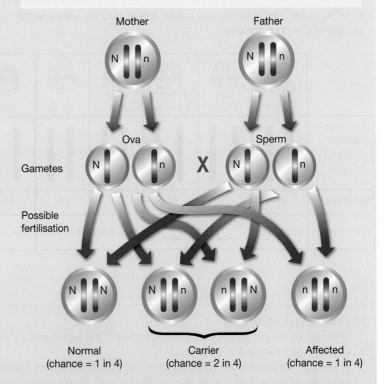

A chance event?

Genetic counselling can help parents-to-be who are both carriers of the CF allele with their decision about whether to have a child. If they decide to go ahead, genetic tests can be used to determine the genotype of the embryo. If both parents are heterozygous for cystic fibrosis, as shown in figure 2.90, each child has a 25 per cent chance of having cystic fibrosis (through being homozygous recessive), and a 75 per cent chance of not having the disease. It is important to note that the chance is independent for each child. If the parents already had one child with cystic fibrosis, the next child would still have a 25 per cent chance of having it. Additionally, each child has a 50 per cent chance of not having the disorder but being a carrier with one CF allele. You can see how these chances are calculated in the Punnett square shown in figure 2.90.

If the genetic test shows that the child will have cystic fibrosis, the parents need to consider many important factors around this. Genetic counselling may also help the parents with this very difficult decision.

FIGURE 2.90 If these two heterozygous parents had five children, would it be possible for none of them to have the disease?

Punnett square for $Nn \times Nn$

N = normal allele
n = cystic fibrosis allele

	N	n
N	NN	Nn
n	Nn	nn

Offspring probabilities

Genotype: $\frac{1}{4}$ NN: $\frac{1}{2}$ Nn: $\frac{1}{4}$ nn

Phenotype: $\frac{3}{4}$ normal: $\frac{1}{4}$ cystic fibrosis

---

EXTENSION: Using Punnett squares for incomplete dominance and multiple alleles

Do you know which type of blood you have flowing through your capillaries? The inheritance of your blood type involves multiple alleles, and both codominant inheritance and complete dominance.

The ABO gene — multiple alleles

The inheritance of blood types A, B, AB and O are determined by the ABO gene. There are three different alleles for the ABO gene. Two of these carry instructions to make a particular type of protein called an antigen; the other does not.

One of the alleles codes for antigen A and the other codes for antigen B. If you possess both alleles, you have the instructions to produce both antigen A and antigen B. This is an example of codominant inheritance because both blood types are expressed in the heterozygote. If you refer to the ABO gene as I, the allele that codes for making:

- antigen A can be referred to as I^A
- antigen B can be referred to as I^B.

The ability to make antigen A or B is shown as a capital letter because it is dominant to the inability to make either antigen (which is recessive). If you refer to the ABO gene as I, then the allele that codes for making neither antigen can be referred to as i.

Blood type A, B, AB or O?

The combination of the ABO alleles in your genotype determines your blood type. If your genotype is ii, you will have blood type O. If it is I^Ai, you will have blood type A, and if I^AI^B, you will have blood type AB. Table 2.7 shows the six possible genotypes and the blood groups that they code for.

TABLE 2.7 Genotype and phenotype of blood groups

Genotype	Phenotype
$I^A I^A$	Blood type A
I^Ai	Blood type A
$I^B I^B$	Blood type B
I^Bi	Blood type B
$I^A I^B$	Blood type AB
ii	Blood type O

Apply and analyse

3. Predict the probabilities of the phenotypes and genotypes of the offspring of the following, using the provided Punnett squares.
   a. A homozygous brown-eyed parent and a blue-eyed parent
   b. Two parents heterozygous for brown eyes

Punnett square for BB × bb
B = allele for brown eyes
b = allele for blue eyes

	B	B
b		
b		

Punnett square for Bb × Bb
B = allele for brown eyes
b = allele for blue eyes

	B	b
B		
b		

4. A student completed their Punnett square, but forgot to note down the genotypes and the phenotypes of the parents when examining the earlobe trait. Determine the identity of the parents and enter this into the Punnett square shown.

Punnett square for
A = allele for attached earlobes
a = allele for unattached earlobes

	Aa	aa
	Aa	aa

Offspring probabilities
Genotype: 2 Aa: 2 aa
Phenotype: 2 attached
2 unattached

5. A majority of traits are inherited on autosomes. Some traits are inherited on the sex chromosomes. For an X-linked recessive trait to be expressed, the allele for the dominant trait cannot be present. Suggest why males have a greater chance of showing an X-linked recessive trait than females.

6. Two individuals are carriers for cystic fibrosis, a recessive trait that leads to issues with a membrane protein. They are planning to have a child.
   a. What is the chance that their child will have cystic fibrosis?
   b. The two individuals have a baby daughter. At birth, screening shows that she does not have cystic fibrosis. What is the chance this child is a carrier for cystic fibrosis?

7. Three experiments are conducted to investigate the inheritance of traits in pea plants. The following table shows the experimental results.

TABLE Results of monohybrid crosses with peas

Set	Treatment	Experimental outcomes
A	A pure breeding round pea plant and a hybrid breeding wrinkled pea plant are crossed.	All pea plants in the first generation had round-shaped peas.
B	A pure breeding yellow pea plant and a pure breeding green pea plant are crossed.	All pea plants in the first generation had yellow-coloured peas.
C	Two hybrid breeding tall stem pea plants are crossed.	75 per cent of the pea plants in the first generation had tall stem whereas 25 per cent of the pea plants in the first generation had short stem.

   a. Two pea plants from the first generation of Set B are then crossed. Predict the genotypic and phenotypic rations of the second generation.
   b. Two short stem pea plants from the first generation of Set C are then crossed. Predict the genotypic and phenotypic rations of the second generation.

8. A type blood is dominant to O type blood. B type blood is also dominant to O type blood. The alleles for each are represented as I^A (for A type), I^B (for B type) and i (for O type).
   a. A mother who is homozygous for A type blood has a child with a father who has O type blood. Using a Punnett square, show the likelihood for the different blood types in her child.
   b. When an individual has both the I^A and I^B alleles, they have AB blood (as neither A or B are dominant over each other; this is known as co-dominance). Construct a Punnett square showing the probabilities of the different blood types if two AB parents have a child.
   c. Determine the chance of a couple with blood types AB and A having a child with the following.
      i. Blood type A   ii. Blood type B   iii. Blood type AB   iv. Blood type O
   d. Can a father with blood type A and a mother with blood type B have a child with blood type O? Explain.

9. Construct Punnett squares and use the provided table to assist you in answering the following questions.

TABLE Examples of dominant and recessive traits

Dominant trait	Recessive trait
Unattached earlobes	Attached earlobes
Mid-digital hair present	Mid-digital hair absent
Normal skin pigmentation	Pigmentation lacking (albinism)
Non-red hair	Red hair
Rhesus-positive (Rh +ve) blood	Rhesus-negative (Rh −ve) blood
Dwarf stature (achondroplasia)	Average stature
Widow's peak	Straight hairline

a. Find the probability of Sally (who is homozygous for dwarf stature) and Tom (who has average stature) having a child with dwarf stature.

b. Find the probability of Fred (who is heterozygous for dwarf stature) and Susy (who has average stature) having a child with dwarf stature.

c. What is the chance of two parents who are both heterozygous for free earlobes having a child with attached ear lobes?

d. Michael is heterozygous for mid-digital hair, whereas Debbie does not have mid-digital hair. What is the chance of their children having mid-digital hair?

Evaluate and create

10. **SIS** Some experts suggest that the major histocompatibility complex can influence mate selection. Research this claim and discuss your findings.

11. a. What is sexual selection? Give two examples.

b. How is sexual selection different from natural selection?

c. Suggest implications of sexual selection for our species. (You should describe inheritance in your response.)

d. Suggest the possible impact of sexual selection on your future reproductive life.

12. While the science of love is still in its infancy, advances in molecular biology and technology have increasingly allowed us to peer through its window.

a. Find out examples of research on the chemistry of love or love potions.

b. Do you believe that this research should be continued? Justify your response.

c. Suggest possible issues that may arise with the knowledge obtained and its possible applications.

d. Discuss if, how and who should regulate or control this type of research.

13. **SIS** A farmer is tracking the size of the farm's crops over generations. The farmer knows that tall crops are dominant to short crops. Each generation, the farmer crosses the plants to a short crop, and observes the phenotypes in the successive generation. The results of the crosses are shown in the following graph.

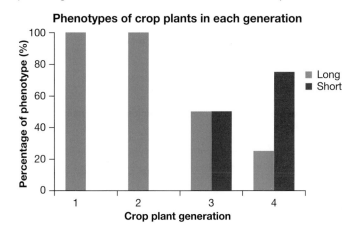

Phenotypes of crop plants in each generation

a. Discuss the trends seen in this graph.

b. Why does the short crop trait only appear for the first time in offspring in the third generation? Justify your response.

c. The farmer decides they only want tall plants. What genotype plant do they need to breed from the short plant crop in order to allow for this?

d. What would you expect to see in the fifth generation? Show this using Punnett squares.

Fully worked solutions and sample responses are available in your digital formats.

LESSON
2.10 Interpreting pedigree charts

LEARNING INTENTION

At the end of this lesson you will be able to interpret pedigree charts and describe patterns of inheritance shown within them.

2.10.1 What are pedigree charts?

A diagram that shows a family's relationships and how characteristics are passed on from one generation to the next is a **pedigree chart**. A pedigree chart for Linda and Geoff's family is shown in figure 2.94.

pedigree chart diagram showing the family tree and a particular inherited characteristic for family members

FIGURE 2.94 An example of a pedigree chart

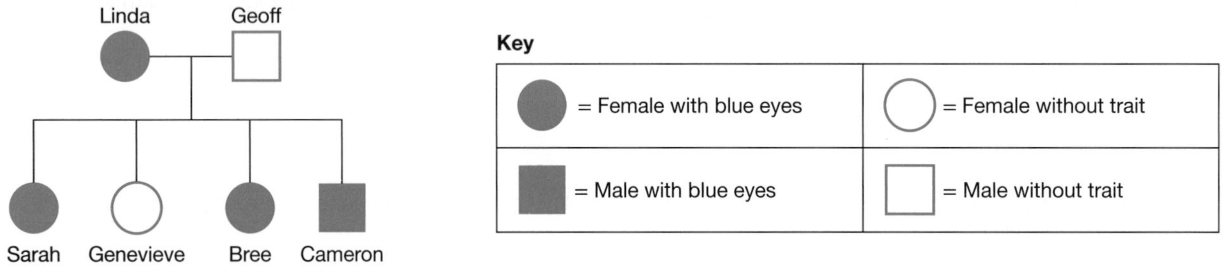
Linda
Geoff
Sarah
Genevieve
Bree
Cameron
Key
= Female with blue eyes
= Female without trait
= Male with blue eyes
= Male without trait

Some rules for drawing pedigree charts are outlined in figure 2.97.

FIGURE 2.95 Different symbols represent different aspects of a pedigree.

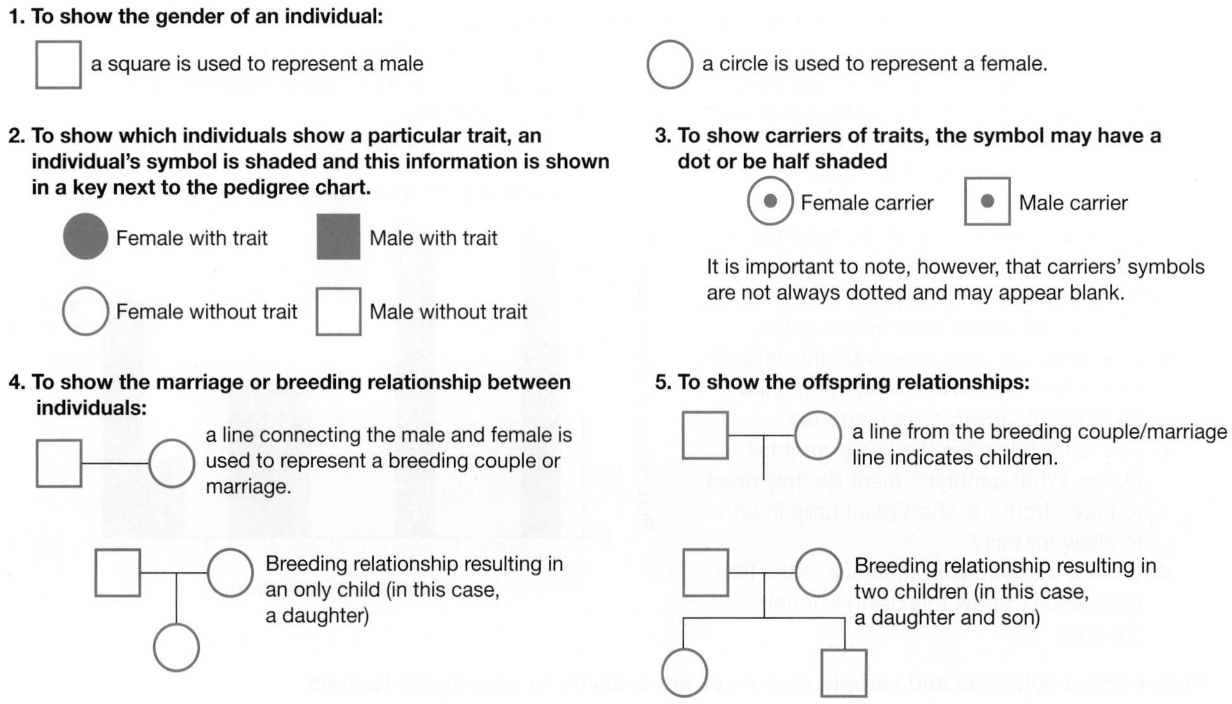
1. To show the gender of an individual:
a square is used to represent a male
a circle is used to represent a female.
2. To show which individuals show a particular trait, an individual's symbol is shaded and this information is shown in a key next to the pedigree chart.
Female with trait
Male with trait
Female without trait
Male without trait
3. To show carriers of traits, the symbol may have a dot or be half shaded
Female carrier
Male carrier
It is important to note, however, that carriers' symbols are not always dotted and may appear blank.
4. To show the marriage or breeding relationship between individuals:
a line connecting the male and female is used to represent a breeding couple or marriage.
Breeding relationship resulting in an only child (in this case, a daughter)
5. To show the offspring relationships:
a line from the breeding couple/marriage line indicates children.
Breeding relationship resulting in two children (in this case, a daughter and son)

2.10.2 Who's who in a pedigree chart?

Pedigree charts can be used to observe patterns and predict the inheritance of traits within families. Patterns in the inheritance of these traits can also show whether the trait is dominant or recessive and whether it is carried on an autosome or sex chromosome.

The pedigree chart in figure 2.96 shows how individuals and generations can be identified so interpretation of patterns can be more effectively communicated. The shaded individual at the top of the chart is identified as I-1 (individual 1 in the first generation) and the shaded individual in the bottom row is identified as III-3 (individual 3 in the third generation). The daughters of individual I-1 are identified as II-3 and II-4.

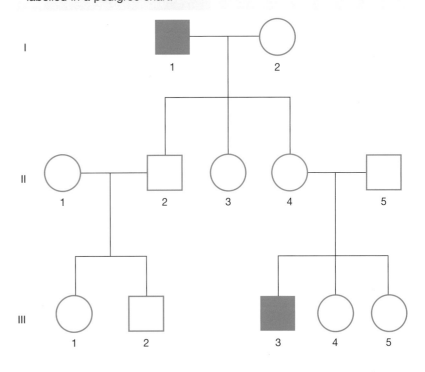

FIGURE 2.96 Each individual is numbered and each generation labelled in a pedigree chart.

2.10.3 Naming inheritance

The inheritance of various traits can be described in terms of the location of the gene responsible and whether the inheritance is dominant or recessive. For traits located on the sex chromosomes (X or Y), the trait is considered to be sex-linked. For traits located on the autosomes (1 to 22 in humans), the trait is considered to be autosomal.

Some different types of inheritance are:
- autosomal recessive
- autosomal dominant
- X-linked recessive
- X-linked dominant
- Y-linked.

Pedigrees are a useful way to help determine how a trait may be inherited.

Some patterns in pedigrees include the following.
- Dominant traits cannot 'skip' generations — in order to have a dominant trait, at least one parent of an affected individual must also have it (unless it is a new mutation). The trait can be seen in each generation. Once it vanishes from a family, it does not suddenly reappear.
- Recessive traits can 'skip' generations; that is, a child can show the trait without either parents having it.
- Autosomal traits do not usually show an obvious gender skew.
- Sex-linked traits often have a gender skew.

FIGURE 2.97 Determining type of inheritance from a pedigree

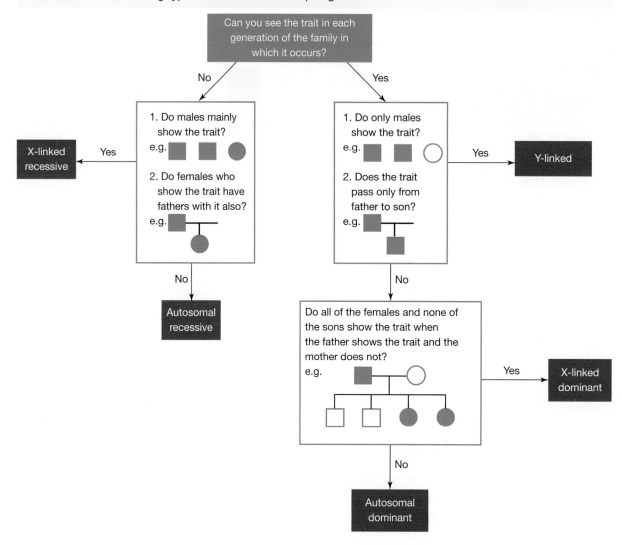

2.10.4 What can be inherited?

Many traits are inherited in various ways, whether this be sex-linked, autosomal, recessive or dominant. This is also the case with many genetic diseases and disorders.

TABLE 2.8 Some diseases that can be inherited

Inherited disorder	Type of inheritance	Symptoms of disorder
Duchenne muscular dystrophy	X-linked recessive	Inability to walk from a young age due to the lack of a functional protein dystrophin, which is vital for muscle function
Haemophilia A and B (HEMA, HEMB)	X-linked recessive	Bleeding disorders
Fragile X syndrome	X-linked dominant	Intellectual disabilities and lower than average IQ
Alport syndrome	X-linked dominant	Kidney disease and hearing loss
Huntington's disease (HD)	Autosomal dominant	Usually mid-life onset; progressive, lethal degenerative neurological disease
Intestinal polyposis	Autosomal dominant	Many small bulges in the colon form; may lead to colon cancer

Dwarfism	Autosomal dominant	Inhibited growth
Sickle-cell disease	Autosomal recessive	Red blood cells become deformed into a sickle shape when oxygen levels are low, which can lead to impaired mental function, paralysis and organ damage
Thalassaemia (THAL)	Autosomal recessive	Reduced red blood cell levels
Cystic fibrosis	Autosomal recessive	Very thick mucus and phlegm produced in lungs and pancreas, leading to respiratory issues and blockage of many traits

ACTIVITY: Researching genetic disorders

a. In a group, make a list of human genetic disorders. Each person in the group is to write a report on one. Your report could take the form of a poster or brochure.
b. Include which gene or chromosomal abnormality is responsible for the disorder and some of the characteristics the affected person would show.
c. Find out whether organisations are available to support people who have the disorder and their families.

The regulation and timing of when genes are switched on and off is important in the onset of certain genetic diseases. Many genes, such as those controlling the production of enzymes necessary for respiration, are active throughout the life span of a person. Some are switched on only at particular times and in specific tissues. This regulates development. Late onset genetic disorders, such as Huntington's disease and a form of Alzheimer's disease, result from particular defective genes becoming activated later in life. Duchenne muscular dystrophy or muscle deterioration is a disease that gradually develops from late childhood.

The inheritance of different traits

figures 2.98 to 2.101 show four pedigrees for four different inherited traits.

FIGURE 2.98 Pedigree for cystic fibrosis: autosomal recessive. It is able to skip generations, and does not have a gender skew.

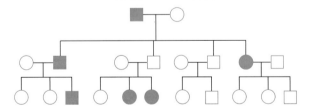

FIGURE 2.99 Pedigree for Huntington's disease: autosomal dominant. It does not skip generations, and does not have a gender skew.

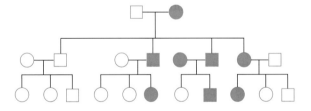

FIGURE 2.100 Pedigree for Duchenne muscular dystrophy: X-linked recessive. It is able to skip generations, and shows an obvious gender skew.

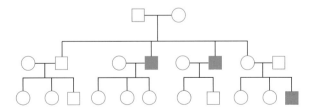

FIGURE 2.101 Pedigree for Fragile X: X-linked dominant. It does not skip generations, and shows an obvious gender skew.

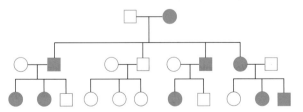

Apply and analyse

6. Huntington's disease is an autosomal dominant condition. Refer to the pedigree provided to answer the following questions.

   a. If *H* = Huntington's disease and *h* = normal, state the genotype(s) and phenotypes of the following.

i. I-1	**iv.** II-4
ii. I-2	**v.** II-5
iii. II-1	**vi.** II-3

   b. Use a Punnett square to predict the chances of the following.
   - **i.** I-1 and I-2 having a child with Huntington's disease
   - **ii.** II-4 and II-5 having a child with Huntington's disease

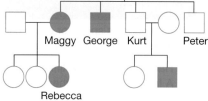

Huntington's pedigree

7. Refer to the pedigree of the family in the following diagram. The inheritance of broad lips (*B*; unshaded individuals) is dominant to the inheritance of thin lips (*b*; shaded individuals).

   a. How many females are shown in the pedigree chart?

   b. How many males are shown in the pedigree chart?

   c. How many females have the thin lips trait?

   d. Suggest the genotype of Maggy's parents.

   e. Suggest how Maggy inherited thin lips when her parents did not.

   f. Suggest the genotypes of (i) Peter, (ii) Kurt, (iii) George and (iv) Rebecca.

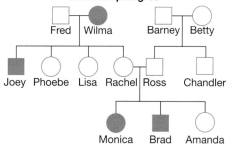

Pedigree of lip shape

8. The pedigree provided traces the recessive trait of albinism in a family. The shaded individuals lack pigmentation and are described as being albino.

   a. List any observations from the pedigree that support albinism being a recessive trait.

   b. If the albinism allele was represented as *n* and normal skin pigmentation as *N*, state the possible genotypes for each of the individuals in the pedigree.

   c. Is this trait sex-linked or autosomal? Explain your response.

   d. If Rachel and Ross choose to have another child, what is the chance that this child is albino?

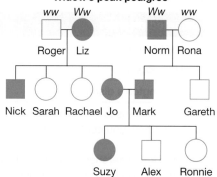

Albinism pedigree

9. The pedigree provided traces the dominant trait, a widow's peak, in a family.

   a. List any observations from the pedigree that support the widow's peak being a dominant trait.

   b. If the widow's peak allele is represented as *W* and the straight hairline as *w*, state the possible genotypes for each of the individuals in the pedigree.

   c. If Jo and Mark were to have another child, what is the chance of it having a widow's peak?

   d. If Ronnie were to have a child with a man who did not have a widow's peak, what is the probability that their child would have a widow's peak?

   e. If Norm and Rona were to have another child, what is the probability that they would have a child without a widow's peak?

Widow's peak pedigree

10. Being colour blind is an X-linked recessive trait. Refer to Chris and Heather's family pedigree chart and information on their inheritance of colour blindness in the following figure.

    a. State the genotype for:
    - **i.** Chris
    - **ii.** Heather.

    b. State the phenotype for:
    - **i.** Chris
    - **ii.** Heather.

    c. Is it possible for Peter to have a child who is colour blind? Explain.

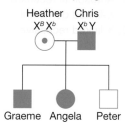

Colour-blind pedigree

Evaluate and create

11. A couple, Doris and Joseph, have a daughter, Vicki, and two sons, Joey and Robert.
Joey and Robert do not have children, but Vicki has two daughters, Belinda and Alanna. Belinda is married to
Paul and has two children, a girl, Allira, and a boy, Hunter. Vicki is married to Chris. Chris's parents are Doreen
and Brian. Chris has five brothers and one sister. His sister, Maree, is married to Wayne, and has four children:
three boys, David, Daniel and Johnno, and a daughter, Cassie.
The following individuals have blue eyes: Joseph, Joey, Belinda, Paul, Hunter, Allira, Wayne, David, Johnno
and Cassie.
Show this family as a pedigree and shade all those with blue eyes.

Hypertrichosis pedigree

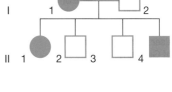

12. Jacob has hypertrichosis or werewolf syndrome, as does his mother. His
father, however, does not. Hypertrichosis is an X-linked dominant trait
that is characterised by increased hair growth on the face and upper
body. Jacob is shown in the following pedigree chart as individual II-4.
For the following questions, assume that X^H = hypertrichosis and
X^h = normal hair growth.
   a. Use the Punnett square provided to determine the chances of
   Jacob's parents having children with the syndrome and state the
   chance of their having the following.
      i. A daughter with hypertrichosis
      ii. A son with hypertrichosis
      iii. A child with hypertrichosis
   b. If Jacob had children with Bella (who does not have hypertrichosis),
   use a Punnett square to determine the chances of their child being the
   following.
      i. A daughter who has hypertrichosis
      ii. A son who has hypertrichosis
      iii. A daughter who does not have hypertrichosis
      iv. A son who does not have hypertrichosis
   c. Since Jacob is affected with hypertrichosis, do his sons and daughters have the same chance of inheriting
   the condition? Explain.
   d. How does this compare to a father who has an autosomal recessive trait? If he shows the trait and his wife
   does not, what are the chances that his daughters will show the trait? Explain.
   e. How does this compare to a father who has an autosomal dominant trait? If he shows the trait and his wife
   does not, what are the chances that his daughters will show the trait? Explain.

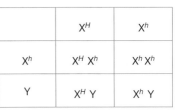

	X^H	X^h
X^h	$X^H X^h$	$X^h X^h$
Y	$X^H Y$	$X^h Y$

13. Queen Victoria was a carrier of the X-linked recessive trait haemophilia. This trait affects blood clotting. Queen
Victoria carried the allele for haemophilia on one of her X chromosomes. This germline mutation was passed
on to other members in her family. Use the pedigree provided to answer the following.
   a. What evidence is there from the
   pedigree that this trait is X-linked
   recessive?
   b. If X^H = normal trait and X^h =
   haemophilia, state the genotype of:
      i. Queen Victoria
      ii. her husband
      iii. her daughter Beatrice
      iv. her son Leopold (Duke of Albany)
      v. her granddaughter Alexandra
      vi. her great grandson Alexei.

Prenatal screening

Can tests be done to detect abnormalities of a baby when the mother is still pregnant?

Prenatal screening tests can be performed to detect abnormalities in the fetus during pregnancy. These include biochemical tests and analysis of the number and structure of chromosomes. Depending on the stage of development, different procedures are used to collect fetal cells for analysis.

Chorionic villus sampling (CVS) (as shown in figure 2.103) is a procedure in which genetic information about the embryonic cells may be obtained. Usually performed around six to eight weeks of pregnancy, CVS involves removal of tissue from the chorion (a membrane surrounding the fetus which is genetically identical to the embryo).

Another procedure, called **amniocentesis** (also shown in figure 2.103), is used to collect a sample of the amniotic fluid surrounding the fetus. Usually performed around 16 to 18 weeks, the enzyme function of the fetal cells within the fluid may be tested. The amniotic fluid may also be tested because the presence of some substances within it may suggest a metabolic disorder.

Amniocentesis and CVS can be used in the identification of genetic and chromosomal abnormalities.

However, these types of techniques do involve some risk of miscarriage or damage to the fetus.

More recently, less invasive tests have been developed. One such example is the aptly named non-invasive prenatal testing (NIPT) which involves first and second trimester serum screening using a sample of the mother's blood. Cell-free DNA testing is another example that looks at the baby's DNA in the mother's blood.

FIGURE 2.103 Chorionic villus sampling (CVS) and amniocentesis are two processes in prenatal screening.

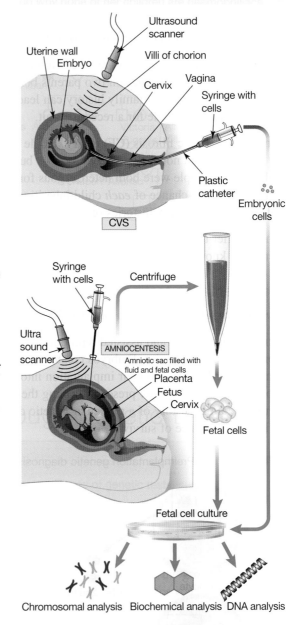

prenatal screening testing a fetus during pregnancy to detect any abnormalities

chorionic villus sampling (CVS) a type of prenatal screening in which tissue from the chorion is tested during pregnancy

amniocentesis a type of prenatal screening in which a sample is collected from the amniotic fluid around a fetus

Newborn screening

Does a newborn baby have PKU, hypothyroidism or cystic fibrosis?

Australian state screening laboratories test newborn babies for metabolic disorders such as phenylketonuria (PKU) and galactosaemia, and other disorders such as cystic fibrosis and hypothyroidism, as outlined in table 2.10. One way these tests are performed is through a heel prick test (or Guthrie test). This early testing allows intervention that may increase the chance of successful treatment. The problem, untreated symptoms and incidence of these disorders is also shown in table 2.10.

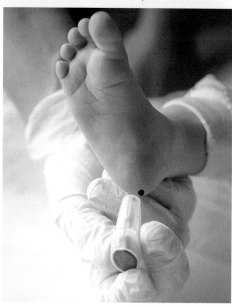

FIGURE 2.104 The heel prick test

TABLE 2.10 Screening tests for genetic disorders

Genetic disorder	Problem	Symptoms untreated	Incidence
Phenylketonuria (PKU)	Enzyme that metabolises phenylalanine to tyrosine defective or not produced	Excessive levels of phenylalanine in brain tissue can cause brain damage	1 in 10 000
Galactosaemia	Enzyme that metabolises galactose defective or not produced	Excessive levels of galactose in the blood can result in failure to thrive	1 in 40 000
Congenital hypothyroidism	Absent or poorly developed thyroid gland	Slowed growth and brain development	1 in 3500
Cystic fibrosis	Altered form of a transport protein that is not able to do its job	Respiratory and digestive problems; shortened life expectancy	1 in 2500

2.11.3 Predictive testing

If a family member has been diagnosed with a genetic condition, are you likely to develop this condition later in life?

Predictive testing can be used to detect genetic disorders that may be evident after birth — often later in life. This type of testing can help identify mutations that increase your chance of developing disorders with a genetic basis, such as breast cancer (BRCA1 or BRCA2 mutations) and Alzheimer's disease (variant of APOE gene).

Another example of this testing can be performed for Huntington's disease. Huntington's disease (HD) is an autosomal dominant inherited brain disorder that causes specific cells in the brain to die. The HD mutation results from a three-base sequence (for example, C-A-G) repeated many times (triplet repeat expansion). Although the mutation is present at birth, symptoms do not usually appear until adulthood. By this time, the person may have already had children of their own. Although the disease still has no cure, predictive testing provides affected individuals with knowledge of their status for the disease.

Genetic counselling involves helping individuals, couples and families understand and adjust to the implications of a genetically related health condition. Genetic counsellors can work directly with patients and their families. They also often work in hospitals in collaboration with medical specialists. Genetic counsellors may help with questions such as:

- We're thinking about starting a family, should we be genetically tested?
- What is the genetic condition and what are its related risks?
- What does the genetic test entail?
- What will the genetic test tell me?
- What do the results of the genetic test mean to me — and to my family?
- Now I have the results, what next?
- What sort of health care decisions do I need to make?
- We need help dealing with this. What are our options? What types of support are available?

FIGURE 2.105 Genetic counsellors are important in the processes around genetic testing.

on Resources

Weblinks Cystic fibrosis and carrier screening
What is genetic counselling?
Can genetic testing services really predict your future?
New genetic testing technology for IVF embryos

2.11.4 Tools for genetic testing

Scientists use various tools in laboratories to analyse samples for genetic testing.

Gene probes

Gene probes can be used to find out if an individual is carrying an allele for a particular genetic disease. These are short sequences of DNA that contain complementary bases to the allele under investigation.

DNA from the individual being tested is extracted and the DNA strands are separated. These single DNA strands are then mixed with the appropriate gene probe. If the sample DNA and the gene probe pair up, the result is positive for the presence of that particular allele.

FIGURE 2.106 Probes help detect a specific sequence in DNA.

Radioactive probe (DNA)

ATCCGA

1 The probe is mixed with single-stranded DNA from various bacterial clones.

Single-stranded DNA

ATGCGCTTATC   AGCCTTATGCA
AGGTAGGCTAA

2 The probe tags the desired gene by binding to its complementary DNA sequence.

ATCCG
AGGTAGGCTA

DNA profiling and DNA fingerprinting

DNA fingerprinting has been useful in forensic investigations, paternity tests, evolutionary studies (to determine the relatedness of different organisms), and to search for the presence of a particular variations of a gene.

Both DNA fingerprinting and DNA profiling use patterns in repeating base sequences of DNA for identification.

DNA profiling is more commonly used today and involves the use of gel electrophoresis to sort DNA fragments on the basis of their size and charge. In DNA profiling, PCR is used to produce multiple copies of the specific repeating base sequence. This is because it requires less DNA, is faster, provides more easily interpreted patterns, uses fluorescent labelled probes rather than radioactively labelled probes, and is able to distinguish between allele sequences of just one pair.

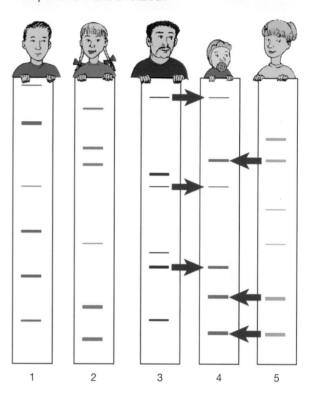

FIGURE 2.107 Who are the parents of individual 4? Are persons 1 and 2 related?

1 2 3 4 5

EXTENSION: DNA chips

DNA chips or DNA microarrays enable the presence of a range of gene variations to be investigated at the same time. DNA chips can measure expression levels of a large number of genes simultaneously, and determine the genotype in multiple regions of a genome.

DNA chips are made up of probes consisting of short fragments of selected genes attached to a wafer. Adding a sample of an individual's DNA to a particular type of DNA chip will result in a 'light up' response in the presence of one of the genes being tested for. This positive response is caused by the matching DNA locking onto the relevant probe on the chip. For example, a chip that contains probes for cystic fibrosis will only lock onto the cystic fibrosis gene (allele) within the DNA sample.

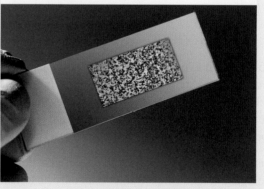

FIGURE 2.108 A DNA chip

 Resources

 Weblink DNA chip

2.11.5 Implications of genetic testing

Genetic tests can be used for the diagnosis, prediction or predisposition to particular genetic diseases or other inherited traits.

> **DISCUSSION**
>
> Should embryos be screened for their potential of carrying specific genetic disorders? If they are, what happens if the genetic test is positive? If prenatal diagnosis of the fetus during pregnancy identified the presence of an allele that codes for a genetic disorder, what are the consequences of knowing this information? Likewise, if the blood of a newborn baby tests positive for PKU, how may this information be used?

The construction of DNA databases, applications of bioinformatics and increased availability of other types of DNA profiling techniques will open up many new questions, problems and issues. Who owns the resulting genetic information and decides what is done with it? Who should make these decisions? Who should access the information? Could insurance companies or employers reject individuals based on genetic testing results?

> **DISCUSSION**
>
> Read the following headlines and discuss your thoughts related to these. You may wish to research the topics.
>
> **FIGURE 2.109** Headlines around genetic testing
>
> *A Harvard scientist is developing a DNA-based dating app to reduce genetic disease. Critics call it eugenics.*
>
> **Should you get tested for the breast cancer genes?**
>
> How to protect your DNA data before and after taking an at-home test
>
> **DIY genetic testing can unveil the mystery of your ancestry — but what happens to your data?**
>
> Can genetic testing services really predict your future?
>
> **Harmful or helpful? Researchers urge caution on genetic testing for children**
>
> *Employees jump at genetic testing. Is that a good thing?*

2.11.6 Genetic ancestry testing

How much do you want to know about your family and its history? Family historians used to search through documents in archives in their quest to find out more about their ancestral roots. Advances in both information technology and genetic technology have changed not only the information that they can now gain, but also the manner in which it is obtained — and shared. As a result of increased access to DNA tests, digitised documents, shared databases and pedigree charts, the landscape of identity of genealogy itself is changing.

Genetic genealogy involves the use of genetic tests (as well as traditional genealogical methods) to infer biological relationships (rather than only historical) between individuals. Because we do not inherit all of the genes of all of our ancestors, our genetic lines of descent can differ from the previous identity stories told by family historians. This pattern is expected due to the way in which DNA is inherited between generations.

Implications of genetic ancestry testing

While the number of people using genetic ancestry testing continues to increase, so do the issues that surround it. Some issues relate to ownership and privacy. For example, what is done with the genetic information and who has access to it? Also, if a family member is tested, much of the genetic information of other family members can also be inferred. Unexpected results also create issues. Some individuals, for example, have found out that

genetic genealogy the use of DNA along with other genealogical tests to infer relatedness between individuals

they were not biologically related to the family that they thought they were, or have found family that they didn't know that they were related to. Is this technology regulated? What, if any, obligations do the companies that perform genetic ancestry testing have towards the individual, their families and global communities?

on Resources

eWorkbooks	Genetic testing (ewbk-13060)	
	Genetic testing and life insurance (ewbk-13062)	
Weblinks	The limits of ancestry DNA tests, explained	
	Genetic testing and life insurance	

2.11 Activities

learn on

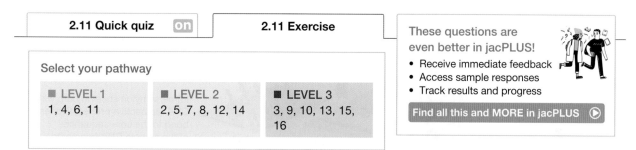

2.11 Quick quiz **on** | **2.11 Exercise**

Select your pathway

■ **LEVEL 1**
1, 4, 6, 11

■ **LEVEL 2**
2, 5, 7, 8, 12, 14

■ **LEVEL 3**
3, 9, 10, 13, 15, 16

These questions are even better in jacPLUS!
• Receive immediate feedback
• Access sample responses
• Track results and progress

Find all this and MORE in jacPLUS ⓘ

Remember and understand

1. **a.** State whether the following statements are true or false.

Statements	True or false?
i. Genetic tests can provide information that can be used in the diagnosis, prediction or predisposition to particular genetic diseases or other inherited traits.	
ii. Genetic tests can be used in carrier screening for genetic mutations, but not for gender determination.	
iii. Carrier testing can be used to determine the presence for the allele for a recessive trait.	
iv. Prenatal screening tests are used to detect genetic disorders in newborn babies.	
v. Amniocentesis and chorionic villus sampling can be used in the identification of chromosomal abnormalities.	
vi. If you receive a positive result for a genetic disorder using predictive testing, you will develop the disorder.	

b. Rewrite any false statements to make them true.
2. Match the genetic test to the question it may resolve in the following table.

Genetic test	Question
a. Carrier testing	**A.** Although I'm still pregnant, can I check if my baby has any chromosomal abnormalities?
b. Newborn screening	**B.** Am I a carrier of an allele for a recessive genetic disorder?
c. Predictive screening	**C.** Before implanting my IVF embryo, can I check if it has any significant genetic disorders?
d. Preimplantation screening	**D.** Does my newborn baby have any genetic disorders associated with beneficial early intervention?
e. Prenatal screening	**E.** My grandmother had breast cancer. Can I check if it had a genetic basis and if I carry the BRCA mutation?

3. Use the information in the text on DNA fingerprinting and profiling, and in the following DNA fingerprinting figure to explain how this process may allow for genetic disorders to be diagnosed.

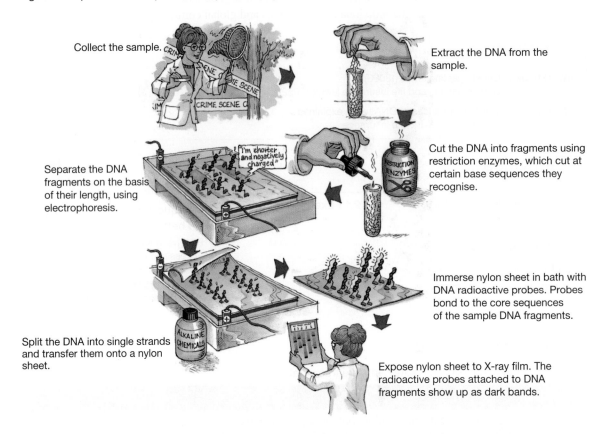

Collect the sample.

Extract the DNA from the sample.

Cut the DNA into fragments using restriction enzymes, which cut at certain base sequences they recognise.

Separate the DNA fragments on the basis of their length, using electrophoresis.

"I'm shorter and negatively charged"

Immerse nylon sheet in bath with DNA radioactive probes. Probes bond to the core sequences of the sample DNA fragments.

Split the DNA into single strands and transfer them onto a nylon sheet.

Expose nylon sheet to X-ray film. The radioactive probes attached to DNA fragments show up as dark bands.

4. Suggest six reasons for using genetic tests.
5. State three genetic disorder screening tests performed on newborn babies and explain why these tests are important.
6. Outline similarities and differences between chorionic villus sampling and amniocentesis.
7. a. List four reasons for using DNA fingerprinting.
   b. Distinguish between DNA fingerprinting and DNA profiling.

Apply and analyse

8. **SIS** Suggest ways in which information from genetic tests may be used by organisations such as insurance companies, medical facilities and workplaces.

9. **SIS** The symptoms of the autosomal dominant inherited disorder Huntington's disease (HD) don't usually appear until the affected person is over 30 years old. Predictive testing can be carried out to determine an individual's genetic status. In the following figure, H represents the faulty HD allele and h the normal allele. The mother, I-1, is currently showing the symptoms of HD and has the genotype Hh.
   a. Suggest the genotype of the following.
      i. Individual II-3
      ii. Individual I-2
   b. Suggest whether the children are likely to develop HD. Justify your response.
   c. Find out more about HD and current related research and issues.

Analysis of a HD pedigree

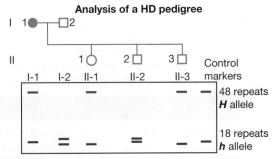

I 1●──□2

II   1○   2□   3□   Control
                              markers

   I-1   I-2   II-1   II-2   II-3
   —   —   — —   48 repeats
                                        H allele

   —   ═   —   ═   — —   18 repeats
                                        h allele

10. **SIS** Duchenne muscular dystrophy (DMD) is an inherited X-linked recessive disorder. The provided figure shows a pedigree and RFLP patterns that were obtained using a direct gene probe. Those individuals affected are shaded.

    a. State the individuals with DMD.
    b. Describe the RFLP pattern of those individuals with DMD.
    c. If the mother, I-2, is a carrier of the DMD gene, which of her daughters is also a carrier?
    d. Is the father, I-3, a carrier of the DMD gene? Explain.

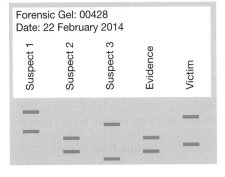

Analysis of a DMD pedigree

11. The provided diagram shows the DNA fingerprint of a victim, the DNA fingerprint from evidence taken from her body after an attack, and the DNA fingerprints of three suspects.

    a. Using the information in the DNA fingerprints, which of the three suspects is most likely to be guilty of the crime against the victim?
    b. Give reasons for your response to part **a**.
    c. Suggest why a sample was taken from the victim as well as the foreign DNA sample being collected from her body.
    d. i. State some other forensic diagnostic tools that exist to identify those guilty of crimes.
       ii. How do these compare with DNA fingerprinting?

Evaluate and create

12. What is thalassaemia? Research the screening and diagnostic tests for this genetic disorder. Find out more about the Thalassaemia Society and its involvement in genetic counselling. Create a visual representation of your findings.

13. **SIS** Research what a patent is and suggest implications of patenting any of the following.

    a. Genes
    b. Gene products
    c. Specific drugs that target a gene or gene products

14. What are bioethics and how do they relate to genetic testing? Research the use of a particular type of genetic testing. When involved in a bioethical discussion, you need to consider the questions listed in the following figure.

15. a. Find out what types of genetic testing occur in Australia.
    b. Are there any laws, rules or regulations associated with genetic testing? If so, what are they?
    c. List examples of different views and perspectives on genetic testing.
    d. Suggest why there are differing views.

> 1 What is the issue?
> 2 Who will be affected by the issue?
> 3 What are the positive points of view? (Who or what benefits, and why and how?)
> 4 What are the negative points of view? (Who or what is disadvantaged, and why and how?)
> 5 What are some of the possible alternatives?
> 6 How may these alternatives affect those involved?
> 7 What is a possible solution that may be acceptable to those involved?
> 8 What is your opinion on the issue?

16. Use the **New genetic testing technology for IVF embryos** weblink (available in the Resource panel) to answer the following questions.

    a. What does PGD stand for?
    b. What does the PGD test identify in embryos? Include four specific examples in your response.
    c. Outline the opportunity that this test offers families with histories of genetic disorders.
    d. Outline a negative aspect of the PGD test.
    e. At which stage is the embryo when a single cell is removed from it?
    f. Are you aware of any bias in the article? How many different perspectives were included? If you were to write the article, what other information or details might you include?

Fully worked solutions and sample responses are available in your digital formats.

LESSON
2.12 Biotechnological tools to edit the genome

LEARNING INTENTION

At the end of this lesson you will be able to provide examples of how advances in our scientific understanding have led to the development of a variety of biotechnological tools, techniques and applications.

Are we in the middle of a molecular biology and biotechnology revolution? What new discoveries will the future bring and what implications will they have on our lives?

2.12.1 Biotechnological revolution

Human genes in bacteria? Insect genes in plants? Cotton plants producing granules of plastic for ultra-warm fibre? While these may sound bizarre, they already exist.

We are living in a time of biological and technological revolution. Advances in our scientific understanding are often a result of developments in technology and technological advances. Such advances in biotechnology are gathering momentum so fast that our lives will never be the same.

2.12.2 Just science fiction?

If you saw the movie *Jurassic Park*, you may recall the scene in which scientists extract dinosaur DNA from mosquitoes that had been trapped in amber. They place this prehistoric DNA (with a mix from some other living organisms to fill the gaps) into surrogate eggs. While the science in *Jurassic Park* has more than a few holes in it, we do have (and are still developing) technologies to mix the DNA from different species together, and clone genes, some types of tissues and organs, and entire organisms. We are also able to manipulate the DNA of organisms so that they make proteins and possess features that they previously did not.

2.12.3 Genetic engineering — tools and techniques

Genetic engineering is one type of biotechnology that involves working with DNA, the genetic material located within cells. **Genetic engineers** use specialised tools and techniques to cut, join, copy and separate DNA. These are outlined in table 2.11.

genetic engineering one type of biotechnology that involves working with DNA

genetic engineers scientists who use special tools to cut, join, copy and separate DNA

TABLE 2.11 Different tools in genetic engineering

Tool	Job
CRISPR-Cas9	Edits sections of DNA
Bacterial plasmids	Gene cloning
Restriction enzymes	Cuts DNA into fragments at precise locations
Electrophoresis	Separates DNA fragments by their size
DNA probes	Finds specific DNA fragments
Ligase	Joins DNA fragments
Vectors	Transports DNA into cells
Polymerase chain reaction (PCR)	Amplify (make large quantities of) specific DNA

Restriction enzymes and DNA ligases

Need to cut and paste DNA? Genetic engineers use two types of enzymes as tools to achieve this.

Restriction enzymes (also known as endonucleases) act like scissors, cutting the double-stranded DNA molecule at a specific recognition site. Each restriction enzyme has its own particular DNA sequence that it cuts at. As a result of this cutting, two overhanging ends ('sticky ends') that expose nucleotide bases at each end may be produced.

If the same restriction enzyme is used to cut another sample of DNA, the DNA fragments will have complementary 'sticky ends'. Like glue, **DNA ligase** can join these DNA fragments together. If the fragments were from different sources, **recombinant DNA** has been produced.

FIGURE 2.110 Cutting with restriction enzymes

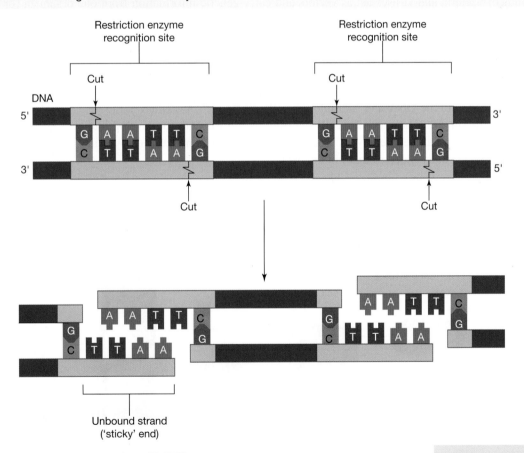

Unbound strand
('sticky' end)

Polymerase chain reaction (PCR)

Need more DNA? No problem! As introduced in section 2.7.6 and shown in figure 2.69, PCR is a technique that can be used to produce large quantities from an original DNA sample. PCR takes place in a thermal cycler. Three steps (denaturation, annealing and elongation) are repeated each cycle. While one cycle can yield two identical copies, 30 cycles can produce over a billion identical copies of the original DNA.

restriction enzymes enzymes that cut DNA at specific base sequences (recognition sites)

DNA ligase an enzyme that joins DNA fragments together

recombinant DNA a molecule of DNA that contains fragments from more than one source

Gel electrophoresis

Gel electrophoresis is a technique that is used to separate molecules based on their physical properties such as their size and electric charge. This technique can be used to separate DNA fragments that have been cut into different sizes by restriction enzymes. DNA markers (DNA molecules with a known size) can be used as standards to estimate the size of the DNA fragments.

An electrical current forces the DNA fragments to move through the pores of the gel. As the negatively charged DNA moves towards the positive electrode, the smaller DNA fragments move faster than the larger DNA fragments.

FIGURE 2.111 Gel electrophoresis separates DNA based on size.

Vectors and gene guns

Agrobacterium tumefaciens is a soil bacterium. This bacterium is able to get inside and infect many plants, such as vines and fruit trees. In doing so, it transfers a tiny piece of its DNA into the host cell. This programs the host cell to make chemical compounds for the sneaky bacterium to feed on. Genetic engineers saw the possibility of using this bacterium as a vector to carry the genes they wanted from one plant into another.

Other kinds of bacteria and viruses act as vectors and carry genetic information from one organism (or one synthesised to be like that organism) to another organism.

Vectors can be used to carry genes for producing protein in soybean and sunflower plants, enzymes to control chemical processes, or compounds that keep insects or pathogenic viruses at bay.

Gene guns have also been used to insert foreign genes into plant tissues. Fine particles of gold are coated with the DNA and shot into the cells. Some cells are killed in the process, but some survive. The surviving cells multiply by mitosis and can develop into complete plants with an altered genotype. Many plant crops such as maize and soybean have had favourable genes added to them in this way.

gel electrophoresis a technique used to separate molecules based on their size through an agarose gel

FIGURE 2.112 Inserting foreign DNA into plants

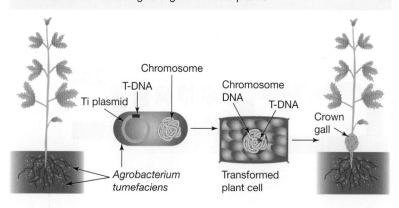

FIGURE 2.113 Gene guns used to insert DNA into cells.

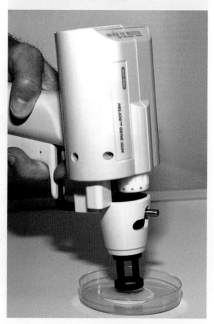

2.12.4 Some applications of gene technology

Genetically modified organisms

Do you want your tomatoes to be frost resistant? Just add the genes from Arctic fish! The addition of genes from Arctic fish to the genome of tomato plants is an example of **recombinant DNA technology**. The DNA has been recombined. The feature coded for by the foreign gene is then expressed by its new host — so no more tomato 'frostbite'!

Genetically modified organisms (GMO) contain artificially altered DNA. They may have had an existing gene altered, deleted or 'switched off' — or a foreign gene added. If the foreign gene has come from a different species, such as the frost resistant tomato just mentioned, the result would be a **transgenic organism**.

Table 2.12 provides some examples of ways in which the genomes of genetically modified organisms may be altered, and some reasons for doing it.

> **recombinant DNA technology**
> technology that can form DNA that does not exist naturally, by combining DNA sequences that would not normally occur together
>
> **genetically modified organism**
> a organism where the genome has been altered
>
> **transgenic organism** an organism with genetic information from another species in its genome

TABLE 2.12 Why do we genetically modify?

What's the change?	Why do it?	What's an example?
Existing gene altered	To alter the expression of a gene	Plants making more protein; gene therapy; growth hormone
Existing gene deleted or 'switched off'	Prevents the expression of a particular trait	Control ripening in fruit (e.g. Flavr-Savr tomato) to extend shelf life
Foreign gene added	A trait associated with the donor gene may be expressed	Bacteria with human gene inserted into their DNA make human insulin

Genetic manipulation techniques are now widely used. For example:
- Crop plants can be genetically engineered to grow in a variety of conditions, or even to make their own insecticide so that they are pest resistant!
- The shelf life of fruit such as tomatoes can be extended by switching off the ripening gene in tomatoes.
- The addition of the genes that code for proteins such as insulin, clotting factor VIII or growth hormone can be put into microbes (such as bacteria and yeast) so that they become tiny protein-producing factories.

DISCUSSION

a. Use the internet to identify five examples of genetically modified food.
b. Research these GM foods and summarise your findings.
c. Would you eat genetically manipulated food? Why or why not?
d. Discuss your views with others.

Cloning

As can be seen in figure 2.115, a variety of types of cloning are possible depending on its purpose.

FIGURE 2.114 Various techniques are involved in cloning.

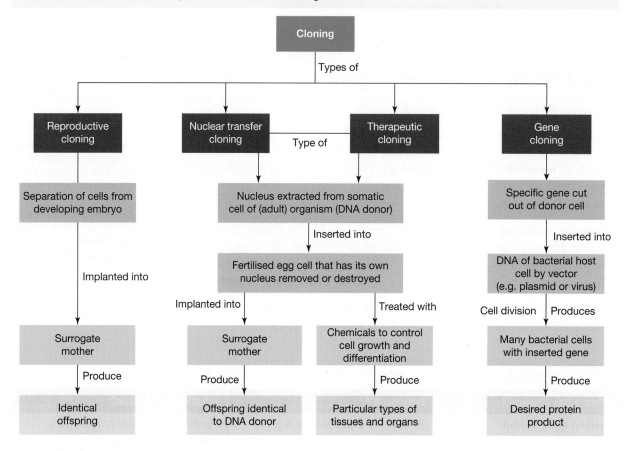

Gene cloning

As can be seen in figure 2.115, a variety of types of cloning are possible. Gene cloning involves making multiple copies of genes. More copies of the gene means more potential production of the protein that it codes for.

An example of gene cloning is when the gene that codes for a specific protein is inserted into circular pieces of DNA, known as plasmids, in bacteria.

Bacteria with these altered plasmids can then act as micro factories and produce considerable quantities of proteins, such as insulin for diabetics and the clotting factor VIII for haemophiliacs.

FIGURE 2.115 Bacteria with the human gene inserted into their DNA make human insulin.

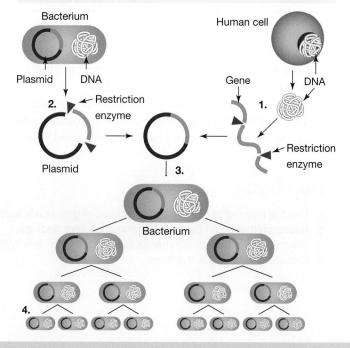

SCIENCE AS A HUMAN ENDEAVOUR: The story of Dolly the sheep

The story of Dolly the sheep began at the Roslin Institute in Scotland or, more specifically, as a single cell from the udder of a ewe. Follow her story in the diagram in figure 2.116.

Dolly made history as the first mammal to be cloned from a single adult cell using somatic cell nuclear transfer. Until then, biologists did not believe that once a cell had developed and become specialised, it could be reprogrammed to become different.

A group of cells that come from a single cell by repeated mitosis have the same genetic coding as each other. They are clones of each other. All of Dolly's cells came from the original fusion of an unfertilised egg and DNA from an udder cell. Because no genetic input came from another sheep, Dolly was a clone of the parent ewe from which the udder cell came.

During her life, Dolly gave birth to six lambs. Other clones of Dolly were also created, many of which are still alive today. While Dolly was the first cloned mammal, this paved the way for the cloning of other mammals including pigs and horses.

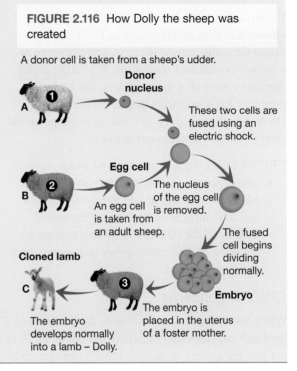

FIGURE 2.116 How Dolly the sheep was created

A donor cell is taken from a sheep's udder.

Donor nucleus

These two cells are fused using an electric shock.

Egg cell

An egg cell is taken from an adult sheep.

The nucleus of the egg cell is removed.

The fused cell begins dividing normally.

Embryo

The embryo is placed in the uterus of a foster mother.

Cloned lamb

The embryo develops normally into a lamb – Dolly.

DISCUSSION

While the cloning of mammals has been performed since the 1990s, no human has yet been cloned.

Do you think human cloning should be used? Outline and discuss advantages and disadvantages around this, and your thoughts about this as a possible technology.

---

on Resources

▶ **Video eLesson**   Ancient resurrection (eles-1070)

🔗 **Weblink**   Click and clone

2.12.5 Gene therapy and CRISPR-Cas9

The goal of **gene therapy** is to use genes to treat or prevent disease. This experimental technique could be used to introduce a new gene, or to inactivate or replace an allele responsible for a genetic disorder.

While there has been success in some cystic fibrosis trials, there have also been complications. Issues have arisen with the use of some types of viruses as vectors to deliver the new genetic material into humans, and serious health risks to the human patients involved have also developed.

Because gene therapy has the potential to alter an individual's DNA, it has raised a variety of ethical concerns. Some of these relate to the decision regarding which genes can or should be tampered with — and also whether changes should only be performed on somatic cells.

Although gene therapy is still experimental, other technologies are being developed that may alter this in the near future. One of these is called CRISPR-Cas9.

gene therapy altering genes with the intention to treat or prevent disease

CRISPR — need your genome edited?

CRISPR — clustered regularly interspaced short palindromic repeats — is an example of a technology that has been developed for genome editing. This technology exploits a quirk in the immune systems of bacteria.

Bacteria capture snips of DNA from invading viruses and then display them as DNA segments called CRISPR arrays. These essentially serve as a gallery of criminal mug shots. If a future attack matches a mug shot in the gallery, the bacteria produces RNA segments that target the invader's DNA. An enzyme, such as Cas9, then chops up the viral DNA and neutralises the threat.

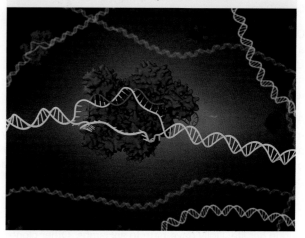

FIGURE 2.117 A target sequence (in purple) allows for the DNA to be cut at a specific location.

CRISPR technologies have allowed us to:
- edit genomes quickly and cheaply, with the potential to delete or add genes to genomes and to control which genes get expressed
- determine the function of different genes by targeting and silencing genes and seeing what traits are affected.

CRISPR concerns

While much excitement surrounds the potential of CRISPR technology, concerns also exist. In some situations the Cas9 enzymes have misfired and edited DNA in unexpected places. Could the use of CRISPR result in not just the editing of an individual's genome, or even just the next generation, but the altering of genomes of entire species?

ACTIVITY: Media articles and gene technologies

In a group, compile a folio of about ten journal and other media articles relating to gene technology. Make sure you note the date and source of each one. Each person will choose an article to analyse. In your analysis include:
a. the kind of gene technology being reported on
b. a simple description or explanation of the particular example of technology
c. any issues relating to the example.

DISCUSSION: Do-it-yourself creation kits

Imagine if everyone was able to access genetic engineering tools to create and develop new forms of life. Imagine being able to create new plants for your garden, and pets that you can currently only dream about. Will designing genomes become a personal thing — a type of art form or expression of creativity?

What sorts of biotechnology games will be designed and produced? What sorts of lessons and learning may young children get from creating their own organisms to watch grow and interact with? Should rules and regulations be put in place for this new biotechnology? What sort of rules and regulations? Who should make them? How can they be enforced?

FIGURE 2.118 Should you be able to use do-it-yourself creation kits?

Resources

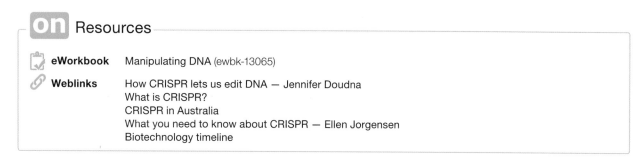

eWorkbook	Manipulating DNA (ewbk-13065)	
Weblinks	How CRISPR lets us edit DNA — Jennifer Doudna	
	What is CRISPR?	
	CRISPR in Australia	
	What you need to know about CRISPR — Ellen Jorgensen	
	Biotechnology timeline	

2.12 Activities

learn on

2.12 Quick quiz on	2.12 Exercise

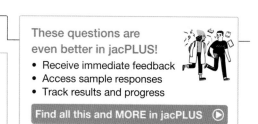

These questions are even better in jacPLUS!
- Receive immediate feedback
- Access sample responses
- Track results and progress

Find all this and MORE in jacPLUS ▶

Select your pathway

■ LEVEL 1	■ LEVEL 2	■ LEVEL 3
1, 2, 4	3, 6, 8, 9	5, 7, 10, 11

Remember and understand

1. a. State whether the following statements are true or false.

Statements	True or false?
i. Genetic engineering is a type of biotechnology that involves working with DNA, the genetic material located in cells.	
ii. The human gene for insulin can be inserted into bacteria so that they make human insulin.	
iii. Restriction enzymes are used to paste DNA fragments together.	
iv. Vectors can be used to transport DNA into cells.	
v. All transgenic organisms are genetically modified.	
vi. All genetically modified organisms are transgenic.	
vii. The goal of gene therapy is to use genes to treat or prevent disease.	
viii. Gene therapy is currently being used as an effective and safe treatment for genetic diseases.	
ix. CRISPR technology has been developed for genome editing.	
x. Gene cloning can be used to make multiple copies of genes.	

b. Rewrite any false statements to make them true.

2. Match the tool with its job in the following table.

Tool	Job
a. DNA ligase	**A.** Separate DNA fragments by their size
b. DNA probe	**B.** Amplify (make large quantities of) specific DNA
c. Electrophoresis	**C.** Cut DNA into fragments at precise locations
d. PCR	**D.** Find specific DNA fragments
e. Restriction enzyme	**E.** Join DNA fragments
f. Vector	**F.** Transport DNA into cells

3. Describe two methods that can be used to transfer foreign genes into other organisms. ▶

4. **a.** What is CRISPR the abbreviation for?
   b. Describe the link between bacteria and CRISPR.
   c. Identify the key use of CRISPR technology.

Apply and analyse

5. Distinguish between the following terms.
   a. Genetically modified organism and transgenic organism
   b. Gene cloning and gene therapy
   c. Nuclear transfer cloning and reproductive cloning
   d. Gel electrophoresis and PCR
6. Draw a simple flow diagram to summarise how bacteria are used to produce insulin that people with diabetes can use.
7. Refer to figure 2.116 showing Dolly's creation.
   a. Why Dolly was famous?
   b. Describe what had to be done to the unfertilised egg taken from ewe number 2.
   c. Describe how the donor cell DNA (from ewe 1) got inside the empty cell from ewe 2.
   d. State ewe 3's role.
   e. Explain why Dolly is called a clone.
   f. Identify which ewe was the original Dolly.
   g. Does Dolly have any genetic material from a ram (male sheep)? Explain.

Evaluate and create

8. **SIS** Clearly outline an investigation on how CRISPR may be used to edit out a faulty gene. Ensure your method is clear.
9. Since Dolly, other animals have also been cloned. Find out more about the cloning of at least one other animal. Report on the following.
   a. The reasons for cloning that particular animal
   b. How the animal was cloned
   c. The advantages and disadvantages associated with the cloning of the animal
10. **SIS** Mutants or miracles? Scientists have genetically modified cattle so they do not have horns. Use the internet to find out how this was achieved, and the advantages and disadvantages of using gene technology for this purpose. Report your findings on this topic, and explain and justify your opinion.
11. **SIS** Imagine you are a genetic engineer who is tasked to design an organism that you think would have a good chance of surviving on Mars.
    a. Find out the environmental conditions on Mars and think about what sort of life form could survive there.
    b. Provide detailed plans of the design of your organism.

Fully worked solutions and sample responses are available in your digital formats.

LESSON
2.13 Interwoven stories of DNA and the genome

LEARNING INTENTION

At the end of this lesson you will be able to provide examples of how our understanding of DNA and our genome have changed over time.

2.13.1 Hologenome theory of evolution

Did you know that your DNA contains a wonderful tapestry of not just your human ancestry but also other genomes?

Although we may consider ourselves to be 'purely human', our definition of 'human' may require some revision. As with other plants and animals, we contain DNA from a variety of sources. The **hologenome** theory of evolution emphasises the role that microorganisms have within our evolution.

hologenome the sum of genetic information of a host and its microbiota

Microorganisms (microbiota) living within and on their hosts (such as plants or animals) can be described as symbionts. Symbionts are organisms that live in a **symbiotic** relationship — such as when different species live together and function in close association with each other. Collectively, a host and all their microbiota make up a **holobiont**. Some scientists propose this to be a unit of selection in evolution.

The combined genomes of the host and its microbes make up the hologenome. Genetic variation within the holobiont may occur in both the host and the microbial symbiont genomes; this variation can be inherited by the offspring. Some of this variation may occur within existing microbes; other variation may be due to microbes newly acquired from the environment.

Some scientists consider this view as Lamarckian, because the inheritance of variation via microbes follows Lamarck's idea that traits acquired within the lifetime of the parent can be transmitted to the next generation.

FIGURE 2.119 The relationship between the hologenome, microbiome and host geome

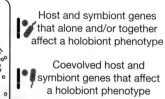

Host and symbiont genes that alone and/or together affect a holobiont phenotype

Coevolved host and symbiont genes that affect a holobiont phenotype

Host genes and symbionts that do not affect a holobiont phenotype

Environmental microbes that are not part of the holobiont

symbiotic a very close relationship between two organisms of different species

holobiont a host and their associated microbiota

SCIENCE AS A HUMAN ENDEAVOUR: Climate change and holobionts

Some scientists have suggested that reef corals and possibly some other multicellular organisms may alter the microbial communities within their bodies to cope with environmental stresses such as those caused by climate change. More studies on this may provide strategies we can use to help reduce the loss of biodiversity in ecosystems threatened by environmental changes related to climate change.

FIGURE 2.120 The relationship between various components in an ecosystem

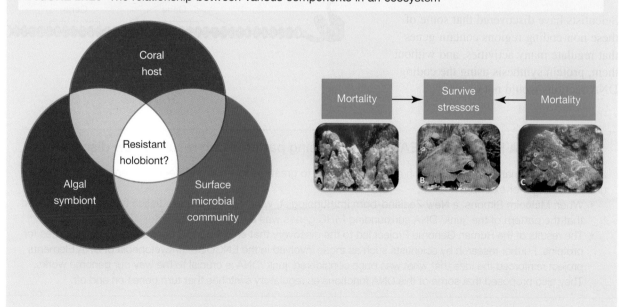

Scientists such as Dr Emmanuel Buschiazzo focus on marine environments and DNA. Dr Emmanuel Buschiazzo's research involved blending genomic approaches and population genetics to unravel the biology of two Caribbean coral holobionts, *Montastraea faveolata* and *Acropora palmata*.

FIGURE 2.121 **a.** Dr Emmanuel Buschiazzo **b.** *Montastraea faveolata* coral **c.** *Acropora palmata* coral

DISCUSSION

If the human genome has evolved as a holobiontic union of vertebrate and virus, could plagues be considered a vital evolutionary survival tool for our descendants? What are the implications of this theory in terms of how we treat diseases and think about viral infections? Do viruses have a place not just in our present, but also in shaping our future evolution?

2.13.2 Coding and non-coding of nuclear DNA

DNA in your nucleus is made up of coding and non-coding sections. It was long thought that the non-coding DNA served no purpose. Because this DNA contained highly repetitive sections and did not code for amino acids, it was often referred to as 'junk' DNA. New technologies and discoveries, however, are fast changing our views about the non-coding DNA.

Scientists have discovered that some of these non-coding regions contain genes that regulate many activities, and without them, protein synthesis using the coding DNA sections would not occur.

FIGURE 2.122 Coding and non-coding regions

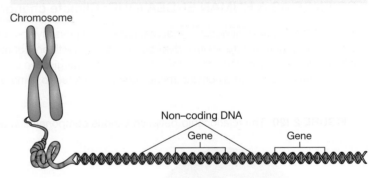

SCIENCE AS A HUMAN ENDEAVOUR: Analysing patterns can result in new discoveries

Scientists often analyse patterns. Such analysis can lead to great discoveries and deeper scientific understanding. For example:

- When Malcolm Simons, a New Zealand-born immunologist, was testing people's tissue types, he discovered that the pattern of the 'junk' DNA surrounding MHC genes was a very good predictor of tissue type.
- The results of the Human Genome Project led to the discovery that only around 2 per cent of our DNA codes for proteins. Further research by scientists such as those involved in the ENCODE: Encyclopedia of DNA Elements project reinforced the idea that what was once considered 'junk' DNA is crucial to the way our genome works. They also proposed that some of this DNA functions as regulatory switches that turn genes on and off.

2.13.3 Viruses and jumping genes in our genome

The human immunodeficiency virus (HIV) is an example of a retrovirus. Retroviruses convert their RNA genome into DNA before implanting it into host chromosomes. If the viral genome is incorporated into the chromosomes in the host's sex cells (referred to as germline), it can become a part of the genome of future generations.

Are you aware that the incorporation of viral DNA has happened repeatedly in our own lineage? This mechanism helps to explain the varied sources of the DNA in the human genome. It also explains why about 8 per cent of our genome is composed of sequences with a viral origin.

SCIENCE AS A HUMAN ENDEAVOUR: HERVs and health

Human endogenous retroviruses (HERVs) are the most common virus-derived sequences within our genome. Bioinformatics is being used to research HERVs, HERV variations, molecular mimicry and links to human disease. Evidence suggests that HERV variations may contribute to disease susceptibility. For example, some patients with motor neuron disease (MND), multiple sclerosis, and some types of cancer and autoimmune diseases have exhibited over-expressed HERV-encoded genes. Some other HERV variations, however, can have beneficial effects. These include protection against infection by some viruses and production of several essential proteins during reproduction. For example, HERVs are involved in the production of various viral proteins that are crucial in fetal development (such as synctins) and in dampening the mother's immune system, so that it does not attack the developing fetus.

SCIENCE AS A HUMAN ENDEAVOUR: Jumping genes and Barbara McClintock

Barbara McClintock (1902–1992) was a cytogeneticist whose scientific theories (and possibly her gender) clashed with the scientific community of her early research years. McClintock investigated how chromosomes change during reproduction in maize (*Zea mays*). Maize proved to be the perfect organism for the study of transposable elements (TE) or 'jumping genes'.

McClintock contributed to our understanding of the mechanism of crossing over during meiosis and produced the first genetic map for maize. She also linked regions of the chromosome with physical traits and demonstrated the roles of telomeres and centromeres. McClintock also discovered **transposition** and outlined how it could be involved in turning the expression of genes on and off.

FIGURE 2.123 Barbara McClintock

Changing scientific paradigms can take time

McClintock pioneered the study of **cytogenetics** in the 1930s. Before the structure of DNA was even discovered, McClintock was the first scientist to outline the basic concept of epigenetics, recognising that genes could be expressed and silenced.

McClintock's theories were revolutionary because they suggested that an organism's genome was not a stationary entity, but something that could be altered and rearranged. This view was highly criticised by the scientific community at the time. She was eventually awarded the Nobel Prize in Physiology or Medicine in 1983, when she was over 80, for research she had done many years before.

Your jumping genes

Some of the repeating sequences within our non-coding DNA are known as **transposons** or 'jumping genes'. They may have originated from invading viruses. These sections have the ability to copy themselves independently of the rest of the genome and then randomly insert themselves in other sections of the genome. Some scientists have suggested that these transposons have shaped our evolution.

transposition the ability of a gene to change position on the chromosome

cytogenetics the study of heredity at a cellular level, focusing on cellular components such as chromosomes

transposons a section of chromosome that moves about the chromosome within a cell through the method of transposition

FIGURE 2.124 a. DNA sequences known as transposons or 'jumping genes' can copy themselves into other sections of the genome through two different methods. **b.** An example of a corn in which transposons have acted

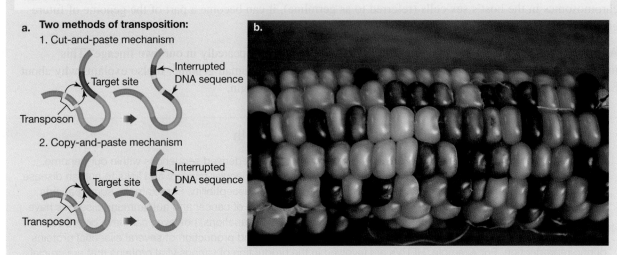

a. Two methods of transposition:
1. Cut-and-paste mechanism
2. Copy-and-paste mechanism

FIGURE 2.125 Jumping genes in marsupials

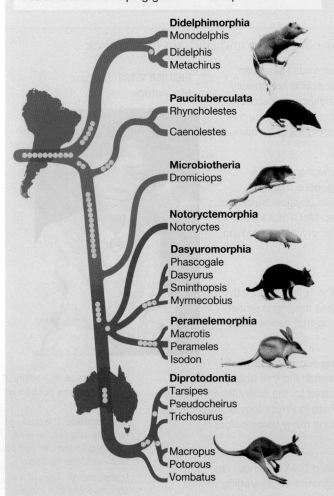

Didelphimorphia
- Monodelphis
- Didelphis
- Metachirus

Paucituberculata
- Rhyncholestes
- Caenolestes

Microbiotheria
- Dromiciops

Notoryctemorphia
- Notoryctes

Dasyuromorphia
- Phascogale
- Dasyurus
- Sminthopsis
- Myrmecobius

Peramelemorphia
- Macrotis
- Perameles
- Isodon

Diprotodontia
- Tarsipes
- Pseudocheirus
- Trichosurus
- Macropus
- Potorous
- Vombatus

Australian marsupials and jumping genes

Scientists have argued about how marsupials such as kangaroos, opossums and Tasmanian devils evolved in South America and Australia. DNA sequencing and the fossil record tell two different stories. Do jumping genes hold the answer to the mystery?

German evolutionary biologist Maria Nilsson has been investigating this mystery by looking at strange bits of DNA called **retroposons**. Retroposons have the ability to break off chromosomal DNA and then copy and paste themselves elsewhere in the genome. Once they copy and paste themselves, their locations are stable, making them a reliable marker for determining evolutionary relationships.

Nilsson's retroposon data suggest that the Australian and South American marsupials could be divided into two distinct groups that had little contact as they evolved. This supports the DNA sequencing data that they share a single South American ancestor that travelled to Australia before the continents drifted apart, and that they evolved separately afterwards.

retroposons segments of DNA that can break off a chromosome and paste themselves elsewhere in the genome

2.13.4 Our bacterial origins

Much of what we inherit is thought to have a bacterial origin.

The origin of mitochondria and chloroplasts

As well as the genetic material within your nucleus, genetic material is also found within mitochondria in your cells. This is called mitochondrial DNA (mtDNA). Interestingly, mtDNA is circular rather than linear and resembles bacterial DNA rather than eukaryotic DNA, suggesting a bacterial origin.

The endosymbiotic theory states that mitochondria (alongside chloroplasts) originated as prokaryotic microbes that had been ingested by an ancestral eukaryotic cell, and not only survived within it but also developed a symbiotic relationship with it.

SCIENCE AS A HUMAN ENDEAVOUR: Our protein assassin

Australian and British scientists have identified the process through which our natural killer cells (part of our immune system) puncture and destroy virus-infected or cancerous cells. A protein called perforin is responsible for forming a pore in the diseased cell. The natural killer cells can then inject toxins into the diseased cell to kill it from within.

By using powerful technologies such as the Australian Synchrotron and cryo-electron microscopy, scientists have determined the structure of perforin and how it creates pores. This protein resembles cellular weaponry used by bacteria, and it is possible that our immune system may have incorporated the genetic information from bacteria within our evolutionary past.

FIGURE 2.126 Where did perforin, our protein assassin, come from?

DISCUSSION

Select one of the following options for discussion with your peers.
a. If it weren't for viruses, ocean ecosystems would stop.
b. How do you feel about the possibility of having viral DNA in your DNA? Construct a PMI chart for a discussion on this question with your peers.
c. Do you think that life (as we know it) could have occurred on planets other than Earth? Justify your response and discuss it with your peers.
d. Do you think that all life on Earth descended from a common origin? Discuss this and justify your opinion.
e. If living things did not share an ancestor that shared ribosomes, ATP and the triplet code, why are these found universally among all living things? Discuss and justify your reasons

---

 Resources

    Weblinks Barbara McClinton Nobel Prize lecture
                    ENCODE: Encylopedia of DNA Elements

2.13 Activities

learn**on**

| 2.13 Quick quiz **on** | 2.13 Exercise |

These questions are
even better in jacPLUS!
- Receive immediate feedback
- Access sample responses
- Track results and progress

Find all this and MORE in jacPLUS ▶

Select your pathway

■ LEVEL 1	■ LEVEL 2	■ LEVEL 3
1, 2, 4	3, 5, 7, 10	6, 8, 9, 11

Remember and understand

1. a. State whether the following statements are true or false.

Statements	True or false?
i. Genetic variation within the holobiont may occur in both the host and the microbial symbiont genomes.	
ii. The human genome contains viral DNA.	
iii. Human DNA is made up of coding and non-coding sections.	
iv. It is possible that our immune system may have incorporated genetic information from bacteria within our evolutionary past.	
v. Some of the repeating sequences within our non-coding DNA are known as transposons or 'jumping genes' and may have originated from invading viruses.	
vi. DNA sequences known as transposons have the ability to copy themselves independently of the rest of the genome and then randomly insert themselves in other sections of the genome.	
vii. If a viral genome is incorporated into the chromosomes in the host's germline, it does not become a part of the genome of future generations.	
viii. Once retroposons copy and paste themselves into the genome, their locations are stable, making them a reliable marker for determining evolutionary relationships.	

b. Rewrite any false statements to make them true.
2. State another name for 'jumping genes'.
3. Provide an example of a retrovirus.
4. Match the term to its description in the following table.

Term	Description
a. Holobiont	**A.** Organism that lives in a symbiotic relationship
b. Hologenome	**B.** Close association between two or more organisms of different species, which may benefit one or both members
c. Retroposon	**C.** DNA sequences that can copy themselves into other sections of the genome
d. Symbiont	**D.** Piece of DNA that can break off chromosomal DNA and then copy and paste itself elsewhere in the genome
e. Symbiosis	**E.** The combined genomes of the host and the microbes within it
f. Transposon	**F.** The host organism and all its symbiont microbiota

Apply and analyse

5. **SIS** **a.** State three contributions that Barbara McClintock (1902–1999) made to our scientific understanding.
   b. Describe why McClintock's theories were initially highly criticised by the scientific community and then later considered revolutionary.
6. **a.** State what the hologenome theory of evolution emphasises.
   b. Outline why we no longer call regions in our DNA that do not code for proteins 'junk' DNA.
   c. Suggest how we know that germline endogenisation has occurred within our human lineage.
   d. Explain how research on retroposons has contributed to our knowledge about the evolution of Australian marsupials.
7. If viruses kill about 20 per cent of all living material in the oceans every day, releasing their contents for other organisms to grow, does that mean that they drive ocean ecosystems? Explain.
8. The genetic code consists of 64 possible triplet DNA combinations that code for one or more of the 20 different natural amino acids; all species on Earth use the same code. Suggest why this might be used as evidence of only one genesis on Earth.

Evaluate and create

9. Provide three pieces of evidence that show mitochondria have a bacterial origin.
10. Research one of the following scientific claims and then construct an evidence-based argument that either negates or supports the claim.
    • What was once referred to as 'junk' DNA has played a key role in making us human, because it distinguishes primates from other mammals.
    • What was once referred to as 'junk' DNA in humans is due to the invasion of a million copies of jumping genes.
    • If viruses had not invaded the genomes of our ancestors, mammals as we know them today would not have evolved.
11. **SIS** Carl Woese relied on RNA sequences rather than structural features to determine evolutionary relationships among prokaryotes. He discovered that prokaryotes were actually composed of two very different groups: bacteria (cyanobacteria and heterotrophic bacteria) and archaea (halophiles and thermophiles). Due to research by Carl Woese in the 1970s, many scientists now accept that prokaryotes can be divided into two distinct groups and that those in the archaea group are older than bacteria. These ancient ancestors of life on our planet were riddled with viruses. Does that mean that life on Earth originated from viruses? Discuss your thoughts on this theory and find out what evidence allowed for this to be supported.

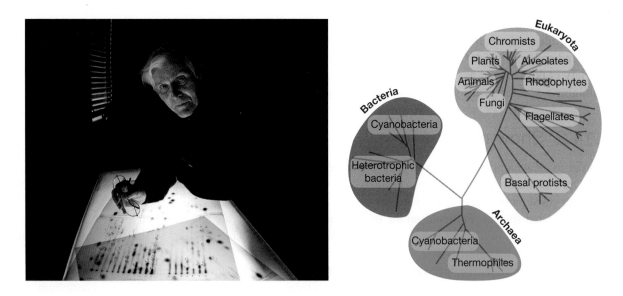

Fully worked solutions and sample responses are available in your digital formats.

LESSON
2.14 Thinking tools — Priority grids

2.14.1 Tell me

What is a priority grid?

Priority grids compare two aspects of an idea using two vertical axes. They allow you to discover the best option to follow and why, in turn helping you make decisions and see how views and judgements compare with others.

You can create a priority grid in a few different ways, with some examples shown in figures 2.127 and 2.128.

Priority grids allow you to determine how important different choices are in terms of their impact and the effort required when undertaking them, allowing you to easily see tasks that are quick wins, rewarding yet time-consuming, or thankless (tasks that take time with little impact).

FIGURE 2.127 Two examples of ways priority grids may be constructed

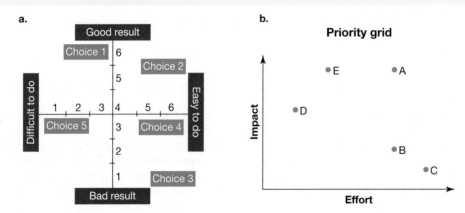

Why use a priority grid?

Priority grids compare different choices to aspects of the idea, allowing for pros and cons to be easily compared and prioritised. Other comparative types of grids or analyses, such as a SWOT analysis, consider a variety of positive and negative aspects, with a focus on opportunity and threats. This can be useful when examining a specific focus, whereas a priority grid provides a quick insight into different points of focus.

2.14.2 Show me

In order to create a priority grid, you should complete the following.
1. Draw two continuums that cross through each other at right angles (you may wish to split the two lines into equal parts to help guide you).
2. Put a label such as 'Difficult' on the left end of the horizontal scale and one such as 'Easy' on the right end (as shown in figure 2.127A). You might also use a term such as 'Effort' to show this as a continuum (as shown in figure 2.127B).
3. Put a label such as 'High reward' on the top of the vertical scale and one such as 'Low reward' at the bottom of the vertical scale. You might also use a term such as 'mpact' to show this as a continuum.
4. Write a list of choices, activities or tasks appropriate to the topic you are examining.
5. Assess each choice using the two lines, and mark the location where you think it fits, ensuring you label this.
6. Discuss your choices with others and share your ideas, values and judgements, enabling you to make your final priority grid.

figure 2.128 shows an example of using different types of priority grids when determining who to test for genetic diseases (the letters A to D represent the same variable in each grid).

FIGURE 2.128 Using priority grids when determining who to test for genetic diseases

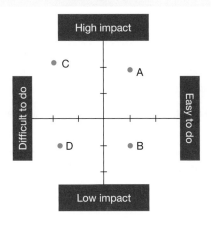

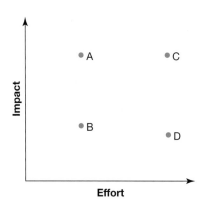

A. Testing at risk individuals where treatment can occur with early intervention
B. Testing at risk individuals where early intervention will not affect the course of a genetic disease
C. Testing every individual for diseases which can be treated with early intervention
D. Testing every individual for all diseases, where early intervention will not affect the course of a genetic disease

2.14.3 Let me do it

2.14 Activity

1. In your team, brainstorm statements or choices related to the genetic testing of embryos. Select five of these statements and position them on a priority grid with the following labels.
   Horizontal
   Left side: Difficult decision
   Right side: Easy decision
   Vertical
   Top: Good result
   Bottom: Bad result
2. a. Identify issues related to gene therapy.
   b. Identify different possible uses of gene therapy.
   c. Construct a priority grid to map out different potential uses of gene therapy.
3. Advances in technology may enable us to achieve wonders that were once only accessible in science fiction stories. What do you know about these advances? What benefits and issues can you identify? What is your opinion? Explore these questions by answering the following.
   a. Identify reasons people would want to clone humans.
   b. Research media stories that raise issues related to cloning, particularly of humans. Summarise these issues and identify some implications.
   c. Identify reasons people would want to edit their DNA.
   d. Research scientific technologies that may enable the editing of the human genome (for example, CRISPR).
   e. Outline your opinions on the editing of the human genome using a priority chart.
   f. Share and discuss your opinion with others.

Fully worked solutions and sample responses are available in your digital formats.

These questions are even better in jacPLUS!
• Receive immediate feedback
• Access sample responses
• Track results and progress

Find all this and MORE in jacPLUS ▶

LESSON
2.15 Project — Genetic discrimination

Scenario

Consider the following article on genetic discrimination in Australia.

Australians need more protection against genetic discrimination: health experts

Genomic testing — the ability to read an individual's genetic code and identify their risk of conditions such as cancer — has opened up huge possibilities in personalised medicine.

But it has also introduced serious ethical challenges. Particularly, there is the danger of life insurance companies using such information to discriminate against those at higher risk of conditions.

Canada, Britain and most European countries have already banned or restricted life insurers from using genetic test results.

Australia's response so far has been mostly to leave it to industry self-regulation. But our research suggests most health professionals don't think this is enough. More than 90% of the experts we surveyed agreed more government oversight is required.

Australia's regulatory approach

Australia's federal Private Health Insurance Act (2017) prohibits health insurers from using genetic information to discriminate against customers. But there is no legal prohibition against life insurers using results to charge people higher premiums or deny them coverage altogether. This applies to death cover, total and permanent disability, critical illness/trauma and income-protection cover.

In 2018 a joint parliamentary inquiry recommended a prohibition against life insurers using the outcomes of predictive genetic tests, at least in the medium term. It also recommended the government maintain a watching brief and consider legislation in future.

The federal government did not respond to the inquiry's report, leaving it to the industry to self-regulate. In 2019 the financial services industry's peak body, the Financial Services Council, introduced a five-year moratorium on insurers using applicants' genetic test results up to certain financial limits.

Life insurers can only ask for or use genetic test results for policies worth more than A$500,000 for death cover or total and permanent disability cover, A$200,000 for critical illness/trauma cover, and $4000/month for income protection.

Given the median yearly household income is about A$122,000, these thresholds are arguably too low to prevent insurers from using genetic test results in many cases.

Our survey results

With the moratorium now half over (it will end in 2024), we surveyed health professionals to gauge their views about Australia's approach. The survey was part of a federal government-funded research project to evaluate the moratorium.

Of 166 respondents, 121 were genetic specialists — geneticists and genetic counsellors who help people make sense of and make decisions about genetic testing. There are 480 such specialists in Australia registered with the Human Genetics Society of Australasia. With genetic testing increasingly being offered outside genetics clinics, we also invited specialists such as oncologists to take part.

Not everyone answered every question, so the following percentages are based on those that answered specific questions. While 93% agreed consumers are better protected under the moratorium, 88% remained concerned about genetic discrimination.

The most common complaints were that the financial thresholds were too low, there was no certainty for patients beyond 2024, and the insurance industry couldn't be trusted to regulate itself.

More than 90% said the Australian government should introduce legislation to regulate life insurers.

Canada's legislation, for example, bans insurers and other service providers from using genetic test results to discriminate against applicants.

The British government, meanwhile, has a hybrid regulatory model. This involves a Code on Genetic Testing and Insurance agreed to between the government and life insurance industry. In our survey, 95% said a similar approach is required for Australia.

Safeguarding Australia's genomic future

Genetic technology is transforming health care. Precision medicine relies on genomic testing to personalise therapeutic treatments. Genomic research is also critical to understanding disease, improving diagnostic methods and guiding the selection of the most effective drugs for treatment.

To maximise its potential and ensure public trust in genomics, it seems clear more must be done to prevent genetic discrimination and ensure all Australians — particularly those most at risk from genetic conditions — can benefit from the genomics revolution.

Source: https://theconversation.com/australians-need-more-protection-against-genetic-discrimination-health-experts-168563

Your task

Complete a research project in the form of an infographic or a short video to create awareness among your community about the importance of understanding the danger of genetic discrimination.

Guidelines:
- Present your information in a clear and concise format.
- Have a solid story line.
- Make sure your information is up to date.
- Explain your scientific content accurately.
- Be unique and creative.

The following questions might guide you in your task:
1. Explain what genetic discrimination is, based on the article provided.
2. List the key points, in a table similar to the one below, the reasons to allow or disallow the use of genetic information in health insurance settings. Debate this topic in class.

Reasons to allow	Reasons to disallow

Note that you can refer to the marking rubrics online for assessment criteria.

on Resources

💡 **ProjectsPLUS** Genetic discrimination (pro-0259)

LESSON
2.16 Review

Hey students! Now that it's time to revise this topic, go online to:

Review your results

Watch teacher-led videos

Practise questions with immediate feedback

Find all this and MORE in jacPLUS

Access your topic review eWorkbooks

 Resources

■ Topic review Level 1	■ Topic review Level 2	■ Topic review Level 3
ewbk-13068	ewbk-13070	ewbk-13072

2.16.1 Summary

Chromosomes, genes and DNA

- DNA (deoxyribonucleic acid) is a nucleic acid made up of nucleotides.
- Genes are specific segments of DNA that contain the codes for the production of specific proteins that can result in the expression of particular traits.
- Genes are organised into larger structures called chromosomes, which are located within the nucleus of eukaryotic cells.
- The location of a gene on a specific chromosome is called its locus, whereas genes that are located on the same chromosome are said to be linked.
- Chromosomes can be divided into two main types: autosomes and sex chromosomes (the X and Y chromosomes).
- The differences in chromosome pair size, shape and banding can be seen in a karyotype.

Discovering the structure of DNA

- Each nucleotide is made up of three parts: sugar (deoxyribose in DNA, ribose in RNA), phosphate and a nitrogenous base (adenine, guanine, cytosine, thymine or uracil).
- Chargaff's rule states adenine (A) binds to a thymine (T) and cytosine (C) binds to a guanine (G).
- DNA has the shape of a double helix made up of two anti-parallel strands of nucleotides, with the outside part of the 'ladder' made up of a sugar–phosphate frame (alternating sugar and phosphate).

Unlocking the DNA code

- Proteins are made up of amino acids. The instructions for making proteins are coded for in the sequence of nitrogenous bases in DNA.
- RNA is another type of nucleic acid made up of nucleotides. RNA differs from DNA in that it is single stranded, contains ribose instead of deoxyribose, and contains uracil instead of thymine. Messenger RNA (mRNA) and transfer RNA (tRNA) are two types of RNA.
- Three nucleotides in DNA that code for a particular amino acid, or a start or stop instruction, are called a triplet, in mRNA a codon, and in tRNA an anti-codon.
- Transcription is the process of making a complementary mRNA copy of the DNA message to make a protein. mRNA passes through the pores of the nuclear membrane into the cytoplasm to take its genetic copy of the protein instruction message to ribosomes.

- Translation is the process of reading the mRNA message and making the protein it codes for.
- tRNA transfers the appropriate amino acid to its matching code on the mRNA, and then these amino acids are joined together by peptide bonds to make a protein.

Mitosis and meiosis

- All cells come from pre-existing cells.
- Mitosis and meiosis are two types of cell division involved in cell production.
- Mitosis is the type of cell division involved in growth, development and repair of tissues. It results in cells that are clones because they are genetically identical to each other and to the original cell.
- Meiosis results in sex cells (or gametes) that contain half the number of chromosomes (haploid). The process of meiosis is a key source of variation due to independent assortment and crossing over of chromosomes.

Changing the code through mutations

- DNA replication is the process by which DNA makes identical copies of itself.
- Although the process of DNA replication includes checkpoints, sometimes mistakes get through, which can result in a mutation — a change in the DNA.
- In an induced mutation, the cause can be identified; in a spontaneous mutation, it cannot. A factor that triggers mutations in cells is called a mutagen or mutagenic agent.
- Point mutations may be due to the substitution of a nucleotide with a different nucleotide, or the addition or deletion of a nucleotide.

Exploring patterns in the genome and genetic sequences

- The development of automated DNA sequencers has reduced the cost and time required to sequence DNA.
- Gene sequencing identifies the order of nucleotides along a gene.
- A genome is the complete complement of genetic material in a cell or organism. The study of genomes is called genomics. The genome size is often described in terms of the total number of base pairs (or bp).
- The key findings from the Human Genome Project (HGP) were that the human genome has a size of around three billion base pairs (or 3000 Mb) and contains around 20000 to 25000 genes, and that all humans share about 99.9 per cent of their DNA.
- Genome maps describe the order of genes and the spacing between them on each chromosome.

Inheritance of genes

- The passing on of characteristics from one generation to the next is called inheritance and the study of inheritance is called genetics.
- These alternative forms or expressions of a gene are called alleles.
- The combination of the alleles that you have for a particular gene is called your genotype.
- Your characteristics or features are examples of your phenotype.
- If the allele for a dominant trait is present, it will be expressed. An allele for a dominant trait is denoted with a capital letter.
- The recessive trait is hidden in the presence of the dominant trait and can be expressed only if the allele for the dominant trait is not present. An allele for a recessive trait is denoted with a lower case letter.
- Individuals may be heterozygous and have different alleles at a gene (for example, *Bb*) or homozygous, and have the same alleles (*BB* or *bb*).
- The term 'carrier' is used to describe someone who is heterozygous for a particular trait.

Punnett squares and predicting inheritance

- A Punnett square is a diagram that is used to predict the outcome of a genetic cross for each offspring.
- There are different types of inheritance including complete dominance, codominance, incomplete dominance or sex-linked inheritance.

Interpreting pedigree charts

- A diagram that shows a family's relationships and how characteristics are passed on from one generation to the next is a pedigree chart.
- Pedigree charts can be used to observe the mode of inheritance, including autosomal dominant, autosomal recessive, X-linked dominant and X-linked recessive.

Exposing your genes through genetic testing

- Genetic tests can be used to help identify specific genetic variations within your DNA that may relate to your genetic family relationships or to your health.
- As well as being used to determine gender, genetic tests can be used for the diagnosis, prediction or predisposition to particular genetic diseases or other inherited traits.

Biotechnological tools to edit the genome

- Genetic engineering is a type of biotechnology that involves working with DNA, the genetic material located in cells.
- Genetically modified organisms have had their genetic information altered in some way. Transgenic organisms contain a genetic material (such as a gene) from a different species.
- DNA fingerprinting and DNA profiling use variations in the patterns of base sequences in DNA between individuals to identify them.
- CRISPR technology has been developed for genome editing.

Interwoven stories of DNA and the genome

- Organisms contains DNA from a variety of sources. This combination of the hosts genome and its microbes make up the hologenome.
- Non-coding regions contain genes that regulate many activities, where previously it was thought they served no purpose.

2.16.2 Key terms

allele alternate form of a gene for a particular characteristic
alleles alternate forms of a gene for a particular characteristic
amino acid an organic compound that forms the building blocks of proteins
amniocentesis a type of prenatal screening in which a sample is collected from the amniotic fluid around a fetus
asexual reproduction reproduction that does not involve fusion of sex cells (gametes)
autosomes non-sex chromosomes
bioinformatics the science of analysing biological data through computers, particularly around genomics and molecular genetics
carrier an individual heterozygous for a characteristic who does not display the recessive trait
cell the smallest unit of life and the building blocks of living things
centromere a section of a chromosome that links sister chromatids
Chargaff's rule a rule that states the pairing of adenine with thymine and cytosine with guanine
chorionic villus sampling (CVS) a type of prenatal screening in which tissue from the chorion is tested during pregnancy
chromatid one identical half of a replicated chromosome
chromosomes tiny thread-like structures inside the nucleus of a cell that contain the DNA that carries genetic information
clones genetically identical copies
codon sequence of three bases in mRNA that codes for a particular amino acid
complementary base pairs in DNA, specific base pairs will form between the nitrogenous bases adenine (A) and thymine (T) and between the bases cytosine (C) and guanine (G)
complete dominance a type of inheritance where traits are either dominant or recessive
cytogenetics the study of heredity at a cellular level, focusing on cellular components such as chromosomes
deoxyribonucleic acid (DNA) a substance found in all living things that contains its genetic information
deoxyribose the sugar in the nucleotides that make up DNA
diploid the possession of two copies of each chromosome in a cell
DNA ligase an enzyme that joins DNA fragments together
DNA replication process that results in DNA making a precise copy of itself
dominant a trait (phenotype) that requires only one allele to be present for its expression in a heterozygote
epigenetics the study of the effect of the environment on the expression of genes
fertilisation penetration of the ovum by a sperm
gametes reproductive or sex cells such as sperm or ova

gel electrophoresis a technique used to separate molecules based on their size through an agarose gel

gene segment of a DNA molecule with a coded set of instructions in its base sequence for a specific protein product; when expressed, may determine the characteristics of an organism

gene therapy altering genes with the intention to treat or prevent disease

genetic engineering one type of biotechnology that involves working with DNA

genetic engineers scientists who use special tools to cut, join, copy and separate DNA

genetic genealogy the use of DNA along with other genealogical tests to infer relatedness between individuals

genetically modified organism a organism where the genome has been altered

genetics study of inheritance

genome the complete set of genes present in a cell or organism

genome maps maps that describe the order and spacing of genes on each chromosome

genomics the study of genomes

genotype genetic instructions (contained in DNA) inherited from parents at a particular gene locus

haploid the possession of one copy of each chromosome in a cell

heterozygous a genotype in which the two alleles are different

holobiont a host and their associated microbiota

hologenome the sum of genetic information of a host and its microbiota

homologous chromosomes with matching centromeres, gene locations, sizes and banding patterns

homozygous a genotype in which the two alleles are identical

homozygous dominant a genotype where both alleles for the dominant trait are present

homozygous recessive a genotype where both alleles for the recessive trait are present

induced mutation a mutation of DNA in which the cause can be identified

inheritance genetic transmission of characteristics from parents to offspring

karyotype an image that orders chromosomes based on their size

kinetochore a region on a chromosome associated with cell division

linkage analysis use of markers to scan the genome and map genes on chromosomes

locus position occupied by a gene on a chromosome

maternal chromosomes chromosomes from the ovum

meiosis cell division process that results in new cells with half the number of chromosomes of the original cell

Mendelian inheritance an inheritance pattern in which a gene inherited from either parent segregates into gametes at an equal frequency

messenger RNA (mRNA) single-stranded RNA transcribed from a DNA template that then carries the genetic to a ribosome to be translated into a protein

mitosis cell division process that results in new genetically identical cells with the same number of chromosomes as the original cell

molecular genetics study of genetics at a molecular level

monohybrid ratio the 3:1 ratio of a particular characteristic for offspring produced by heterozygous parents, controlled by autosomal complete dominant inheritance

monomers molecules that are the building blocks of larger molecules known as polymers

monosomy a condition where there is only one copy of a particular chromosome instead of two

mutagen agent or factor that can induce or increase the rate of mutations

mutations changes to DNA sequence, at the gene or chromosomal level

nitrogenous base a component of nucleotides that may be one of adenine, thymine, guanine, cytosine or uracil

nucleic acids molecules composed of building blocks called nucleotides, which are linked together in a chain

nucleotides compounds (DNA building blocks) containing a sugar part (deoxyribose or ribose), a phosphate part and a nitrogen-containing base that varies

nucleus roundish structure inside a cell that contains DNA and acts as the control centre for the cell

ova female reproductive cells or eggs

paternal chromosomes chromosomes carried in the sperm

pedigree chart diagram showing the family tree and a particular inherited characteristic for family members

phenotype characteristics or traits expressed by an organism

point mutation a mutation at one particular point in the DNA sequence, such as a substitution or single base deletion or insertion

polymerase chain reaction (PCR) a process which amplifies small amounts of DNA

polymers molecules made of repeating subunits of monomers joined together in long chains

prenatal screening testing a fetus during pregnancy to detect any abnormalities

proteins molecules, such as enzymes, haemoglobin and antibodies made up of amino acids

Punnett square a diagram used to predict the outcome of a genetic cross

recessive a trait (phenotype) that will only be expressed in the absence of the allele for the dominant trait

recombinant DNA a molecule of DNA that contains fragments from more than one source

recombinant DNA technology technology that can form DNA that does not exist naturally, by combining DNA sequences that would not normally occur together

restriction enzymes enzymes that cut DNA at specific base sequences (recognition sites)

restriction fragment length polymorphisms (RFLPs) variations in the lengths of DNA fragments in individuals with different alleles of a gene

retroposons segments of DNA that can break off a chromosome and paste themselves elsewhere in the genome

ribonucleic acid (RNA) a type of nucleic acid that contains ribose sugar

ribose the sugar found in nucleotides of RNA

ribosomal RNA (rRNA) a special type of RNA that forms the structure of ribosomes

ribosome organelle found in the cells of all organisms in which translation occurs

sex chromosomes chromosomes that determine the sex of an organism

sex-linked inheritance an inherited trait coded for by genes located on sex chromosomes

sexual reproduction reproduction that involves the joining together of male and female gametes

single nucleotide polymorphisms (SNPs) genetic differences between individuals that can result from single base changes in their DNA sequences

sister chromatids identical chromatids on a replicated chromosome

somatic cells cells of the body that are not sex cells

sperm male reproductive cell

spontaneous mutation a mutation of DNA that cannot be explained or identified

symbiotic a very close relationship between two organisms of different species

telomerase an enzyme involved in maintaining and repairing a telomere

telomere a cap of DNA on the tip of a chromosome that enables DNA to be replicated safely without losing valuable information

transcription the process by which the genetic message in DNA is copied into a mRNA molecule

transfer RNA (tRNA) molecules located in the cytosol that transport specific amino acids to complementary mRNA codons in the ribosome

transgenic organism an organism with genetic information from another species in its genome

translation the process in which amino acids are joined in a ribosome to form a protein

transposition the ability of a gene to change position on the chromosome

transposons a section of chromosome that moves about the chromosome within a cell through the method of transposition

triplet a sequence of three nucleotides in DNA that can code for an amino acid

trisomy a condition where there are three copies of a particular chromosome instead of two

variation differences between cells or organisms

zygote a cell formed by the fusion of male and female reproductive cells

on Resources

eWorkbooks

Study checklist (ewbk-13074)
Literacy builder (ewbk-13075)
Crossword (ewbk-13077)
Word search (ewbk-13079)
Reflection (ewbk-13067)

Solutions Topic 2 Solutions (sol-1125)

Practical investigation eLogbook Topic 2 Practical investigation eLogbook (elog-2424)

Digital document Key terms glossary (doc-40481)

2.16 Activities

2.16 Review questions

Select your pathway

■ LEVEL 1	■ LEVEL 2	■ LEVEL 3
2, 3, 7, 10, 14	1, 4, 5, 8, 12, 15	6, 9, 11, 13, 16, 17

These questions are even better in jacPLUS!
- Receive immediate feedback
- Access sample responses
- Track results and progress

Find all this and MORE in jacPLUS ▶

Remember and understand

1. **a.** Identify whether the following statements are true or false.

Statements	True or false?
i. All nucleotides that make up DNA contain the same type of nitrogenous base.	
ii. According to Chargaff's base-pairing rule, adenine binds to cytosine and guanine binds to thymine.	
iii. The process of making a complementary mRNA copy of the DNA message is called translation.	
iv. Meiosis prevents doubling of the chromosomes at fertilisation because it produces cells with half the chromosome number of the original cell.	
v. The recessive trait can only be expressed if the allele for the dominant trait is not present.	
vi. Alleles on chromosomes inherited from each of your parents contribute to your genotype.	
vii. Mutations in both germline and somatic cells are passed on to the next generation.	
viii. Errors or changes in DNA, genes or chromosomes are called mutations.	
ix. Pedigree charts can be used to observe patterns and to predict the inheritance of traits within families.	
x. CRISPR technology has been developed for genome editing.	
xi. Gel electrophoresis can be used to separate DNA fragments.	
xii. Some of the repeating sequences within our non-coding DNA may have originated from invading viruses.	

b. Rewrite any false statements to make them true.

2. Arrange the sentence fragments provided to complete the sentence that has been started for you. A living organism _____.
   - is made up of
   - cells
   - DNA
   - which contain in the nucleus
   - which are made up of
   - chromosomes
   - which contain
   - genes

3. Copy and complete the following figures using the terms provided.
   - Homozygous dominant
   - Cytosine
   - Sugar
   - Nitrogenous base
   - Adenine
   - Phosphate
   - Heterozygous
   - Thymine
   - Homozygous recessive
   - Guanine

a.

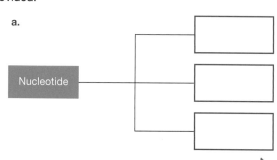

b.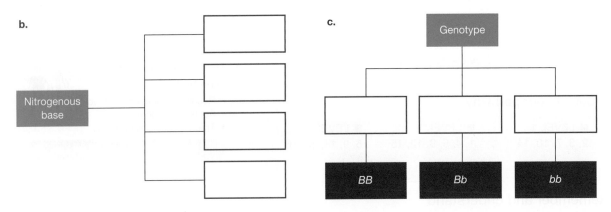

c.

4. Distinguish between the following.
   a. Mitosis and meiosis
   b. Fertilisation and meiosis
   c. Autosomes and sex chromosomes
   d. Dominant and recessive

5. Match the term with its description in the following table.

Term	Description
a. Diploid	**A.** A type of cell division important for growth, repair, and replacement
b. Fertilisation	**B.** The fusion of gametes
c. Gamete	**C.** The number of chromosomes in human gametes
d. Haploid	**D.** The number of chromosomes in normal human somatic cells
e. Meiosis	**E.** A type of cell division to produce gametes
f. Mitosis	**F.** Also known as a sex cell
g. Allele	**G.** Alternative form of a gene
h. Gene	**H.** Characteristics or features determined by genotype and influenced by environment
i. Genetics	**I.** Combination of alleles for a particular trait
j. Genotype	**J.** Segment of DNA that codes for the production of a particular protein
k. Inheritance	**K.** The branch of science that involves the study of inheritance
l. Phenotype	**L.** The passing on of characteristics from one generation to the next

6. Observe the following figure and use your understanding from this topic to answer the questions.

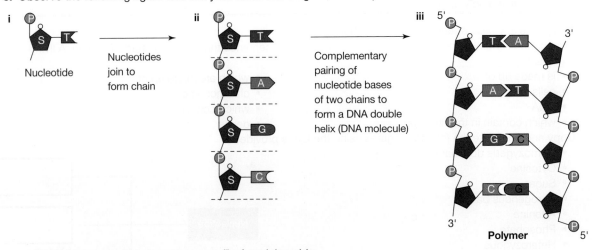

   a. State the name given to a monomer (i) of nucleic acids.
   b. If a mistake were made during DNA replication that resulted in thymine being replaced by cytosine in the DNA strand (ii), describe the effect this would have on the structure and function of the double-stranded DNA (iii).

c. State the name of the term used to describe changes in DNA (or chromosomes).

d. Outline two factors that can contribute to causing changes in DNA.

e. Provide an example of a genetic disease that results from a change in the DNA or chromosome.

7. Observe the following figure and use your understanding from this topic to answer the questions.

a. Identify labels for A to G.

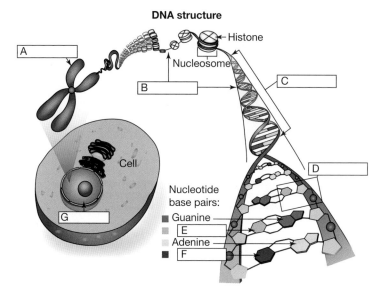

DNA structure

b. Describe the relationship between nucleotides, DNA, genes and chromosomes.

Apply and analyse

8. Mitosis and meiosis are both processes involved in the production of new cells.

a. Outline two differences between these processes.

b. If an organism has 50 chromosomes, how many chromosomes would you expect in the following?

   i. A cell produced through mitosis

   ii. A cell produced through meiosis

c. In which of these does crossing-over occur? What occurs during this process?

d. Summarise the stages of mitosis.

9. Use the mutation mind map provided and your own knowledge from this topic to answer the following questions.

a. Define the following terms.

   i. Mutation

   ii. Mutagenic agent

b. State the name of the following.

   i. A physical mutagenic agent

   ii. A chemical mutagenic agent

   iii. Genetic disorder characterised by a change in the number of autosomes

   iv. Genetic disorder characterised by a change in the number of sex chromosomes

   v. Genetic disorder in which a base substitution mutation has occurred

c. Distinguish between the following.

   i. Induced mutations and spontaneous mutations

   ii. Physical and chemical mutagenic agents

   iii. Base insertion, base substitution and base deletion mutations

d. Describe what is meant by the following.

   i. A frameshift mutation

   ii. The type of base mutation(s) that can cause it

   iii. The effect that it may have

e. Suggest two more pieces of information that could be incorporated into the mind map on mutation.

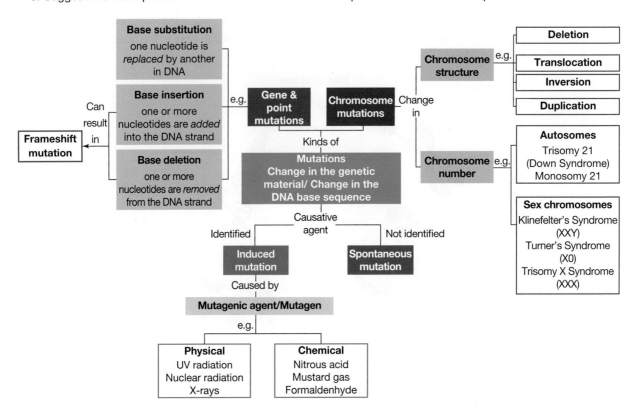

10. Examine the pedigree chart provided. Let the dark hair allele be represented by *B* and the red hair allele by *b*.
    a. Redraw the pedigree using pedigree symbols and shade all individuals with red hair.
    b. Write the genotypes for the individuals B, G, H and K.
    c. Write the phenotypes for the individuals D and F.
    d. If individuals G and H had another child, what is the chance that it would have red hair?
    e. If individuals F and I had a child, do you think it might be possible for it to have red hair? Explain your reasoning.

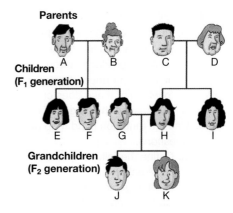

11. Use the pedigree chart provided, showing inheritance of cystic fibrosis, to determine the following.
    a. Which type of inheritance is responsible for cystic fibrosis? Justify your response.
    b. Is individual II-3 a male or a female? Justify your response.
    c. If *N* = normal and *n* = cystic fibrosis allele, state the genotype and phenotype with respect to cystic fibrosis for the following.
    i. Individual I-1
    ii. Individual I-2
    iii. Individual II-4
    iv. Individual III-3

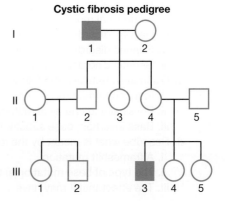

12. Assume that coat colour in mice is inherited by autosomal inheritance and that black coat (*B*) is dominant to white coat (*b*).
    Use a Punnett square to predict the phenotypes of the offspring of a cross between the following.
    a. A homozygous black mouse with a heterozygous black mouse
    b. A homozygous white mouse with a heterozygous black mouse
    c. Two heterozygous black mice
    d. A homozygous black mouse with a homozygous white mouse

13. The pedigree chart provided shows the inheritance pattern of Huntington's disease within a family. Huntington's disease is an autosomal dominant trait. Individuals with Huntington's disease are shaded in the provided pedigree chart.

Huntington's disease pedigree

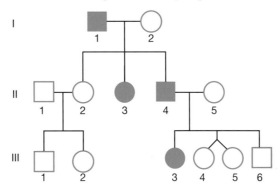

    a. State how many generations are shown.
    b. How many females are in the pedigree?
    c. How many males are in the second generation?
    d. Identify three individuals who have Huntington's disease.
    e. If *H* represents the allele for Huntington's disease, state the genotypes of the following.
        i. Individuals I-1 and I-2
        ii. Individuals II-2 and II-4
    f. Identify the chance that the next child of the following couples will have Huntington's disease. (Use a Punnett square to help you.)
        i. Individuals II-4 and II-5
        ii. Individuals II-1 and II-2

14. Summarise each of the different types of genetic testing and provide a clear example of each.

Evaluate and create

15. Construct Venn diagrams and add shading to pedigrees like those shown in the following figures to illustrate the similarities and differences between each of the types of inheritance.

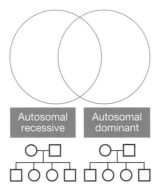

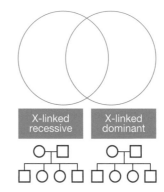

16. **a.** **MC** Use the following diagram to decide which of the statements is correct.
   A. Pig A and pig B are genetically identical.
   B. Pig D and pig C are genetically identical.
   C. Pig A and pig C are genetically identical.
   D. None of the pigs are genetically identical because the environment affects their genotype.

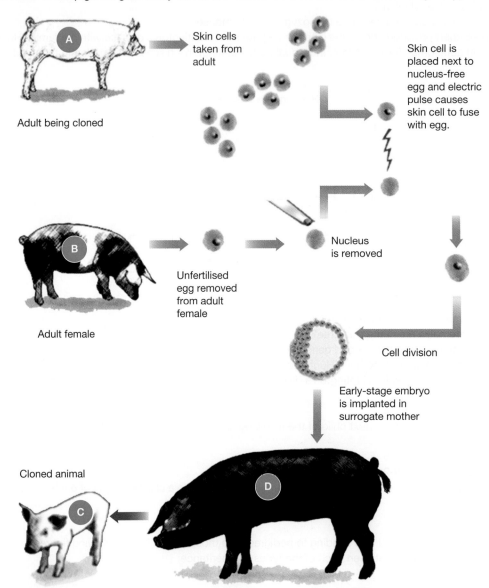

Adult being cloned

Skin cells taken from adult

Skin cell is placed next to nucleus-free egg and electric pulse causes skin cell to fuse with egg.

Adult female

Unfertilised egg removed from adult female

Nucleus is removed

Cell division

Early-stage embryo is implanted in surrogate mother

Cloned animal

b. Outline ethical implications of this process. Evaluate these implications and state your opinion about the use of this gene technology.

17. Use the provided figures showing the relationship between DNA and chromosomes and your own knowledge from this topic to answer the questions.
   a. State the number of the following.
      i. DNA strands in the double helix
      ii. Molecule(s) of DNA in each chromosome
   b. Suggest a reason DNA is wound around histone proteins.
   c. Outline the relationship between the following.
      i. DNA, histone proteins and chromosomes
      ii. DNA, genes and chromosomes
   d. Suggest why the artist used two different colours to represent the chromosomes in figure 2.
   e. Are chromosomes always visible in the nucleus? Explain.
   f. Suggest two improvements for each figure.

Figure 1

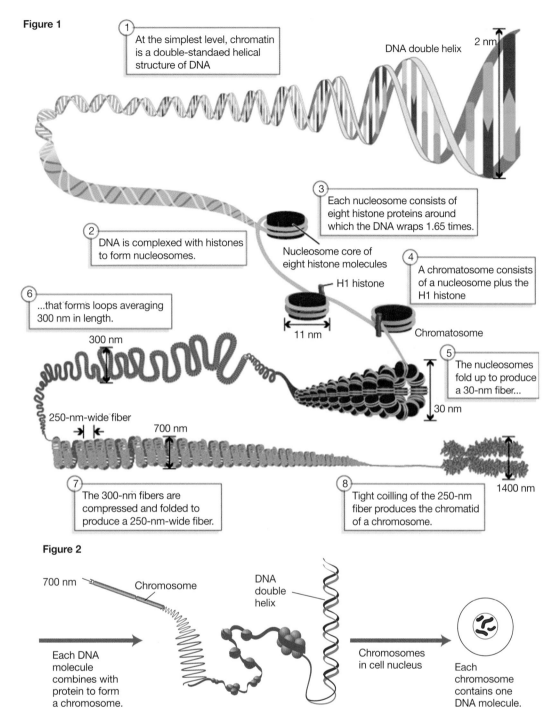

① At the simplest level, chromatin is a double-standaed helical structure of DNA

DNA double helix

2 nm

③ Each nucleosome consists of eight histone proteins around which the DNA wraps 1.65 times.

Nucleosome core of eight histone molecules

② DNA is complexed with histones to form nucleosomes.

④ A chromatosome consists of a nucleosome plus the H1 histone

H1 histone

Chromatosome

11 nm

⑥ ...that forms loops averaging 300 nm in length.

300 nm

⑤ The nucleosomes fold up to produce a 30-nm fiber...

30 nm

250-nm-wide fiber

700 nm

1400 nm

⑦ The 300-nm fibers are compressed and folded to produce a 250-nm-wide fiber.

⑧ Tight coilling of the 250-nm fiber produces the chromatid of a chromosome.

Figure 2

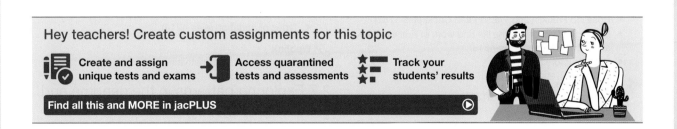

700 nm

Chromosome

DNA double helix

Each DNA molecule combines with protein to form a chromosome.

Chromosomes in cell nucleus

Each chromosome contains one DNA molecule.

Fully worked solutions and sample responses are available in your digital formats.

Online Resources

 Resources

Below is a full list of **rich resources** available online for this topic. These resources are designed to bring ideas to life, to promote deep and lasting learning and to support the different learning needs of each individual.

2.1 Overview

 eWorkbooks
- Topic 2 eWorkbook (ewbk-13017)
- Starter activity (ewbk-13019)
- Student learning matrix (ewbk-13021)

 Solutions
- Topic 2 Solutions (sol-1125)

Practical investigation eLogbooks
- Topic 2 Practical investigation eLogbook (elog-2424)
- Investigation 2.1: Do you fit into your genes or do they fit into you? (elog-2426)

Video eLesson
- Epigenetics — switching genes on and off (eles-4210)

2.2 Chromosomes, genes and DNA

 eWorkbooks
- Genes and chromosomes (ewbk-13024)
- Labelling a chromosome (ewbk-13022)
- Karyotype (ewbk-13026)

 Practical investigation eLogbook
- Investigation 2.2: Extracting DNA (elog-2428)

Teacher-led video
- Investigation 2.2: Extracting DNA (tlvd-10808)

Interactivities
- Cell structures (int-5870)
- Labelling a chromosome (int-8123)

Weblinks
- Karyotype - case studies
- Virtual karyotype

2.3 Discovering the structure of DNA

eWorkbooks
- DNA (ewbk-13030)
- Labelling DNA (ewbk-13028)

 Practical investigation eLogbook
- Investigation 2.3: Constructing a model of DNA (elog-2438)

Video eLessons
- Rosalind Franklin and Watson and Crick (eles-1782)
- Montage of images of the life and work of Francis Crick and James Watson (eles-4213)
- DNA structure (eles-4211)

Interactivities
- Complementary DNA (int-0133)
- Labelling DNA (int-8124)

Weblink
- The history of DNA timeline

2.4 Unlocking the DNA code

 eWorkbooks
- Protein synthesis (ewbk-13034)
- Labelling the components of protein synthesis (ewbk-13032)

Video eLessons
- Transcription (eles-4167)
- Translation (eles-4168)
- Protein synthesis (eles-4295)

Interactivities
- Transcription (int-8125)
- Translation (int-8126)
- Labelling the components of protein synthesis (int-8127)

2.5 Mitosis and meiosis

 eWorkbooks
- Mitosis (ewbk-13038)
- Labelling the stages of mitosis (ewbk-13036)
- Meiosis (ewbk-13042)
- Labelling the stages of meiosis (ewbk-13040)

Practical investigation eLogbooks
- Investigation 2.4: Observing cell division (elog-2432)
- Investigation 2.5: What's the chance? (elog-2434)

Video eLessons
- Amoeba (eles-2694)
- Euglena (eles-2695)
- Stages of mitosis (eles-4215)
- Stages of meiosis (eles-4216)

Interactivities
- Mitosis (int-3027)
- The stages of mitosis (int-3028)
- Labelling the stages of mitosis (int-8128)
- Labelling the stages of meiosis (int-8129)

2.6 Changing the code through mutations

eWorkbook
- Mutations (ewbk-13044)

Video eLessons
- Polydactyl cat (eles-2698)
- DNA and Hiroshima (eles-1781)
- Types of mutation (eles-4214)

Weblink
- Scientists warn against vitamins

2.7 Exploring patterns in the genome and genetic sequences

eWorkbooks
- Blood samples (ewbk-12770)

Weblinks
- Epigenetic transformation — you are what your grandparents ate
- The Human Genome Project
- Culture and science

2.8 Inheritance of genes

eWorkbooks
- Alleles and inheritance (ewbk-13047)
- Mendel's experiments (ewbk-13049)
- Mendel's peas (ewbk-13051)

Practical investigation eLogbooks
- Investigation 2.6: How does the environment affect phenotype? (elog-2436)
- Investigation 2.7: Genetics database (elog-2438)

Video eLessons
- Genotype (eles-4222)
- Phenotype (eles-4223)
- Co-dominance (eles-4224)

Interactivities
- Making families (int-0681)
- Genotype (int-0668)
- Generating the phenotype (int-0178)

2.9 Punnett squares and predicting inheritance

eWorkbook
- Practising Punnett squares (ewbk-13053)

Interactivity
- Punnett squares (int-8131)

2.10 Interpreting pedigree charts

eWorkbooks
- Pedigrees (ewbk-13055)
- Haemochromatosis (ewbk-12772)
- First Nations Australians kinship system (ewbk-13057)

Video eLesson
- Autosomal recessive disorders (eles-4221)

Interactivity
- Pedigrees and genotypes (int-8122)

Weblinks
- First Nations Australians kinship system - Nations, Clans and family groups
- First Nations Australians kinship system - Moiety
- First Nations Australians kinship system - Totems
- First Nations Australians kinship system - Skin Names

2.11 Exposing your genes through genetic testing

eWorkbook
- Genetic testing (ewbk-13060)
- Genetic testing and life insurance (ewbk-13062)

Weblinks
- Cystic fibrosis and carrier screening

- What is genetic counselling?
- Can genetic testing services really predict your future?
- New genetic testing technology for IVF embryos
- DNA chip
- The limits of ancestry DNA tests, explained
- Genetic testing and life insurance

2.12 Biotechnological tools to edit the genome

eWorkbook
- Manipulating DNA (ewbk-13065)

 Video eLessons
- Sample preparation for DNA electrophoresis (eles-2699)
- DNA amplification (eles-4164)
- Restriction enzymes (eles-4212)
- Ancient resurrection (eles-1070)

Weblinks
- PCR
- Electrophoresis
- Click and clone
- How CRISPR lets us edit DNA — Jennifer Doudna
- What is CRISPR?
- CRISPR in Australia
- What you need to know about CRISPR — Ellen Jorgensen
- Biotechnology timeline

2.13 Interwoven stories of DNA and the genome

Weblinks
- ENCODE: Encylopedia of DNA Elements
- Barbara McClinton Nobel Prize lecture

2.15 Project — Genetic discrimination

ProjectsPLUS
- Genetic discrimination (pro-0259)

2.16 Review

eWorkbooks
- Topic review Level 1 (ewbk-13068)
- Topic review Level 2 (ewbk-13070)
- Topic review Level 3 (ewbk-13072)
- Study checklist (ewbk-13074)
- Literacy builder (ewbk-13075)
- Crossword (ewbk-13077)
- Word search (ewbk-13079)
- Reflection (ewbk-13067)

Digital document
- Key terms glossary (doc-40481)

 To access these online resources, log on to **www.jacplus.com.au**

3 Evolution

CONTENT DESCRIPTION

Use the theory of evolution by natural selection to explain past and present diversity and analyse the scientific evidence supporting the theory (AC9S10U02)

Source: F–10 Australian Curriculum 9.0 (2024–2029) extracts © Australian Curriculum, Assessment and Reporting Authority; reproduced by permission.

LESSON SEQUENCE

SCIENCE INQUIRY AND INVESTIGATIONS

Science inquiry is a central component of Science curriculum. Investigations, supported by a **Practical investigation eLogbook** and **teacher-led videos**, are included in this topic to provide opportunities to build Science inquiry skills through undertaking investigations and communicating findings.

LESSON
3.1 Overview

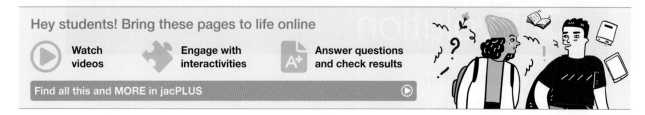
3.1.1 Introduction

Five minutes to midnight? If we compressed Earth's 4.5-billion-year history into a single year, Earth would have formed on 1 January and the present time would be represented by the stroke of midnight on 31 December. Using this timescale, the first primitive microbial life forms appeared in late March, followed by more complex photosynthetic micro-organisms towards the end of May. Land plants and animals emerged from the seas in mid-November. Dinosaurs arrived early on the morning of 13 December and then disappeared forever in the evening of 25 December. Although human-like creatures appeared in Africa during the evening of 31 December, it was not until about five minutes before the New Year that our species, *Homo sapiens*, appeared on Earth.

The great diversity of living things may be explained by the theory of evolution by natural selection. Variations upon which natural selection acts may be determined by both genetic and environmental factors. The selection of some variations over others is related to their possible effects on increasing the chances of survival and reproduction of individuals that possess them, such as a caterpillar that is able to camouflage and hide from predators. In this way, favourable variations may be passed from one generation to the next. But what is the evidence for this theory and how can it be evaluated and interpreted?

FIGURE 3.1 All environments change over time.

3.1.2 Think about evolution

1. Why do humans still have a tailbone despite being born without a tail?
2. What has dating got to do with rocks?
3. How much Neanderthal DNA do you think you have in your genome?
4. How can one species become two?
5. What does a clock have to do with your ancestors?
6. Why should we celebrate our differences?

3.1.3 Science inquiry

Understanding the relatedness between species

When you look at living things around you, have you considered how closely related you are?

Life on earth is classified into three domains: Bacteria, Archaea and Eukarya. Every living thing on Earth is thought to have descended from one single entity. This was a sort of primitive cell that floated around in the primordial soup over three billion years ago. It has been named the last universal common ancestor or LUCA (as shown in figure 3.2). Considerable controversy surrounds this ancestor, because it has left no fossil remains or any other physical clues of its identity. Researchers, however, are comparing genes from all forms of life and have put together a portrait of this cell that could be the ancestor of us all.

FIGURE 3.2 The theory of the last universal common ancestor

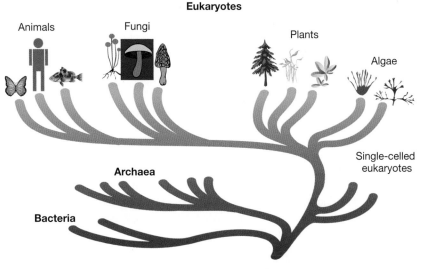

Scientific evidence suggests that LUCA was anaerobic and autotrophic and may have lived in deep iron-sulfur rich hydrothermal sea vents. Advances in technologies over the last couple of decades have enabled scientists to sequence genomes and to produce genome databanks. Interestingly, the expression of some of LUCA's 355 genes may have enabled it to metabolise hydrogen as a source of energy and to tolerate very high temperatures.

elog-2217

INVESTIGATION 3.1

Exploring relatedness and the last universal common ancestor

Aim

To gain an understanding of common ancestors and how features are retained or changed across species

PART A: Investigating similarities in closely related species

Materials

- Different species of plants

Method

1. Carefully observe the features of the four different species of possums in the figure. Make lists of how the possums are similar and how they are different in your results.
2. Go outside and explore four different species of plant. You may wish to take photos of each. Make lists of how the plants are similar and how they are different in your results.

Results

1. Construct a table outlining ways in which the possums are different and ways in which they are similar.
2. Construct a table outlining ways in which the plants are different and ways in which they are similar.

Discussion

1. Suggest reasons for the differences.
2. Suggest how species may have become different.
3. Would plants and possums have common ancestors? What features do they have in common that may suggest this.

Conclusion

Summarise the differences and similarities between species and why these might be important in evolution.

PART B: LUCA — your ancestor

Method

1. Write a hypothesis about what you think Earth was like three billion years ago. Find out whether your hypothesis supports current evidence.
2. Suggest the features of organisms that could survive on Earth three billion years ago.
3. Suggest processes or features that may have increased the chance of organisms passing on their traits to their offspring at this time.
4. Design an organism that could survive these conditions.

Results

Provide a labelled diagram of your organism that includes descriptions of all of its features.

Discussion

1. Do any organisms living today share any of these features?
2. Explain why the organism you designed would be able to survive in conditions from three billion years ago.
3. State what LUCA is an abbreviation for and outline why it is important.
4. **a.** All forms of life are coded by nucleic acids (DNA or RNA). Suggest a reason for this.
   b. Compare your organism with others in your class. What differences would you expect in their DNA and RNA?
5. Based on figure 3.2, identify the following.
   a. Which group branched off the earliest from LUCA?
   b. Are humans more closely related to fungi, plants, algae, bacteria or archaea? Explain how you drew this conclusion.
6. If scientists discovered evidence of a species such as the one you created, how do you think they might show that this is the last universal common ancestor? What processes would they follow to support this?

7. Scientific evidence suggests LUCA possessed the gene necessary to produce the enzyme reverse gyrase. Find out this trait may have enabled LUCA to survive in its harsh environment.

8. Research what is meant by the terms *anerobic* and *autotrophic*. How might these traits relate to LUCA's survival?

9. Phylogenetists study the genetic relationships and evolutionary history of organisms. Discoveries from investigations of the historical relatedness of genomes has resulted in some scientists shifting their support for a two-domain phylogenetic tree. In the two-domain tree, scientists hypothesise that eukaryotes evolved from endosymbiosis between archaea and bacteria, as shown in the following figure.

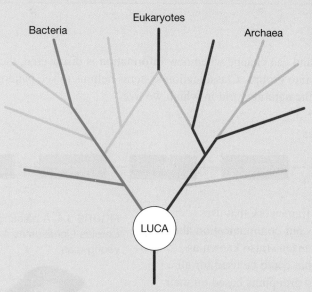

A schematic of the two-domain tree, with eukaryotes evolving from endosymboisis between members of the two original trunks of the tree, archaea and bacteria

Will further advances in technology support this new hypothesis, or will it lead to the synthesis of a new line of thinking?

10. a. Research scientific evidence for and against the two-domain tree hypothesis. Summarise the secondary data you found (include the sources of your information).
    b. Was the research you explored valid and reliable? How do you know?
    c. On the basis of your research, which phylogenetic LUCA tree do you support? Explain.
    d. Discuss your findings and share your views with others.

Conclusion

Summarise what is meant by the last universal common ancestor and what features it might have had to survive in a very different environment from today.

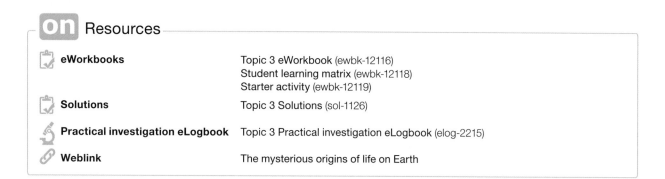

Resources

eWorkbooks	Topic 3 eWorkbook (ewbk-12116)
	Student learning matrix (ewbk-12118)
	Starter activity (ewbk-12119)
Solutions	Topic 3 Solutions (sol-1126)
Practical investigation eLogbook	Topic 3 Practical investigation eLogbook (elog-2215)
Weblink	The mysterious origins of life on Earth

LESSON
3.2 Classification

LEARNING INTENTION

At the end of this lesson you will be able to define the term 'species' and outline the binomial naming system of classification.

3.2.1 Binomial nomenclature

Although classification systems are not fixed and can change when new information is discovered, they are particularly useful for categorising organisms into groups. Classification systems help us to see patterns and order, so we can make sense and meaning of the natural world in which we live.

FIGURE 3.3 The naming system from Linnaeus

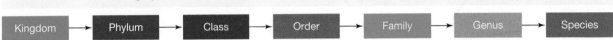

Kingdom → Phylum → Class → Order → Family → Genus → Species

Classifying organisms into groups provides a framework that uses specific criteria and terminology and improves our communication about organisms. The Swedish naturalist Carolus Linnaeus (also known as Carl von Linné) developed a naming system that could be used for all living organisms. It involved placing them into groupings based on their similarities. He called the smallest grouping **species**.

Linnaeus' naming system was called the **binomial system of nomenclature** because it involved giving each species a particular name made up of two words (The word *binomial* comes from the Latin terms *bi-*, meaning 'two', and *nomen*, meaning 'name'). The scientific names given to organisms were often Latinised. In this system, the species name is made up of a genus name as the first word and a descriptive or specific name as the second word. A capital letter is used for the genus name and lower case for the descriptive name. If handwritten, the species name should be underlined; if typed, it should be in *italics*.

FIGURE 3.4 A painting of Carolus Linnaeus as a young man

FIGURE 3.5 Naming a species involves using the genus and descriptive name

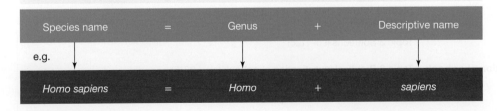

Species name	=	Genus	+	Descriptive name
e.g.				
Homo sapiens	=	*Homo*	+	*sapiens*

species taxonomic unit consisting of organisms capable of mating and producing viable and fertile offspring

binomial system of nomenclature system devised by Carolus Linnaeus giving organisms two names, the genus and another specific name

on Resources

🔗 **Weblink** A higher level of classification of all living things

3.2.2 Species

Our current definition of species

Although genetic technologies have blurred the lines of our classification system, our usual definition of species refers to individuals that can interbreed to produce fertile offspring.

Species also fit into another grouping in terms of where they belong within an ecosystem. Ecosystems consist of a number of different communities. Within these communities are **populations** of individual organisms of a species living together in a particular place (habitat) at a particular time.

Understanding what defines a species is vital in understanding **biodiversity**, and how biodiversity within different species is driven by the process of **evolution**.

population members of one species living together in a particular place at the same time

biodiversity total variety of living things on Earth

evolution the process in which traits in species gradually change over successive generations

FIGURE 3.6 The interrelationship between ecosystems and species

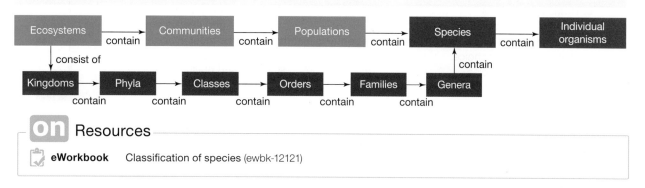

On Resources

📝 **eWorkbook** Classification of species (ewbk-12121)

3.2 Activities

learn on

| 3.2 Quick quiz on | 3.2 Exercise |

These questions are even better in jacPLUS!
- Receive immediate feedback
- Access sample responses
- Track results and progress

Find all this and MORE in jacPLUS ⊙

Select your pathway

| ■ LEVEL 1 1, 2, 7, 9 | ■ LEVEL 2 3, 5, 8, 11 | ■ LEVEL 3 4, 6, 10, 12 |

Remember and understand

1. **a.** State whether the following statements are true or false.

Statements	True or false?
i. Classification is fixed and does not change.	
ii. Classifying organisms into groups provides a framework that uses specific criteria and terminology and improves our communication about organisms.	
iii. As you move from kingdoms to species, there are more differences between the members of the group.	
iv. Individuals that can interbreed and produce fertile offspring belong to the same species.	
v. Binomial nomenclature involves giving each species a particular name made up of its phylum and its order.	

b. Rewrite any false statements to make them true. ▶

2. Name the three domains that organisms can be categorised into.

3. What naming system did Carolus Linneaus devise and what did it do?

4. a. Define the term *species*.

   b. Outline the naming system used in the binomial system of nomenclature.

5. Match the term to its description in the following table.

Term	Description
a. *Homo*	**A.** Descriptive name
b. *Homo sapiens*	**B.** Genus
c. *sapiens*	**C.** Species name

Apply and analyse

6. **SIS** **a.** Outline why scientists classify organisms into groups.

   b. Explain why classification of organisms is not fixed.

   c. Suggest criteria that are used to divide organisms into different kingdoms.

7. Describe the relationships between the two terms in each of the following pairs.

   a. Genus and genera
   b. Phyla and phylum
   c. Homo and sapiens
   d. Genus and species
   e. Species and populations
   f. Species and ecosystems

8. **MC** Identify which one of the following pairs contains members with the most in common and then justify your choice.

   A. Kingdom and species
   B. Order and phylum
   C. Genus and family
   D. Class and order

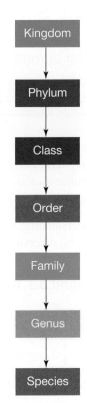

Evaluate and create

9. Create a flowchart using the following terms to show their connections.

- Populations
- DNA
- Multicellular
- Ecosystems
- Organism
- Systems
- Tissues
- Species
- Cells
- Classes
- Communities
- Kingdoms
- Families
- Phyla
- Orders
- Genera
- Nucleus
- Organs

10. Research and report on one of the following contributors to classification. Evaluate how their contributions affected our understanding of classification.

- Carolus Linnaeus
- Robert H. Whittaker
- Carl Woese

11. Research and construct a table showing the key similarities and differences between the three domains, Archaea, Bacteria and Eukarya.

12. **SIS** **a.** Investigate recent research in different kingdoms of living things.

   b. Outline evidence being used to suggest new classifications, or modifications to our current classifications.

   c. State and justify your opinion on whether changes should be made based on the evidence you have outlined.

Fully worked solutions and sample responses are available in your digital formats.

LESSON
3.3 Processes of natural selection

3.3.1 Best suited

Survival of the fittest is more than having muscles, being tough or working out at the gym. It's about being better suited to a particular environment and having an increased chance of surviving long enough to be able to have offspring that will take your genes into the next generation.

Consider the moth in figure 3.7. It is able to camouflage and it is less visible to predators. Therefore, it is more likely to survive and pass on its genes to its offspring. This enables it to be 'fitter' in its environment.

FIGURE 3.7 Can you see the moth in this image?

3.3.2 The mechanism for evolution

Vive la différence!

We are all different! Our differences increase the chance of survival of our species. If our environmental conditions were to change, some of us might have an increased chance of surviving over others. Those who survive might then pass on any genetically inherited advantage to their children, who would also have an increased chance of survival — at least unless the environment changed to their disadvantage. If this happened, other variations may have increased chances of survival. This is what the theory of evolution and **natural selection** is all about.

The process of natural selection

Darwin and Wallace's theory of evolution included the suggestion that the mechanism for evolution was natural selection. Natural selection relies on two main factors: *variation* and *selection* (refer to figure 3.8 and sections 3.3.3 and 3.3.4).

The steps of natural selection

1. *Variation* of inherited characteristics occurs in a species and some of these variations will increase the chances of surviving in a particular environment.
2. Under certain selective pressures, members with favourable traits will have an increased chance of survival over others. This is known as *selection*.
3. Surviving members have an increased chance of reproducing and passing on their inherited traits to their offspring.
4. Over time and many generations, organisms will possess traits that are better suited to their environment and so increase their chances of survival.

natural selection process in which organisms better adapted for an environment are more likely to pass on their genes to the next generation

FIGURE 3.8 How selection and variation influence natural selection and evolution

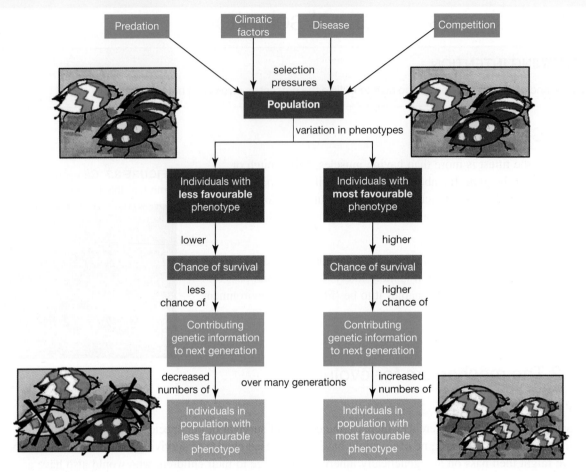

3.3.3 Variation

The theory of natural selection starts with the observations that
- more individuals are produced than their environment can support
- individuals within populations are usually different from each other in some way — they show variations.

According to this theory, some of these variations will provide an increased chance of survival over other variations within a particular environment. In other words, some variations will provide individuals with a competitive advantage.

Favourable variations

Individuals that possess a favourable variation (or phenotype) will have an increased chance of reproducing and passing on this variation (through their genes) to their offspring, such as the frog in figure 3.9. If a predator comes into this environment, it would have difficulty finding and eating the frog. Inheritance of this variation (or phenotype) will also increase the chances of survival of the offspring and, hence, the possibility that they will contribute their genes into the next generation. Over time and many generations, if this variation continues to provide a selective advantage, the number of individuals within a population that show the favourable variation will increase.

FIGURE 3.9 This frog is able to camouflage and hide from predators.

Less favourable variations

Individuals that have less favourable variations or are not as well suited to their environment (such as the frog in figure 3.10) will not be able to compete as effectively. For example, if a predator that eats frogs comes into the environment, this frog will be easily spotted. They may die young or produce few or no offspring. Therefore, they will have a limited contribution to the gene pool of the next generation. This will lead to a decrease in the number of individuals with that particular variation within the population.

FIGURE 3.10 The extinct golden toad (*Incilius periglenes*) could not survive changing climates and possible infections.

 **Resources**

▶ **Video eLessons** Camouflage (eles-2702)
Natural selection in peppered moths (eles-4233)

3.3.4 Selection

Organisms live within ecosystems, which are made up of various living (biotic) and non-living (abiotic) factors. These factors contribute to selecting which variations provide the individual with an increased chance of surviving. It is for this reason that these factors may be referred to as **selective pressures** or **selective agents**.

selective pressures factors that contribute to selecting which variations will provide the individual with an increased chance of surviving over others

selective agents the different living (biotic) and non-living (abiotic) agents that influence the survival of organisms

> Selective pressures determine which variations provide a selective advantage and which have a disadvantage. They are a driving force for natural selection. These pressures may be biotic or abiotic.

Biotic factors that may act as selective agents include:
- predators
- disease
- competitors
- prey
- mating partners.

Abiotic factors that may act as selective agents include:
- temperature
- shelter
- sunlight
- water
- nutrients.

Competition

Individuals within a population compete with each other for resources such as food, shelter or mates. Those with a selective advantage over other individuals are better able to compete for the resource.

FIGURE 3.11 Predators are a selective agent for their prey; prey that are faster or better at hiding have an advantage.

FIGURE 3.12 Can you suggest different selective pressures that may affect these organisms? What features might give them an increased chance of survival?

There may be situations in which competing is not about resources, but about competing to not be eaten by a predator or killed by a particular disease. In this case, individuals with a particular variation that reduces their chance of being eaten or killed will have a higher chance of survival. For example, organisms that can better camouflage in their environments have a higher chance of survival than those with variations that stand out to predators. In other cases, some organisms may have genes that provide resistance to a certain disease, whereas others variants are killed.

3.3.5 Isolation

int-5877

In 1835, Charles Darwin's expedition on the HMS Beagle saw him travel to the Galapagos Islands, an archipelago consisting of 13 major islands and 7 smaller ones (127 islands, islets and rocks in total). Here, he observed that species of finches had unique beak shapes and believed they originated from a mainland species.

How did this difference occur? Geographical isolation allowed the finches to develop adaptations specific to their environment and food sources. Because they could only breed with finches on their island, they were able to evolve and become different.

Other types of isolation that can occur include:
- Temporal isolation – when two or more species reproduce at different times.
- Behavioral isolation – when there is a difference in mating rituals.
- Ecological isolation – when two species, who live in different areas, do not breed because they are isolated from each other.

FIGURE 3.13 The finches of the Galapagos Islands look different from each other due to geographical isolation and the food sources available.

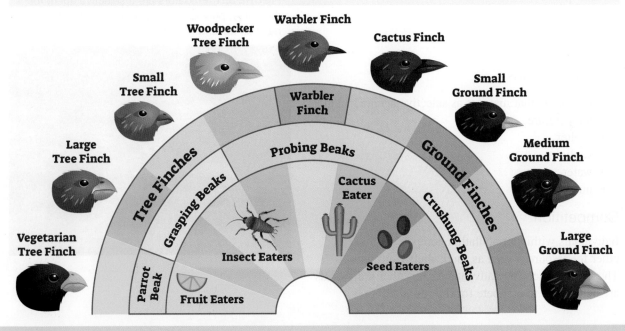

3.3.6 Tales of resistance

Many mutations are harmful to the organism and decrease its chances of survival. Some, however, may actually increase chances of survival.

> If a mutation results in a characteristic that gives the organism an increased chance of survival, it is more likely that the organism will survive long enough to reproduce and pass this trait on to the next generation.

If the organism's offspring inherit the genetic information for this new 'increased survival' trait, over time an increased proportion of the population may possess this trait.

This is the way in which populations of organisms can become resistant to methods that humans have used to kill them or control their population sizes. Those individuals within the population with the mutation that confers resistance to the control method live long enough to produce offspring who also possess the resistant characteristic. Over time, future generations are likely to contain increased numbers of individuals with resistance against the control method, making it no longer effective.

Resistant bacteria

Mutation is the key source of genetic variation in asexually reproducing organisms. Unless a mutation occurs, organisms that reproduce asexually produce clones of each other and are genetically identical. Errors or changed sequences in their DNA during cell division can be the source of new alleles (alternative forms of a gene).

FIGURE 3.14 Mutations in bacteria can result in some individuals having resistance against antibiotics. When these bacteria reproduce, their offspring also show antibiotic resistance.

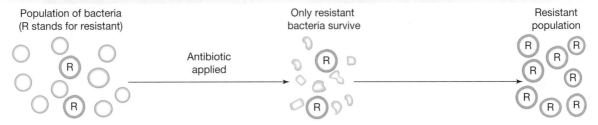

Prior to the discovery and use of penicillin, many people died from a variety of infections that can currently be treated with antibiotics. Penicillin and other antibiotics are drugs that can kill or slow the growth of bacteria. Referred to as 'magic bullets', these drugs revolutionised medicine. Although still widely used today, they are not as effective as they once were. Many bacteria have evolved to be resistant to them.

CASE STUDY: Multidrug-resistant tuberculosis

Multidrug-resistant tuberculosis (MDR TB) is a type of tuberculosis (TB). TB is a severe disease that is one of the world's greatest killers. Caused by the bacterium *Mycobacterium tuberculosis*, TB is transmitted through the air and usually targets the lungs of its hosts. MDR TB is of even more concern because it can result from being infected with bacteria that are resistant to two of the most effective drugs (isoniazid and rifampicin) used to treat TB. As well as working on the development of new TB treatments, scientists are focusing on improving communicable disease control programs.

FIGURE 3.15 *Mycobacterius tuberculosis* is a rod-shaped bacterium that causes tuberculosis.

Resistant bugs and bunnies

Variations due to mutation can also be advantageous in sexually reproducing organisms. The mutation may lead to a resistance to a particular pesticide, such as:

- insects developing resistance to the pesticide DDT (see figure 3.16)
- resistances to a particular viral disease, as in the case of some rabbits being resistant to the myxomatosis virus that was used to try to control their populations.

If some individuals within the population have a mutation that enables them to survive and reproduce, they may pass this trait on to their offspring, which will also be resistant. An increase in the number of organisms within the population resistant to the pesticide or virus reduces the effectiveness of the control method.

FIGURE 3.16 Mutations in flies can result in some individuals having resistance against a particular pesticide. When these flies reproduce, their offspring also show pesticide resistance.

 Resources

Weblinks   Multidrug-resistant tuberculosis
Multidrug resistant tuberculosis in Australia and our region
Resistance in mosquitoes

3.3.7 Structural and physiological adaptations of First Nations Australians

Genetic studies and archaeological data indicate that modern humans colonised Australia around 65 000 years ago. One scientific theory suggests First Nations Peoples descended from Africa around 70 000 years ago but First Nations beliefs assert that they came from this Earth 65 000 years ago. Archaeological evidence including stone tools, rock art, charcoal deposits and human skeletal fossils provide information about the age and complex nature of the cultures of First Nations Australians.

The human fossils of the people living in Australia between 40 000 and 10 000 years ago suggest that they had much larger bodies and showed great variation in their physical structure. For example, evidence suggests that First Nations Australians living in desert areas are able to tolerate sub-zero temperatures without an increase in metabolic activity.

INVESTIGATION 3.2

Investigating some of the structural and physiological adaptations of First Nations Australians to the Australian environment

Aim

To gain understanding about the structural and physiological adaptations of First Nations Australians to their environment.

Materials

Internet access, computer.

Method

1. Research some structural and physiological adaptations of First Nations Australians to the Australian environment.
2. Record your data in a table.

Discussion

Describe the structural variations in First Nations Australians suggested by your research.

Conclusion

Summarise the findings of this investigation about the structural and physiological adaptations of First Nations Australians to their environment.

ACTIVITY: Creating a natural selection cartoon

Search the internet for cartoons and simulations on natural selection. Then use the best ideas to create your own cartoon, comic strip, picture story book or animation.
a. Design and construct your own organism.
b. Give this organism a name and describe the environment in which it lives.
c. Use this organism as the common ancestor for four other variations of the organism.
d. Construct each variation, giving each a name and describing how the variation increases its chances of survival in its environment.

 Resources

 eWorkbook Natural selection (ewbk-12123)

3.3 Activities

learnon

| 3.3 Quick quiz   on | 3.3 Exercise |

Select your pathway

■ LEVEL 1	■ LEVEL 2	■ LEVEL 3
1, 4, 6, 11	2, 3, 7, 9, 13	5, 8, 10, 12, 14

These questions are
even better in jacPLUS!
- Receive immediate feedback
- Access sample responses
- Track results and progress

Find all this and MORE in jacPLUS ▶

Remember and understand

1. **a.** State whether the following statements are true or false.

Statements	True or false?
i. Natural selection is about being better suited to a particular environment and having an increased chance of surviving long enough to be able to have offspring that will take your genes into the next generation.	
ii. The mechanism for evolution is natural selection.	
iii. Individuals within a population that have fewer favourable variations or that are not as well suited to their environment will not be able to compete as effectively as those with more favourable variations.	
iv. Biotic and abiotic factors within ecosystems may contribute to selecting which variations provide the individual with an increased chance of surviving, and may be referred to as selective pressures or selective agents.	
v. Mutation is the key source of genetic variation in asexually reproducing organisms.	
vi. Mutations in bacteria can result is some individuals having resistance against antibiotics, but when these bacteria reproduce, their offspring will not show antibiotic resistance.	

   b. Rewrite any false statements to make them true.
2. **a.** Identify three resources for which individuals within a population may compete.
   b. Identify three examples of **(i)** biotic and **(ii)** abiotic selective pressures or selective agents.
3. Explain how variation within populations increases the chances of survival of the species.
4. Describe what is meant by the term *natural selection*.
5. Outline the link between natural selection and evolution.
6. Describe a link between mutation, variation and resistance to a pesticide or antibiotic.
7. Using diagrams, explain how bacterial resistance may rapidly increase in a population of bacteria.

Apply and analyse

8. A population of bears has natural variation in the thickness of their fur. The average temperature in the area in which the bears live dropped by 12 °C due to changed environmental conditions. Over time, the number of bears with thicker fur increased in the population.
   a. What was the selective pressure in this scenario?
   b. Explain the process that allows the bears with thicker fur to increase in number.
9. **SIS** Myxomatosis (caused by Myxoma virus) was used as a method to control rabbit populations in Australia. However, it is no longer effective. Suggest reasons for the ineffectiveness of a previously effective method of control.
10. Treatments are no longer effective for several infectious diseases (for example, MDR-TB and golden staph) because the pathogens (for example, bacteria) that cause them are resistant to it. Find out more and report on an example of this.

Evaluate and create

11. Construct a flowchart to describe natural selection using the following terms.
- Predation
- Competition
- Variation in phenotypes
- Climatic factors
- Less favourable
- Disease
- Over many generations
- Genetic contribution to next generation
- Decreased numbers
- Increased numbers
- Selection pressures
- Population
- Lower chance
- Survival
- Phenotype
- Individuals
- Higher chance
- Most favourable

12. DDT was a pesticide used to kill mosquitoes. It is no longer effective and has caused some unexpected environmental and ecological issues.
- **a.** Research the pesticide DDT and its history and use, and outline your findings.
- **b.** Identify ecological and environmental concerns about the use of DDT.
- **c.** Suggest (and, if possible, discuss) reasons for the gradual decrease in DDT's effectiveness as a pesticide.

13. Penicillin is a very effective antibiotic against a number of different types of bacteria.
- **a.** Research penicillin and describe its history and its use.
- **b.** Outline reasons some bacteria are now resistant to penicillin.
- **c.** Explain how this resistance may affect humans.
- **d.** Should we still use penicillin to treat bacterial infections? Justify your response.

14. **SIS** The English peppered moth, *Biston betularia*, rests on tree trunks during the day. Prior to 1850, this species had a speckled pale grey colour that effectively camouflaged it from predators as it rested on the pale lichen-covered trunks. In about 1850, a black version of this moth appeared. By 1895, these black moths made up about 98 per cent of the population.
- **a.** Research this species of moth. Outline your findings.
- **b.** Identify the source of the new variation.
- **c.** Investigate and describe what was happening in England between 1850 and 1895 that may have had an impact on the survival of these moths.
- **d.** Suggest why and how the number of black moths in the population increased so dramatically.

Fully worked solutions and sample responses are available in your digital formats.

LESSON
3.4 Biodiversity as a function of evolution

LEARNING INTENTION

At the end of this lesson you will be able to provide examples of biodiversity and variation. You should also be able to describe causes of this and explain their relevance to the survival of a species.

3.4.1 It's great to be different

Look at the dogs in the photograph in figure 3.17. What differences can you see? How did these come about when all the individuals belong to the same species?

Variation in humans

Have a look at the people around you. How many differences do you notice? How can you explain your observations? One part of your response might deal with genetics and inherited traits; another part might deal with the environment. The **variation** of characteristics or phenotypes within populations has contributed to the survival of our species.

> **variations** differences between cells or organisms

FIGURE 3.18 Lots of variation exists between individuals.

3.4.2 Genetic diversity

Biodiversity (or biological diversity) has to do with variation within living things. Biodiversity can exist on a number of different levels. For example, it can be described in terms of an **ecosystem**, at the level of species, or even at the level of individual genes.

Species diversity is the number of different species within an ecosystem.

In contrast, **genetic diversity** is the range of genetic characteristics within a single species. The most important level in terms of evolution is that of the gene. Variation at a genetic level is an important factor that allows for evolution to occur.

ecosystem community of living things that interact with each other and with the environment in which they live

species diversity the number of different species within an ecosystem

genetic diversity variation in genes between members of the same species

Causes of genetic diversity

Genetic diversity is important because it codes for variations of phenotypes, some of which may better suit the individual organism to a particular environment than others, giving it an increased chance of survival. If this individual survives, the chance of it reproducing to pass the advantageous gene to its offspring is increased, also giving the offspring an increased chance of surviving. Overall, this genetic advantage will increase the survival of the species within that particular environment. So, how does this genetic diversity come about?

FIGURE 3.19 Causes of variation

```
                          Causes of variation
                                  |
            ┌─────────────────────┴─────────────────────┐
      Within individual                            Within population
            |                                             |
   ┌────────┼────────┐                         ┌──────────┼──────────┐
Mutations  Meiosis  Gamete                  Gene flow  Random      Natural
                    combinations                       genetic     selection
                                                        drift
   |                    |                                  |            |
Changes in        Random joining                    Change occurs   Some variations
genetic material  of sperm and ova                  by chance       provide increased
                  at fertilisation                                   chance of survival
        ┌──────────┴──────────┐          ┌──────────┴──────────┐            |
  Independent           Crossing over:   Immigration:      Emigration:   Pass on genetic
  assortment:           exchanging       possible introduction  possible loss of  Over time,
  random assortment     parts of         of alleles into    alleles from   increase in
  of chromosomes        chromosomes      population         population    favourable variation.
```

3.4.3 Mutation

Mutation can occur in all organisms and is the source of new genetic variation. A change in the genetic code (DNA) can lead to a change in the protein that is coded for and produced by that segment of DNA. Examples of mutations were explored in lesson 2.6. This can change the organism's characteristics.

In figure 3.20, for example, a change in DNA has led to the production of a protein that changes the colour of a mouse from white to black. Mutations that occur in germline cells (such as sperm and eggs) are the source of new alleles (alternative forms of genes) within populations.

mutation change to DNA sequence, at the gene or chromosomal level

FIGURE 3.20 Differences in DNA and genes can lead to variation between individuals.

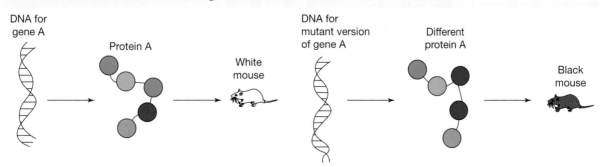

Adaptations

Variations that increase chances of survival may lead to new **adaptations**. An adaptation may be considered to be a special feature or characteristic that improves an organism's chance of survival in its environment.

Different types of adaptations are possible; for example:
- structural adaptations (such as hair to keep warm or the colour of a moth)
- behavioural adaptations (such as courtship display to attract a mate)
- physiological adaptations (such as the ability to produce concentrated urine to conserve water).

Can you think of adaptations that you possess that increase your chances of survival in your current environment?

on Resources

▶ **Video eLesson** Blue footed booby mating dance (eles-2701)

🔗 **Weblink** The five fingers of evolution

3.4.4 Variation between individuals

Variation between individuals that reproduce sexually may also be the result of several other factors besides mutation. Variation between individuals can be described in terms of **alleles** — the alternative forms of genes. These alleles can be passed on to future generations.

Variation can occur during **meiosis** due to:
- the independent assortment of the **chromosomes** into the **gametes** (figure 3.21)
- **crossing over** of sections of maternal and paternal chromosomes (figure 3.22).

These processes were introduced in lesson 2.5.

The combination of gametes that fuse together during fertilisation provides another source of variation, as does the selection of a particular mate.

FIGURE 3.21 Independent assortment and crossing over during meiosis are two causes of variation.

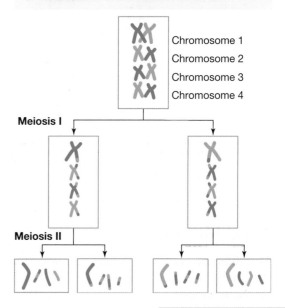

Chromosome 1
Chromosome 2
Chromosome 3
Chromosome 4

Meiosis I

Meiosis II

FIGURE 3.22 The process of crossing over

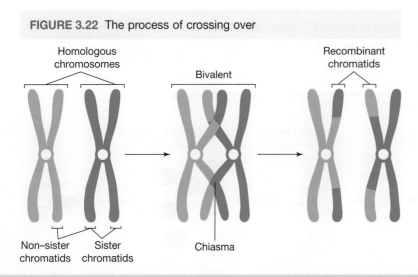

Homologous chromosomes

Bivalent

Recombinant chromatids

Non–sister chromatids

Sister chromatids

Chiasma

adaptation a feature that aids in the survival of an organism

alleles alternative forms of a gene for a particular characteristic.

meiosis cell division process that results in new cells with half the number of chromosomes of the original cell

chromosomes tiny thread-like structures inside the nucleus of a cell that contain the DNA that carries genetic information

gametes reproductive or sex cells such as sperm or ova

crossing over exchange of alleles between maternal and paternal chromosomes

3.4.5 Variation within populations

Genetic variation within populations can be referred to in terms of the frequency and number of different alleles within the population. While the genotype describes the combination of alleles for a particular trait within an individual, a **gene pool** describes the alleles for a particular trait within a population.

Within a population, gene mutations, sexual reproduction and **gene flow** may all promote genetic biodiversity or variation within a gene pool. **Genetic drift** (changes due to chance events) and natural selection (changes due to some organisms having a survival and reproductive advantage under certain pressures) can also have an impact on allele frequencies — and can result in changes that reduce biodiversity within populations.

gene pool the total genetic information of a population, usually expressed in terms of allele frequency

gene flow the movement of individuals and their alleles between populations

genetic drift changes in allele frequency due to chance events such as floods and fires

FIGURE 3.23 Examples of genetic drift and natural selection. **a.** In genetic drift, changes in the gene pool are due to chance events (in this case, all beetles had an equal chance of being stepped on). **b.** In natural selection, specific organisms have a trait that gives them an advantage against a certain selective pressure (in this case, orange beetles are more camouflaged).

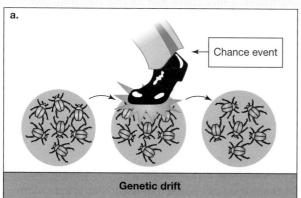

Genetic drift

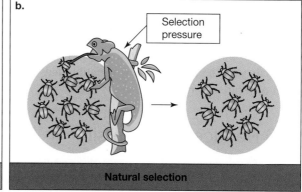

Natural selection

CASE STUDY: Frogs and gene pools

When individuals of a species of frog mate, they recombine their genetic material to produce offspring that show a wide variety of characteristics. Such variability within a species is important because it may enhance the chances of survival of the individual's offspring in a changing environment. Some individuals may have genes (or alleles) that assist in their survival. They may then pass these favourable genes on to their offspring. On the other hand, if a large number of frogs emigrated or were removed from a particular habitat without mating, their genes (or alleles) would be removed from the gene pool. Once removed, they are gone forever.

FIGURE 3.24 A species of frog might show a wide variety of characteristics; this variability enhances the chances of survival.

Gene flow

Movement of individuals between populations provides another possible source of diversity. **Emigration** (moving out) may result in the loss of particular alleles; **immigration** (moving in) may result in the addition of new alleles into the population (particularly when the individuals reproduce in the new population).

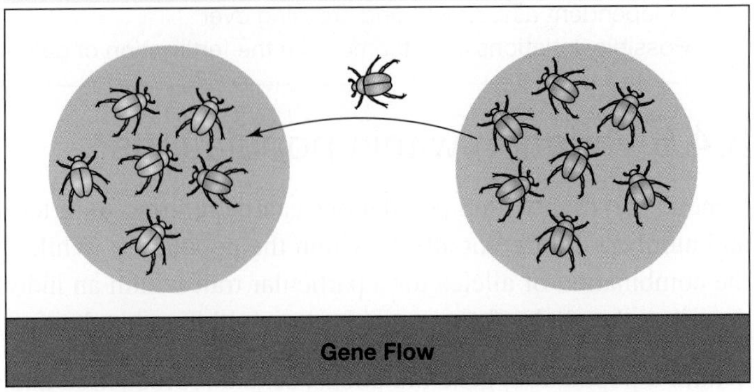

Before advances in technology provided humans with relatively easy long-distance travel, our species was split into small groups. The separate identities of these groups was maintained by geographical barriers such as mountains and oceans, and by attitudinal and social barriers.

FIGURE 3.25 Gene flow is the movement of individuals.

Gene Flow

With the advent of faster and more accessible means of transport and improved communication technologies, these barriers are now starting to break down, and migration and interbreeding between human groups is widespread.

Sometimes the variation introduced into a population is not beneficial. An inherited anaemic disease, thalassaemia, is common among people living along the Mediterranean coast. As people from this part of the world migrated to Australia, they brought with them the thalassaemia allele. For an individual to have thalassaemia, they need two parents who carry the allele. Therefore, an increasing number of people within the Australian population have this disease.

In other cases, the introduction of a genetic trait into a population may increase the chances of survival of individuals with the trait. This new variation may contribute to increasing the fitness of the population to the current or future environment.

3.4.6 Reduced biodiversity

The use of reproductive technologies such as artificial selection, artificial insemination, IVF and cloning has the potential to unbalance natural levels of biodiversity. These technologies can be used in horticulture, agriculture and animal breeding to select which particular desired characteristics will be passed on to the next generation.

Artificial selection

For thousands of years, humans have used selective breeding techniques to breed domestic animals and plants. We have selected which animals will mate together based on their possession of particular features, to increase the chances that their offspring will have features that suit our needs. This type of selective breeding is called **artificial selection**. Artificial selection is different from natural selection in that it is not random. Once the organisms with desired traits are selected, new varieties of the species of plants and animals are generated.

When fewer individuals are selected for breeding, the genetic diversity is reduced, and inbreeding may result. As well as decreasing variation in the traits of offspring, inbreeding can increase the chances of inherited diseases.

emigration a type of gene flow in which an individual leaves a population

immigration a type of gene flow in which an individual enters a population

artificial selection the process in which humans breed animals or plants in such a way to increase the proportion of desired traits

Artificial insemination

The sperm of a prize-winning racehorse may be used to inseminate many mares, increasing the chance of offspring that also possess the race-winning features of their father. This leads to a larger contribution of alleles from this horse than would naturally be possible. It can also lead to reduced genetic diversity within the populations of horses in which this occurs.

Artificial insemination is also used to impregnate dairy cows. Straws of semen collected from bulls can be tested or kept in liquid nitrogen to be stored for future use.

FIGURE 3.26 A liquid nitrogen bank filled with straws of frozen bull semen used for impregnating dairy cows

IVF and embryo screening

In-vitro fertilisation (IVF) techniques allow the testing and selection of embryos for particular characteristics prior to their implantation. This can also have an impact on genetic diversity. Imagine the effect of implanting only female embryos or only those with a particular recessive trait.

DISCUSSION

In small groups, discuss the theoretical implications for human populations if humans were able to allow for the selective implantation of embryos that:

a. are male
b. possess recessive traits such as red hair or blue eyes
c. have a potential for a higher IQ
d. have a potential for paler skin.

Cloning

Imagine the production of a population of genetically identical individuals. Although they may be well suited to a particular environment, what might happen when the environment changes to one that they are not suited to? Cloning can occur in many ways, as outlined in topic 2. (Refer to section 2.12.4 for some examples of this, including information about the cloning of the first mammal, Dolly the sheep.)

FIGURE 3.27 Cloning in plants

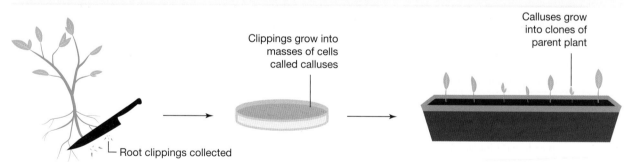

Clippings grow into masses of cells called calluses

Calluses grow into clones of parent plant

Root clippings collected

Consequences of reduced biodiversity

Reducing variation in genetic diversity can lead to the eradication of populations or entire species. This is known as **extinction**. If the population or species is exposed to an environmental change or threat (for example, disease, climate change or lack of a particular resource), the reduced variation may mean that there is less chance that some of the species will survive to reproduce.

> **extinction** complete loss of a species when the last organism of the species dies

SCIENCE AS A HUMAN ENDEAVOUR: Seeds of inheritance

Gregor Mendel (1822–1884), an Austrian monk, used peas of different colours and shapes in his experiments and is responsible for the development of the fundamentals of the genetic basis of inheritance (see lesson 2.8 for more information).

His research into inheritance patterns began in 1856 and ended in 1865, when he presented the results of his experiments with nearly 30 000 pea plants.

Having such a large data set can validate scientific findings by allowing trends to be more easily seen and analysed more effectively and efficiently. It is also likely to result in a smaller margins of error.

Would Mendel's theory of inheritance have the same effect if he only had 10 plants? Can your scientific findings be validated if you have only completed the experiment a few times?

FIGURE 3.28 Mendel conducted his studies with edible peas.

On Resources

📋	**eWorkbook**	Variation, biodiversity and adaptations (ewbk-13081)
🔗	**Weblinks**	Zoos Victoria case study: the Eastern Barred Bandicoot How we'll resurrect the Gastric Brooding frog and the Tasmanian tiger

3.4 Activities

learn on

3.4 Quick quiz on	3.4 Exercise

Select your pathway

■ LEVEL 1 1, 4, 6, 9, 11	■ LEVEL 2 2, 3, 7, 10, 13	■ LEVEL 3 5, 8, 10, 12, 14, 15

These questions are even better in jacPLUS!
- Receive immediate feedback
- Access sample responses
- Track results and progress

Find all this and MORE in jacPLUS ▶

Remember and understand

1. a. State whether the following statements are true or false.

Statements	True or false?
i. Biodiversity (or biological diversity) has to do with variation within living things.	
ii. The similarity of characteristics or phenotypes within populations has contributed to the survival of our species.	
iii. The use of reproductive technologies such as artificial selection, artificial insemination, IVF and cloning has the potential to select which particular characteristics will be passed on to the next generation, and hence may result in reduced biodiversity.	
iv. In artificial selection, because fewer individuals are selected for breeding, genetic diversity is increased.	
v. Inbreeding may result in increased variation in traits of offspring and decreased chances of inherited diseases.	
vi. Reduced variation in genetic diversity increases the chances of survival of the species.	

b. Rewrite any false statements to make them true.

2. Match the type of diversity to its description in the following table.

Type of diversity	Description
a. Biodiversity	**A.** Diversity in ecosystems
b. Ecosystem diversity	**B.** A range of genetic characteristics within a single species
c. Genetic diversity	**C.** The number of different species within an ecosystem
d. Species diversity	**D.** The total variety of living things on Earth, their genes and the ecosystems in which they live

3. Select the appropriate word to complete the following sentences.
   - meiosis
   - protein
   - sperm
   - independent assortment
   - phenotypes
   - alleles
   - sex
   - crossing over
   - DNA
   - increasing

   a. Genetic diversity is important because it codes for variations of _____, some of which may better suit the individual organisms to a particular environment than others, _____ its chances of survival.

   b. A change in the genetic code in _____ can lead to a change in the _____ that is coded for and produced by that segment of DNA, which can change the organism's characteristics.

   c. Mutations that occur in _____ cells (such as _____ and eggs) are the source of new _____ (alternative forms of genes) within population.

   d. Variation between individuals that reproduce sexually can also occur during _____ due to _____ of sections of maternal and paternal chromosomes, and also due to the _____ of the chromosomes into gametes.

4. Match the cause of variation within populations with its description in the following table.

Cause of variation	Description
a. Emigration	**A.** Change in variation that occurs by chance
b. Gene flow	**B.** Individuals move into population — possible gain of alleles into population
c. Immigration	**C.** Individuals move out of population — possible loss of alleles from population
d. Natural selection	**D.** Movement of individuals between (in or out of) populations
e. Random genetic drift	**E.** Where some variations provide individuals with increased chances of survival so they survive and pass genetic information on to offspring, meaning over time, the number of those in population with the favourable variation increases

5. Identify each of the following examples as either **(i)** behavioural adaptation, **(ii)** physiological adaptation or **(iii)** structural adaptation.
   a. Hair to keep warm
   b. Courtship display to attract a mate (as pictured)
   c. Ability to produce concentrated urine to conserve water

Apply and analyse

6. **a.** Explain why genetic diversity is important to the survival of a species.
   b. Predict what will happen if a species does not possess the genes that allow it to adapt to new environmental conditions.

7. Describe and provide examples of sources of genetic variation for the following.
   a. An individual
   b. A population

8. **a.** Outline the relationship between mutation and genetic variation.
   b. Explain why mutations are important to asexually reproducing organisms.
   c. Are all mutations bad? Justify your response.

9. Distinguish between the following pairs of terms.
   a. Genetic diversity and species diversity
   b. Immigration and emigration
   c. Alleles and the gene pool
10. a. Describe how IVF and related technologies could reduce human genetic diversity.
    b. Discuss the advantages and disadvantages of artificial insemination.

Evaluate and create

11. a. Describe the role of meiosis in generating genetic variation.
    b. Draw a diagram or create a model to summarise this.
12. a. Describe how humans have achieved artificial selection. Outline two clear examples.
    b. State the desired outcome of artificial selection.
    c. Predict possible consequences of artificial selection on genetic diversity. Justify your response.
13. **SIS** a. List the differences you can see among different breeds of dogs.
    b. Explain how these variations likely came about.
    c. Find out more about artificial selection in dogs and how it is currently used in Australia. Construct a graph to display your findings.

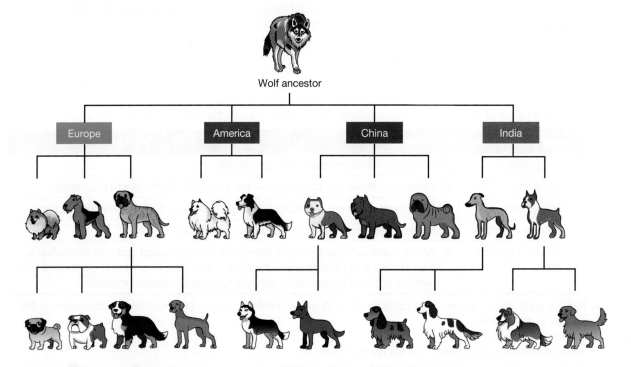

14. *Brassica oleracea* is a common ancestor to a number of vegetables.
    a. Identify differences between *Brassica oleracea* and each of the species in the following figure.
    b. Explain how these differences came about.
    c. All of these vegetables were produced by artificial selection and share a common ancestor. Could this also have happened by natural selection? Justify your response.

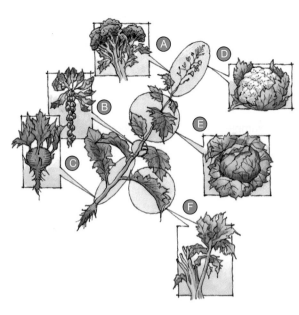

(A) Broccoli (inflorescence) (*Brassica oleracea* var. *cymosa*)

(B) Brussels sprouts (lateral buds) (*Brassica oleracea* var. *gemmifera*)

(C) Kohlrabi (stem) (*Brassica oleracea* var. *gonglyoides*)

(D) Cauliflower (flower) (*Brassica oleracea* var. *botrytis*)

(E) Cabbage (terminal buds) (*Brassica oleracea* var. *capitata*)

(F) Kale (leaf) (*Brassica oleracea* var. *acephala*)

15. **SIS** Read the article and answer the following questions.

Bacteria don't sunbake!

The ozone layer helps to filter out many harmful UV rays. However, in Precambrian times, over 540 million years ago, there was no ozone layer. How then did life survive? To answer this question, a team of biologists from NASA's Ames Research Centre in California has been studying microbial mats of bacteria and blue–green algae, the types of organisms that lived in Precambrian times.

The natural production of DNA in each organism was studied by placing some of the mat into a plastic bag that was transparent to UV light. Some radioactive phosphate was added into the bag. (Phosphate is used by cells to produce DNA.) Every few hours, the amount of phosphate in the DNA of some of the cells was measured. The results for both bacteria and blue–green algae showed the same pattern of DNA production. At sunrise, the amount of phosphate in the DNA was high. DNA production then ceased at noon for three to six hours, resuming just before sunset. Photosynthesis occurred throughout the whole day.

Lynn Rothschild, head of the research team, concluded that the cells cease DNA production at noon because of the harmful UV light. The cells use this time to repair any DNA damage before they begin to divide again. This mechanism might give some unicellular organisms a natural advantage if the Earth's ozone layer continues to be destroyed.

  a. What question were the researchers trying to answer?
  b. Describe the experiment they set up.
  c. Suggest improvements to their experiment design.
  d. What were their results? What were their conclusions?
  e. Often, the DNA damage caused by harmful UV light can lead to mutation. Are all mutations likely to affect the survival of an organism and its offspring? Explain your response.
  f. What implications do these conclusions have for life on Earth if the ozone layer continues to break down?
  g. Research and report on one of the following.
    • Australian research on the ozone layer
    • UV radiation and skin cancer
    • The ozone depletion and our health

Fully worked solutions and sample responses are available in your digital formats.

LESSON
3.5 Genetics and evolution

LEARNING INTENTION

At the end of this lesson you will be able to describe and provide examples of speciation, divergent evolution, convergent evolution and coevolution.

3.5.1 The formation of new species

Variation, struggle for survival, selective advantage and inheritance of advantageous variations form the basis for the theory of evolution by natural selection. They also provide an explanation for how new species arise. New species are continually arising. In 2018 alone, more than 200 new species were discovered. The formation of new species is called **speciation**. Organisms are classified as different species when they are unable to produce fertile and viable offspring through reproduction.

The process of speciation

1. Variation exists in the initial population.
2. A barrier separates an initial population into two or more smaller populations that cannot interbreed (reproductive isolation).
3. Each population is exposed to different selection pressures, resulting in the selection of different variations or phenotypes (through natural selection).
4. Over time, the populations may become so different that even if they were brought back together, they might not be able to produce fertile offspring.

FIGURE 3.29 Speciation is a process by which a new species develops from another.

| Variation of characteristics is present in a population. | The breeding population becomes isolated. | Different characteristics arise through random genetic drift, mutation and environmental pressures. | The environment changes. Through selection, some characteristics are favoured over others. Those best suited to the environment survive. | Survivors reproduce and pass on favourable genes and features to offspring. | The frequency at which the genes for the new characteristics appear increases. | The isolated population is now quite different, producing a new species. |

Examples of barriers

Barriers dividing species come in many different forms. The easiest way to visualise this is often through geographical barriers, such as being separated by a mountain or a stream (as shown in figure 3.30).

speciation formation of new species

FIGURE 3.30 During speciation, populations are exposed to different selective pressures.

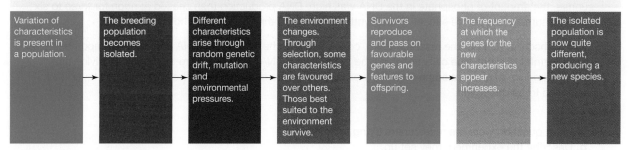

Single population — one species

A geographic barrier separates population into two

Isolated populations subjected to different selection pressures; different phenotypes favoured

When populations come together again, they can no longer interbreed — two species

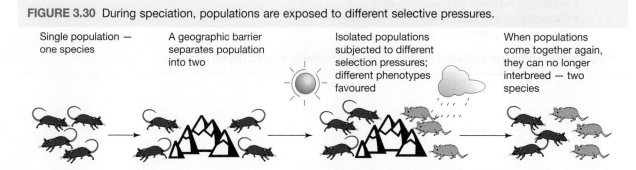

3.5.2 Patterns of evolution

Phyletic evolution occurs when a population of a species progressively changes over time to become a new species (this is much less common).

Branching evolution or **divergent evolution** is more common; in this case, a population is divided into two or more new populations that are prevented from interbreeding.

Other patterns of evolution can also be observed. These are not mechanisms of speciation, but rather ways in which species evolve similar traits due to similarities in selective pressures. Examples are convergent evolution and coevolution:

- **Convergent evolution** occurs when different organisms develop traits with similar functions due to being in similar selective pressures. These traits evolve independently and are not seen in a common ancestor.
- **Coevolution** occurs when the evolution of one organism occurs in response to changes in another organism.

Divergent evolution

Divergent evolution is a type of evolution in which new species evolve from a shared ancestral species.

At some point in history a barrier has divided the population into two or more populations. This barrier has interfered with interbreeding between the populations, causing them to become reproductively isolated. This process is shown in figure 3.31, and is a mechanism of speciation.

Adaptive radiation

Adaptive radiation is said to have occurred when divergent evolution of one species results in the formation of many species that are adapted to a variety of environments. Darwin's finches (figure 3.32) and Australian marsupials (figure 3.33) are two examples. Australian marsupials are thought to have evolved from a common possum-like ancestor.

Homologous structures

Species that share a common ancestor often have characteristics that result from this. These structures are known as homologous structures. **Homologous structures** provide important evidence for evolution, and will be further explored in section 3.6.3.

allopatric speciation a type of speciation where populations are separated by a geographical barrier

sympatric speciation a type of speciation where populations are separated by a barrier that is not geographical

phyletic evolution when a population of a species progressively changes over time to become a new species

divergent evolution when a population is divided into two or more new populations that are prevented from interbreeding

convergent evolution tendency of unrelated organisms to acquire similar structures due to similar environmental pressures

coevolution the process in which two species evolve in partnership so that they depend on each other

adaptive radiation when divergent evolution of one species results in the formation of many species that are adapted to a variety of environments

homologous structures body structures that can perform a different function but have a similar structure due to their evolutionary origin

FIGURE 3.31 How new species come about as a result of divergent evolution.

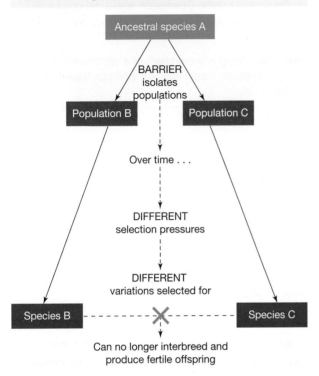

Ancestral species A

BARRIER isolates populations

Population B   Population C

Over time . . .

DIFFERENT selection pressures

DIFFERENT variations selected for

Species B   Species C

Can no longer interbreed and produce fertile offspring

FIGURE 3.32 Darwin's finches are examples of adaptive radiation.

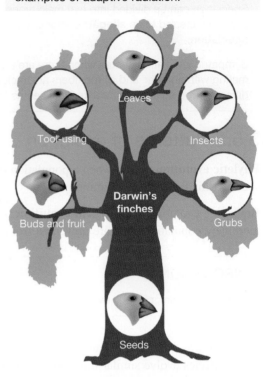

Leaves

Tool-using   Insects

Darwin's finches

Buds and fruit   Grubs

Seeds

FIGURE 3.33 Marsupials show adaptive radiation because they have evolved from a common ancestor but, due to different selection pressures, different characteristics have been selected.

Resources

▶ **Video eLesson** These finches are very different but have a common ancestor (eles-2919)

It is important to note that homologous structures are as a result of divergent evolution. These structures (for example, the forelimb in various organisms) have similar structures. However, they have evolved from a common ancestor to have different functions, as a result of species adapting to different niches within ecosystems.

Convergent evolution

Convergent evolution is the opposite to divergent evolution. While it is also an evolutionary process, it does not result in the formation of new species and is therefore not a mechanism of speciation.

FIGURE 3.34 Convergent evolution

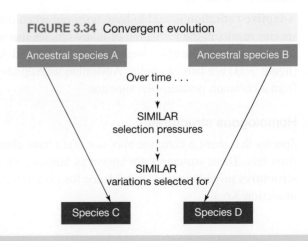

Ancestral species A   Ancestral species B

Over time . . .

SIMILAR selection pressures

SIMILAR variations selected for

Species C   Species D

FIGURE 3.35 **a.** The Australian echidna, **b.** African aardvark, **c.** South-East Asian pangolin and **d.** South American anteater share similar features. These features were selected for because they gave a selective advantage in obtaining food supply in their environment, rather than because of a recent common ancestry.

Convergent evolution is the result of similar selection pressures in the environment selecting for similar features or adaptations. These adaptations have not been inherited from a common ancestor, but rather have evolved independently.

Figure 3.35, for example, shows four species. All have snouts suited to eating similar food, but none of them are related.

Analogous structures

Unrelated species living in very similar environments (with similar selection pressures) in different parts of the world have evolved similar structures. For example, the fins of a dolphin and a shark (seen in figure 3.36), or the wings of a bat and a butterfly are the result of convergent evolution. Structures that perform the same role but have different evolutionary origins are called analogous structures.

FIGURE 3.36 Convergent evolution can be seen in the fins of sharks, wings of penguins and flippers of dolphins.

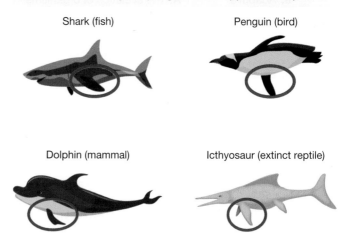

Shark (fish)　　　　　　Penguin (bird)

Dolphin (mammal)　　　　Icthyosaur (extinct reptile)

3.5 Activities

learn on

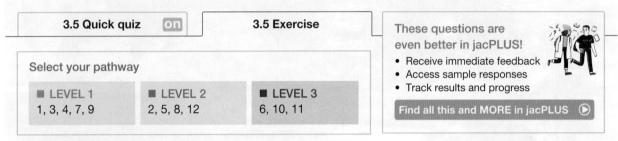

Remember and understand

1. **a.** State whether the following statements are true or false.

Statements	True or false?
i. Australian marsupials show adaptive radiation because they have evolved from a common ancestor but, due to different selection pressures, different characteristics have been selected.	
ii. Kangaroos and koalas belong to the same species.	
iii. Convergent evolution can result from different selection pressures in the environment selecting for different variations in evolution from a common ancestor.	
iv. Divergent evolution can result from similar selection pressures in the environment selecting for similar features or adaptations that have not been inherited from a common ancestor.	

   b. Rewrite any false statements to make them true.
2. State four key ideas that formed the basis of the theory of evolution by natural selection.
3. Match the term to its most appropriate description in the following table.

Term	Description
a. Adaptive radiation	**A.** The evolution of organisms in response to each other, such as plants and their pollinators
b. Coevolution	**B.** The result of different selection pressures in the environment selecting for different variations in evolution from a common ancestor
c. Divergent evolution	**C.** When a population of a species progressively changes over time to become a new species
d. Phyletic evolution	**D.** When divergent evolution of one species results in the formation of many species that are adapted to a variety of environments

4. **a.** Distinguish between the terms *speciation* and *evolution*.
   b. Describe how a new species can be formed.
5. **a.** Compare and contrast *divergent evolution* and *convergent evolution*.
   b. Describe an example of divergent evolution.
   c. Describe an example of convergent evolution.

6. **a.** Outline the relationship between *adaptive radiation* and *divergent evolution*.
   b. Describe an example of adaptive radiation.
7. Use flowcharts to describe the following.
   a. Divergent evolution
   b. Convergent evolution

Apply and analyse

8. **SIS** The following figures show two species of North American hares that are closely related and share a common ancestor. The snowshoe hare, *Lepus americanus* (left), lives in northern parts of North America where it snows in winter. The black-tailed jack rabbit, *Lepus californicus* (right), lives in desert areas.

   a. Identify differences between these animals.
   b. Outline reasons for these differences.
   c. Describe how these differences came about.
   d. Is this an example of convergent evolution or divergent evolution? Justify your response.

9. Identify where each of the following figures belong in the convergent evolution table provided.

| Numbat | Bobcat | Spotted cuscus | Flying squirrel |
| Sugar glider | Lemur | Anteater | Spotted-tailed quoll |

TABLE Examples of different placental mammals and Australian marsupials

Niche	Placental mammal	Australian marsupial
Anteater		
Cat		
Climber		
Glider		

10. **a.** Carefully examine each of the pairs of organisms i. to iv. as shown. Identify whether the following features of each pair is an example of convergent evolution or divergent evolution.
  i. Fins of the dolphin and shark
  ii. Nose of the numbat and anteater
  iii. Shape of the sea dragon and seahorse
  iv. Wing structure in the European goldfinch and pine siskin (finch)
b. Justify your responses.

Evaluate and create

11. **SIS** The following figure shows how one ancestral species can undergo evolution and give rise to a number of new species.

    a. Select a species that is currently alive on Earth.

    b. Use the internet and other resources to find information about other species this organism is related to.

    c. How long ago was your selected species thought to have diverged from each species you listed?

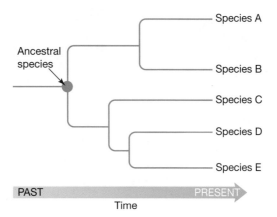

12. Honeycreepers are found only in the Hawaiian Islands and share a common ancestry. Examples of four species of honeycreepers are shown in the following figures.

    a. Suggest reasons for their different appearance.

    b. Create a flowchart outlining the events that may have occurred to make honeycreepers look so different.

Fully worked solutions and sample responses are available in your digital formats.

LESSON
3.6 Evidence of the theory of evolution

LEARNING INTENTION

At the end of this lesson you will be able to describe how the fossil record and biogeography provide evidence to support the theory of evolution.

3.6.1 Biogeography

A long time ago, long before humans inhabited Earth, the continents were joined together. If you could travel back in time 10 million years, not only would the continents look different, but life on Earth would also be very different from what it is today.

Earth's age is estimated to be around 4.6 billion years. Geologists have constructed a geological timeline that divides this time into five **eras**, some of which are further divided into **periods**. This timeline, with its divisions and information from fossil records, is shown in figure 3.38.

> **eras** divisions of geological time defined by specific events in the Earth's history, divided into periods
>
> **periods** subdivisions of geological time, divided into epochs

FIGURE 3.37 Dinosaurs roamed Earth millions of years ago.

FIGURE 3.38 The fossil record provides an incomplete picture of life as it has existed on Earth.

Millions of years ago	Era	Period	Approximate origin of life forms	Approximate occurrence of other key events
Present	Cenozoic	Quaternary		
50	Cenozoic	Tertiary		
100	Mesozoic	Cretaceous		South America separates from South Africa
200	Mesozoic	Jurassic		
200	Mesozoic	Triassic		
300	Palaeozoic	Permian		Pangaea moving apart
300	Palaeozoic	Carboniferous		Pangaea formed
400	Palaeozoic	Devonian		
	Palaeozoic	Silurian		
500	Palaeozoic	Ordovician		
	Palaeozoic	Cambrian		
600	Proterozoic	Precambrian		
700	Proterozoic	Precambrian		
800	Proterozoic	Precambrian		
900	Proterozoic	Precambrian		
1000	Proterozoic	Precambrian		
1500	Proterozoic	Precambrian		
2000	Proterozoic	Precambrian		Oxygen-rich atmosphere formed
2500	Proterozoic	Precambrian		
3000	Archaeozoic	Precambrian		Anaerobic atmosphere formed
3500	Archaeozoic	Precambrian		
4000	Archaeozoic	Precambrian		
4500	Archaeozoic	Precambrian		Earth and solar system formed

Life forms labelled in figure: Bacteria, Seaweeds, Fungi, Land plants, Seed plants, Single-celled animals, Worms, Crustaceans, Insects, Marine vertebrates, Land vertebrates, Birds, Primates

INVESTIGATION 3.3

4.6 billion years of history

Aim

To construct a timeline of the history of Earth

Materials

- roll of toilet paper, cash register tape or similar
- pens or markers

Method

1. Use the roll of paper to create a timeline of the history of Earth. Begin by choosing an appropriate scale to represent the 4.6 billion years of history.
2. Indicate the events shown in figure 3.37 on your timeline.

Results

Create a copy of your timeline, outlining the events that you showed and the scale you used.

Discussion

1. A student was describing the evolution of life on Earth and wrote, 'For much of Earth's history, not much happened'. Is this statement justified?
2. Explain why a long roll of paper is necessary to construct this timeline.

Conclusion

Summarise the findings for this investigation about the history of Earth.

 Resources

 eWorkbook   Geological time (ewbk-12131)

 Weblinks   Geological time
The Pangaea pop-up TED talk

Fossil evidence

Fossil evidence suggests that the continents were once joined. Figure 3.39 shows some examples of how fossils provide this evidence.

Fossils of the land-dwelling dinosaur *Mesosaurus* (which lived about 270 million years ago) have been discovered in only two places in the world — the eastern side of South America and the western side of South Africa (as shown in figure 3.39). These continents are now separated by 6600 kilometres of ocean, but the fossil evidence suggests that the continents were once joined.

Fossil evidence also suggests that many of Australia's marsupials originated in South America.

EXTENSION: Microfossil evidence of life on Mars

Evidence that life may once have existed on Mars has been found inside a meteorite that landed in Antarctica. The meteorite has been dated at about 13 000 years old. Examination of thin slices of the meteorite under an electron microscope has suggested the presence of microfossils of single cells. This is significant because it is thought that life on Earth also started out as single cells.

FIGURE 3.39 Fossil patterns across continents

AFRICA

INDIA

Fossil evidence of the massive land reptile lystrosaurus

SOUTH AMERICA

ANTARCTICA

AUSTRALIA

Fossil remains of cynognathus, a massive land reptile approximately 3 metres long.

Fossil remains of the freshwater reptile mesosaurus.

Fossils of the fern glossopteris, found in all of the southern continents, show that they were once joined.

3.6.2 Fossils

What are fossils?

To gain insight into what life was like in the past, you need look no further than rocks. Within rocks you may find fossils — evidence of past life. The study of organisms by their fossil remains is called palaeontology.

Fossils can be parts of an organism, such as bones, teeth, feathers, scales, branches or leaves. They can also be footprints, burrows and other evidence that an organism existed in an area. For example, a dinosaur track has been discovered in the Otway Range in southern Victoria. By observing the footprints in the track, palaeontologists can work out the size, weight and speed of the dinosaur that made them.

Types of fossils

Many types of fossils are possible. These can be physical fossils (direct evidence such as body fossils) or trace fossils (indirect evidence of an organism's presence). Physical (body) fossils include casts (figure 3.41), moulds (figure 3.42), imprints (figure 3.43), petrified organisms (figure 3.44) and whole organisms that have been frozen (figure 3.45) or trapped in sap or amber (figure 3.46). Trace fossils include footprints (figure 3.47) and coprolite (or fossilised dung) (figure 3.48).

FIGURE 3.40 This skull provides evidence of past life.

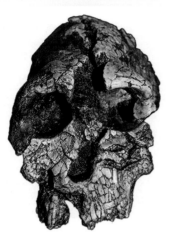

FIGURE 3.41 Cast fossil: a rock with the shape of an organism protruding (sticking out) from it

FIGURE 3.42 Mould fossil: a rock that has an impression (hollow) of an organism

FIGURE 3.43 Carbon imprint fossil: the dark print of an organism that can be seen on a rock

FIGURE 3.44 Petrified fossil: organic material of living things that has been replaced by minerals, such as petrified wood

FIGURE 3.45 Whole organism fossil: larger organisms that have been preserved whole by being mummified or frozen, such as this baby mammoth found in 2007 in Siberia

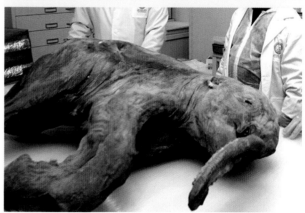

FIGURE 3.46 Amber fossil: parts of plants, insects or other small animals that have been trapped in a clear substance called amber.

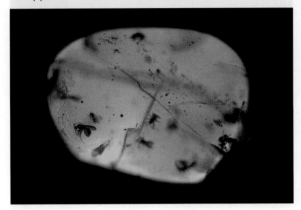

FIGURE 3.47 Footprint fossil: imprints of evidence of past life

FIGURE 3.48 Fossiled dung (coprolite): evidence that an organism lived in an area

ACTIVITY: Picturing fossils

Use an image search engine to locate images of each of the types of fossils described in this section. Cut and paste the pictures into a Word document. Write a caption for each image. The caption should include the name of the fossilised organism shown in the picture, the location the fossil was found and the type of fossil (for example, cast).

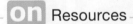 Resources

▶ **Video eLesson** Fossilised dinosaur tracks in mud (eles-2706)

✦ **Interactivity** Revelation: 'Fossils' (int-1018)

🔗 **Weblinks** Digital atlas of ancient life
What does it mean to be human?

How are fossils formed?

Fossilisation is a rare event. Usually when an organism dies, microorganisms are involved in its decomposition so that eventually no part of it remains. However, if an organism is covered shortly after its death by dirt, mud, silt or lava (as can happen if it becomes trapped in a mudslide or in the silt at the bottom of the ocean), the microorganisms responsible for decomposition cannot do their job because of the lack of oxygen. Over millions of years, the material covering the dead organism is compressed and turned into rock, preserving the fossil within it.

int-5874

FIGURE 3.49 Fossil formation

1. A dinosaur dies and is quickly covered by sediment.

2. Over time, the sediment turns into rock. The remains of the dinosaur turn into a fossil.

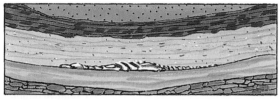

3. The fossil is flattened by the layers of rock.

4. The rock is folded and eroded and the fossil can be seen on the surface.

elog-2223

tlvd-10815

INVESTIGATION 3.4

Creating casts of leaves

Aim

To make a cast of a leaf fossil

Materials

- leaf
- clay (or plasticine)
- petroleum jelly
- mixed plaster

Method

1. Roll out a rectangular piece of clay and cover it with petroleum jelly.
2. Press a ribbed leaf into the clay to make an impression.
3. Remove the leaf and build some clay walls about 1 centimetre high at the edge of the rectangle.
4. Cover the walls with petroleum jelly and pour in some mixed plaster.
5. When the plaster has set, remove the clay and examine the cast for the leaf impression.

Results

Draw a clear sketch of your cast.

Discussion

1. How is a cast different from a mould?
2. Describe how you might create a mould instead of a cast.
3. Outline two ways in which you may improve this investigation.
4. How does your cast differ from others in your class?

Conclusion

Summarise your findings for this investigation about leaf fossils.

Dating fossils

The age of fossils is estimated in two main ways. One is called **relative dating** and the other is called **absolute dating**. The key difference between these two types of dating can be outlined using the following analogy:

- If you were to ask me 'What is your **relative age**?', I would reply, 'I am the eldest of three daughters'.
- If you were to ask me 'What is your **absolute age**?', I would tell you I am 16 years old.

> Relative dating gives a relative age that provides a comparison only.
> Absolute dating provides an absolute age, which is a numeric age.

Relative dating

Relative dating is used to determine the relative age of a fossil. Because the layers of sedimentary rock are usually arranged in the order they were deposited, the most recent layers are near the surface and the older layers are further down. The position or location of a fossil in the strata, or layers, of rock provides an indication of the time in which the animal lived. Relative dating can also provide information about which other species were living at the same time and the order in which they appeared in the area.

Interpretation of the relative dating method requires considering the movement of tectonic plates in which the rocks lie. It is possible that a layer (or layers) containing fossils could have been thrust upwards by a sideways force to form a **fold**, or broken and moved apart in opposite vertical directions to form a **fault**.

FIGURE 3.50 Fossils can be dated by observing their position in the rock layers.

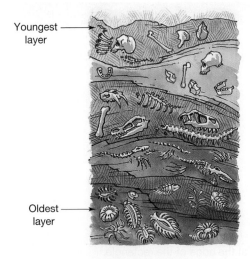

Youngest layer

Oldest layer

FIGURE 3.51 The formation of folds and faults can cause changes in the rock layers.

a. Folds

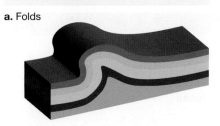

b. Faults

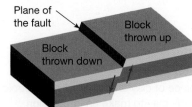

Plane of the fault

Block thrown up

Block thrown down

relative dating method of dating that determines the age of a rock layer by relating it to another layer using superposition and the fossils contained

absolute dating determining the age of a fossil and the rock in which it is found using the remaining amount of unchanged radioactive carbon

relative age age of a rock compared with the age of another rock

absolute age number of years since the formation of a rock or fossil

fold a layer of rock bent into a curved shape, which occurs when rocks are under pressure from both sides

fault a break in a rock structure causing a sliding movement of the rocks along the break

 **Resources**

▶ **Video eLesson** Folds (eles-2707)

Absolute dating

Fossils or the rocks in which they were located can also be dated by various radiometric techniques, which are based on the rate of decay or **half-life** of particular **isotopes**. The half-life of an isotope is the amount of time it takes for its radioactivity to halve. The use of these techniques to determine the absolute age of rocks and fossils is called **radiometric dating**.

Carbon dating is a specific type of radiometric dating and can be used to date fossils up to about 60 000 years old. Most of the carbon contained in living things is carbon-12, but a small amount of the radioactive isotope carbon-14 is also present. Organisms incorporate this into their bodies from the small amount of radioactive carbon dioxide that is naturally present in the air. When an organism dies, the unstable carbon-14 decays, but the carbon-12 does not. The ratio between carbon-12 and carbon-14 can be used to determine the absolute age of the fossil.

Radiometric dating can also be used to determine the age of inorganic materials (materials not containing carbon), such as the rocks surrounding fossils. **Potassium–argon dating** is commonly used to determine the absolute age of ancient rocks. Uranium–lead dating (see figure 3.52) can also determine the age of rocks.

half-life time taken for half the radioactive atoms in a sample to decay; that is, change into atoms of a different element

isotopes atoms of the same element that differ in the number of neutrons in the nucleus

radiometric dating a technique in which radioactive substances are used to calculate the age of rocks or dead plants and animals

carbon dating a radiometric dating technique that uses an isotope of carbon-14 to determine the absolute age of fossils

potassium–argon dating a radiometric dating technique based on the measure ratio of potassium-40 to argon-40

int-5876

FIGURE 3.52 Over time, the uranium-235 present in rocks decays into lead-207. The half-life of uranium is the amount of time it takes for half the uranium initially present in the rock to decay into lead.

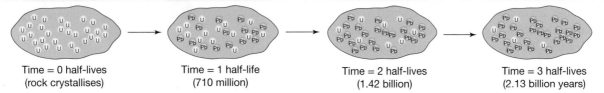

| Time = 0 half-lives (rock crystallises) | Time = 1 half-life (710 million) | Time = 2 half-lives (1.42 billion) | Time = 3 half-lives (2.13 billion years) |

TABLE 3.1 Some radioactive isotopes and their daughter products of decay

Radioactive parent isotope	Daughter product	Half-life (years)	Uses
Carbon-14	Nitrogen-14	5730	Used for dating organic (carbon-based) remains up to about 60 000 years old
Uranium-235	Lead-207	710 000 000	Used for dating igneous rocks containing uranium-based minerals in the range from about 1000 to 1 000 000 years
Potassium-40	Argon-40	1 300 000 000	Used for dating igneous rocks containing potassium-bearing minerals from 500 000 years and older
Rubidium-87	Strontium-87	47 000 000 000	Used to date the most ancient igneous rocks

Note: The longer the half-life of a radioactive isotope, the older the material that can be dated.

 Resources

eWorkbook Ages of fossils (ewbk-12135)

Fossils telling tales

The fossil record gives us evidence that species have changed over time. For example, a fossilised skeleton of a bird (*Archaeopteryx*) found in Bavaria has been dated at 150 million years old. It clearly shows feathers, which are a feature of all modern birds; however, it also has dinosaur characteristics such as teeth, claws on its wings and a long, jointed bony tail. From this, scientists have deduced that birds evolved from a dinosaur ancestor. The evolution, or change over time, of other species can also be followed by studying the fossil record.

Horsing around in time

The fossil record gives us evidence of gradual change occurring over time. An example can be seen in fossils of horse species from different times.

Fossils indicate that horses have become taller, their teeth are now better suited to grazing than eating leaves and fruit, and their feet have a single hoof rather than spread-out toes.

Over time, environmental changes have led to some variations having a selective advantage over others. Reduced availability of fruit and leaves but increased availability of tough grasses resulted in selection for teeth better suited to grazing. As forests were replaced by open plains, longer legs and hoofs may have been selected for to provide a better chance of escaping predators.

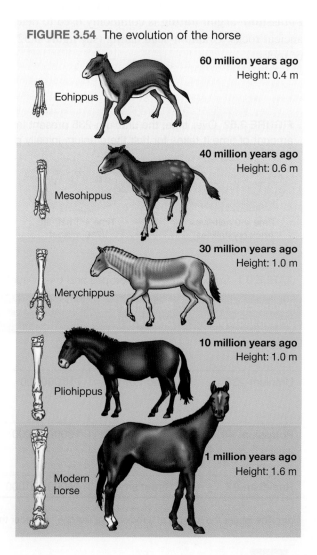

FIGURE 3.54 The evolution of the horse

Eohippus — **60 million years ago** Height: 0.4 m

Mesohippus — **40 million years ago** Height: 0.6 m

Merychippus — **30 million years ago** Height: 1.0 m

Pliohippus — **10 million years ago** Height: 1.0 m

Modern horse — **1 million years ago** Height: 1.6 m

FIGURE 3.53 The fossilised skeleton of *Archaeopteryx*

3.6.3 More evidence for evolution

A variety of evidence supports the theory of evolution. Biogeography and the fossil record have been addressed earlier, let us now focus on three other examples of evidence — **comparative anatomy**, **comparative embryology** and **molecular biology**.

FIGURE 3.55 Examples of different evidence for evolution

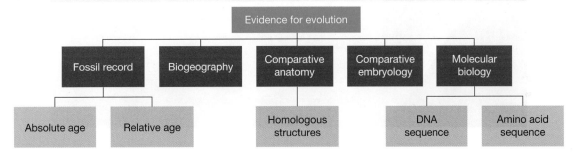

Comparative anatomy

The forearms of vertebrates (such as mammals, amphibians, reptiles and birds) are remarkably similar in structure. Each, however, is used for a different function, such as swimming, walking or flying. The structure of the vertebrate forearm (also referred to as the pentadactyl limb) can be traced back to the fin of a fossilised fish from which amphibians are thought to have evolved.

FIGURE 3.56 These structures have the same basic structure since they are all derived from a vertebrate forelimb. Do they have identical functions?

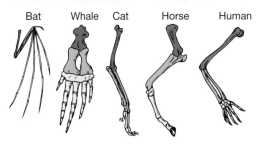

Homologous structures

Similarity in characteristics that result from common ancestry is known as **homology**. Anatomical signs of evolution such as the similar forearms of mammals are called homologous structures, and these provide evidence for divergent evolution.

For example, in figure 3.56, you can see that each limb has a similar number of bones arranged in the same basic pattern. Even though their functions are different, the similarity of the basic structure still exists.

Vestigial structures

Vestigial structures are structures that no longer serve a purpose in an organism, but are remnants from ancestral species in which the structure had an important function.

comparative anatomy exploring similarities and differences in the anatomical structures of various species

comparative embryology exploring similarities and differences between embryos of various species

molecular biology the study of the structure and composition of molecules within a cell

homology similar characteristics that result from common ancestry

vestigial structures structures with no apparent function that are remnants from a past ancestor

These provide evidence of evolution in a similar way to homologous structures, showing common ancestry and evidence of divergent evolution.

Some examples of vestigial structures:
- the hip bone present in whales
- the palmaris longus muscle present in the wrists of some humans (remnants of a muscle that used to assist our tree-climbing ancestors)
- nipples in human males
- the wings of flightless birds.

FIGURE 3.57 The **a.** palmaris longus muscle and **b.** hip bone of a whale no longer serve a purpose, but are remnants from ancestral species.

ACTIVITY: Testing for the palmaris longus muscle

Touch your thumb and little finger together, with your inner wrist facing towards you. Flex your wrist. Do you have the palmaris longus muscle? Compare this with others in your class.

Test your other hand. Do you have the muscle in this wrist? Some individuals have it in both wrists, some have it in one and others do not have this muscle at all!

Comparative embryology

Organisms that go through similar stages in their embryonic development do so because they have evolved from a common ancestor.

FIGURE 3.58 Examples of embryonic stages of different organisms

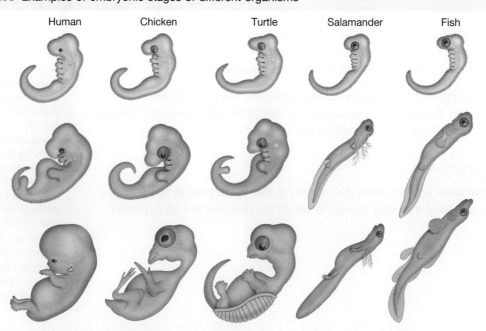

During the early stages of development, the human embryo and the embryos of other animals appear to be quite similar (as seen in figure 3.58). For example, the embryos of fish, amphibians, reptiles, birds and mammals all initially have gill slits. As the embryos develop further, the gill slits disappear in all but fish. It is thought that gill slits are a characteristic that all these animals once shared with a common ancestor.

Molecular biology

How amazing is it that all living things share the same overall genetic coding system? Although the nucleotide sequences in their DNA may vary, the 'language' used and the rules to read it are the same. This commonality had enabled genetic engineers to cut DNA out of one organism and paste it into another. By doing this, they have been able to get an organism to make a protein that they previously could not.

We can use this concept of a universal genetic code to determine the evolutionary relationships between species. Similarities and differences between their DNA sequences and amino acid sequences in proteins can be used to determine how closely they are related and to estimate when they shared a common ancestor.

Proteins are universally important chemicals that are essential to the survival of organisms. In topic 2, we looked at the coding and synthesis of proteins (see lesson 2.4 and figure 3.59).

The genetic message to make proteins is stored in DNA. A section of the DNA (gene) is transcribed into messenger RNA (mRNA), which is then translated into proteins. Each of the DNA triplets and mRNA codons code for a specific amino acid, and the sequence of the nucleotides determines the sequence of the amino acids that will make up a specific protein.

FIGURE 3.59 Reviewing how DNA sequences are used to code for proteins

DNA sequences

By comparing the DNA sequences, we can determine the relatedness with species and observe how DNA has changed over time. DNA can be obtained not only from living organisms, but often also from fossilised remains.

DNA sequences have been described as 'documents of evolutionary history'. Comparisons of DNA from different species may be made through:
- direct comparison of DNA base sequences
- comparing whole genomes
- DNA hybridisation.

DNA molecules consist of a series of nucleotides containing nitrogenous bases that form a base sequence (a combination of A, C, G and T). We can predict that species with a more recent common ancestor have more similarity in their base sequences.

Comparing DNA base sequences

Comparisons can be performed using DNA sequencing, which was outlined in topic 2. The process of sequencing is automated and is completed using instruments known as DNA sequencers.

Comparing whole genomes

Sometimes comparing just a small section of DNA does not give a full understanding of the evolutionary history of species. Therefore, comparing the entire genome can be more useful. This can be performed through exploring the number of shared genes between species, and not just within a DNA sequence.

TABLE 3.2 A section of DNA, in which comparisons are made to humans. Based on this sequence, it can be assumed that Species A is more closely related to humans than species D.

Human	A	C	G	G	T	T	C	A	G
Species A	A	C	G	C	T	T	C	A	G
Species B	A	C	G	C	T	T	C	T	G
Species C	T	C	G	C	T	T	C	T	G
Species D	T	C	G	C	T	T	C	T	A

DNA hybridisation

Another technique that can be used is **DNA hybridisation**. While less commonly used in recent years (due to the advancement of technologies in DNA sequencing), it is an easy way to compare similarities between species. DNA hybridisation is a technique that can be used to compare DNA sequences in different species to determine how closely related they are.

> **DNA hybridisation** a technique that can be used to compare the DNA in different species to determine how closely related they are

The tree diagram in figure 3.61 illustrates the inferred evolutionary relationships within a group of primates, based on DNA hybridisation. Which primate is most closely related to humans and which is least closely related?

FIGURE 3.60 The process involved in DNA hybridisation

int-5880

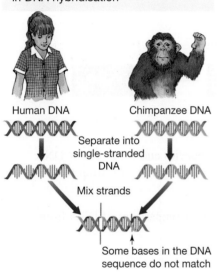

Human DNA   Chimpanzee DNA

Separate into single-stranded DNA

Mix strands

Some bases in the DNA sequence do not match

FIGURE 3.61 The branches of the phylogenetic tree shows when a species diverged from a common ancestor.

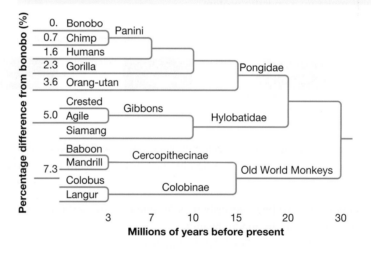

Amino acid sequences

As mentioned earlier, the specific protein produced by organisms depends on the sequence of nucleotides in the DNA (DNA sequence) in a gene. A change in the DNA sequence may lead to a change in the type and sequence of amino acids in the protein produced. This idea is used when comparing the amino acid sequences of specific proteins in organisms.

Cytochrome *c* is a common protein used to determine evolutionary relationships. It is involved in the conversion of energy into a form the cell can use during cellular respiration. This protein is found in many species, but different species have slightly or significantly different versions of this protein. Although a part of the cytochrome

FIGURE 3.62 The amino acids in cytochrome *c* differ between different organisms.

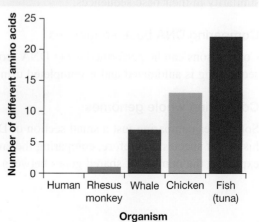

molecule maintains a specific shape, over time other parts of the molecule have mutated. The graph shown in figure 3.62 provides an example of the number of differences in the amino acid sequence of cytochrome *c* in various vertebrates.

Another example of comparing amino acid sequences is shown in figure 3.63.

FIGURE 3.63 Comparing amino acid sequences between different species. Each letter represents an amino acid (for example, H is histadine and V is valine).

```
honeycreepers (10)  ...CRDVQFGWLIRNLHANGASFFFICIYLHIGRGIYYGSYLNK--ETWNIGVILLLTLMATAFVGYVLPWGQMSFWG...
      song sparrow  ...CRDVQFGWLIRNLHANGASFFFICIYLHIGRGIYYGSYLNK--ETWNVGIILLLALMATAFVGYVLPWGQMSFWG...
  Gough Island finch ...CRDVQFGWLIRNIHANGASFFFICIYLHIGRGLYYGSYLYK--ETWNVGVILLLTLMATAFVGYVLPWGQMSFWG...
         deer mouse  ...CRDVNYGWLIRYMHANGASMFFICLFLHVGRGMYYGSYTFT--ETWNIGIVLLFAVMATAFMGYVLPWGQMSFWG...
   Asiatic black bear ...CRDVHYGWIIRYMHANGASMFFICLFMHVGRGLYYGSYLLS--ETWNIGIILLFTVMATAFMGYVLPWGQMSFWG...
      bogue (a fish)  ...CRDVNYGWLIRNLHANGASFFFICIYLHIGRGLYYGSYLYK--ETWNIGVVLLLLVMGTAFVGYVLPWGQMSFWG...
             human  ...TRDVNYGWIIRYLHANGASMFFICLFLHIGRGLYYGSFLYS--ETWNIGIILLLATMATAFMGYVLPWGQMSFWG...
 thale cress (a plant) ...MRDVEGGWLLRYMHANGASMFLIVVYLHIFRGLYHASYSSPREFVWCLGVVIFLLMIVTAFIGYVLPWGQMSFWG...
       baboon louse  ...ETDVMNGWMVRSIHANGASWFFIMLYSHIFRGLWVSSFTQP--LVWLSGVIILFLSMATAFLGYVLPWGQMSFWG...
       baker's yeast  ...MRDVHNGYILRYLHANGASFFFMVMFMHMAKGLYYGSYRSPRVTLWNVGVIIFTLTIATAFLGYCCVYGQMSHWG...
```

SCIENCE AS A HUMAN ENDEAVOUR: The molecular clock

In 1966, biochemists Vincent M. Sarich and Allan Wilson noticed that changes in the amino acid sequences of particular proteins in related species appeared to occur at a steady rate. They found more amino acid sequence differences the longer that two species had existed separately. From these observations, the concept of the **molecular clock** arose.

The molecular clock concept uses the known rate of amino acid changes (through mutation) and the number of differences in the amino acid sequences of two species to estimate the time since the species diverged. Based on their analysis of immunological proteins, Sarich and Wilson concluded that humans and African apes shared a common ancestor a lot later (no more than five million years ago) than was suggested by palaeontologists (who suggested 15 to 25 million years ago).

FIGURE 3.64 Vincent M. Sarich and Allan Wilson

FIGURE 3.65 The molecular clock concept suggests, on the basis of amino acid sequence differences, species A and B are more closely related than A and C, or B and C. It also suggests that species A and B diverged from a common ancestor just over 4 million years ago and that species A, B and C shared a common ancestor about 10 million years ago.

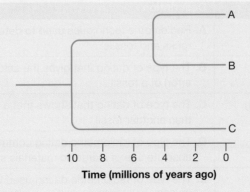

molecular clock a technique that uses known mutation rates of biomolecules to determine the time two species diverged

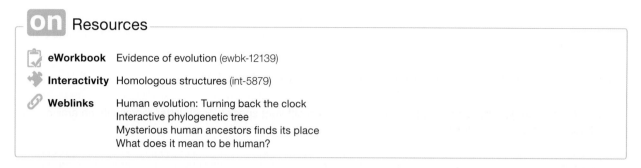

On Resources

📋 **eWorkbook**	Evidence of evolution (ewbk-12139)
🧩 **Interactivity**	Homologous structures (int-5879)
🔗 **Weblinks**	Human evolution: Turning back the clock
	Interactive phylogenetic tree
	Mysterious human ancestors finds its place
	What does it mean to be human?

3.6 Activities

learnon

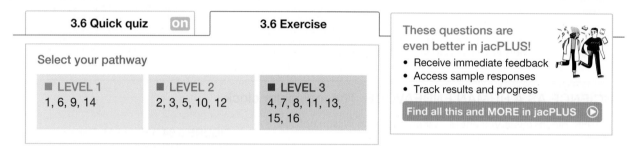

3.6 Quick quiz on	**3.6 Exercise**

Select your pathway

■ LEVEL 1	■ LEVEL 2	■ LEVEL 3
1, 6, 9, 14	2, 3, 5, 10, 12	4, 7, 8, 11, 13, 15, 16

These questions are even better in jacPLUS!
- Receive immediate feedback
- Access sample responses
- Track results and progress

Find all this and MORE in jacPLUS ▶

Remember and understand

1. **a.** State whether the following statements are true or false.

Statements	True or false?
i. Fossilisation is a common event.	
ii. The formation of folds and faults has no effect on the rock layers.	
iii. Fossils are approximately the same age as the layer of rock in which they are found.	
iv. The position or location of a fossil in the strata, or layers, of rock gives an indication of the time in which the animals lived.	
v. The half-life of an isotope is the amount of time it takes for half of the nuclei in the sample to undergo radioactive decay.	

   b. Rewrite any false statements to make them true.
2. Match the term to its description in the following table.

Term	Description
a. Absolute dating	**A.** Radiometric techniques used to determine the absolute age of rocks and fossils
b. Carbon dating	**B.** The type of dating that gives the actual age (plus or minus some error) of a fossil
c. Potassium–argon dating	**C.** The type of dating that shows that a fossil is younger or older than another fossil
d. Radiometric dating	**D.** The type of radiometric dating commonly used to determine the absolute age of inorganic materials such as ancient rocks
e. Relative dating	**E.** The type of radiometric dating used to date fossils up to about 60 000 years old

3. **a.** Define the following terms.
   i. Homology
   ii. Homologous structures
   b. Provide an example of a homologous structure.

4. Describe how the following can be used as a source of evidence for evolution.
   a. Homologous structures
   b. Comparative embryology
   c. DNA sequences
   d. Amino acid sequences
5. What is DNA hybridisation and why is it useful?

Apply and analyse

6. a. What does biogeography refer to?
   b. Describe the relationship between biogeography and evolution. Use three examples to support this.
7. Biogeography is often used as evidence of evolution for numerous species.
   a. Three species of flightless birds are located in Australia, New Zealand and South America. Which would you expect to be more closely related? Explain your answer.
   b. Explain why biogeography does not provide strong evidence for the evolution of flying birds.
8. a. State the relationship between carbon dating and radiometric dating.
   b. Describe how carbon dating is used to estimate the age of fossils.
   c. A fossil of an extinct archaic species of human was found in a cave in China. It was thought to be around 20 000 years old. Identify if carbon dating or potassium–argon dating would be most appropriate for this fossil, and justify your response.
9. Outline the role of geologists and palaeontologists. Why are their roles important in understanding fossils?
10. **SIS** The phylogenetic tree shown in figure 3.61 shows the inferred evolutionary relationship between primates based on their DNA.
    a. **MC** Identify which of the following species is most closely related to bonobos.
       A. Baboon
       B. Chimp
       C. Gorilla
       D. Human
    b. **MC** Identify how long ago humans and chimps shared a common ancestor.
       A. About 3 million years ago
       B. About 7 million years ago
       C. About 10 million years ago
       D. About 15 million years ago
    c. Identify whether humans are more closely related to chimps or gorillas.
    d. Identify whether humans or baboons are more closely related to orangutans or mandrills.
    e. Identify how long ago humans and baboons shared a common ancestor.
       A. About 3 million years ago
       B. About 10 million years ago
       C. About 20 million years ago
       D. About 30 million years ago
11. **SIS** Select one of the following topics to research and report on. Provide examples in your response.
    • What is the pentadactyl limb and how is it relevant as evidence for evolution?
    • What is the difference between homologous structures and analogous structures?
    • What vestigial structures are found in humans other than those already mentioned in lesson 3.6?

Evaluate and create

12. The Venn diagram provided can be used to include some analogies as well as key differences to distinguish the relative age from the absolute age of fossils.
    Copy and complete the Venn diagram using the following terms: order, 16 years old, eldest of three daughters, carbon dating, potassium–argon dating, youngest layer, strata, oldest layer.
13. Australia has a number of great fossil sites. Use the internet and other sources to investigate one of the following fossil sites: Naracoorte, Riversleigh, Bluff Downs, Murgon and Lightning Ridge. Summarise information about the fossil site you have chosen under the following headings.

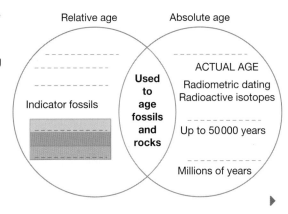

a. Why the area is rich in fossils
b. Examples of fossils that have been found here
c. Age of the fossils found in this area
d. Important information revealed by the fossils found in this area

14. Examine the following picture of the fossilised Saurischian dinosaur.
   a. Write a brief description of the animal, including what it may have eaten and how it may have moved.
   b. Suggest a reason it had so many large openings in its skull.

15. **SIS** Examine the following table showing the DNA sequence from part of a haemoglobin gene from four mammalian species.

TABLE A comparison of the DNA sequence in four species

Species				DNA sequence
Human	TGACAAGAACA	–	GTTAGAG	–	TGTCCGAGGACCAACAGATGGGTACCTG GGTCCCAAGAAACTG
Orangutan	TCAC GAGAACA	–	GTTAGAG	–	GTCCGAGGACCAACAGATGGGTACCTG GGTCTCCAAGAAACTG
Rhesus monkey	TGAC GAGAACA	A	GTTAGAG	–	TGTCCGAGGACCAACAGATGGGTACCTG GGTCTCCAAGAAACTG
Rabbit	TGGTGA TAACA	A	GACAGAG	A	TATCCGAGGACCAGCAGATAGG AACCTG GGTCTCTAAGAAGCTA

Differences between the human DNA sequence and those of other species are shown by coloured letters.
The dash (–) Is used to keep the sequences alligned.

   a. T, G, C and A represent nitrogenous bases. Are these the same four bases in DNA in all other species?
   b. In terms of the first 11 nitrogenous bases, which mammalian species is most similar to humans?
   c. In terms of the first 11 nitrogenous bases, which mammalian species is least similar to humans?
   d. On the basis of the data in the table, rank these species in terms of how long ago they may have shared a common ancestor with humans.
   e. Suggest how these differences in the sequence of nitrogenous bases in DNA may have arisen.

16. Cytochrome *c* is known primarily for its function in mitochondria and its involvement in ATP synthesis. Scientists are also researching its involvement in the process of programmed cell death (apoptosis) and to determine evolutionary relationships.
   a. Investigate the structure and function of cytochrome *c*.
   b. Draw a clear diagram of a cytochrome *c* protein.

Fully worked solutions and sample responses are available in your digital formats.

LESSON
3.7 Thinking tools — Gantt charts

3.7.1 Tell me

What is a Gantt chart?

Gantt charts (also known as harmonograms) are a type of bar chart split into different columns to show sequences of events. They have a variety of purposes, and can be used not only in science, but also for general purposes such as for scheduling, as seen in figure 3.66.

Why use a Gantt chart?

Gantt charts provide an easy way to visualise not only the order of events, but also the length of time of those events. They are quick to read and analyse.

In many situations, they are preferable to other visual thinking tools such as storyboards, because they do not rely on additional factors such as diagrams or sketches.

FIGURE 3.66 A Gantt chart used as a schedule

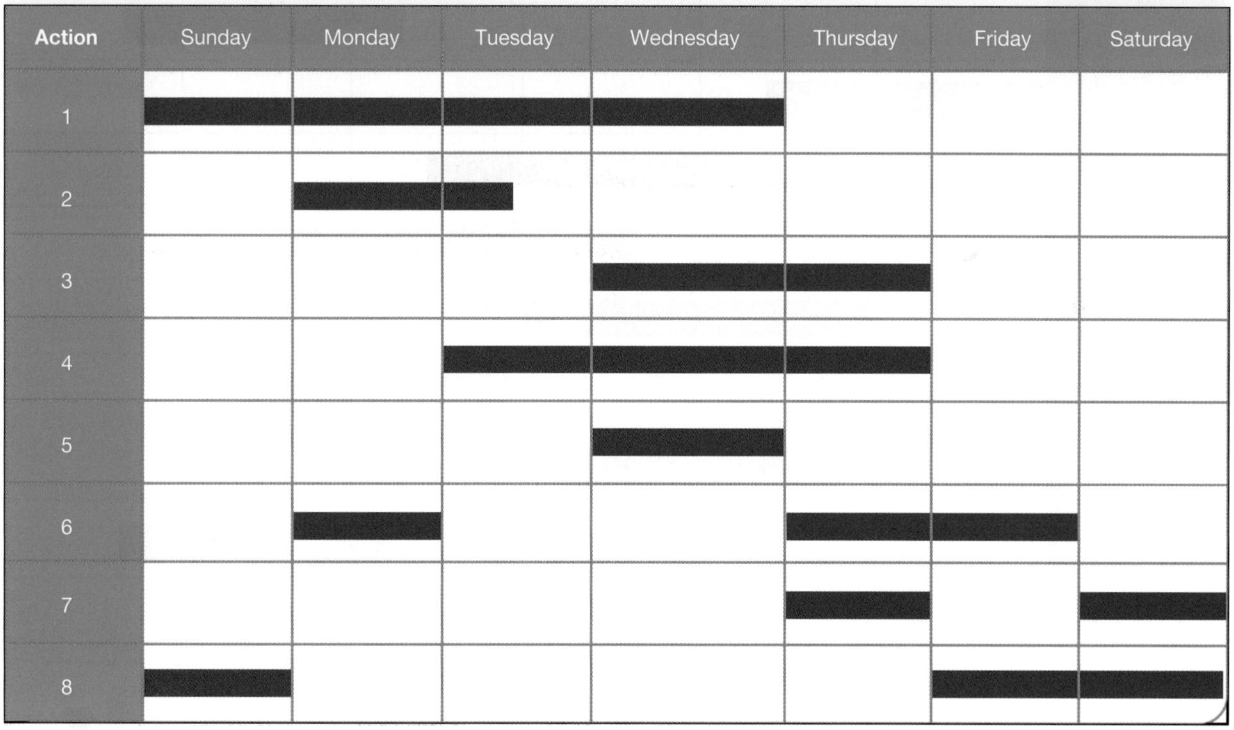

Action	Sunday	Monday	Tuesday	Wednesday	Thursday	Friday	Saturday
1	███	███	███	███			
2		███	███				
3				███	███		
4			███	███	███		
5				███			
6		███			███	███	
7					███		███
8	███					███	███

3.7.2 Show me

In order to create a Gantt chart you should:
1. Write a list of events or scenarios involved in your topic.
2. Note down the timelines for each of these; for example, this may be in millions of years (as is the case with evolution), years or days of the week.
3. Create an appropriate time scale for your chart and place this at the top of your chart, running vertical lines to separate periods.
4. Add your different events, individuals or scenarios down the left of the time scale.
5. Add bars in each row to show the timeframes for each event, individual or scenario.

An example is shown in figure 3.67, showing some of the main individuals involved in adapting the different theories of evolution.

FIGURE 3.67 A Gantt chart showing the lifetimes of some of the main individuals involved in adapting the different theories of evolution

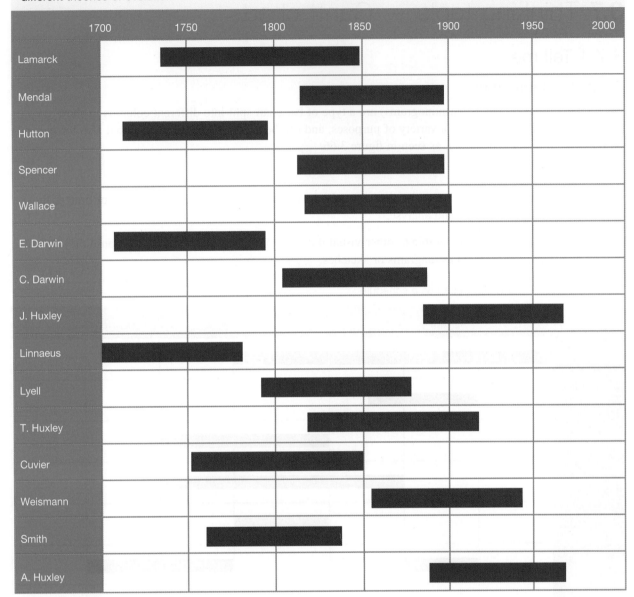

3.7.3 Let me do it

3.7 Activity

1. Using figure 3.38 from lesson 3.6, create a Gantt chart, sharing the fossil record of life as it has existed on Earth.
2. Research current theories on human evolution and the appearance of different hominin species over time. Use a Gantt chart to summarise some of your findings.
3. Find out more about an aspect of human evolution that interests you and construct a Gantt chart to share what you have found out.
4. a. Find out more about the 'hobbit-like' human ancestor (*Homo floresiensis*) and the various scientific discussions about its lineage.
   b. Use a Gantt chart to summarise your findings.

Fully worked solutions and sample responses are available in your digital formats.

These questions are even better in jacPLUS!
- Receive immediate feedback
- Access sample responses
- Track results and progress

Find all this and **MORE** in jacPLUS

LESSON
3.8 Project — Natural selection board game

3.8.1 Scenario

Most people in Australia today have played a board game such as *Monopoly, Scrabble* or *Snakes and Ladders* sometime in their life. Australia's love for an old-fashioned board or card game is still going strong — evident during the 2020 pandemic when, alongside toilet paper, board and card games were sold out in record speed! Apart from being a great choice when there's no electricity and that they can be played and enjoyed by people from different generations, psychologists suggest that their continued popularity can also be attributed to the winner being determined as much by luck as by skill. In this way, board games are much like real life!

The effectiveness of using game play as a way of teaching concepts is the stock-in-trade of the educational game company BrainGames, which produces computer games that teach science, maths, history and geography concepts. Games such as *The Revenge of Pavlov's Dogs* and *Where in the World is Amerigo Vespucci?* have made them the leader in the educational games market. However, keen to exploit the non-computer-equipped market sector, BrainGames now want to branch out into board games, and the first board game they want to produce is to be based on one of the key ideas of biology.

FIGURE 3.68 The popularity of board games is still going strong.

3.8.2 Your task

As part of the Games Development Division at BrainGames, you and your team are to develop a prototype board game based on the idea of natural selection and evolution. In this game, players will be able to select a variety of characteristics to give an organism and then, over the course of the game, see whether these organisms survive intact as their environment is changed. Your prototype must include:

- a game board
- game pieces
- a rule book.

You may also choose to include game mechanics such as cards, spinners or dice.

FIGURE 3.69 Will the organism you create be more successful than the dodo?

  Resources

ProjectsPLUS Natural selection — the board game! (pro-0112)

LESSON
3.9 Review

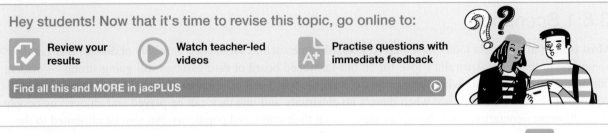
Access your topic review eWorkbooks

on Resources

■ Topic review Level 1
ewbk-12150

■ Topic review Level 2
ewbk-12152

■ Topic review Level 3
ewbk-12154

3.9.1 Summary

Classification

- The main way we name species is through the binomial system of nomenclature, which involves giving species a two-word name. The first word is the genus and the second is a descriptive term; for example, *Homo sapiens.*
- Although genetic technologies have blurred the lines of our classification system, our usual definition of species refers to individuals within a group that can interbreed to produce fertile offspring.

Processes of natural selection

- Alfred Wallace and Charles Darwin are jointly credited with developing the theory of evolution.
- Natural selection is the mechanism through which evolution occurs and refers to the selection of particular phenotype (variation) within a population due to biotic and/or abiotic selection pressures.
- The four key ideas that form the basis of Darwin's theory of evolution by natural selection are as follows.
  - Variation — variation occurs within populations.
  - Selection — some individuals within the population have adaptations that help them compete successfully under certain selective pressures.
  - Inheritance — successful individuals reproduce and pass their characteristics on to their offspring.
  - Time — many generations will be produced and the organisms will possess the traits that increases their chances of survival.
- Many mutations generally decrease the survival of an organism, however, it sometimes increases the chances of survival. Drug resistant diseases and bugs will produce offspring that are resistant to anti-biotics or particular pesticides.

Biodiversity as a function of evolution

- Biodiversity (or biological diversity) relates to variation within living things. It can be described in terms of an ecosystem, at the level of species, or even at the level of individual genes.
- Genetic diversity refers to differences between individuals (due to differences in genes and DNA), while species diversity refers to the number of different species that inhabit a particular community.
- Mutation can occur in all organisms and is the source of new genetic variation.
- Mutations are important to asexually reproducing organisms because this is the only way that variation can arise within the species.
- A mutation may result in a changed trait that confers a survival advantage because it makes an organism more suited to its environment.

- Variation between individuals that reproduce sexually may also occur due to
  - crossing over of sections of maternal and paternal chromosomes during meiosis
  - the independent assortment of the chromosomes into the gametes during meiosis
  - the random combination of gametes that fuse together during fertilisation
  - the random selection of a particular mate.
- Genetic drift (changes due to chance events), gene flow (immigration and-emigration), and natural selection can have an impact on allele frequencies and, hence, variation within populations.
- The use of reproductive technologies such as artificial insemination, IVF and cloning has the potential to unbalance natural levels of biodiversity.

Genetics and evolution

- Speciation is a process by which a new species develops from another.
- An organism's adaptations will aid their survival if the adaptations are suited to the particular environment of the organism and the selection pressures it is subject to.
- Divergent evolution is when new species evolve from a shared ancestral species. Over time, these species diverge as they develop different characteristics due to different selection pressures.
- Adaptive radiation is when divergent evolution of one species results in the formation of many species that are adapted to a variety of environments.
- Convergent evolution is when unrelated species develop similar features due to their exposure to similar selection pressures.
- All environments change over time. If they change too rapidly, the genes required for survival in the changed environment may not be present in the gene pool and that species may become extinct.

Evidence of the theory of evolution

- Evidence for the theory of evolution includes the fossil record, biogeography, comparative anatomy, comparative embryology and molecular biology (DNA sequences and amino acid sequences).
- Some types of evidence can provide information about shared common ancestors.
- The fossil record shows how different species have lived, survived and become extinct, and how species have changed over time, while fossil layers allow relative dating of fossils.
- Biogeography supports the idea that each species evolved in one location and then spread out to other areas because closely related species are distributed across continents that were once joined together.
- Homologous structures are body structures that can perform a different function but have a similar structure (for example, the vertebrate pentadactyl limb) due to their evolutionary origin, such as a shared common ancestor.
- Comparative embryology is where a similarity of features during stages of embryonic development can be seen due to shared common ancestors.
- Comparing the DNA sequences and amino acid sequences of proteins between organisms allows for similarities and differences to be used to determine how closely the organisms are related to each other.

3.9.2 Key terms

absolute age   number of years since the formation of a rock or fossil

absolute dating   determining the age of a fossil and the rock in which it is found using the remaining amount of unchanged radioactive carbon

adaptation   a feature that aids in the survival of an organism

adaptive radiation   when divergent evolution of one species results in the formation of many species that are adapted to a variety of environments

alleles   alternative forms of a gene for a particular characteristic.

allopatric speciation   a type of speciation where populations are separated by a geographical barrier

artificial selection   the process in which humans breed animals or plants in such a way to increase the proportion of desired traits

binomial system of nomenclature   system devised by Carolus Linnaeus giving organisms two names, the genus and another specific name

biodiversity   total variety of living things on Earth

carbon dating   a radiometric dating technique that uses an isotope of carbon-14 to determine the absolute age of fossils

chromosomes   tiny thread-like structures inside the nucleus of a cell that contain the DNA that carries genetic information

coevolution   the process in which two species evolve in partnership so that they depend on each other

comparative anatomy   exploring similarities and differences in the anatomical structures of various species

comparative embryology   exploring similarities and differences between embryos of various species

convergent evolution   tendency of unrelated organisms to acquire similar structures due to similar environmental pressures

crossing over   exchange of alleles between maternal and paternal chromosomes

divergent evolution   when a population is divided into two or more new populations that are prevented from interbreeding

DNA hybridisation   a technique that can be used to compare the DNA in different species to determine how closely related they are

ecosystem   community of living things that interact with each other and with the environment in which they live

emigration   a type of gene flow in which an individual leaves a population

eras   divisions of geological time defined by specific events in the Earth's history, divided into periods

evolution   the process in which traits in species gradually change over successive generations

extinction   complete loss of a species when the last organism of the species dies

fault   a break in a rock structure causing a sliding movement of the rocks along the break

fold   a layer of rock bent into a curved shape, which occurs when rocks are under pressure from both sides

gametes   reproductive or sex cells such as sperm or ova

gene flow   the movement of individuals and their alleles between populations

gene pool   the total genetic information of a population, usually expressed in terms of allele frequency

genetic diversity   variation in genes between members of the same species

genetic drift   changes in allele frequency due to chance events such as floods and fires

half-life   time taken for half the radioactive atoms in a sample to decay; that is, change into atoms of a different element

homologous structures   body structures that can perform a different function but have a similar structure due to their evolutionary origin

homology   similar characteristics that result from common ancestry

immigration   a type of gene flow in which an individual enters a population

isotopes   atoms of the same element that differ in the number of neutrons in the nucleus

meiosis   cell division process that results in new cells with half the number of chromosomes of the original cell

molecular biology   the study of the structure and composition of molecules within a cell

molecular clock   a technique that uses known mutation rates of biomolecules to determine the time two species diverged

mutation   change to DNA sequence, at the gene or chromosomal level

natural selection   process in which organisms better adapted for an environment are more likely to pass on their genes to the next generation

periods   subdivisions of geological time, divided into epochs

phyletic evolution   when a population of a species progressively changes over time to become a new species

population   members of one species living together in a particular place at the same time

potassium–argon dating   a radiometric dating technique based on the measure ratio of potassium-40 to argon-40

radiometric dating   a technique in which radioactive substances are used to calculate the age of rocks or dead plants and animals

relative age   age of a rock compared with the age of another rock

relative dating   method of dating that determines the age of a rock layer by relating it to another layer using superposition and the fossils contained

selective agents   the different living (biotic) and non-living (abiotic) agents that influence the survival of organisms

selective pressures   factors that contribute to selecting which variations will provide the individual with an increased chance of surviving over others

speciation   formation of new species

species   taxonomic unit consisting of organisms capable of mating and producing viable and fertile offspring

species diversity   the number of different species within an ecosystem

sympatric speciation   a type of speciation where populations are separated by a barrier that is not geographical

variations   differences between cells or organisms

vestigial structures   structures with no apparent function that are remnants from a past ancestor

on Resources

eWorkbooks	Study checklist (ewbk-12141) Literacy builder (ewbk-12143) Crossword (ewbk-12145) Word search (ewbk-12147) Reflection (ewbk-12149)
Solutions	Topic 3 Solutions (sol-1126)
Practical investigation eLogbook	Topic 3 Practical investigation eLogbook (elog-2215)
Digital document	Key terms glossary (doc-40102)

3.9 Activities

learnon

3.9 Review questions

Select your pathway

■ LEVEL 1	■ LEVEL 2	■ LEVEL 3
1, 2, 5, 7, 13	4, 6, 8, 10, 12, 15	3, 9, 11, 14, 16, 17

These questions are even better in jacPLUS!

- Receive immediate feedback
- Access sample responses
- Track results and progress

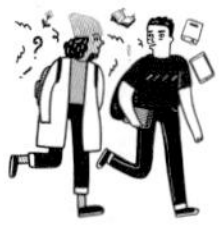

Find all this and MORE in jacPLUS

Remember and understand

1. **a.** Identify whether the following statements are true or false.

Statements	True or false?
i. As you move from kingdoms to species, there are fewer differences between the members of the group.	
ii. Organisms of the same species resemble each other and can interbreed to produce fertile offspring.	
iii. In binomial nomenclature, the name of each species is made up of two words, with the genus name beginning with a lower-case letter and the descriptive name beginning with a capital letter.	
iv. The variation of phenotypes within populations has contributed to the survival of our species.	
v. Radiometric dating determines the age of a fossil and the rock in which it is found based on the rate of decay of particular isotopes.	
vi. DNA hybridisation is a technique used to compare the similarity of DNA.	
vii. Biogeography relates to the geographical distribution of species.	
viii. Gene flow describes change in variation within populations due to chance events.	
ix. Speciation involves the loss of a species.	
x. An adaptation is a special characteristic or feature that improves an organism's chance of surviving in its environment.	

   b. Rewrite false statements to make them true.

2. Select the correct term from the following list to complete the sentences.
   - protein
   - selective
   - population
   - sperm
   - germline
   - frequency
   - phenotypes
   - DNA
   - struggle
   - increasing
   - evolution
   - new

    a. Genetic diversity is important because it codes for variations of_____, some of which may better suit the individual organism to a particular environment than others, _____ its chances of survival.

    b. A change in the genetic code in _____ can lead to a change in the _____ that is coded for and produced by that segment of DNA, which can change the organism's characteristics.

    c. Genetic variation within populations can be referred to in terms of the _____ of particular alleles within the population, and the source of new alleles are mutations that occur in _____ cells (such as _____ and eggs).

    d. Over time and many generations, if a particular variation continues to provide a _____ advantage, the number of individuals within a _____ that show the favourable variation will increase.

    e. Variation, _____ for existence, selective advantage and inheritance of advantageous variations form the basis for the theory of _____ by natural selection and also provide and explanation for how _____ species arise.

3. Recall the names or terms for the following.

    a. The scientific name for humans

    b. A type of evolution in which organisms face similar selective pressures and so evolve similar traits

    c. The southern supercontinent

    d. Person's name included in a system of nomenclature

    e. Around 30 of these make up the surface of the Earth

    f. A permanent change in DNA

    g. The type of dating that assumes that lower layers of rock are older than the ones above

    h. The rank between phylum and order

    i. Members of the same species living together in the same place at the same time

    j. Body structures that perform a different function but have a similar basic structure due to a shared common ancestry

4. Match the following terms to their definitions.

Term	Definition
a. Variation	**A.** The process by which the individuals with the most advantageous variation survive and reproduce more successfully than others
b. Fossil	**B.** A special characteristic that improves an organism's chance of surviving in an environment
c. DNA hybridisation	**C.** The range of structural and behavioural differences in a species
d. Comparative anatomy	**D.** Evidence of past life
e. Convergent evolution	**E.** Structures that may look similar due to comparable selection pressures rather than shared ancestry
f. Competition	**F.** A technique used to compare the similarity of DNA
g. Biodiversity	**G.** The struggle for resources between members of the same species
h. Natural selection	**H.** The geographical distribution of species
i. Radiometric dating	**I.** Comparing the structure of organisms
j. Clone	**J.** A process by which unrelated organisms living in similar environments develop similar features
k. Adaptation	**K.** A technique that uses measurements of isotopes to determine the age of rocks and fossils
l. Biogeography	**L.** Variation among organisms at the ecosystem, species and gene level
m. Analogous structures	**M.** The process in which organisms with particular features are selected and bred together
n. Artificial selection	**N.** A genetically identical organism
o. Protein	**O.** Made up of amino acids

5. Complete the following table, outlining differences between analogous and homologous structures.

TABLE Comparing analogous and homologous structures

	Analogous structures	Homologous structures
Structure (similar or different)		
Function (similar, different or either)		
Selective pressures (similar or different)		
Type of evolution (convergent or divergent)		
Origin (common ancestor or independent)		

6. Sequence the following into the correct evolutionary order.
   a. Flowering plants evolve
   b. Early dinosaurs evolve
   c. Mammals, flowering plants, insects, fish and birds dominate
   d. Bacteria evolve
   e. All living things are in the ocean; massive increase in multicellular organisms
   f. Most dinosaurs become extinct
   g. Greatest mass extinction of all time
   h. Dinosaurs dominate the planet
7. Outline the steps in which natural selection brings about genetic change in a population.
8. Distinguish between the following terms.
   a. Speciation and natural selection
   b. Divergent evolution and convergent evolution
   c. Primates and hominins
   d. Genetic variation and species variation
   e. Crossing over and independent assortment

Apply and analyse

9. Examine the table provided and identify the epoch in which the following occurred.
   a. First Nations Australians arrived in Australia
   b. The first marsupials appeared in Australia
   c. Swimming and flying mammals appeared
   d. The dinosaurs became extinct

TABLE Mammals and marsupials observed in different epochs

Some marsupial fossil finds and events	Epoch (millions of years ago)		Major mammal events
Present	**HOLOCENE** 0.01–present	Quaternary period	Humans investigate Earth's history
Most of the large Pleistocene marsupials became extinct about 15 000–30 000 years ago.	**PLEISTOCENE** 1.64–0.01 mya	Quaternary period	First Nations Australians arrived in Australia about 65 000 years ago.
Many giant burrowing marsupials became extinct; there were grazing kangaroos and lots of diprotodons.	**PLIOCENE** 5.2–1.64 mya	Cenozoic era	*Homo habilis*, the earliest known human, appeared in East Africa.
Primitive marsupial 'mice' and 'tapirs' were found at Lake Eyre, South Australia, and diprotodons at Bullock Creek, Northern Territory.	**MIOCENE** 23.5–5.2 mya	Cenozoic era	Lots of marsupial mammals were living in Australia and South America.
First Australian marsupials occurred about 23 million years ago. Fossils of diprotodons and a relative of the pygmy possum were found in Tasmania.	**OLIGOCENE** 35.5–23.5 mya	Tertiary period	First marsupials appeared in Australia. First primates appeared.
Lots of marsupial fossils of this age were found in South and North America.	**EOCENE** 56.5–35.5 mya	Tertiary period	Swimming and flying mammals appeared.
Dinosaurs became extinct about 65 million years ago.	**PALAEOCENE** 65–56.5 mya	Tertiary period	More mammals appeared after dinosaurs became extinct.

10. Refer to the animal kingdom evolutionary tree provided to answer the following questions.
    a. Identify which animal group (or phylum) is most closely related to the Chordata.
    b. Identify which is more closely related to the Mollusca phylum: Arthropoda or Nematoda.
    c. Do you think that Platyhelminthes would have more or less in common with Annelida than with Cnidaria? Justify your response.
    d. Explain the significance of adaptations of organisms in relation to their survival.

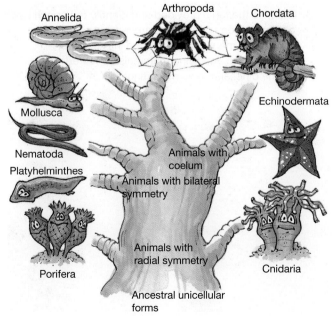

11. Outline the difficulties that the following scientists had in being able to express their scientific opinions because of society during their lifetime.
    a. Gregor Mendel
    b. Jean-Baptiste de Lamarck
    c. Charles Darwin

Evaluate and create

12. Examine the diagram provided showing the strata of rock, and deduce the answers to the following questions.
    a. Write down the names of the fossils in order from youngest to oldest.
    b. Which layer is the same age as layer 3, layer 4 and layer 5 respectively? How can you tell?
    c. Out of all the fossil layers numbered 1–11, which layer is the oldest?

13. Examine the figures of a. the *Archaeopteryx* and b. a modern flying bird.
    a. How are they similar?
    b. How are they different?
    c. Suggest reasons for the similarities and differences.

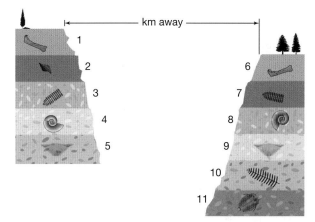

14. **SIS** A 2006 study showed that, because Australia had banned the use of one particular group of new antibiotics (the fluoroquinolones) in livestock, the level of resistance to the human antibiotic cyprofloxacin was only 2 per cent. In countries where these antibiotics are used in livestock, the human resistance level was around 15 per cent. Explain these findings.

15. When Darwin visited the Galápagos Islands, he found that longer-beaked species of finches were found in areas where insects were plentiful and shorter-beaked finches were found where seeds were the main food source. Explain this observation using the theory of evolution by natural selection.

16. Are humans still evolving? Outline your reasons and provide evidence to justify your response.

17. Biologists make inferences about evolutionary relationships based on comparative anatomy. Molecular biology techniques such as DNA hybridisation and protein sequencing may support or contradict these inferences. Explain how molecular biology techniques can be used to work out evolutionary relationships.

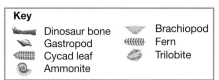

Key

Dinosaur bone		Brachiopod	
Gastropod		Fern	
Cycad leaf		Trilobite	
Ammonite			

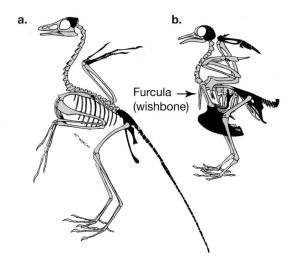

Fully worked solutions and sample responses are available in your digital formats.

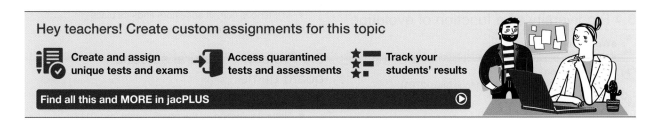

Hey teachers! Create custom assignments for this topic

Create and assign unique tests and exams

Access quarantined tests and assessments

Track your students' results

Find all this and MORE in jacPLUS

Online Resources

 Resources

Below is a full list of **rich resources** available online for this this topic. These resources are designed to bring ideas to life, to promote deep and lasting learning and to support the different learning needs of each individual.

3.1 Overview

 eWorkbooks
- Topic 3 eWorkbook (ewbk-12116)
- Starter activity (ewbk-12119)
- Student learning matrix (ewbk-12118)

Solution
- Topic 3 Solutions (sol-1126)

Practical investigation eLogbooks
- Topic 3 Practical investigation eLogbook (elog-2215)
- Investigation 3.1: Exploring relatedness and the last universal common ancestor (elog-2217)

Video eLesson
- Ancient DNA (eles-1069)

Weblink
- The mysterious origins of life on Earth

3.2 Classification

 eWorkbook
- Classification of species (ewbk-12121)

Weblink
- A higher level of classification of all living things

3.3 Processes of natural selection

eWorkbooks
- Labelling influences on natural selection (ewbk-12125)
- Natural selection (ewbk-12123)

Practical investigation eLogbook
- Investigation 3.2: Investigating some of the structural and physiological adaptations of First Nations Australians to the Australian environment (elog-2219)

Video eLessons
- Camouflage (eles-2702)
- Natural selection in peppered moths (eles-4233)

Interactivity
- How selection and variation influence natural selection and evolution (int-8162)

Weblinks
- Multidrug-resistant tuberculosis
- Multidrug resistant tuberculosis in Australia and our region
- Resistance in mosquitoes

3.4 Biodiversity as a function of evolution

 eWorkbook
- Variation, biodiversity and adaptations (ewbk-13081)

Video eLesson
- Blue footed booby mating dance (eles-2701)

Weblinks
- The five fingers of evolution
- Zoos Victoria case study: the Eastern Barred Bandicoot
- How we'll resurrect the Gastric Brooding frog and the Tasmanian tiger

3.5 Genetics and evolution

 eWorkbooks
- Isolation and new species (ewbk-12127)
- Patterns of evolution (ewbk-12129)

Video eLessons
- These finches are very different but have a common ancestor (eles-2919)
- How a new species evolves (eles-0162)

Weblinks
- Reproductive barriers
- Mysterious human ancestor finds it place

3.6 Evidence of the theory of evolution

 eWorkbooks
- Biogeography and flightless birds (ewbk-12133)
- Geological time (ewbk-12131)
- Ages of fossils (ewbk-12135)
- Fossils (ewbk-12137)
- Evidence of evolution (ewbk-12139)

Practical investigation eLogbooks
- Investigation 3.3: 4.6 billion years of history (elog-2221)
- Investigation 3.4: Creating casts of leaves (elog-2223)

Teacher-led videos
- Investigation 3.4: Creating casts of leaves (tlvd-10815)

Video eLessons
- Fossilised dinosaur tracks in mud (eles-2706)
- Folds (eles-2707)

Interactivities
- Revelation: 'Fossils' (int-1018)
- Fossil formation (int-5874)
- Uranium-235 decays into lead-207 (int-5876)
- The evolution of the horse (int-5878)
- The process involved in DNA hybridisation (int-5880)
- Homologous structures (int-5879)

Weblinks
- Geological time
- Digital atlas of ancient life
- Human evolution: Turning back the clock
- Interactive phylogenetic tree
- The Pangaea pop-up TED talk
- What does it mean to be human?
- Mysterious human ancestors finds its place
- What does it mean to be human?

3.8 Project — Natural selection board game

ProjectsPLUS
- Natural selection — the board game! (pro-0112)

3.9 Review

 eWorkbooks

- Topic review Level 1 (ewbk-12150)
- Topic review Level 2 (ewbk-12152)
- Topic review Level 3 (ewbk-12154)
- Study checklist (ewbk-12141)
- Literacy builder (ewbk-12143)
- Crossword (ewbk-12145)
- Word search (ewbk-12147)
- Reflection (ewbk-12149)

Digital document

- Key terms glossary (doc-40102)

To access these online resources, log on to **www.jacplus.com.au**

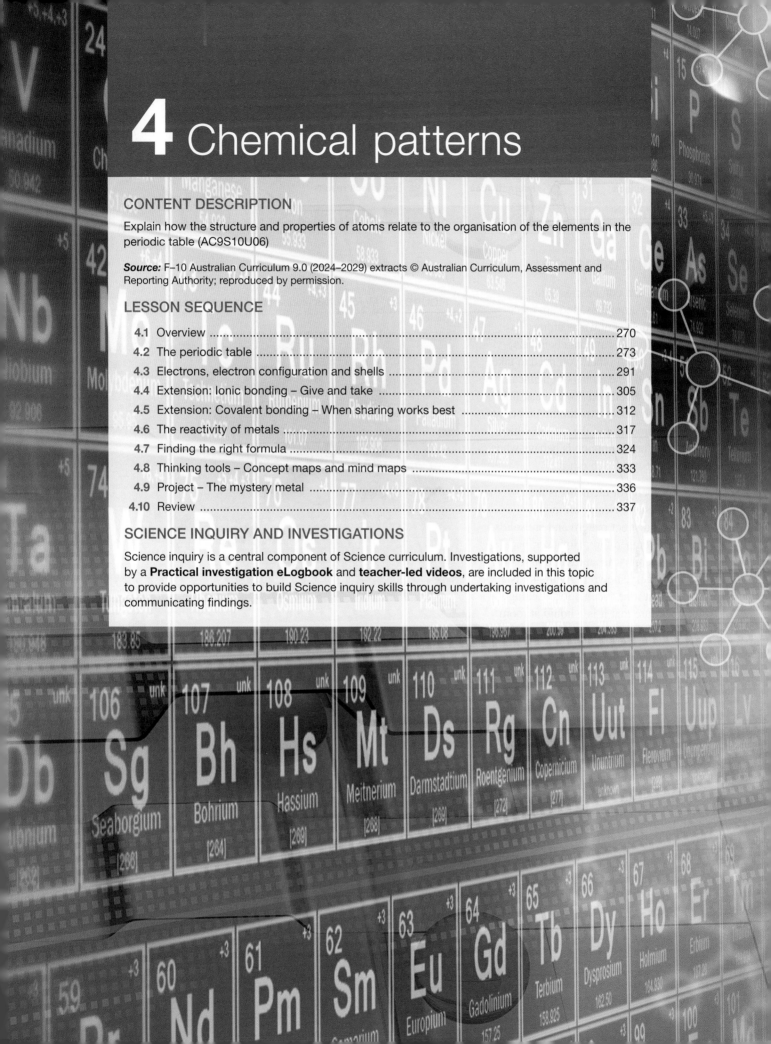

4 Chemical patterns

CONTENT DESCRIPTION

Explain how the structure and properties of atoms relate to the organisation of the elements in the periodic table (AC9S10U06)

Source: F–10 Australian Curriculum 9.0 (2024–2029) extracts © Australian Curriculum, Assessment and Reporting Authority; reproduced by permission.

LESSON SEQUENCE

SCIENCE INQUIRY AND INVESTIGATIONS

Science inquiry is a central component of Science curriculum. Investigations, supported by a **Practical investigation eLogbook** and **teacher-led videos**, are included in this topic to provide opportunities to build Science inquiry skills through undertaking investigations and communicating findings.

LESSON
4.1 Overview

4.1.1 Introduction

Elements and compounds form everything around us, and are made up of particles known as atoms. To understand the way that chemicals react with each other, you need to take a look inside the atoms of chemical elements. When you do, you can find patterns that help explain the properties of the elements and the way in which elements and compounds behave when they react with each other.

One property of elements is their physical state. Nearly all metals follow a distinct pattern and are solids at room temperature. However, one metal, mercury, is an exception to this. The mercury shown in figure 4.1 is a metal, but it has such a low melting temperature that it is a liquid at room temperature.

FIGURE 4.1 Mercury has unique properties compared to other metals

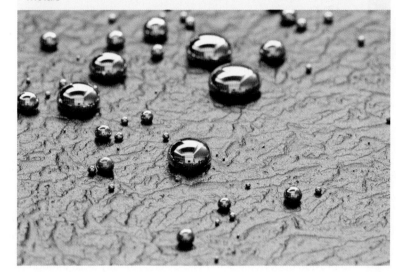

 Resources

 Video eLesson The atom (eles-1775)

Watch this video from *The story of science* about the development of the model of the atom.

4.1.2 Think about chemical patterns

1. Who was Dmitri Mendeleev and how was he able to predict the future?
2. What are metalloids?
3. Why do cars no longer use petrol containing lead?
4. Why do we talk about shells when describing electrons?
5. Why are you more likely to find pure gold on or near the Earth's surface than pure copper or iron?
6. What is the connection between the reactivity of metals and the ancient Roman Empire?

4.1.3 Science inquiry

Inside the elements

Atoms are the building blocks of the chemical **elements**. They are, therefore, also the building blocks of compounds and mixtures. For thousands of years, alchemists and scientists have searched for patterns in the substances that make up the universe. Many of them succeeded to some extent. But the discovery by Lord Ernest Rutherford in 1911 that most of the atom was empty space, and subsequent discoveries about the particles inside that atom by Niels Bohr and other scientists, provided the missing links in the patterns.

Atoms have the same general structure shown in figure 4.2. The main difference in their structures is in the numbers of each particle within the atom (the sub-atomic particles).

atoms very small particles that make up all things

elements pure substances made up of only one type of atom

FIGURE 4.2 Parts of the atom

int-5797

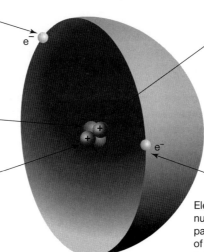

Electrons are about 1/2000th the size of protons and neutrons. Electrons have an electrical charge of negative one (−1). An atom has the same number of electrons as protons. The charges balance out so an atom has no electrical charge. It is said to be neutral.

Protons and neutrons are almost the same size. A proton has an electrical charge of positive one (+1). The number of protons in an atom determines what type of atom it is. For example, every atom with 79 protons is a gold atom, and every gold atom has 79 protons. Substances that are made up of only one type of atom are called elements.

A neutron has no electrical charge.

Protons and neutrons make up the nucleus. They are held together by very strong nuclear forces. Almost all of the mass of an atom is in the nucleus.

Electrons move rapidly around the nucleus. Although they follow no set paths, electrons are arranged in a series of energy levels around the nucleus. These energy levels are referred to as electron shells.

---

elog-2442

INVESTIGATION 4.1

Modelling an atom

Aim

To model an atom and observe what makes up most of an atom

Materials

- Calculator
- Semolina grains (a small number)
- Two different coloured pieces of plasticine
- Ruler
- Map of local area (showing scale)
- Mathematical compass

Method

1. Use the periodic table or other means of research to determine the number of protons and the number of neutrons that make up the nucleus of an atom of beryllium.
   How many of each type will there be?
2. Using one colour of plasticine to represent the protons, construct the required number of small balls. Make each ball the size of a pea.
3. Repeat step 2 with the other colour of plasticine to represent the required number of neutrons.
4. Now join all the balls of plasticine to represent the nucleus. Press them together just hard enough to make them stick. Try not to squash them too much. The total diameter of your nucleus should be no larger 4 cm (about the size of a ping pong ball or golf ball).

5. Measure the diameter and calculate the radius of the nucleus you have made. Try to do this as accurately as possible.
6. In this model, the semolina grains represent the electrons. How many will you require for this model? What do you notice about their size in comparison to each of the protons and neutrons in the nucleus?
7. The electrons are about 10 000* times further out than the protons and neutrons. Calculate where it will be necessary to place the semolina grains in this model. Do you think it is practical to complete your model?
Tip: Make sure you use the radius, not the diameter, of your nucleus in this calculation.

Extension

8. Now imagine that the size of the nucleus is about the size of a basketball.
   Work out where the electrons would need to be placed on a model of this scale.

   Obtain a map of your local area and using the map scale and a compass, draw a circle with this radius, with your school at the centre. Do you live within this circle?

* This figure is an approximation. It is difficult to measure the distances involved due to their exceedingly small size as well as limitations within the methods used. The properties of the particles involved also make this difficult.

Results

Answer the questions that are asked in the method above.

Make sure you answer in a way that is easy for someone else to understand.

Discussion

1. Explain what your results allow you to understand about the structure of an atom.
2. Draw a labelled diagram of the nucleus of beryllium, ensuring you show the correct number of protons and neutrons.
3. Explain how you would calculate the number of electrons in an atom of beryllium.
4. The beryllium atom loses two electrons to become an ion. What would you expect the charge of this ion to be?
5. Identify the sub-atomic particle or particles that:
   a. orbit the nucleus
   b. can be found inside the nucleus
   c. has/have no electric charge
   d. has/have a positive electric charge
   e. has/have a negative electric charge
   f. is/are lightest.
6. The atom shown in figure 4.2 belongs to a single chemical element.
   a. What is the atomic number of the element?
   b. Which particles are counted to determine the atomic number of the element?
   c. Identify the element in the diagram.

Extension

7. If you did the extension, survey how many students in your classroom live in the area you circled on the map.
8. Research the contributions of Lord Ernest Rutherford, Niels Bohr and Sir James Chadwick to our knowledge of the atom. What features from each of these people's work are shown in your model?

Conclusion

Summarise the findings of your investigation, linking this to the structure of an atom.

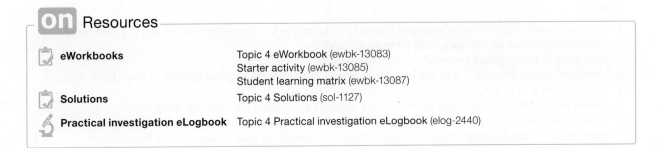

on Resources

eWorkbooks
Topic 4 eWorkbook (ewbk-13083)
Starter activity (ewbk-13085)
Student learning matrix (ewbk-13087)

Solutions
Topic 4 Solutions (sol-1127)

Practical investigation eLogbook
Topic 4 Practical investigation eLogbook (elog-2440)

LESSON
4.2 The periodic table

LEARNING INTENTION

At the end of this lesson you will be able to explain the atomic structure and properties of elements that are used to organise them in the periodic table and how new elements are being made and added to it.

4.2.1 The patterns emerge

Two thousand years ago, only ten elements had been identified and only a few more were discovered by the early eighteenth century. During this period, attempts to classify them and identify patterns proved difficult. It was much like attempting a jigsaw puzzle when most of the pieces are missing.

With the progress of science and chemistry, in particular during the seventeenth and eighteenth centuries, many more elements were discovered and the need to organise them in some way soon became apparent. Over time, many scientists (of many different nationalities) attempted and contributed to this task, culminating with the work of Russian chemist Dmitri Mendeleev.

In 1869 he presented an arrangement of the known elements into rows and columns based on their properties and atomic weights. By following these patterns closely, he was able to leave gaps for undiscovered elements and confidently predicted the properties of the element germanium 15 years before it was discovered. Today, the properties of new elements are predicted before their discovery, just as they were in Mendeleev's time. A timeline of the discovery of all the different elements is shown in figure 4.3.

FIGURE 4.3 A timeline of the discovery of elements

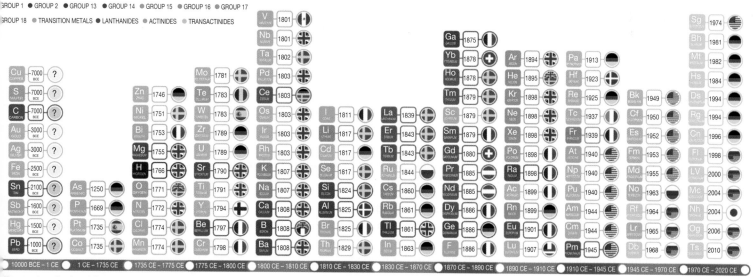

The years shown for element discoveries subsequent to those elemets which were known since antiquity are those in which the element in question was isolated for the first time. The flages identify the country in which the discovery was made, rather than the nationality of the discovers (s).

FIGURE 4.4 The periodic table of elements

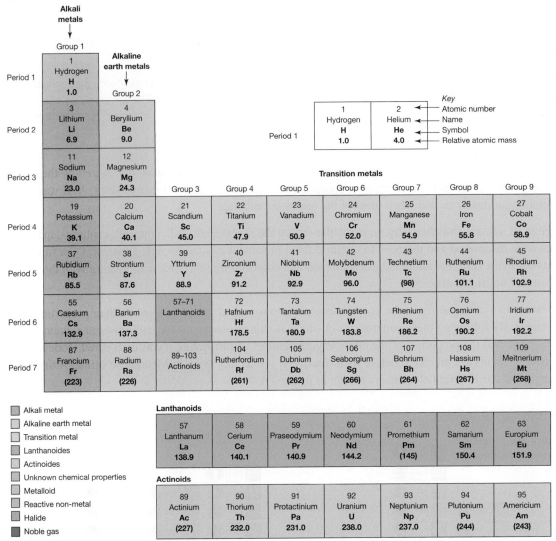

4.2.2 The periodic table

Mendeleev's work formed the basis of what we now know as the **periodic table** and it has been able to accommodate all the elements that have been discovered since.

The elements are arranged into a grid with rows and columns.
- Rows contain elements in order of increasing mass or atomic weight. The rows of elements are called **periods**.
- Columns contain families of elements with similar properties, and are called **groups**.

Figure 4.4 shows the periodic table with all currently discovered elements. At the time of publication, scientists have reported the discovery of elements with atomic numbers up to 118. The discoveries of four elements (113, 115, 117 and 118) were only confirmed by the International Union of Pure and Applied Chemistry (IUPAC) in 2016.

This arrangement is called the periodic table because elements with similar properties occur at regular intervals or periods.

periodic table the table listing all known elements, grouped according to their properties and in order of atomic number (number of protons in their nucleus)

periods rows of elements on the periodic table

groups columns of the periodic table containing elements with similar properties

Noble gases → Group 18

Halogens → Group 17

Non-metals →

Group 10	Group 11	Group 12	Group 13	Group 14	Group 15	Group 16	Group 17	Group 18
								2 Helium **He** 4.0
			5 Boron **B** 10.8	6 Carbon **C** 12.0	7 Nitrogen **N** 14.0	8 Oxygen **O** 16.0	9 Fluorine **F** 19.0	10 Neon **Ne** 20.2
			13 Aluminium **Al** 27.0	14 Silicon **Si** 28.1	15 Phosphorus **P** 31.0	16 Sulfur **S** 32.1	17 Chlorine **Cl** 35.5	18 Argon **Ar** 39.9
28 Nickel **Ni** 58.7	29 Copper **Cu** 63.5	30 Zinc **Zn** 65.4	31 Gallium **Ga** 69.7	32 Germanium **Ge** 72.6	33 Arsenic **As** 74.9	34 Selenium **Se** 79	35 Bromine **Br** 79.9	36 Krypton **Kr** 83.8
46 Palladium **Pd** 106.4	47 Silver **Ag** 107.9	48 Cadmium **Cd** 112.4	49 Indium **In** 114.8	50 Tin **Sn** 118.7	51 Antimony **Sb** 121.0	52 Tellurium **Te** 127.6	53 Iodine **I** 126.9	54 Xenon **Xe** 131.3
78 Platinum **Pt** 195.1	79 Gold **Au** 197.0	80 Mercury **Hg** 200.6	81 Thallium **Tl** 204.4	82 Lead **Pb** 207.2	83 Bismuth **Bi** 209.0	84 Polonium **Po** (210)	85 Astatine **At** (210)	86 Radon **Rn** (222)
110 Darmstadtium **Ds** (271)	111 Roentgenium **Rg** (272)	112 Copernicium **Cn** (285)	113 Nihonium **Nh** (280)	114 Flerovium **Fl** (289)	115 Moscovium **Mc** (289)	116 Livermorium **Lv** (292)	117 Tennessine **Ts** (294)	118 Oganesson **Og** (294)

64 Gadolinium **Gd** 157.3	65 Terbium **Tb** 158.9	66 Dysprosium **Dy** 162.5	67 Holmium **Ho** 164.9	68 Erbium **Er** 167.3	69 Thulium **Tm** 168.9	70 Ytterbium **Yb** 173.1	71 Lutetium **Lu** 175.0

96 Curium **Cm** (247)	97 Berkelium **Bk** (247)	98 Californium **Cf** (251)	99 Einsteinium **Es** (254)	100 Fermium **Fm** (257)	101 Mendelevium **Md** (258)	102 Nobelium **No** (255)	103 Lawrencium **Lr** (256)

The observation that the physical and chemical properties of the elements recur at regular intervals when elements are listed in order of atomic weight is known as the **periodic law**.

> **periodic law** elements with similar properties occur at regular intervals when all elements are listed in order of atomic weight

SCIENCE AS A HUMAN ENDEAVOUR: The development of the periodic table

Dmitri Mendeleev

The organisation of the chemical elements into the periodic table is regarded as one of the most important developments in modern chemistry. This is almost always credited to the Russian chemist Dmitri Mendeleev, who published his findings in 1869, although another scientist, Lothar Meyer, could be considered unlucky to have not also been given recognition. Like many important scientific advances, both then and now, Mendeleev's achievement built upon the work of preceding scientists and the knowledge and technology of the time. It also highlights the international nature of science.

Some of the other scientists who laid the foundations for Mendeleev include the following.

- Antoine Lavoisier (French)

In 1789 Lavoisier published an arrangement of the elements known at the time based upon their chemical properties. Although primitive by today's standards and containing some errors (some 'elements' were later found to be compounds), it was able to distinguish between metals and non-metals.

- Johann Döbereiner (German)

In 1817, Döbereiner grouped some of the then known elements into groups of three, or 'triads', based upon their chemical properties and atomic weights. (Atomic weight is an old term that is not used much today. Today the term relative atomic mass is preferred. This gives an indication of how the masses of atoms of different elements compare to each other.) Döbereiner found that in these triads, the element in the middle had an atomic weight that was roughly equal to the average of the other two. Also, its properties and appearance were in between the other two. For example, he placed chlorine, bromine and iodine into a triad in that order. Their known atomic weights at the time were 35.5, 80 and 127. He then noticed that the properties of bromine were in between those of chlorine and iodine. Although he was able to do this with two other triads, he could not explain why and his work went largely unnoticed until after Mendeleev published his table. It did, however, provide the important clue that atomic weights were somehow important.

- John Newlands (English) and the Law of Octaves

Newlands arranged all the elements known at that time in ascending order based on atomic weight and then numbered them. He then arranged them into seven columns based on similar chemical properties. He noted that each eighth element shared similar chemical properties. He likened this to a musical scale where each eighth note is similar to the first. Newlands published his findings in 1864.

- Lothar Meyer (German)

Meyer worked independently on organising the chemical elements at the same time as Mendeleev. He graphed various properties of the then known elements against atomic weight, thus showing their periodic variation. When Mendeleev published his work in 1869, Meyer was able to publish his version of the periodic table just two months later. The two versions were remarkably similar.

FIGURE 4.5 **a.** Antoine Lavoisier **b.** Johann Döbereiner **c.** John Newlands **d.** Lothar Meyer

a.   b.   c.   d.

How did Mendeleev do this?

Mendeleev spent years collecting information about each of the 63 elements known at that time. On each card he recorded the atomic weight and a summary of both physical and chemical properties. Using these cards, he was able to produce an arrangement similar to that of Newlands', based on increasing atomic weight. Mendeleev's table, however, had two significant differences to that of Newlands'. It possessed an eighth column, into which Mendeleev placed a number of elements that were required to maintain patterns in the rest of the table and whose properties appeared to be significantly different. It also contained gaps and Mendeleev speculated that these represented elements that were yet to be discovered.

FIGURE 4.6 This stamp, issued in 2019, celebrates the 150th anniversary of Mendeleev's periodic table

Why were Mendeleev's ideas accepted?

Mendeleev's arrangement was an elegant representation of the then known elements. However, it went further than just being a clever arrangement. In science, a theory must not only be able to explain known facts, but it must also be able to make predictions that can be experimentally tested. Using the gaps inserted into his arrangement, Mendeleev made predictions about the elements yet to be discovered. The subsequent discovery of a number of these elements confirmed his predictions with remarkable accuracy (see section 4.2.3).

The periodic table today

The modern periodic table has some similarities, but also a number of differences to the one proposed by Mendeleev. Once again, scientists of many different nationalities have contributed to it. The main differences are detailed below.

- It is based on atomic number (see section 4.2.4) rather than atomic weight. The existence of protons had yet to be discovered in Mendeleev's time.
- It contains many more elements.
- It has a region containing the transition metals, as well as regions representing the lanthanoids and actinoids (see figure 4.4). These new regions contain many of the elements that Mendeleev placed in his eighth column (because he could not fit them anywhere else!)
- There are eighteen columns in today's table.
- Seven of these columns (labelled groups 1–2 and 13–17) represent groups from Mendeleev's table. Group 18, however, is a totally new column, which contains elements that were almost 'surprise' discoveries. Their existence was not even predicted before their discovery.

ACTIVITY: Atomic number versus relative atomic mass

1. Observe argon (Ar) and potassium (K) in the periodic table in figure 4.4. Which one has the highest relative atomic mass? Which one has the highest atomic number? What would happen to their positions if the periodic table was based on atomic mass instead of atomic number?
2. Look carefully at the first 92 elements in a modern periodic table and identify three other pairs of elements that would be 'round the wrong way' if relative atomic mass rather than atomic number were used.

4.2.3 An educated guess

Mendeleev was so confident about the periodic law that he deliberately left gaps in his periodic table and was able to predict the properties of the unknown elements that would fill the gaps. Mendeleev predicted the existence of germanium, which he called eka-silicon. This element was discovered in 1886, 15 years later. table 4.1 shows the accuracy of Mendeleev's predictions.

TABLE 4.1 Properties of eka-silicon and germanium

Properties of eka-silicon as predicted by Mendeleev	Properties of germanium, discovered in 1886
A grey metal	A grey-white metal
Melting point of about 800 °C	Melting point of 958 °C
Relative atomic mass of 73.4	Relative atomic mass of 72.6
Density of 5.5 g/cm^3	Density of 5.47 g/cm^3
Reacts with chlorine to form compounds with four chlorine atoms bonded to each eka-silicon atom	Reacts with chlorine and forms compounds in a ratio of four chlorine atoms to every germanium atom

Mendeleev's work led many scientists to search for new elements. By 1925, scientists had identified all 92 naturally existing elements.

The periodic table in figure 4.4 includes the names, **symbols** and **atomic numbers** of all currently known elements. The symbols are a form of shorthand for writing the names of the elements and are recognised worldwide.

symbols one- or two-letter code(s) used for the elements

atomic number the number of protons in the nucleus of an atom

Some periodic tables describe the properties of each element, including its physical state at room temperature, melting point, boiling point and **relative atomic mass**. Most elements exist as solids under normal conditions and a few exist as gases. Only two elements exist as liquids at normal room temperature — bromine and mercury.

4.2.4 Counting sub-atomic particles

The periodic table is organised on the basis of atomic numbers.
- The atomic number of an element is the number of **protons** present in each atom. Atoms with the same atomic number have identical chemical properties.
- Because atoms are electrically neutral, the number of protons in an atom is the same as the number of **electrons** (protons are positively charged and electrons are negatively charged, so they cancel each other out).
- The **mass number** of an atom is the sum of the number of protons and **neutrons** in the atom (the particles within the **nucleus** of the atom). The number of neutrons in an atom can therefore be calculated by subtracting the atomic number from the mass number.

The notation for the mass and atomic number is shown in figure 4.7. It shows an example for carbon, which has an atomic number of 6 and a mass number of 12. Because this is an atom, it is electrically neutral, and also has six electrons.

FIGURE 4.7 Notation used for atoms of different elements

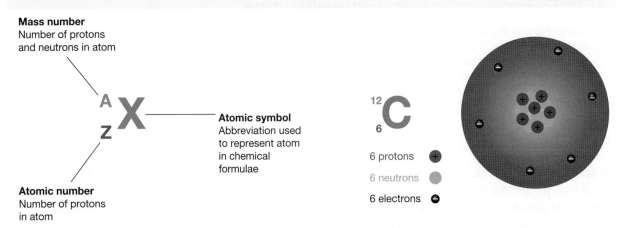

Calculating the number of protons and neutrons in an atom

Where Z is the atomic number and A is the mass number:

A = number of protons + number of neutrons

Z = number of protons

A – Z = number of neutrons

relative atomic mass mass of the naturally occurring mixture of the isotopes of an element

proton a positively charged particle found in the nucleus of an atom

electron very light, negatively charged particle inside an atom that moves around the central nucleus

mass number the number of protons and neutrons in the nucleus of an atom

neutron uncharged particle found in the nucleus of an atom

nucleus the central part of an atom, made up of protons and neutrons

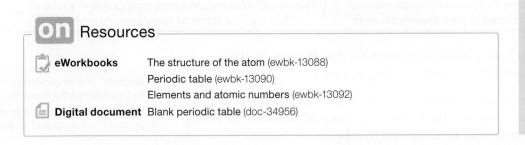

Resources

📋 **eWorkbooks**   The structure of the atom (ewbk-13088)
Periodic table (ewbk-13090)
Elements and atomic numbers (ewbk-13092)

📄 **Digital document**   Blank periodic table (doc-34956)

FIGURE 4.8 Unleaded petrol was introduced to Australia in 1986.

Lead poisoning was a common occurrence in ancient Rome because the lead the Romans used to make water pipes and cooking utensils slowly dissolved into the water. Acute lead poisoning causes mental impairment and personality changes. The effects are not immediately noticeable, but occur gradually as the amount of lead in the body accumulates. Some historians attribute the strange behaviour of several Roman emperors to lead poisoning.

In the Middle Ages, plates, cups and other drinking vessels were often made from pewter, an alloy of lead and tin. The acids in food and drinks caused lead to leach out and cause poisoning.

Until 1986, lead was added to petrol to stop 'knocking' in car engines. Unleaded fuel allows a catalytic converter to prevent pollutants such as nitrous oxides and carbon monoxide from being emitted. With lead in the petrol, these devices couldn't work. Lead emissions from cars were possibly causing a build-up of lead in humans in built-up areas.

The word *plumber* is derived from the Latin word *plumbum*, meaning 'lead'. Look up the symbol for lead in the periodic table. Where do you think this symbol came from?

4.2.5 How heavy are atoms?

Measuring and comparing the masses of atoms is difficult because of their extremely small size. Chemists solve this problem by comparing equal numbers of atoms, rather than trying to measure the mass of a single atom.

A further problem arises because not all atoms of an element are identical. Although all atoms of a particular element have the same atomic number, they can have different numbers of neutrons. Hence, some elements contain atoms with slightly different masses. Two atoms of the same element with different numbers of neutrons are called **isotopes**.

If you were to take a spoonful of most elements, they would contain a mix of the different isotopes. To determine what an atom would weigh 'on average', the relative abundance and mass of each isotope is used to calculate a **weighted mean**. This number is referred to as the relative atomic mass, which has replaced the older term of 'atomic weight', and is usually not a whole number. The mass number (A) of an element can usually be found by rounding the relative atomic mass.

> **isotopes** atoms of the same element that differ in the number of neutrons in the nucleus
>
> **weighted mean** the average mass of an element that is calculated from the percentage of each isotope in nature

FIGURE 4.9 Examples of different isotopes of carbon

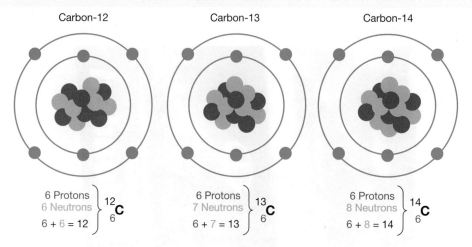

Carbon-12
6 Protons
6 Neutrons $\Big\}$ $^{12}_{6}C$
6 + 6 = 12

Carbon-13
6 Protons
7 Neutrons $\Big\}$ $^{13}_{6}C$
6 + 7 = 13

Carbon-14
6 Protons
8 Neutrons $\Big\}$ $^{14}_{6}C$
6 + 8 = 14

4.2.6 Families of elements

The periodic table contains nine groups (or families) of elements, some of which have been given special names, as seen in figure 4.10.

- Group 1 elements are known as **alkali metals**. The alkali metals all react strongly with water to form basic (alkaline) solutions.
- Group 2 elements are referred to as **alkaline earth metals**.
- Group 15 elements are known as pnictogens or the nitrogen family.
- Group 16 elements are knows as chalcogens or the oxygen family.
- Group 17 elements are known as **halogens** (or halides). The halogens are brightly coloured elements. Chlorine is green, bromine is red-brown and iodine is silvery-purple.
- Group 18 elements are known as **noble gases**. The noble gases are inert, meaning they do not readily react with other substances.
- The block of elements in the middle of the table is known as the **transition metal block**. Elements in the Lanthanoids and Actinoids blocks are also transition metals.

alkali metals very reactive metals in group 1 of the periodic table

alkaline earth metals reactive metals in group 2 of the periodic table

halogens non-metal elements in group 17 of the periodic table

noble gases elements in the last column of the periodic table that are unreactive

transition metal block a block of metallic elements in the middle of the periodic table

FIGURE 4.10 The named families of elements in the periodic table

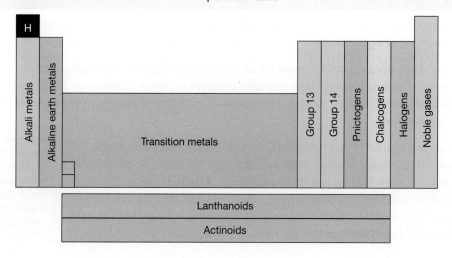

4.2.7 Metals, non-metals and metalloids

The line that zigzags through the periodic table (as seen in figure 4.11) separates the **metals** from the **non-metals**. About three-quarters of all elements are classified as metals, which are found on the left-hand side of the table. The non-metals are found on the upper right-hand side of the table. Eight elements that fall along the line between metals and non-metals have properties belonging to both. They are called **metalloids**.

FIGURE 4.11 The periodic table can be broken into metals, non-metals and metalloids. Polonium (*) and astanine (**) are sometimes disputed as being metalloids.

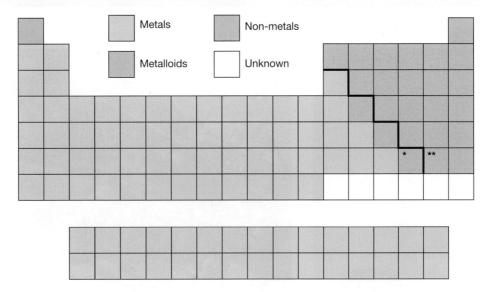

Some of the newly discovered elements are yet to have their properties exactly determined, but based on the periodic table, predictions can be made as to their likely properties. Due to having extremely short half-lives, it is difficult to confirm their exact properties.

Metals

Metals (with the exception of mercury) have several features in common.

- They are solid at room temperature.
- They can be polished to produce a high shine or **lustre**.
- They are good **conductors** of electricity and heat.
- They can be beaten or bent into a variety of shapes. We say they are **malleable** (the word malleable comes from the Latin word *malleus*, meaning 'hammer').
- They can be made into a wire. We say they are **ductile**.
- They usually melt at high temperatures. Mercury, which melts at −40 °C, is one exception.

FIGURE 4.12 Copper wire exhibits many common metallic properties.

metals elements that conduct heat and electricity and are shiny solids that can be made into thin wires and sheets

non-metals elements that do not conduct electricity or heat and usually have a lower boiling and melting point than metals

metalloids elements that have features of both metals and non-metals

lustre the shiny appearance of many metals

conductors materials that allow electric charge to flow through them

malleable able to be beaten, bent or flattened into shape

ductile capable of being drawn into wires or threads; a property of most metals

on Resources

▶ **Video eLesson** The unusual properties of mercury (eles-4184)

Non-metals

Only 22 of the elements are non-metals. At room temperature, eleven of them are gases, ten are solid and one is liquid. The solid non-metals have most of the following features in common.

- They cannot be polished to give a shine like metals; they are usually dull.
- They are **brittle**, which means they shatter when they are hit.
- They cannot be bent into shape.
- They are usually poor conductors of electricity and heat.
- They usually melt at low temperatures.

> **brittle** breaks easily into many pieces

FIGURE 4.13 Common examples of non-metals are sulfur, carbon and oxygen.

FIGURE 4.14 Metalloids are important materials often used in electronic components of computer circuits.

Metalloids

Some of the elements in the non-metal group look like metals. One example is silicon. While it can be polished like a metal, silicon is a semi-conductor of heat and electricity and cannot be bent or made into wire. Elements that have features of both metals and non-metals are called metalloids. There are eight metalloids altogether: boron, silicon, arsenic, germanium, antimony, polonium, astatine and tellurium.

Polonium (* in figure 4.11) is sometimes disputed, and is sometimes classed as a metal rather than a metalloid. Astanine (** in figure 4.11) can also be classified as a halogen in the non-metal group. Astanine is the rarest occurring natural element in the Earth's crust and, due to rapid decay, its exact properties are not known.

DISCUSSION

Silicon is used to make computer chips. This is the reason why the home of the technology industry in California is called Silicon Valley. Computer chips are wafers of silicon with grooves cut in them for electrons to flow through, called circuits.

- What properties of silicon make it ideal for making computer chips?
- In pursuit of faster computers, manufacturers are trying to fit more circuits onto the same sized chip of silicon. Is there a limit to how many circuits can be cut into a chip? If so, why?

INVESTIGATION 4.2

Properties of metals and non-metals

Aim

To investigate and compare the properties of metals and non-metals

Materials

- safety glasses, gloves and laboratory coat
- 1 M hydrochloric acid
- water
- magnesium
- iron filings
- copper filings
- sulfur powder

- universal indicator
- 4 test tubes
- 4 gas jars filled with oxygen gas
- 4 deflagrating spoons
- dropping pipette
- spatula
- Bunsen burner, heatproof mat and matches

Method

Before you begin write a hypothesis for each reaction, ask yourself what you expect to observe for the metals and the non-metals in each of the reactions.

CAUTION

The heating part of this experiment should be done in a fume cupboard. Safety glasses, gloves and laboratory coats must be worn at all times. Do not look at the flame during the investigation.

Part A: Physical appearance

1. Describe the appearance and physical state of each element.
2. Which samples look like metals? Which samples look like non-metals?

Part B: Reactions with hydrochloric acid

1. Place a small quantity of magnesium in a test tube.
2. Add about 2 mL of hydrochloric acid.
3. Record your observations in a suitable table in your results.
4. Repeat the above steps with iron filings.
5. Repeat the above steps with copper filings.
6. Repeat the above steps with sulfur powder.

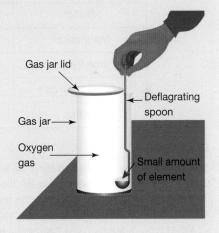

Part C: Burning substances in oxygen

1. Place a small amount of magnesium in a deflagrating spoon and heat it. Do not look directly at the flame.
2. When hot, place it into the gas jar full of oxygen gas (as shown).
3. Record your observations in a suitable table in your results.
4. Repeat the above steps using iron filings.
5. Repeat the above steps using copper filings.
6. Using a fume cupboard, observe a demonstration done by your teacher of the preceding steps using sulfur.

Part D: Determining pH

1. Tip the substance from the deflagrating spoon into the jar. Add about 10 mL of water to each jar from part B and carefully shake. (Alternatively, you can use an additional beaker for this step.)
2. Add 3 drops of universal indicator.
3. Record colour and determine the pH of the solution.
4. Repeat for each substance. Your teacher will demonstrate this for sulfur.

Results

Collect qualitative data for this investigation. (Examples of qualitative data include a colour change, production of a gas, a substance disappearing or changing state, an odour or production of heat.) Ensure you gather data before, during and after each process as outlined in the following examples. These changes are clues as to the chemical processes that are occurring.

TABLE Observations that should be made in this investigation

Before	During	After
Observations of substances (reactants) before the experiment	Observations of substances (reactants and products) during the experiment	Observations of substances (reactants and products) after the experiment

Copy this qualitative data table to record your data.

TABLE Observations of the reaction of different elements for the three different tests

Part	Substance	Before	During	After
B	Magnesium			
	Iron filings			
	Copper filings			
	Sulfur			
C	Magnesium			
	Iron filings			
	Copper filings			
	Sulfur			
D	Magnesium			
	Iron filings			
	Copper filings			
	Sulfur			

Discussion

1. Use the periodic table to determine which of the elements tested were metals and which were non-metals.
2. Describe any differences between the effect of acids on metals and non-metals.
3. Describe what happened when the metals and non-metals reacted with oxygen.
4. The metal or non-metal oxides formed in the gas jars dissolved in water to form acidic and basic solutions. What type of solution did the metals form? What type of solution did the non-metals form?

Conclusion

Summarise the findings for this investigation about the properties of metals and non-metals.

INVESTIGATION 4.3

Comparing the properties of two metal families

elog-2446

Aim

To investigate and compare the properties of metals from two different groups of the periodic table

Materials

- small samples of magnesium, iron, calcium and copper
- 'rice grain' equivalent amounts of calcium chloride, magnesium chloride, iron (II) chloride and copper (III) chloride
- spatula
- 5 test tubes and a test-tube rack

- 2-volt power supply
- 3 connecting leads
- 2 alligator clips
- light globe and holder
- 2 M hydrochloric acid
- water
- matches
- stirring rod
- safety glasses, gloves and laboratory coat

CAUTION

Safety equipment should be worn for the duration of this experiment.

Any experiments that include calcium metal are recommended to be performed as a demonstration by your teacher in a fume cupboard. Calcium metal is highly reactive. It should only be conducted using very small quantities.

Method

1. Design a table to record your results for this investigation. Predict the properties of the different metals before you begin.
2. Describe the physical state (solid, liquid or gas) of each element.
3. Describe the physical appearance of each element.
4. Set up the circuit as shown and determine whether each of the elements conducts electricity.
5. Determine whether any of the elements react with water by placing a small sample in 2 mL of water in a test tube. Calcium should be completed as a demonstration by your teacher. Record any changes that occur in your table.
6. Determine whether the metals react with acid by placing a small sample of each metal in 1 mL of 2 mol L^{-1} hydrochloric acid in a test tube. If a gas is produced, test it by holding a lit match at the mouth of the test tube. Make sure the test tube is pointed away from you.
   - If hydrogen is present, you will hear a 'pop'.
   - If oxygen is present, the match should burn more brightly.
   - If carbon dioxide is present, the match should go out.
7. Add a small amount of each of the metal compounds (magnesium chloride, calcium chloride, iron (II) chloride and copper (III) chloride) to 5 mL of water. Comment on their solubility and the colour of any solution made.

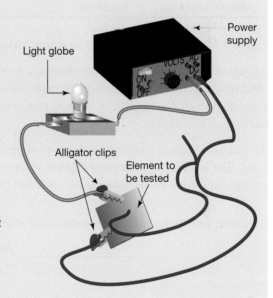

Results

Record the results of each of the observations in an appropriate table. Include the group that each element belongs to.

Discussion

1. What are the properties of copper and iron? Are there any similarities?
2. What are the properties of calcium and magnesium? Are there any similarities?
3. List the metals in order of reactivity with water and acids. List them from most reactive to least reactive.
4. Were there any differences between solubilities of the metal compounds or the colours of the solutions they formed? Describe these differences.
5. What could you infer about the properties of elements in the same group? Give reasons for your answer.

Conclusion

Summarise the findings for this investigation about the properties of metals from two different groups of the periodic table.

4.2.8 Following a trend

The periodic table has a number of repeating patterns. The most obvious is the change from metals on the left of each period to non-metals on the right. Other patterns exist in the physical and chemical properties of elements in the same group or period. Some of these trends are shown in table 4.2.

TABLE 4.2 Patterns in the periodic table

Characteristic	Pattern down a group	Pattern across a period
Atomic number and mass number	Increases	Increases
Atomic radius	Increases	Decreases
Melting points	Decreases for groups 1 to 5 and increases for groups 15 to 18	Generally, increases then decreases
Reactivity	Metals become more reactive and non-metals become less reactive	High reactivity, that decreases and then increases; group 18 elements are inert and do not react
Metallic character	Increases	Decreases

Figure 4.15 shows the differences in atomic radius of various atoms and the trends in these differences.

As you move down a group, more electrons, protons and neutrons are present. The electrons are positioned in electron shells (electron shells are discussed more fully in lesson 4.3). Elements in the same period have the same number of electron shells. As you move down a group, the number of shells increases and, thus, so too does the atomic radius.

As you move across a period, the number of electrons and protons increases. The number of shells remains the same. The increase of negatively charged electrons and positively charged protons across the group means a stronger attraction exists between the shell and the nucleus, causing the atomic radius to decrease.

FIGURE 4.15 Atomic radius increases down a group and decreases across a period.

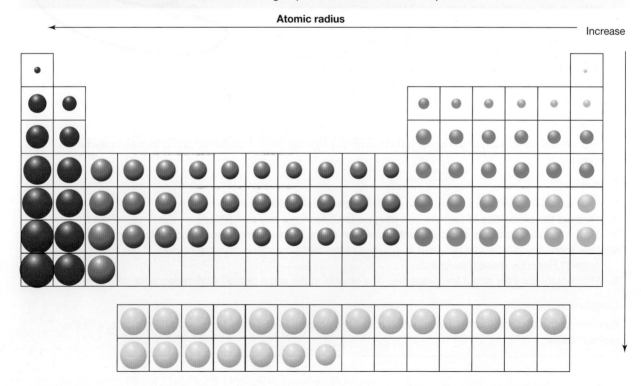

4.2.9 Is the periodic table finished?

As mentioned earlier, four new elements were added to the periodic table in 2016. Are there still more elements to be discovered?

Of all the elements on today's periodic table, 91 of the first 92 occur in nature to some extent or other. The exception is technetium, element number 43, symbol Tc. The elements after number 92, uranium, have all been produced artificially in one way or another*. This has usually involved firing 'bullets' of either tiny sub-atomic particles or existing smaller nuclei at 'targets' consisting of pre-existing nuclei. The hope is that the target nucleus will absorb the smaller particle and therefore make a larger nucleus corresponding to a new element. This is notoriously difficult for a number of reasons:

- It is difficult to actually hit the target due to the almost impossibly small sizes involved.
- The 'bullets' are positively charged. As they approach the target nucleus, which is also positively charged, the like charges repel each other. The closer the bullet gets, the stronger the repulsion.
- Large amounts of energy are required to overcome this repulsion. This means that expensive machines are required, the cost of which is prohibitive and usually only affordable by famous laboratories and government funded institutions.
- If a new element is formed, it is difficult to detect as sometimes only a few atoms are formed. Also, these new atoms are radioactive and do not last very long before decaying.

FIGURE 4.16 How was element number 117 produced?

HOW TO CREATE AN ELEMENT

A team of Russian and US researchers bombarded berkelium (atomic number 97) with calcium-48 (atomic number 20) to create tennessine (atomic number 117), which splits through radioactive decay into smaller elements.

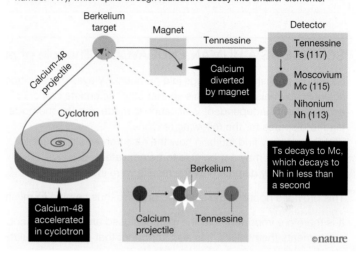

©nature

*Small traces of number 93 (neptunium) and number 94 (plutonium) have been found in nature and in the remnants of atomic explosions. Plutonium can also be produced in a special type of nuclear reactor.

How are new elements named?

The process by which elements are named can be a long one. This is because any new claim must be independently tested and proven. This can take many years. Sometimes disputes can arise as to who has made a claim first and therefore has the right to suggest a name for a new element. A famous case involved Russia and the USA, both of whom claimed to have produced element number 104 first.

Because of this, new elements are given a temporary name and symbol based on the Latin and Greek stubs that relate to their atomic numbers. For example, the new element tennisine (number 117) in figure 4.16 was previously called *ununseptium* (un =1, sept = 7), with the symbol Uus. Once any new element is independently proven and any disputes settled, names are submitted to IUPAC and a name and symbol allocated according to very specific rules.

SCIENCE AS A HUMAN ENDEAVOUR: The role of publication and peer review

Scientists publish their findings in a number of ways. The traditional way is in papers that are published in scientific journals. They do this so that other scientists can criticise and emulate their work. Findings are not accepted until independent verification is made and the methods used found to be valid. This can sometimes lead to disputes for the following reasons:
- People may find faults in how the research was conducted.
- The results may not be able to be emulated.
- People may dispute how the results were interpreted.

This provides rigour to the scientific process thus helping to minimise the spread of faulty knowledge.

It is therefore important that the methods used are clearly described in any paper so that others may do the experiments themselves. It is also important that results, calculations and conclusions are presented logically and clearly so that any faults may be detected.

Often a paper on a particular topic will be followed by further papers on related follow-up topics, either by the original authors or by others working in the same field.

In the case of new elements, this process can often take many years. The new element nihonium (symbol Nh), for example, was first reported in 2004 but was not officially accepted and named until 2016.

Where to from here?

All of this raises an obvious question — are there still new elements to be made and added to the periodic table? The answer is yes. The quest for even heavier elements is an ongoing one and is being conducted in various laboratories in a number of countries. Many of the new elements produced so far have very unstable atoms with very short half-lives. These are sometimes only small fractions of a second. This means that they are difficult to detect, let alone establish their chemical properties. Scientists are being encouraged by a proposed 'island of stability' that might soon be reached. These atoms are predicted to have much longer half-lives and therefore be more stable.

Uses of these new elements

As expected, many of these new elements are only academic curiosities due to their instability and the small number of atoms that have been produced. However, if the 'island of stability' mentioned above proves to be correct, more stable elements with longer half-lives may have new and novel uses.

Despite this, a few of these elements do have some uses. These include:
- Plutonium: Used for nuclear energy and in nuclear weapons. Also used in space exploration for small-scale power generation.
- Americium: Used in smoke detectors (you probably have some in your home!), medicine and in oil wells to measure flow rates.
- Californium: Used in mineral prospecting, medicine, airport detection of certain explosives. Its most important use is in the calibration of instruments used in nuclear reactors.

4.2 Activities

learnon

4.2 Quick quiz on	4.2 Exercise

Select your pathway

■ LEVEL 1 1, 2, 3, 7, 9	■ LEVEL 2 4, 5, 11, 12, 13	■ LEVEL 3 6, 8, 10, 14, 15

These questions are
even better in jacPLUS!
- Receive immediate feedback
- Access sample responses
- Track results and progress

Find all this and MORE in jacPLUS ▶

Remember and understand

1. **a.** State whether the following statements are true or false.

Statements	True or false?
i. The noble gases are found in group 18.	
ii. The non-metals are found in the upper right-hand side of the periodic table.	
iii. There are more non-metals than metals.	
iv. All the elements that are found naturally as liquids are metals.	

 b. Rewrite any false statements to make them true.

2. What is the name of the element in the following?
    a. Group 2, period 3
    c. Group 1, period 4
    b. Group 17, period 2
    d. Group 18, period 3

3. Copy or print a blank periodic table. (A blank periodic table is available to download in the digital documents.)
    Label where you would find the following elements: the noble gases, the alkali metals, the alkaline earth
    metals, the halogens and the transition metals.

4. In the outline of the periodic table shown, some of the elements have been replaced by letters. Using the
    correct chemical symbols, write down which of these elements fit the following categories.

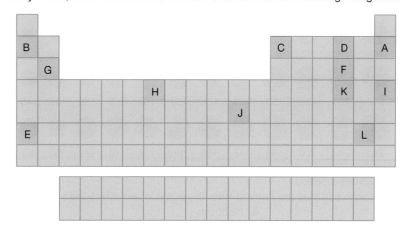

 a. Two elements that are gases at room temperature
 b. Two elements that are metals
 c. Two elements that are transition elements

d. Two elements that are noble gases

e. Two elements that are in the same group

f. Two elements that are in the same period

g. The elements that are alkali metals

h. The element that is a halogen

5. At room temperature, which group of the periodic table consists exclusively of gases?

Apply and analyse

6. Explain why the relative atomic mass of an element is often different from the mass number for an individual atom of that element.

7. Describe what happens to the metallic character of the elements as you go across the periodic table from left to right.

8. **SIS** Explain how Mendeleev was able to predict the properties of elements even before they were discovered.

9. It is almost certain that elements 119, 120 and 121 will be produced at some point in the future. Where would you expect these to be placed on the periodic table?

10. On the periodic table there are a number of elements whose symbols do not appear to match their name. Using a copy of the periodic table that contains symbols and names, make a list of these and explain why you think these elements might have such 'strange' symbols.

11. Complete the following table containing atoms of various elements.

	Nuclide symbol	Name	Atomic number	Mass number	Number of protons	Number of neutrons	Number of electrons
a.	$^{7}_{3}Li$						
b.			30	65			
c.					18	22	
d.				197			79
e.		Uranium		238			

12. **SIS** The diagram provided shows atomic radius (size) across the periodic table.

a. Identify the trend across a period.

b. **MC** Which sub-atomic particles are furthest from the nucleus and affect the atomic radius?

    A. Protons   **B.** Neutrons   **C.** Electrons

c. **MC** Which sub-atomic particles attract your answer to part **b**?

    A. Protons   **B.** Neutrons   **C.** Electrons

d. What happens to the number of protons across a period? Explain how this links to atomic radius.

Evaluate and create

13. **SIS** You have been asked to determine if an element is a metal or a non-metal based on some qualitative data.

TABLE Properties of an unknown element

Test	Observation
State at room temperature	Solid
Electrical conductivity	No

    a. Referring to the data, state whether this is a metal or a non-metal. Explain your response.
    b. Identify the element using the additional information provided in the following table. Explain your answer referring to the characteristics in the table.

TABLE Information and characteristics for the unknown element

Characteristic	Information
Group	In the same group as tellurium
Reactivity	More reactive than selenium

14. Imagine that a new element, number 123, has been reported and subsequently verified by three Australian scientists working at a newly funded laboratory in Canberra. Their surnames are Rutherford, Livermore and Matthews and they have been asked to suggest a possible name for this new element. However, any new element must be named using a set of rules from the IUPAC. Research what these rules are and prepare a report that contains the following:
   - what the IUPAC is
   - what the rules are
   - an illustration of how these rules have been used to name a selection of existing elements
   - a suggestion for the name and symbol for the new element described above.
15. **SIS** It is said that the stars are the 'element factories of the universe'; that is, stars make the elements. Do some research and report on how the stars make elements.

Fully worked solutions and sample responses are available in your digital formats.

LESSON
4.3 Electrons, electron configuration and shells

LEARNING INTENTION

At the end of this lesson you will be able to describe how electrons exist in energy levels around the nucleus called shells and appreciate the experimental evidence for this arrangement.

4.3.1 The influence of electrons

Chemical reactions are fundamentally about electrons moving around, leading to atoms being joined or bonded together. The electrons in each atom account for the chemical behaviour of all matter, because they form the outermost part of the atom.

Shells of electrons

The Bohr model of the atom, proposed by Niels Bohr in 1913, is a representation of atomic structure. According to the Bohr model, an atom consists of a small, positively charged nucleus at the center, surrounded by negatively charged electrons arranged into concentric shells, with each shell having a particular amount of energy. Because of this, shells are also called energy levels. The shells closer to the nucleus are lower in energy, while those further away are higher in energy. The main experimental evidence for this comes from the analysis of atomic spectra. This is discussed in section 4.3.4.

Drawing an accurate picture of an atom using a diagram is difficult because electrons cannot be observed like most particles. Their exact location within the atom is never known — they tend to behave like a 'cloud' of negative charge (see figure 4.17).

An **electron shell diagram** is a simplified model of an atom (see figure 4.18).
- The nucleus of the atom, containing protons and neutrons, is drawn in the middle.
- Electrons are arranged in a series of energy levels around the nucleus.

These energy levels are called **shells** and are drawn as concentric rings around the nucleus. The electrons in the inner shells are more strongly attracted to the nucleus than those in the outer shells.

electron shell diagram a diagram showing electrons in their shells around the nucleus of an atom

shells energy levels surrounding the nucleus of an atom into which electrons are arranged

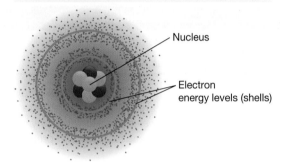

FIGURE 4.17 An atom containing a cloud of electrons arranged into energy levels or shells

Nucleus

Electron energy levels (shells)

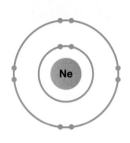

FIGURE 4.18 An electron shell diagram

Ne

SCIENCE INQUIRY SKILLS: Using diagrams

Scientists use diagrams rather than drawings to represent models and processes. Diagrams allow important information to be conveyed in a simple and clear manner and are much easier to create than complex drawings.

Shells can be thought of like a series of 'drawers' that act to store electrons. Each additional shell is further from the nucleus and larger, so can accommodate more electrons.

For example:
- The fourth shell holds a maximum of 2×4^2 electrons, which is 32 electrons.
- The fifth shell holds a maximum of 2×5^2 electrons, which is 50 electrons.

TABLE 4.3 The different electrons per shell

Shell	Number of electrons
1	2
2	8
3	16
4	32
n	$2n^2$

The maximum number of electrons in each shell can be calculated using the rule: the nth shell holds a maximum of $2n^2$ electrons.

Electron configuration

The **electron configuration** of an element is an ordered list of the number of electrons in each shell.

TABLE 4.4 Different electron configurations of four elements

Element	Electron configuration
Lithium	2,1
Silicon	2,8,4
Magnesium	2,8,2
Phosphorus	2,8,5

Determining electron configuration

To determine the electron configuration (for elements with an atomic number up to 18), the following steps should be followed.

1. Identify the number of electrons in the particle. For an atom, this is the same as the atomic number (the number of protons).
2. Add electrons to shell 1 until it is full (two electrons).
3. Add any remaining electrons to shell 2 until it is full (eight electrons).
4. Continue adding the electrons until all electrons are assigned to a shell.
5. Write the list of electrons in each shell separated by commas.

The following steps show how this can be applied to a sodium atom.

1. Sodium has an atomic number of 11 so has 11 protons and 11 electrons.
2. Add two electrons to shell 1 so it is full. This leaves $11 - 2 = 9$ electrons remaining.
3. Add eight electrons to shell 2. This leaves $9 - 8 = 1$ electron remaining.
4. Add the last electron to shell 3.
5. The electron configuration is 2,8,1.

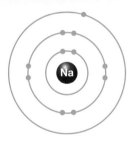

FIGURE 4.19 An electron shell diagram of a sodium atom

DISCUSSION

Electrons and protons have opposite charges so are electrostatically attracted. This poses a question: why don't the electrons spiral into the nucleus?

4.3.2 The periodic table explained

When Mendeleev and Meyer grouped elements on the basis of their similar chemical properties, they were not aware of the existence of electrons. We can now explain many of their observations using our understanding of electron shells.

In lesson 4.2, trends in the periodic table were introduced. The patterns of electron configuration down groups and across periods substantially contribute to these trends.

electron configuration an ordered list of the number of electrons in each electron shell, from inner (low energy) to outer (higher energy) shells

FIGURE 4.20 Patterns of electron configuration in the periodic table.

H Hydrogen 1								He Helium 2
Li Lithium 2,1	Be Beryllium 2,2	B Boron 2,3	C Carbon 2,4	N Nitrogen 2,5	O Oxygen 2,6	F Fluorine 2,7	Ne Neon 2,8
Na Sodium 2,8,1	Ma Magnesium 2,8,2	Al Aluminium 2,8,3	Si Silicon 2,8,4	P Phosphorus 2,8,5	S Sulfur 2,8,6	Cl Chlorine 2,8,7	Ar Argon 2,8,8
K Potassium 2,8,8,1	Ca Calcium 2,8,8,2						

- **Periods (rows):** Moving down a group from one period to the next adds an electron shell.
- **Groups (columns):** Moving across a period from left to right increases the number of outer shell electrons.

The electrons in the outer shell are called **valence electrons** and are important because they are involved in chemical reactions. Atoms in the same group have the same number of valence electrons; therefore, they have similar chemical properties. (In the first 20 elements, helium is an exception to this, but it has a full outer shell.) No more than eight valence electrons can be present, regardless of what shell is the outermost shell. If more than eight valence electrons are present, the atom becomes unstable.

valence electrons outer shell electrons involved in chemical reactions

Filling up in turn

The periodic table contains a great deal of information about electron configuration.

- The first shell can hold up to two electrons, so there are two elements in the first period, with hydrogen containing one electron and helium containing two.
- The second shell holds up to eight electrons, so there are eight elements in the second period.
- The third shell is a bit more complicated, as there are eight elements in the period, even though shell 3 can hold 18 electrons. Let's look at a worked example in table 4.5 to explain this shell.

TABLE 4.5 Worked example of the potassium atom

Principle	Potassium example
The outer shell of an atom can never hold more than eight electrons as the atom would then become unstable.	Electron configuration 2,8,9 \n\n Shell 3 has nine electrons so this would be unstable.
Before the third shell is yet to be filled completely, electrons begin to fill the fourth shell.	Electron configuration 2,8,8,1

This stabilises the atoms because the third shell is no longer the outer shell. The filling of the third shell resumes in the block of elements from scandium to zinc (the transition metals). Once the third shell is full, the fourth shell continues to fill from gallium to xenon.

TABLE 4.6 Electron configurations of different elements

Element	Symbol	Atomic number	Electron configuration
Oxygen	O	8	2,6
Fluorine	F	9	2,7
Neon	Ne	10	2,8
Sodium	Na	11	2,8,1
Magnesium	Mg	12	2,8,2
Sulfur	S	16	2,8,6
Chlorine	Cl	17	2,8,7
Argon	Ar	18	2,8,8
Potassium	K	19	2,8,8,1
Calcium	C	20	2,8,8,2

Note that the fourth shell of the potassium and the calcium atom begins filling before the third shell is full.

FIGURE 4.21 A calcium atom has 20 electrons.

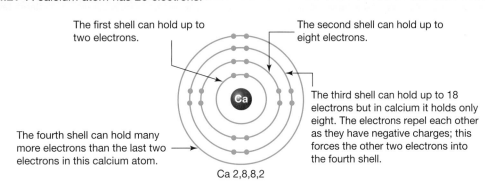

The first shell can hold up to two electrons.

The second shell can hold up to eight electrons.

The third shell can hold up to 18 electrons but in calcium it holds only eight. The electrons repel each other as they have negative charges; this forces the other two electrons into the fourth shell.

The fourth shell can hold many more electrons than the last two electrons in this calcium atom.

Ca 2,8,8,2

SCIENCE AS A HUMAN ENDEAVOUR: Quantum mechanics

Quantum mechanics is the branch of science which focuses on the sub-atomic level, including the behaviour of electrons and shells. It is a famously complicated field, which led the Nobel laureate in physics, Richard Feynman, to state, 'If you think you understand quantum mechanics, then you don't understand quantum mechanics'.

On Resources

📋 **eWorkbook** Electron configuration (ewbk-13098)

🧩 **Interactivity** Shell-shocked (int-0676)

4.3.3 Upwardly mobile electrons

Shells are stable energy levels for electrons, although if enough energy is supplied to an atom, electrons can move from one shell (or energy level) to another (higher) energy level.

FIGURE 4.22 Electrons can absorb energy to move to a higher energy level and emit energy when they fall back to a lower energy level.

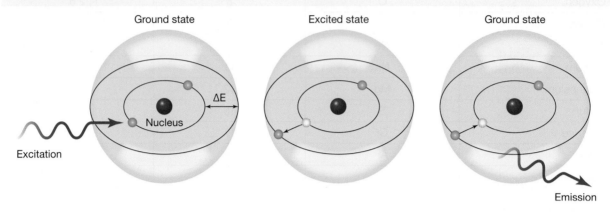

A simple experiment called a **flame test** can be used to demonstrate this process (see figure 4.23).

1. Heat a substance in a flame.
2. Heat energy is absorbed by electrons in the atom, exciting it from the **ground state** to a higher energy level.
3. When the substance cools, electrons fall down in energy levels and the atom changes from the **excited state** to the ground state.
4. When electrons fall back down, they release light energy, which may be visible as colours.

The size of the difference in energy levels determines the colour of the light. Energy levels are slightly different distances apart in different atoms, so flame colours can be used to identify elements.

flame test a type of test in which an element is heated so the atom moves from a ground state to an excited state, and releases coloured light as the electrons fall back down

ground state the lowest energy state of an atom

excited state the highest energy state of an atom, in which electrons jump to a higher shell

eles-2712

FIGURE 4.23 Various metal ions produce characteristic colours when they are volatilised in a flame. Solid (a) or liquid (b) samples can be analysed.

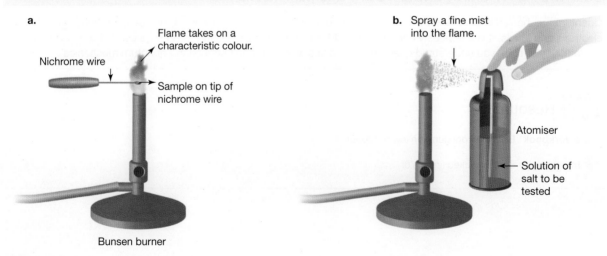

Flame tests

Aim

To observe evidence of electrons dropping from one energy level to another

Materials

- safety glasses and laboratory coat
- 2 M hydrochloric acid
- Bunsen burner, heatproof mat and matches
- 5 evaporating dishes
- barium carbonate
- sodium carbonate
- copper carbonate
- potassium carbonate
- strontium carbonate
- 10 mL measuring cylinder
- spatula

CAUTION

Laboratory coats and safety glasses must be worn at all times.

Method

1. Place 10 mL of 2 M hydrochloric acid in an evaporating dish and place the dish on a heatproof mat.
2. Add a spatula full of the barium carbonate to the evaporating dish.
3. Carefully hold the lit Bunsen burner at an angle over the spray produced by the reacting acid and carbonate as shown in the figure provided. Observe the change in the colour of the flame.
4. Repeat using the other carbonates. Use a different evaporating dish each time.

Note: this is one method of performing flame tests. Other methods involve using nichrome wires and dipping these into the hydrochloric acid and the element being tested.

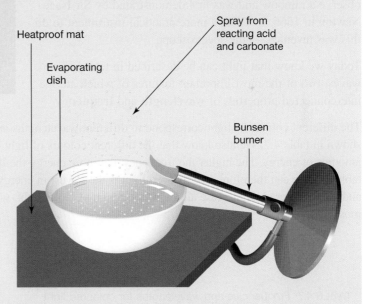

Heatproof mat

Evaporating dish

Spray from reacting acid and carbonate

Bunsen burner

Results

Record the colours produced by the different carbonates in a suitable table.

Discussion

1. Identify the independent (tested), dependent (measured) and controlled (constant) variables in this investigation.
2. Why is it important to only explore one IV at a time?
3. What is the purpose of controlling variables in this investigation?
4. Flame tests provide evidence that electrons do actually occupy different energy levels. Why do different elements produce different colours?
5. Is it the metal part of the compound or the carbonate part (carbon and oxygen) that produces the colour? How do you know?

Conclusion

Summarise the findings for this investigation about electrons dropping from one energy level to another.

4.3.4 Chasing rainbows

The observation that different atoms emit light of different colours has proved to be a very important discovery. It is significant for two reasons:

- It is the main experimental evidence for the existence of electrons in shells.
- It provides a means by which elements may be identified from the light that they give off when energised.

How these connections are made might not seem obvious. Although many elements have distinctly different flame colours as shown in figure 4.24, there are others whose colours are very similar and therefore difficult to distinguish. The link to electron shell structure is even harder to appreciate. To make the required connection, it is necessary to use an instrument called a **spectroscope**.

Seeing the light

You may remember that white light is made up of a range of colours that blend together from red at one end to violet at the other. This is precisely the effect we notice when we observe a rainbow and was first demonstrated by Sir Isaac Newton in 1666. In 1802, a more practical instrument to do this was invented — the spectroscope.

Today we know that light can be described in terms of waves, two of the most important features of which are the interconnected properties of **wavelength** and **frequency**.

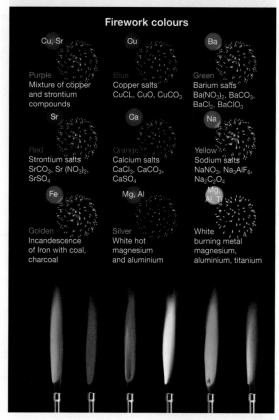

FIGURE 4.24 Fireworks use the principles of the flame test to produce many colours.

The different colours of light correspond to different wavelengths and frequencies as shown in table 4.7. We also know that the different colours of light contain different amounts of energy. The higher the frequency, the more energy the light contains. From table 4.7 we can therefore note that red light is the lowest in energy (lowest frequency) and there is a gradual increase all the way through to violet light, which contains the most energy.

spectroscope an instrument that breaks up the light it receives into distinct frequencies, colours or wavelengths

wavelength the distance between the crest of one wave and the next

frequency the number of waves that pass a particular point each second

TABLE 4.7 Frequencies and wavelengths for coloured light

Colour	Frequency ($\times 10^{12}$Hz)	Wavelength (nm)
Red	400–480	750–625
Orange	480–510	625–590
Yellow	510–530	590–570
Green	530–600	570–500
Blue	600–670	500–450
Indigo	670–700	450–430
Violet	700–750	430–400

FIGURE 4.25 The spectrum of white light seen through a modern spectroscope. This is called a continuous spectrum because the colours merge gradually without any gaps.

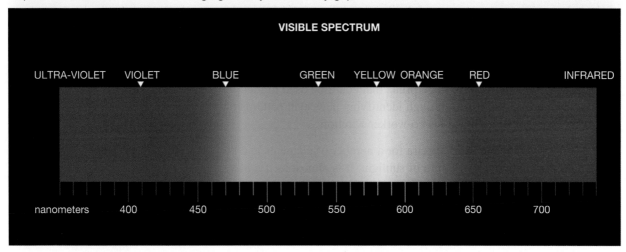

SCIENCE AS A HUMAN ENDEAVOUR: The invention of the spectroscope

What is Robert Bunsen most famous for?

Most people would answer the invention of the Bunsen burner. Bunsen, strictly speaking, did not invent the burner, but he made significant improvements to its design which have lasted to this day. Because of this he is almost universally acclaimed as its inventor.

Bunsen was a brilliant experimentalist and it was his invention of the flame spectroscope, in 1859, that would lead to enormous advances not only in atomic theory, but also in astronomy and chemical analysis. Working with another German chemist, Gustav Kirchoff, Bunsen began to study the light emitted from substances when heated in a hot flame (it was for this reason that he developed the Bunsen burner). To further analyse this light the flame spectroscope was invented. Using this invention, Bunsen and Kirchoff eventually discovered two new elements, rubidium and caesium.

FIGURE 4.26 a. Design of Bunsen's original spectroscope and **b.** a modern version of a spectroscope, similar to that found in many school laboratories

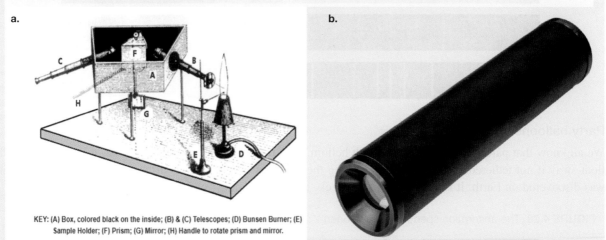

KEY: (A) Box, colored black on the inside; (B) & (C) Telescopes; (D) Bunsen Burner; (E) Sample Holder; (F) Prism; (G) Mirror; (H) Handle to rotate prism and mirror.

The spectroscope has now evolved into many specialised versions. Many of these are complicated instruments important in dedicated fields of scientific research. However, there are a number of derivations that are used in the field of chemical analysis. These have wide usage in areas that benefit society such as food analysis, environmental monitoring, medicine, engineering and pharmaceutical chemistry.

What are spectroscopes used for?

When Bunsen first looked through his spectroscope, what he saw astounded him. Instead of the continuous rainbow, he could observe a series of distinct, brightly coloured lines. It was as if the light from the flame tests that he was observing consisted of just a few colours, with most of the rainbow background missing. Because these patterns were made by light emitted from a hot flame, they became known as **emission spectra**.

Further work established that the pattern of lines in an emission spectrum was different for every element. It was as if these patterns were fingerprints that could be used to identify elements using just the light they emitted when energised. It was through identifying new such patterns that the elements caesium and rubidium were discovered by Bunsen and his co-worker Kirchoff.

Figure 4.27 shows the emission spectra for a number of elements and how these compare to the continuous spectrum of sunlight. It is almost as if a black mask has been placed over the continuous spectrum, and then lines have been cut into it to reveal the colours underneath!

emission spectra the pattern of coloured lines obtained when emitted light from a sample is observed through a spectroscope

FIGURE 4.27 Emission spectra show a pattern of lines which is unique for each element.

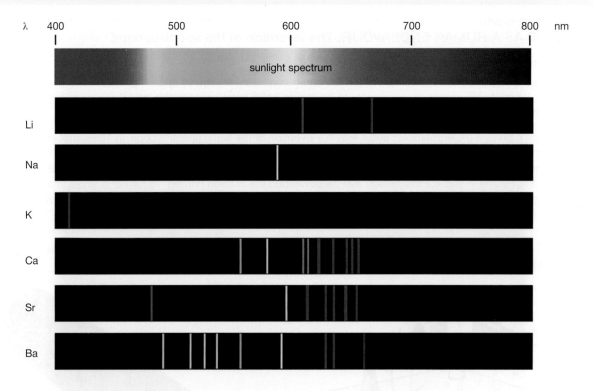

Party balloons and the Sun

We all know that party balloons are filled with helium. This makes them less dense than air, meaning they will float away if not tethered. Helium is unique among the elements because it was discovered in the Sun before it was discovered on Earth. It is named after the Greek 'helios' (meaning 'sun') for this reason.

FIGURE 4.28 The absorption spectrum of hydrogen

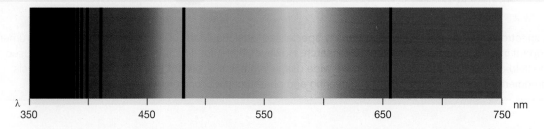

If white light is passed through a gas, a type of spectrum that is the exact opposite of an emission spectrum is obtained. This is a continuous rainbow spectrum, but with dark lines in it. These lines depend on the type of gas that the light has passed through and their pattern exactly matches the corresponding emission spectrum. This type of spectrum is called an **absorption spectrum**. This type of spectrum can therefore be used in exactly the same way as an emission spectrum to identify elements. In 1895, a pattern of such lines was found in sunlight, leading to the discovery of helium.

Starlight, starbright

With the invention of the spectroscope, it was quickly realised that a very important analytical tool was now available, and all it required was light. It meant that starlight could now be analysed to determine the composition of stars, even though these were billions of kilometres away. Because of this we know that the universe is about 90 per cent hydrogen and 9 per cent helium, with all the remaining elements making up the small proportion that is left.

FIGURE 4.29 The composition of stars can be determined by analysing the light they emit.

Another interesting fact observed in starlight is that the lines in the spectra are shifted, very slightly, towards the red end. This is called *red shift*. From this shift, scientists have been able to determine that all the stars are moving away from us. From the *amount* of this shift, they have even been able to calculate the speed with which this is taking place.

Connecting spectral lines to electron shells

An atom possesses many electron shells. However, it will never have enough electrons to fill them all. In the ground state, it is only those shells that represent the lowest energy levels that contain electrons. This means that there is only ever one ground state arrangement that an atom can have. When energy is added to an atom, however, the electrons 'jump up' to higher energy levels (shells). Because there are many of these levels, there are many possibilities for how this may occur, depending on how much energy has been added to the atom. An atom can therefore have a number of excited states.

With this in mind we can make the following points:
- As electrons can only occupy certain shells, and cannot be in between (i.e. they cannot be in a 1½, 2¾ shell etc), there are only a certain number of 'jump downs' possible when they revert from an excited state back to their ground state.
- These 'jump downs' represent transitions from a higher energy level to a lower one. Energy therefore *appears* to be lost.
- However, the law of conservation of energy says that energy *cannot* be lost. It can, however, be transformed from one type to another.
- In this case, the lost energy is transformed into light.
- If the 'jump down' is a large one, the light produced contains more energy and will be further towards the violet end of the spectrum in colour. A distinct line in the spectrum will be produced at the appropriate spot.
- If the 'jump down' is a small one, the light produced contains less energy and will be further towards the red end of the spectrum in colour. A distinct line in the spectrum will once again be produced at the appropriate spot.

absorption spectrum the pattern of dark lines against a rainbow background obtained when white light is passed through a gas

- Elements produce different patterns in their lines because, although the *pattern* of shells is the same, their energy values vary from one element to another. The size of the jumps will therefore vary. Additionally, different atoms have different numbers of electrons that may be undergoing transition and this leads to further differences in their patterns.
- If electron shells did not exist, and electrons could be anywhere around an atom, there would be so many 'jump downs' possible that the lines would blend together and we would not see the sharp pattern of lines that is observed.

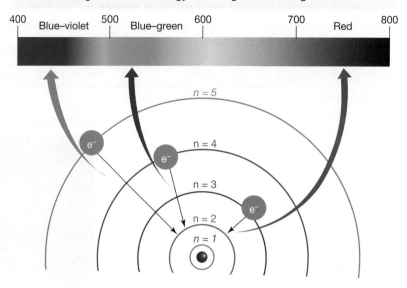

FIGURE 4.30 Each electron transition produces a line of a different colour, with blue being the highest energy and shortest wavelength and red being the lowest energy and longest wavelength.

EXTENSION: Sub-shells

With advances in technology, more and more sophisticated spectroscopes have been produced. These have revealed further features in spectra which cannot be explained by a shell structure alone. It is now realised that shells may further be divided into sub-shells which have small variations in their energy levels. These are labelled using the letters s, p, d, f and g, in conjunction with the major shell number that they are part of. For example, a 2p sub-shell represents a p sub-shell that is part of the second shell. A 3p sub-shell represents a p sub-shell that is part of the third shell.

There are interesting patterns that emerge when this is considered further, as shown in table 4.8. Notice how the electron capacity in the first two sub-shells from shells 2, 3 and 4 is always equal to 8 electrons.

TABLE 4.8 Patterns in shells and sub-shells

Shell number	Number of sub-shells it contains	Sub-shell designations	Electron capacity of each sub-shell
1	1	1s	2
2	2	2s	2
		2p	6
3	3	3s	2
		3p	6
		3d	10
4	4	4s	2
		4p	6
		4d	10
		4f	14

The existence of sub-shells also explains:
- why the pattern of electron shell filling becomes more complicated after calcium (element number 20)
- why the block of transition metals is ten elements wide
- why the blocks of lanthanoids and actinoids (usually placed under the main body of the periodic table) are fourteen elements wide.

elog-2450

INVESTIGATION 4.5

Observing emission spectra

Aim

To observe and record several different types of spectra

Materials

- various vapour lamps set up at stations around the room
- spectroscopes

Method

1. Observe and record the spectrum of sunlight by looking through a window to the sky outside.
Tip: Point the slit of the spectroscope towards the light source and look through the other end. Do not look directly through the slit but slightly to its side to best view the spectrum.
2. Repeat step 1 with any fluorescent light tubes in your classroom.
3. Darken the room. Proceed to each of the vapour lamps in turn, recording the spectrum of each one before moving to the next.

Results

Record your results as labelled diagrams. Make sure you label the colours involved and that you have lines in the correct position corresponding to their colour.

Discussion

1. Comment on the differences between the spectra of sunlight and fluorescent light.
What do you think has caused these differences?
2. How are the spectra of the elements in the vapour lamps different to those of sunlight and fluorescent light?
3. What causes the obvious features in each of the spectra from the vapour lamps?
4. Why was the room darkened prior to step 3?

Conclusion

Summarise the findings for this investigation about different types of spectra.

Resources

📋 **eWorkbook** Electron shells (ewbk-13096)

4.3 Activities

learn on

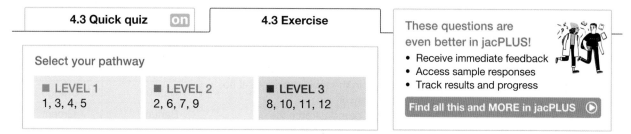

4.3 Quick quiz on   **4.3 Exercise**

These questions are even better in jacPLUS!
- Receive immediate feedback
- Access sample responses
- Track results and progress

Find all this and MORE in jacPLUS ▶

Select your pathway

■ LEVEL 1	■ LEVEL 2	■ LEVEL 3
1, 3, 4, 5	2, 6, 7, 9	8, 10, 11, 12

Remember and understand

1. What is the name given to the different energy levels in which electrons can be found?
2. How many electrons are needed to fill the following?
   a. The first shell   **b.** The second shell
   c. The third shell   **d.** The fourth shell

3. **MC** What element has the electron configuration 2,8,3?
   A. Aluminium
   B. Boron
   C. Sodium
   D. Silicon
4. What information about the electron configuration is given by the group number of an element?
5. What information about the electron configuration is given by the period number of an element?

Apply and analyse

6. Complete the following table of atoms using your periodic table.

TABLE The electron configurations of different elements

Element name	Atomic number	Number of electrons	Electron configuration
a. Carbon			
b.	15		
c.			2
d.			2,8,8,1
e.		5	
f. Neon			
g. Calcium			
h.		14	

7. a. If an atom of an element has one electron in its outer shell, is it a metal or a non-metal? Explain your answer.
   b. If an atom of an element has seven electrons in its outer shell, is it a metal or a non-metal? Explain your answer.
   c. What is special about elements that have eight electrons in their outer shell?
8. Describe the maximum number of occupied electron shells in the known elements. What is its maximum number of electrons?

Evaluate and create

9. **SIS** A student is provided with samples of two elements and asked to identify them. The samples are labelled A and B.
   a. The student performs a series of flame tests for known substances on sample A and records the data in the table. Additionally, they are told that atoms of element A have two valence electrons. Identify the element and explain your answer.
   b. Explain whether the data in the table is quantitative or qualitative.
   c. The student is told that a cold sample of element B absorbs orange light energy. Do you think this observation can be used to identify the element based on the flame test data in the table provided? If so, identify the sample B and explain your answer.
10. **SIS** Identify an experimental technique that demonstrates the existence of electron shells and explain the observations that led to this conclusion.
11. Currently, the element with the atomic number 120 has not officially been discovered and remains hypothetical. Research this element and summarise some of its likely properties. Note down the likely electron configuration for an atom of this element.

TABLE Flame test results

Sample	Colour
A	Green
Sodium	Yellow
Strontium	Red
Potassium	Purple
Calcium	Orange
Barium	Green
Boron	Green

12. The theory of electron shells involves some of the following points:
- Electrons can only occupy certain levels. They cannot be in between these allowed levels.
- They fill in order from the lowest upwards.
- Higher energy levels have more room to accommodate more electrons.
- When electrons drop from a higher energy level to a lower one, they lose energy.

An analogy that is sometimes used compares electron arrangements to people working in a multi-level office building.

Comment on the validity of this analogy, outlining both its strengths and weaknesses.

Fully worked solutions and sample responses are available in your digital formats.

LESSON
4.4 Extension: Ionic bonding — Give and take

LEARNING INTENTION

At the end of this lesson you will be able to describe ionic bonding in terms of the formation of ions and relate it to the number of electrons in the outer electron shells of atoms.

4.4.1 It's the shell structure that counts

Knowledge of the electron shell structures of atoms helps us to understand how compounds such as sodium chloride (table salt) form. When atoms react with each other to form compounds, it is the electrons in the outer shell that are important in determining the type of reaction that occurs.

Two examples of bonds forming between atoms are:
- ionic bonds which form between metals and non-metals
- covalent bonds which form between non-metals.

It's great to be noble

In 1919, Irving Langmuir suggested that the noble gases do not react to form compounds because they have a stable electron configuration of eight electrons in their outer shell. Most other atoms react because their electron arrangements are less stable than those of the noble gases. Atoms become more stable when they attain an electron arrangement that is the same as that of the noble gases. Chemical reactions can allow atoms to obtain this arrangement.

FIGURE 4.31 The outer shells of the noble gases

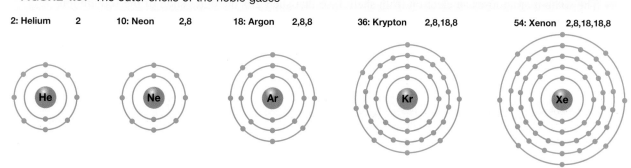

2: Helium   2   10: Neon   2,8   18: Argon   2,8,8   36: Krypton   2,8,18,8   54: Xenon   2,8,18,18,8

The tendency of atoms to gain or lose electrons in order to have eight in their outer shell is known as the **octet rule**. For elements that only contain electrons in shell 1 this is termed the **duet rule**, because shell 1 can only contain a maximum of two electrons.

octet rule the tendency of atoms to gain or lose electrons to have eight in their outer shell

duet rule the tendency of elements that only contain electrons in shell 1 to gain electrons to have two

The atoms of other elements must gain or lose electrons to attain eight in their outer shells. In this way they become more stable, ending up with the electron arrangement of the nearest noble gas in the periodic table. We can predict how many electrons will be gained or lost by an atom using the periodic table, as highlighted in section 4.4.2.

DISCUSSION

Unstable things transforming into more stable things is a fundamental concept in chemistry. The octet rule tells us which particles are more stable than others and allows us to predict changes which could occur to increase stability. Such processes are frequently chemical reactions.

Do you think there are more unstable or stable substances present in the world? Why do you think that is the case?

4.4.2 Some gain, some lose

Atoms that have lost or gained electrons and therefore carry an electric charge are called **ions**. The charge on an ion for any element can be deduced by identifying the number of valence electrons, using the group number from the periodic table.

Metal atoms, such as sodium, magnesium and potassium, have a small number of outer shell electrons.
- They form ions by losing the few electrons in their outer shell. The next shell becomes the outer shell and contains 8 electrons.
- This means that metal ions have more protons than electrons and so are positively charged.
- For example, the magnesium atom loses its two outer shell electrons to become a positively charged magnesium ion. The symbol for the magnesium ion is Mg^{2+}. The '2+' means that two electrons have been lost to form the ion.
- Positively charged ions are called **cations**.

ions atoms that have lost or gained electrons and carry an electrical charge

cations atoms that have lost electrons to become positively charged ions

anions atoms that have gained electrons to become negatively charged ions

Non-metal atoms form ions by gaining electrons to fill their outer shell.
- These ions contain more electrons than protons, so they are negatively charged.
- For example, the chlorine atom gains one electron to obtain eight electrons in its outer shell and therefore it becomes a negatively charged chloride ion. Its symbol is Cl^-. The '−' means that one electron has been gained to form the ion.
- Negatively charged ions are called **anions**.

figure 4.32 shows how sodium and chlorine atoms lose and gain electrons respectively to form ions.
- The sodium atom loses an electron from shell 3, so that shell 2 is the outer shell of the sodium ion.
- Note that the sodium atom becomes a sodium ion and that the chlorine atom becomes a chlor*ide* ion. When non-metals form ions, the suffix '-ide' is used.

TABLE 4.9 A summary of metals and non-metals in ionic bonding

Metal or non-metal	Gain or lose valence electrons	Charge on ion	Name of ion	Example	Name of ion
Metal	Lose	Positive	Cation	Mg^{2+}	Magnesium
Non-metal	Gain	Negative	Anion	Cl^-	Chloride

FIGURE 4.32 How sodium and chlorine atoms form ions

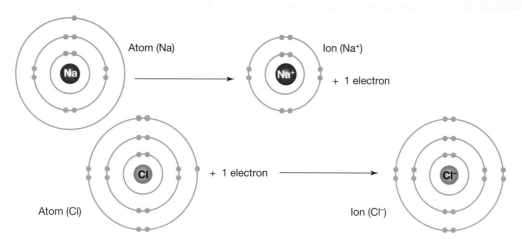

Determining the charge of an ion

1. The last numeral in the group number indicates the number of valence electrons; for example, group 1 has one valence electron and group 17 has seven valence electrons.
2. Determine how many electrons are lost by a metal atom or gained by a non-metal atom to have a full outer shell.
3. Determine the charge on the ion and write the symbol.

FIGURE 4.33 The periodic table can be used to determine the charge on an ion. Group 14 can vary with the ion it forms. Note that the 1 in the 1+ and 1− is often not written (instead the charges are referred to as + and − respectively).

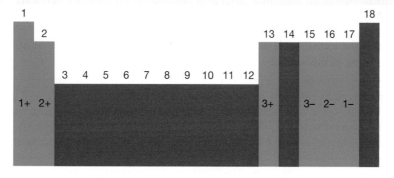

TABLE 4.10 Using the group number to determine the charge on a metal ion

Group	Elements	Atomic electron configuration	Number of valence electrons	Electrons lost to have 8 electrons in the outer shell	Symbol of ion
1	Li	2,1	1	1	Li$^+$
	Na	2,8,1	1	1	Na$^+$
2	Be	2,2	2	2	Be^{2+}
	Mg	2,8,2	2	2	Mg^{2+}
13	Al	2,8,3	3	3	Al^{3+}

Group	Elements	Atomic electron configuration	Number of valence electrons	Electrons gained to have 8 electrons in the outer shell	Symbol of ion
15	N	2,5	5	3	N^{3-}
	P	2,8,5	5	3	P^{3-}
16	O	2,6	6	2	O^{2-}
	S	2,8,6	6	2	S^{2-}
17	F	2,7	7	1	F^{-}
	Cl	2,8,7	7	1	Cl^{-}

Elements in group 14 (such as carbon, lead and tin) are in a unique situation. Do they gain four or lose four? Some of these, such as carbon, tend to do neither and do not usually participate in ionic bonding (instead undergoing covalent bonding, as discussed in lesson 4.5). Others, such as lead and tin, being metals, can lose their outershell electrons and form positive ions.

4.4.3 It's a game of give and take

Chemical bonding between ions is the process of particles 'sticking together' and is always due to **electrostatic attraction** between oppositely charged particles. When cations and anions are close to each other they undergo electrostatic attraction forming an **ionic bond** and a new **ionic compound**.

Ionic bonds most commonly form between metal cations and non-metal anions. In the sodium chloride example:
1. an electron is *transferred* from a sodium atom to a chlorine atom forming two ions
2. an ionic bond forms between the ions.

It is the second step that is the formation of an ionic bond; however, the electron transfer in the first step will be the focus of this section.

electrostatic attraction the attraction between two oppositely charged bodies

ionic bond an attractive force between ions with opposite electrical charge

ionic compounds compounds containing positive and negative ions held together by the electrostatic force

Electron transfer diagrams are a useful way to show the movement of electrons from metal to non-metal atom(s).
1. Draw diagrams of the reactant atoms with electrons arranged in shells. *Note:* it is customary to draw electrons as pairs, as shown in figure 4.34.
2. Draw an arrow to show the movement of an electron from a metal atom to a non-metal atom.

FIGURE 4.34 Electron transfer diagram for the formation of sodium chloride

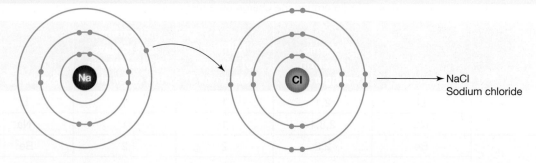

In many situations, more than two atoms may be involved to ensure that all the elements achieve eight electrons in their outer shell, as shown in figure 4.35. For example, when magnesium reacts with chlorine to form magnesium chloride, each magnesium atom loses two electrons. Since each chlorine atom needs to gain only one electron, a magnesium atom gives one electron to each of two chlorine atoms. The resulting Mg^{2+} and Cl^{-} ions are attracted to each other to form the compound $MgCl_2$.

FIGURE 4.35 Electron transfer diagram involving more than two atoms. **a.** Magnesium chloride involves the two chlorine atoms each receiving an electron from magnesium, to form stable chloride ions. **b.** Sodium oxide involves two sodium atoms each giving their electron to oxygen, allowing it to form the stable oxide ion.

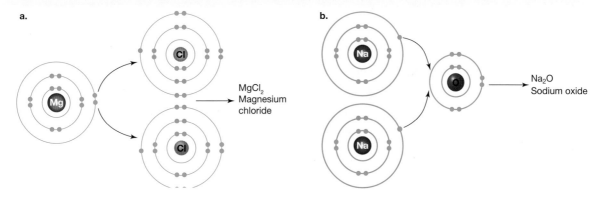

a.

MgCl$_2$
Magnesium chloride

b.

Na$_2$O
Sodium oxide

When an ionic compound forms, the naming of this is based on the ions it contains. The metal, which loses electrons to form a cation, retains its name, forming the first word of the ionic compound. The non-metal, which loses electrons to form an anion, changes its name to end in -ide, forming the second half of the name of the ionic compound.

TABLE 4.12 Different cations and anions and their ion names

Cations (positive ions)			Anions (negative ions)		
Atom name	Ion name	Chemical symbol	Atom name	Ion name	Chemical symbol
lithium	lithium	Li$^+$	iodine	iodide	I$^-$
sodium	sodium	Na$^+$	fluorine	fluoride	F$^-$
potassium	potassium	K$^+$	chlorine	chloride	Cl$^-$
calcium	calcium	Ca^{2+}	oxygen	oxide	O^{2-}
aluminium	aluminium	Al^{3+}	nitrogen	nitride	N^{3-}

4.4.4 What do ionic compounds have in common?

Ionic compounds contain many cations and anions bonded together in a large structure called a **lattice** (see figure 4.36). Within lattices, the cations and anions alternate in three dimensions. They are also present in the required numbers so the overall charge is neutral. When we look at an ionic lattice, we see a crystal structure.

Ionic compounds have the following properties due to the strong ionic bonds and the lattice structure:
- They are usually solids at room temperature.
- They normally have very high melting points because the electrostatic force of attraction between the ions is very strong.
- They are brittle as when a force is applied to the lattice the ions with like charges align and actively repel each other, shattering the lattice.
- They usually dissolve in water to form **aqueous solutions**. (The word aqueous comes from the Latin word *aqua*, meaning 'water'. Other words beginning with the prefixes aque- or aqua- relate to water; for example, aqueduct, aquatic, aqualung).
- Their aqueous solutions normally conduct electricity.
- They conduct electricity in a molten state.

lattice a structure in which particles are bonded to each other to form crystal structures

aqueous solutions solutions in which water is the solvent

FIGURE 4.36 **a.** A stick and ball representation of the lattice structure of sodium chloride; the sticks represent the bonds between the atoms. **b.** The ions in the lattice are effectively held in a tight arrangement. **c.** Individual salt crystals form regular square blocks because of the ionic lattice.

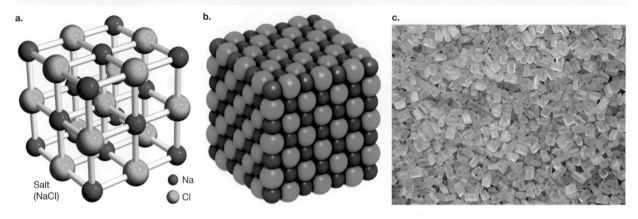

a.

Salt
(NaCl)

● Na
● Cl

b.

c.

on Resources

eWorkbook   Ionic bonding (ewbk-13100)

Interactivities   Pass the salt (int-0675)
                      Ionic models (int-6351)

4.4 Activities

learn on

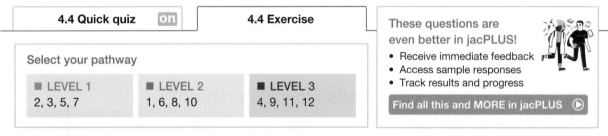

4.4 Quick quiz **on**	**4.4 Exercise**

These questions are even better in jacPLUS!
- Receive immediate feedback
- Access sample responses
- Track results and progress

Find all this and MORE in jacPLUS ▶

Select your pathway

■ LEVEL 1	■ LEVEL 2	■ LEVEL 3
2, 3, 5, 7	1, 6, 8, 10	4, 9, 11, 12

Remember and understand

1. Why do ions form?
2. Identify the name used for the following.
   a. A positively charged ion
   b. A negatively charged ion
3. **MC** Atoms in group 13 are likely to form ions with what charge?
   A. +3
   B. −3
   C. +1
   D. −1
4. Describe three properties of most ionic compounds.
5. What types of elements combine to form ionic compounds?

Apply and analyse

6. Write the symbol for the ion formed by each of the following elements.
   a. Beryllium
   b. Bromine
7. How many electrons have been gained or lost to form the following ions?
   a. Pb^{4+} b. Br^- c. Cr^{3+} d. Se^{2-}
8. a. Complete the following table (you may wish to use a periodic table to assist). The first entry has been completed for you.

TABLE The formation of different ions

Ion symbol	Ion name	Gained or lost electrons?	Number of electrons lost/ gained	Total number of electrons in ion
a. F^-	Fluoride	Gained	1	10
b.	Sodium			
c.	Nitride			
d.		Lost	3	10

   b. Write the electron configuration of a neon atom.
   c. What do you notice about the electron configuration of all the ions in part a. and that of a neon atom? Explain your answer.
9. Draw electron transfer diagrams to show how each of the following ionic compounds form.
   a. Lithium chloride
   b. Magnesium fluoride
   c. Aluminium sulfide
   d. Calcium oxide
10. Describe the type of bonding present in a crystal of potassium fluoride, using a clear diagram.

Evaluate and create

11. **SIS** A student has two metal samples, A and B. They know one is lithium and the other magnesium and have designed an experiment to determine the identify of each sample. The following details and table are listed in their log book.
   1. Burn each metal in oxygen to form a metal oxide.
   2. Measure how much oxygen is required to react with the metal.
   3. Calculate the ratio of metal ions to oxide ions in the metal oxide.
   a. Define the independent variable (IV) and identify the IV for this experiment.
   b. Is the data in the results table quantitative or qualitative data? Justify your answer.
   c. Write the symbol for the following ions.
      i. Lithium
      ii. Magnesium
      iii. Oxide
   d. Identify if sample A is lithium or magnesium. Justify your answer.
12. Research and report on why some transitional metals can form ions of different charges.

TABLE The ratio of ions in different metal samples

Metal sample	Ratio of metal ions to oxide ions in the metal oxide
A	2 : 1
B	1 : 1

Fully worked solutions and sample responses are available in your digital formats.

LESSON
4.5 Extension: Covalent bonding — When sharing works best

LEARNING INTENTION

At the end of this lesson you will be able to describe covalent bonding in terms of the sharing of electrons in the outer shells of atoms.

4.5.1 Covalent bonding

Ionic compounds form when atoms lose or gain electrons. This process explains how a metal atom can transfer electrons to a non-metal atom. However, it cannot explain how two non-metal atoms each *gain* electrons when bonding.

Atoms can also achieve stable electron configurations by *sharing* electrons with other atoms to gain eight electrons in their outer shell. When two or more atoms share electrons, a **covalent bond** is formed. As with all chemical bonding, covalent bonding is the electrostatic attraction between oppositely charged particles, as shown in figure 4.37. In a covalent bond, positive nuclei are attracted to the negative electrons in the bond.

Covalent bonding occurs between non-metal atoms to form **molecules** such as water (H_2O) and carbon dioxide (CO_2), and large **network lattices** such as diamond (C) and quartz (SiO_2).

FIGURE 4.37 Electrostatic attraction in a covalent bond

The electrons experience a force of attraction from both nuclei. This negative-positive-negative attraction holds the two particles together

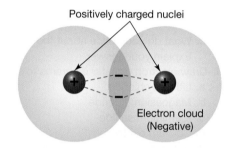

Positively charged nuclei

Electron cloud (Negative)

Molecules can be made of more than one type of atom or made of atoms of the same element. For example, oxygen gas consists of molecules formed when two oxygen atoms share electrons. Water and carbon dioxide are made of different types of atoms.

Network lattices may also be compounds (two or more elements combined) or a single element. The properties of these will be explored in section 4.5.3.

Covalent bonding in water

The octet rule applies to atoms whether they are forming ions and ionic compounds or covalent bonds and molecules (or network lattices).

If we look more closely at a water molecule, we can see that there are covalent bonds between each hydrogen atom and the central oxygen atom, in which electrons of hydrogen and oxygen are shared. In figure 4.38 each hydrogen atom has one valence electron and the oxygen atom has six. When a covalent bond forms between hydrogen and oxygen, *both* the bonding atoms gain access to the electron, so it contributes to the octet or duet rule.

- Hydrogen started with one valence electron and gained one from the bond for a total of two.
- Oxygen started with six electrons and gained one from each hydrogen atom for a total of eight.

Now both hydrogen and oxygen atoms have stable electron configurations, so water is a stable molecule.

covalent bond a shared pair of electrons holding two atoms together

molecules small particles of a chemical substance consisting of one or more types of atoms

network lattice a lattice in which each atom is covalently bonded to adjacent atoms

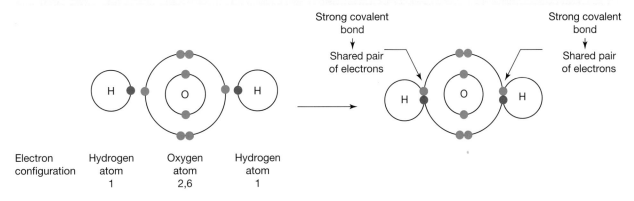

FIGURE 4.38 Formation of covalent bonds between hydrogen and oxygen atoms in a water molecule

Strong covalent bond
Shared pair of electrons

Strong covalent bond
Shared pair of electrons

Electron configuration

Hydrogen atom 1

Oxygen atom 2,6

Hydrogen atom 1

4.5.2 Electron dots: what's the point?

To ensure that each atom in a molecule receives the right number of electrons we use **electron dot diagrams** (also known as Lewis structures). Only valence electrons are included in electron dot diagrams as these are the electrons which are involved in bonding and satisfying the octet (or duet) rule.

electron dot diagrams
diagrams using dots to represent the electrons in the outer shell of atoms and to show the bonds between atoms in molecules

Drawing electron dot diagrams for an atom

1. Draw the element symbol.
2. Determine the number of outer shell electrons for the atom. (An easy way to do this is to look at the last number of the group.)
3. Add one electron to each of four regions around the symbol until all four regions have an electron or all electrons are assigned.
4. Form pairs with any remaining electrons.

TABLE 4.13 Electron dot diagrams for some elements. Only the valence electrons need to be shown.

Symbol	Electron configuration	Electron dot diagram
H	1	H •
C	2,4	• C •
O	2,6	: O •
F	2,7	: F :

Using electron dot diagrams for molecules

When elements combine to form **covalent compounds**, they share electrons to achieve a stable outer shell with eight electrons (the octet rule). Hydrogen has a full outer shell when it has two electrons (the duet rule), but all the other elements in the table need eight electrons to form a stable outer shell.

Drawing electron dot diagrams for a molecule

1. Draw the symbol for the element that will form the most bonds in the centre (the atom that needs the most electrons to gain an outer shell of eight electrons).
2. Draw the other atoms around (or next to) this central atom.
3. Determine how many valence electrons each atom requires. This will be the number of covalent bonds that it forms.
4. Add a pair of electrons between any two atoms that are forming a covalent bond. These are known as **bonding electrons**.
5. Determine how many electrons each atom has left and add these as **non-bonding pairs** (also known as lone pairs).
6. Check to see that the octet or duet rule is satisfied for all atoms and that all the valence electrons have been drawn.

It is also possible to draw a **structural formula**, where a dash is used to represent bonding electrons. The dash represents the covalent bond and the non-bonding electrons are not drawn.

Not all covalent bonds are the same.
- One pair of electrons forms a **single covalent bond**.
- However, some atoms need to share two pairs of electrons to satisfy the octet rule, forming a **double covalent bond**.
- **Triple covalent bonds** may also be formed when three pairs of electrons are shared between two atoms.

The number of each type of atom required to form a molecule determines the chemical formula for that element (for example, O_2) or compound (for example, H_2O).

covalent compounds compounds in which the atoms are held together by covalent bonds

bonding electrons shared electrons holding two atoms together

non-bonding pairs electron pairs not involved in covalent bonding (also known as lone pairs)

structural formula a diagram showing the arrangement of atoms in a substance with covalent bonds drawn as dashes

single covalent bond a covalent bond involving the sharing of one pair of electrons

double covalent bond a covalent bond involving the sharing of two pairs of electrons

triple covalent bond a covalent bond involving the sharing of three pairs of electrons

TABLE 4.14 The formation of covalent molecules with single covalent bonds

Name and formula	Atoms	Compound	Structural formula	Explanation
Chlorine Cl_2	:Cl• + •Cl:	:Cl:Cl:	Cl — Cl Note: The line represents a sharing of two electrons and is called a single covalent bond.	Each chlorine atom needs to share one electron to gain a stable outer shell of 8 electrons.
Hydrogen chloride HCl	H• + •Cl:	H:Cl:	H — Cl	Both the hydrogen and chlorine atom need to share one electron to gain a stable outer shell.
Water H_2O	H• H• + •O:	:O:H H	H—O—H	Each hydrogen atom needs one electron and the oxygen atom needs two to gain a stable outer shell.

TABLE 4.15 The formation of double and triple covalent bonds

Name and formula	Atoms	Compound	Structural formula	Explanation
Oxygen O_2	O̤ ̈+ ̈O̤	O̤ ̈̈ O̤	O = O *Note:* The double line represents a double covalent bond.	Each oxygen atom needs to share two electrons to gain a stable outer shell of 8 electrons.
Nitrogen N_2	:N̈ ̇ + ̇ N̈:	:N ⦂⦂ N:	N ≡ N *Note:* The triple line represents a triple covalent bond.	Each nitrogen atom shares three electrons to gain a stable outer shell of 8 electrons.
Carbon dioxide CO_2	̇C̤ ̇ + :Ö: :Ö:	O̤ ⦂⦂ C ⦂⦂ O̤	O = C = O	Each oxygen atom needs two electrons and the carbon atom needs four electrons to gain a stable outer shell of 8 electrons.

DISCUSSION

Antioxidants are chemicals that stop oxygen radicals damaging important molecules, including DNA in our cells. An oxygen radical is a covalent compound containing oxygen that is highly reactive due to containing unpaired electrons. Oxygen radicals are often formed in the body during chemical processes.

Thinking about the valence electrons, why do you think oxygen atoms that occur in our body are so dangerous, yet 21 per cent of the air we breathe to stay alive is molecular oxygen (O_2)?

4.5.3 Properties of covalent compounds

As outlined in section 4.5.1, covalent bonding can lead to the formation of molecules (figure 4.39) and network lattices (figure 4.40). Though these all contain covalent bonding, the properties between them differ.

Most covalent molecular substances have the following properties:
- They exist as gases, liquids or solids with low melting points because the forces of attraction *between* the molecules are weak.
- They generally do not conduct electricity because they are not made up of ions.

FIGURE 4.39 Different examples of covalent molecules

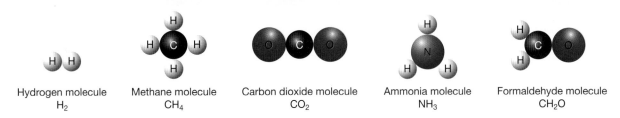

Hydrogen molecule H_2 Methane molecule CH_4 Carbon dioxide molecule CO_2 Ammonia molecule NH_3 Formaldehyde molecule CH_2O

Covalent network lattices (such as those shown in figure 4.40) share the following properties:
- They exist as solids due to the high temperatures required to break the strong covalent bonds in the lattice.
- Their other properties are similar to covalent molecules, as listed in the preceding list.

FIGURE 4.40 Different examples of covalent network lattices

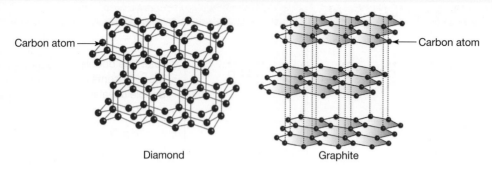

Diamond

Graphite

 **Resources**

📋 **eWorkbook**   Covalent molecules (ewbk-13102)

🧩 **Interactivity**   Making molecules (int-0228)

4.5 Activities

learnon

4.5 Quick quiz on	4.5 Exercise

Select your pathway

■ **LEVEL 1** 1, 3, 6	■ **LEVEL 2** 2, 5, 7, 11	■ **LEVEL 3** 4, 8, 9, 10

These questions are even better in jacPLUS!
- Receive immediate feedback
- Access sample responses
- Track results and progress

Find all this and MORE in jacPLUS ▶

Remember and understand

1. What kinds of elements combine to form covalent compounds?
2. **MC** What does an element's electron dot diagram represent?
   A. The properties of the element.
   B. The order of the number of protons in the atom's nucleus.
   C. The symbol for the atom and the arrangement of the outer shell electrons.
   D. An ordered list of the number of electrons in each electron shell, from inner (low energy) to outer (higher energy) shells.
3. Draw the electron dot diagram for chlorine.
4. Describe two properties of covalent compounds.

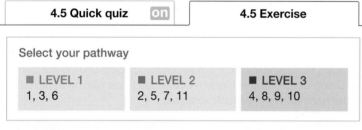

a.

Sulfur trioxide — a gas used to make sulfuric acid

b.

Chloroform — a liquid once used as an anaesthetic

c.

Acetylene — a colourless gas used in welding

5. What is the difference between a single covalent bond and a triple covalent bond in terms of the number of electrons involved?
6. For the covalent compounds shown, state whether their bonds are single, double and/or triple covalent bonds.

Apply and analyse

7. **a.** Draw electron dot diagrams to show how the following covalent compounds form.
   i. Hydrogen fluoride (HF)
   ii. Methane (CH_4)
   iii. Phosphorus trichloride (PCl_3)
   iv. Hydrogen sulfide (H_2S)
   v. Ammonia (NH_3)
   vi. Carbon disulfide (CS_2)

b. What pattern emerges between the structural formula of the compound and the number of electrons involved in bonding?
   c. State whether the covalent bonds in the compounds are single, double or triple bonds.
   d. How many pairs of non-bonding electrons does each compound have?
8. Explain why the noble gases don't form covalent compounds. Use diagrams to justify your response.

Evaluate and create

9. **SIS** Carbon has four valence electrons and chlorine has seven.
   a. Which of these two elements can make the largest variety of different compounds? Explain your answer.
   b. Draw either an electron dot diagram or a structural diagram to show the following.
      i. CH_2Cl_2 ii. CCl_4 iii. C_2H_5Cl iv. $C_2H_2Cl_2$ v. C_2HCl

10. **SIS** The graph provided shows the boiling point of covalent compounds containing hydrogen.
   a. In addition to hydrogen, each compound contains another element. Which two groups in the periodic table are these elements from?
   b. Which compound has the lowest boiling point?
   c. Do the compounds containing elements from periods 2, 3, 4 or 5 have the highest boiling points?
   d. Name all the compounds that will be liquids at room temperature (25 °C).
   e. Which line (pink or blue) includes compounds containing elements from group 16?
   f. Would you predict that H_2Po or HAt would have a higher boiling point? Explain your reasoning.

Boiling point of different covalent compounds

Graph: Boiling point /°C (y-axis from −100 to 100) versus Period 1 to Period 4 (x-axis). Points labelled: H_2O (100, Period 1), HF (~20, Period 1), H_2S (~−60, Period 2), HCl (~−85, Period 2), H_2Se (~−40, Period 3), HBr (~−67, Period 3), H_2Te (~0, Period 4), HI (~−35, Period 4).

11. **SIS** Silicon dioxide, commonly known as silica or sand, is a hard, solid, covalent compound with a very high melting point. Find out about its structure and draw a clear diagram to show this.

Fully worked solutions and sample responses are available in your digital formats.

LESSON
4.6 The reactivity of metals

LEARNING INTENTION

At the end of this lesson you will be able to relate the reactivity of metals to the electron shell structure of their atoms and their location in the periodic table.

4.6.1 Reactivity of metals

Have you ever wondered why gold can be found lying near the surface of the Earth and yet we need to mine iron ore and smelt it in large furnaces before we can obtain iron? The answer lies in the reactivity of the metals. Gold is a very unreactive element. It does not combine readily with other elements to form compounds. Most metals are much more reactive than gold.

When the Earth formed, the more reactive metals — including aluminium, copper, zinc and iron — reacted with other elements to form ionic compounds. These compounds are the **mineral ores** from which the metal elements are obtained. Iron ores include haematite (Fe_2O_3), magnetite (Fe_3O_4), siderite ($FeCO_3$), pyrite (FeS_2) and chalcopyrite ($CuFeS_2$).

FIGURE 4.41 Few metals are found as elements like gold; most are found as compounds or ores.

mineral ores   rocks mined to obtain a metal or other chemical within them

The reactivity of metals is dependent on how easily they are able to give up their outer shell electrons. For example, it is easier for an atom to give up a single electron from an outer shell than to give up two electrons.

The reactivity of metals can be investigated by observing their reactions with acids. A metal reacts with hydrochloric acid according to the equation:

metal + hydrochloric acid → salt + hydrogen gas

In these reactions between metals and acids, electrons are transferred away from the metal atoms to the hydrogen in the acid, forming positive metal ions and hydrogen gas. The metal is said to have displaced the hydrogen from the acid. For this reason, these reactions are also called **displacement reactions**.

displacement reactions chemical reactions involving the transfer of electrons from the atoms of a more reactive metal to the ions of a less reactive metal

smelting the production of a metal in its molten state

alloy a mixture of several metals or a mixture of a metal and a non-metal

SCIENCE AS A HUMAN ENDEAVOUR: Metals in ancient times

The most powerful ancient civilisations succeeded and prospered because they developed better weapons than their enemies by using metals such as copper, tin and iron. The Mesopotamians, who occupied a large region of the Middle East, learned almost 5000 years ago how to separate copper and tin from their ores using a process called **smelting**.

Smelting is a chemical process in which carbon reacts with molten ore to separate the relatively pure metal. In ancient times, charcoal was used in furnaces to provide the carbon. They combined molten copper and tin to produce an **alloy** known as bronze, which is resistant to corrosion and harder than both copper and tin. The ancient Egyptians, Persians and Chinese also used bronze in weapons, ornaments, statues and tools.

FIGURE 4.42 An ancient bronze shield

The ancient Romans used the smelting process to separate iron from iron ore. They strengthened it by pounding it with a hammer and used it to produce weapons, shields and armour that was harder and stronger than bronze. The use of iron weapons allowed the Roman legions to rule the Mediterranean world and beyond for over 400 years.

FIGURE 4.43 The gladius (a short iron sword), together with a long iron shield, gave the Roman army a huge advantage over its enemies. The shields were often used by groups of soldiers to form a protective wall and roof known as a *testudo* (tortoise) around themselves.

SCIENCE INQUIRY SKILLS: Evaluating experiments

No experiments are perfectly designed or performed, so scientists always evaluate experiments to consider limitations and the effect of any limitations on the results. This process allows scientists to determine the quality of the results and, therefore, the strength of their conclusions. Sometimes experiments with significant limitations may need to be repeated or even have their method redesigned before they will generate useful data.

A handy method for considering limitations is an evaluation table. In table 4.16 a simple example has been included for the time taken (DV) to boil 50 mL of water using a Bunsen burner.

TABLE 4.16 An example evaluation table

Limitation	Effect on result (DV)	Improvement
A limitation that may affect the result.	What effect will this limitation have on the result?	A simple way to improve either how the method is designed or performed. Where possible, include how this will affect the result (DV).
Example: Less than 50 mL of water was added as some was left in the measuring device.	*Example: Less heat will be required to boil the smaller mass of water, so the time will be lower.*	*Example: Use a more accurate method of measuring the water volume (e.g. a pipette) to ensure the correct volume of water is added.*

For most experiments, try to think of two to three limitations and list them in order, starting with the greatest effect on the results and working down to the limitation with the smallest effect.

Note that mistakes such as forgetting to add something or dropping an item of equipment aren't listed as limitations. Under such circumstances the results aren't very useful, so it is much better to repeat the experiment.

elog-2452

INVESTIGATION 4.6

Investigating reactivity

Aim

To investigate the reactivity of a range of metals

Materials

- 5 test tubes and a test-tube rack
- safety glasses, gloves and laboratory coat
- 1 cm × 4 cm piece of magnesium ribbon (or equivalent amount)
- 1 cm × 4 cm piece of zinc, copper, aluminium and iron
- 1 M hydrochloric acid
- measuring cylinder, small funnel, thermometer and steel wool

CAUTION

Wear safety glasses, gloves and a laboratory coat.

Method

1. Write a clear hypothesis for this investigation, relating to the observations you expect to see.
2. Polish each of the metal pieces with steel wool to remove any coating.
3. Pour 10 mL of acid into each test tube.
4. Measure and record the initial temperature.
5. Add one piece of metal to each test tube. Look for the presence of bubbles on the surface of the metal.
6. Arrange the test tubes in order of increasing bubble production and record your observations, including any change in temperature.

Results

Record your observations in a suitable table.
- Data should be recorded before, during and after the reaction.
- Describe any changes you observe including but not limited to bubbles, changes in colour of the metal and solution, and any effects on the metal.

Discussion

1. List the metals in order of increasing reactivity.
2. Explain how you know this using evidence from your results.
3. Complete an evaluation table outlining two limitations of this experiment. Use table 4.16 as an example.

TABLE Limitations and improvements for investigating reactivity

Limitation	Effect on result (DV)	Improvement

Conclusion

Summarise your findings from this investigation about the reactivity of metals.

4.6.2 The activity series

The **activity series** (or reactivity series) places the elements in decreasing order of reactivity:

$$Li \rightarrow K \rightarrow Na \rightarrow Ca \rightarrow Mg \rightarrow Al \rightarrow Mn \rightarrow Cr \rightarrow Zn \rightarrow Fe \rightarrow$$

$$Ni \rightarrow Sn \rightarrow Pb \rightarrow H \rightarrow Cu \rightarrow Hg \rightarrow Ag \rightarrow Au \rightarrow Pt.$$

The activity series can be used to predict how a metal will react with an acid. In order to react with acid and release hydrogen gas, the metal must be *before hydrogen* in the activity series.

- Lithium, potassium, sodium and calcium are the most reactive metals. They are so reactive that they will react with water to produce hydrogen gas.
- Magnesium through to lead will react with acid to form hydrogen gas.
- Copper, mercury and silver will not react with acid to form hydrogen gas.
- Gold and platinum are even less reactive than copper and silver.

DISCUSSION

Most of the elements at the start of the activity series were discovered much later than those at the bottom. Gold, silver, mercury and copper were all discovered over 2000 years ago. Potassium, sodium and calcium were not discovered until 1808. Why do you think this is so?

The activity series and displacement reactions

Let's consider two examples (shown in table 4.17):
- reactions between metals and acids
- reactions between metals and metal salts in solution.

TABLE 4.17 Different reactions involving metals

Reaction	Products	Example
Metal and hydrochloric acid	Hydrogen + metal chloride	$2Na(s) + 2HCl(aq) \rightarrow H_2(g) + 2NaCl(aq)$
Metal A and metal B salt	Metal A salt + metal B	$CuSO_4(aq) + Fe(s) \rightarrow Cu(s) + FeSO_4(aq)$

activity series a classification of metals in decreasing order of reactivity

In the first example in table 4.17, the highly reactive metal atom (Na) loses an electron to form a metal ion (Na⁺), which we know is more stable based on the octet rule. The electron is transferred to another atom. In this case, as sodium is more reactive than hydrogen, it displaces the hydrogen atom, which forms hydrogen gas.

In the second example in table 4.17, iron (Fe) is more reactive than copper (Cu). Therefore, it is more likely to become an ion and displace the copper. The iron forms a preferential ionic bond with the sulfate over the copper. This reaction can be observed in figure 4.44.

These are often referred to as displacement reactions, which will be explored further in topic 5.

FIGURE 4.44 The displacement reaction involving a metal and metal salt

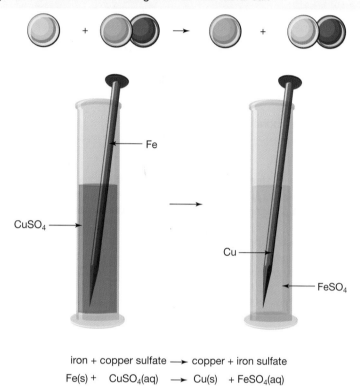

iron + copper sulfate ⟶ copper + iron sulfate

$Fe(s) + CuSO_4(aq) \longrightarrow Cu(s) + FeSO_4(aq)$

The activity series and reactivity

The activity series can also be used to compare the reactivity of different metals. When a piece of metal (composed of atoms) is added to a salt solution (containing metal ions), the more reactive metal will form an ion and the less reactive metal will be forced to receive electrons to form an atom. We observe the formation of atoms as metal crystals. This can be seen in figure 4.45, which shows a displacement reaction between copper atoms in the piece of wire and silver ions in the solution occurs. Copper, the more reactive metal, displaces the silver ions.

FIGURE 4.45 Production of silver crystals by a displacement reaction between copper atoms and silver ions

INVESTIGATION 4.7

Quantified reactivity

Aim

To quantify and measure the reactivity of metals

Materials

- safety glasses, gloves and laboratory coat
- heatproof mat
- steel wool
- gas syringe
- 1 cm × 4 cm piece of zinc, copper, aluminium and iron
- 1 cm × 4 cm piece of magnesium ribbon
- 1 M hydrochloric acid

- retort stand, bosshead and gas syringe clamp
- 1 cm × 6 cm length of plastic tubing
- 250 mL side-arm conical flask
- rubber stopper to fit conical flask
- stopwatch or clock with second hand
- 50 mL measuring cylinder
- distilled water

CAUTION

Wear safety glasses, gloves and a laboratory coat.

Method

1. Polish each of the metal pieces with steel wool to remove any coating.
2. Mount the gas syringe in the clamp as shown in the figure provided. Your teacher will tell you if the syringe needs to be lubricated. Push the plunger fully in and attach the plastic tubing to the nozzle.
3. Pour 50 mL of acid into the flask.
4. Connect the other end of the plastic tubing to the conical flask.
5. Place one of the pieces of metal in the conical flask and quickly seal with the rubber stopper.
6. Have one student act as a timer and another as a recorder.
7. As soon as the metal is dropped in, start timing.
8. Using a suitable table in your results, record the volume of gas in the syringe every 30 seconds until gas is no longer produced, the syringe is full, or 10 minutes has passed (whichever occurs first).
9. Repeat the procedure with the other metals, taking care to rinse out the flask carefully each time with distilled water.

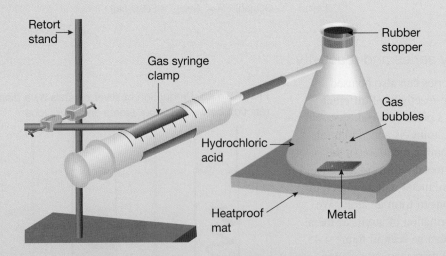

Results

1. Record your results using a suitable table. This will depend on the type of data you have collected (see method, step 8).
2. Using graph paper or software, plot the results for all of the metals on one set of axes. Put the volume of gas on the vertical axis and time on the horizontal axis.

Discussion

1. Identify the independent variable in this experiment.
2. Identify all the dependent variables that you recorded in this experiment.
3. Some of the variables in this investigation were not carefully controlled. Complete an evaluation table (refer back to table 4.16) to explain how these uncontrolled variables may have affected your results and conclusions.
4. Use your graph to list the five metals in increasing order of reactivity and explain your reasoning.
5. Write a word equation for the reaction of each of the metals with the acid. If no reaction occurred, write 'no reaction'.
6. To which general reaction type or types do reactions between metals and acids belong?

Conclusion

Summarise the findings for this investigation about how the reactivity of metals was quantified and measured.

on Resources

☑ **eWorkbook**   Metals and the activity series (ewbk-13104)

▶ **Video eLesson**   Davey and potassium (eles-1773)

4.6 Activities

learn on

4.6 Quick quiz **on**	4.6 Exercise

Select your pathway

■ LEVEL 1	■ LEVEL 2	■ LEVEL 3
1, 3, 4	2, 6, 7	5, 8, 9

These questions are even better in jacPLUS!
- Receive immediate feedback
- Access sample responses
- Track results and progress

Find all this and MORE in jacPLUS ▶

Remember and understand

1. **MC** Name the gas produced in the reaction of a metal with hydrochloric acid.
   A. Hydrogen   **B.** Helium   **C.** Fluorine   **D.** Oxygen
2. Why is iron usually found in the form of a compound in the Earth's crust?
3. Identify the most reactive metal in the following pairs.
   a. Lithium and aluminium
   b. Manganese and zinc
   c. Magnesium and manganese
   d. Calcium and sodium
4. Outline what occurs in a displacement reaction. Provide one example of this type of reaction.

Apply and analyse

5. Explain why the reactivity of metals decreases from left to right across the periods of the periodic table.
6. Three metals are placed in hydrochloric acid: zinc, magnesium and copper. Which would you expect to produce hydrogen the fastest? Justify your response.
7. A piece of magnesium was placed in a solution of copper chloride. Describe the expected observations.

Evaluate and create

8. **SIS** A student runs a series of experiments to investigate the reactivities of metals. The student places a piece of metal in an ionic salt solution and observes whether a reaction occurs.

TABLE The reactivity of different metals and metal salts

	Na	Mg	Fe	Zn
NaCl		✗	✗	✗
MgCl$_2$	✓		✗	✗
FeCl$_2$	✓	✓		✓
ZnCl$_2$	✓	✓	✗	

✓ indicates a reaction occurred

✗ indicates no reaction occurred

A shaded cell indicates that no experiment was performed between a metal and the same metal salt.

a. Identify two observations that could have been made to indicate that a reaction occurred.
b. Which metal is the most reactive? Explain your answer with reference to the results.
c. Write a word equation for the reaction between magnesium and zinc chloride to form metal crystals and a colourless solution, including states.
d. Pieces of the four metals listed in the table are added to a solution of an unknown metal. A reaction is only observed with sodium (Na). Identify the metal using the activity series and explain your answer.
e. The student leaves the reaction test tubes overnight and observes that the pieces of iron (Fe) in the salt solutions now look orange/brown.
   i. Justify whether this indicates that displacement reactions have occurred.
   ii. Can you suggest an alternative explanation for this observation?
9. Silver is a precious metal that is not very reactive. Outline an experiment in which silver can be obtained using silver nitrate solution and copper.

Fully worked solutions and sample responses are available in your digital formats.

LESSON
4.7 Finding the right formula

LEARNING INTENTION

At the end of this lesson you will be able to deduce the formula of a variety of simple covalent and ionic compounds from the valency of their constituent elements.

4.7.1 Chemical ID

Chemicals are usually identified by both a name and a formula. Most people are able to recognise the formula of common compounds such as water (H_2O) and carbon dioxide (CO_2).

A chemical formula (plural *formulae*) is a shorthand way to write the name of an element or compound. It tells us the number and type of atoms that make up an element or compound. Writing the correct formula is of paramount importance in chemistry. Most chemical problems cannot be solved without the knowledge of chemical formulae.

4.7.2 It's elementary

Often the formula of a substance is simply the symbol for the element. Metals such as iron and copper, which contain only one type of atom, are identified simply by the symbol for that element (for example, Fe and Cu). Noble gases such as neon (Ne) have a similar formula.

Some non-metal elements, such as hydrogen, oxygen and nitrogen, exist as simple molecules. These molecules form when atoms of the same non-metal are joined together by covalent bonds. For example, the formula for the element hydrogen is H_2, indicating that two hydrogen atoms are joined together to make each molecule of hydrogen. A **molecular formula** is a way to describe the number and type of atoms that join to form a molecule. Some examples are shown in table 4.18.

TABLE 4.18 Some common non-metal molecules and their molecular formulae

Name	Formula
Hydrogen	H_2
Nitrogen	N_2
Chlorine	Cl_2
Bromine	Br_2
Oxygen	O_2

4.7.3 Formulae of compounds

The formula of a compound shows the symbols of the elements that have combined to make the compound and the ratio in which the atoms have joined together.

For example, the chemical formula for the covalent compound methane, a constituent of natural gas, is CH_4 — one carbon atom for every four hydrogen atoms, shown in figure 4.46a.

The formula for the ionic compound calcium chloride, which is used as a drying agent, is $CaCl_2$ — two chloride ions for every calcium ion, shown in figure 4.46b.

FIGURE 4.46 **a.** A molecule of methane with covalent bonding **b.** Ionic bonding in calcium chloride

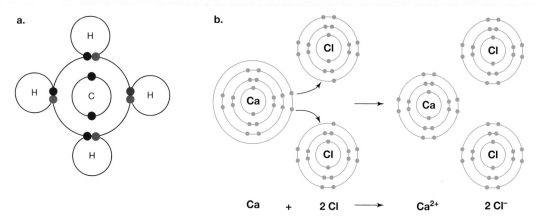

4.7.4 Valency: formulae made easy

Knowledge of the **valency** of an element is essential if we wish to write formulae correctly.

The valency of an element is equal to the number of electrons that each atom needs to gain, lose or share to get eight electrons in its outer shell. For example:

- A chlorine atom has only seven electrons in its outer shell, which can hold eight electrons. By gaining one electron, its outer shell becomes 'full' (with eight electrons). Chlorine therefore has a valency of one.
- A magnesium atom has two electrons in its outer shell. By losing two electrons, it is left with a 'full' outer shell. Magnesium therefore has a valency of two.

A simple guide to remembering the valency of many elements is to remember which group in the periodic table they belong to. table 4.19 provides a simple guide to the valency of many elements.

molecular formula a shorthand statement of the elements in a molecule showing the relative number of atoms of each kind of element

valency the number of electrons that each atom needs to gain, lose or share to 'fill' its outer shell

TABLE 4.19 Valency of groups in the periodic table

Group	Number of valence electrons	Number of electrons to gain/lose	Valency
Group 1	1	Lose 1	1
Group 2	2	Lose 2	2
Group 13	3	Lose 3	3
Group 14	4	Gain/lose 4	4
Group 15	5	Gain 3	3
Group 16	6	Gain 2	2
Group 17	7	Gain 1	1

4.7.5 Writing formulae for and naming covalent compounds

While chemical formulae for covalent compounds can be determined in many ways (such as drawing diagrams or looking at the name), the following method is one technique that can be used.

To write the formula of a non-metal compound made up of only two elements, use the valency of each element and follow the steps shown.

Steps to determine the formulae for covalent compounds

1. Determine the valency of the elements involved.
2. Determine the ratio of atoms that need to combine so that each element can share the same number of electrons.
3. Write the formula using the symbols of the elements and writing the ratios as subscripts next to the element. (The number 1 can be left out because writing the symbol for the element implies that one atom is present.)

CASE STUDY: Examples of determining formulae for covalent compounds

Example 1

Write the formula for carbon dioxide.

Step 1	Determine the valency of the elements involved.
	Carbon has a valency of four; oxygen a valency of two. (That is, carbon needs to share four electrons, while oxygen needs to share two electrons.)
Step 2	Determine the ratio of atoms that need to combine so that each element can share the same number of electrons.
	A ratio of one carbon atom to two oxygen atoms would result in both sharing four electrons.
Step 3	Write the formula using the symbols of the elements and writing the ratios as subscripts next to the element. (The number 1 can be left out because writing the symbol for the element implies that one atom is present.)
	The formula for carbon dioxide is CO_2.

Example 2

Write the formula for phosphorus trichloride.

Step 1	*Determine the valency of the elements involved.*
	Phosphorus has a valency of three; chlorine has a valency of one.
Step 2	*Determine the ratio of atoms that need to combine so that each element can share the same number of electrons.*
	A ratio of one phosphorus atom to three chlorine atoms would result in both sharing three electrons.
Step 3	*Write the formula using the symbols of the elements and writing the ratios as subscripts next to the element.*
	The formula for phosphorus trichloride is PCl_3.

Example 3

Write the formula for dihydrogen monoxide (water).

Step 1	*Determine the valency of the elements involved.*
	Hydrogen has a valency of one; oxygen has a valency of two.
Step 2	*Determine the ratio of atoms that need to combine so that each element can share the same number of electrons.*
	A ratio of two hydrogen atoms to one oxygen atom would result in both sharing two electrons.
Step 3	*Write the formula using the symbols of the elements and writing the ratios as subscripts next to the element.*
	The formula for dihydrogen monoxide is H_2O.

Covalent compounds are usually named using a prefix to denote how many atoms of each element are present. This is important as some elements can combine to form a variety of compounds.

TABLE 4.20 Naming covalent compounds

Number of atoms	Prefix	Examples
1	mono	CO — carbon monoxide NO — nitrogen monoxide
2	di	CO_2 — carbon dioxide NO_2 — nitrogen dioxide
3	tri	NH_3 — nitrogen trihydride PF_3 — phosphorus trifluoride
4	tetra	CCl_4 — carbon tetrachloride SiH_4 — silicon tetrahydride

Some exceptions to these rules exist:
- Mono is often omitted from the first term; for example, carbon monoxide is not usually called monocarbon monoxide.
- Some compounds have trivial (everyday) names that are used more frequently; for example, NH_3 is ammonia and H_2O is water.
- Another naming system is used for carbon compounds; for example, CH_4 is methane. This will be explored in senior Chemistry as part of organic chemistry.

4.7.6 Writing formulae for ionic compounds

The formulae for ionic compounds can be written from knowledge of the ions involved in making up the compound. In ionic compounds, metal ions combine with non-metal ions. tables 4.21 and 4.22 list common positive and negative ions and their names.

Ionic compounds are named by listing the cation (usually metal ion) first and the anion (a non-metal ion or a molecular ion) last. When the anion comprises a single element, the suffix is changed to -ide. Sodium chloride (rather than sodium chlorine) is a handy example to remember and use as a guide.

Metals and cations

Metal atoms usually form cations (positive ions). The size of the positive charge on the ion is called its **electrovalency**. For example:
- a sodium ion has a charge of +1 (Na^+)
- a calcium ion has a charge of +2 (Ca^{2+})
- an aluminium ion has a charge of +3 (Al^{3+}).

Note that the hydrogen ion, although a non-metal ion, exists as a positive ion (H^+).

TABLE 4.21 Electrovalencies of some common cations

Size of the positive charge on an ion		
+1	**+2**	**+3**
Hydrogen (H^+)	Calcium (Ca^{2+})	Aluminium (Al^{3+})
Potassium (K^+)	Copper(II) (Cu^{2+})	Iron(III) (Fe^{3+})
Silver (Ag^+)	Iron(II) (Fe^{2+})	
Sodium (Na^+)	Lead (Pb^{2+})	
Ammonium (NH_4^+)	Magnesium (Mg^{2+})	
	Zinc (Zn^{2+})	

Some of the transition metals (for example, iron) have more than one electrovalency, as shown in table 4.21. The Roman numerals in brackets after the written names of iron and copper identify the electrovalency. Despite being transition metals, zinc and silver only form Zn^{2+} and Ag^+ ions so Roman numerals are not required.

Non-metals and anions

Non-metals usually form anions (negative ions). The size of the negative charges is the electrovalency of the ion. For example:
- a chloride ion has a charge of −1 (Cl^-)
- an oxide has a charge of −2 (O^{2-})
- a phosphide ion has a charge of −3 (P^{3-}).

> **electrovalency** the number of positive or negative charges on an ion

TABLE 4.22 Electrovalencies of some common anions

Size of the negative charge on an ion		
−1	**−2**	**−3**
Bromide (Br^-)	Carbonate (CO_3^{2-})	Phosphate (PO_4^{3-})
Chloride (Cl^-)	Oxide (O^{2-})	Nitride (N^{3-})
Hydrogen carbonate (HCO_3^-)	Sulfate (SO_4^{2-})	
Hydroxide (OH^-)	Sulfide (S^{2-})	
Iodide (I^-)		
Nitrate (NO_3^-)		

Steps to determine the formulae for ionic compounds

1. Determine the electrovalency of the ions that comprise the compound and write down their symbols. List the cation and then the anion.
2. Determine the ratio of ions required in order to achieve electrical neutrality. (Remember compounds have no overall charge.)
3. Write the formula for the compound using the numbers in the ratios as subscripts. (Remember the number 1 does not need to be included.)

CASE STUDY: Examples of determining formulae for ionic compounds

Example 1

Write the formula for sodium chloride.

Step 1	Determine the electrovalency of the ions that comprise the compound and write down their symbols. List the cation then the anion.
	The symbol for the sodium ion is Na^+ and the symbol for the chloride ion is Cl^-.
Step 2	Determine the ratio of ions required in order to achieve electrical neutrality. (Remember compounds have no overall charge.)
	The ratio of negative to positive charges for sodium and chloride ions is 1 : 1. That is, it takes one negatively charged chloride ion to balance the charge of one positively charged sodium ion.
Step 3	Write the formula for the compound using the numbers in the ratios as subscripts. (Remember the number 1 does not need to be included.)
	The formula for the compound is $NaCl$.

Example 2

Write the formula for aluminium oxide.

Step 1	Determine the electrovalency of the ions that comprise the compound and write down their symbols. List the cation then the anion.
	The symbol for the aluminium ion is Al^{3+} and the symbol for the oxide ion is O^{2-}.
Step 2	Determine the ratio of ions required in order to achieve electrical neutrality. (Remember compounds have no overall charge.)
	The ratio of negative to positive charges for aluminium and oxide ions is 2 : 3. That is, it takes three negatively charged oxide ions to balance the charge of the two positively charged aluminium ions.
Step 3	Write the formula for the compound using the numbers in the ratios as subscripts.
	The formula for aluminium oxide is Al_2O_3.

There are also some more complex negative ions called **molecular ions** (or **polyatomic ions**), such as hydroxide ions (OH^-), carbonate (CO_3^{2-}) phosphate (PO_4^{3-}) and sulfate ions (SO_4^{2-}). These groups of atoms have an overall negative charge and are treated as a single entity.

molecular ions groups of atoms that have an overall charge and are treated as an entity; e.g. OH^-, SO4²⁻, NH4⁺

polyatomic ions another term for a molecular ions, in which a group of atoms have an overall charge

CASE STUDY: Examples of determining formulae for ionic compounds with polyatomic ions

Example

Write the formula for calcium phosphate.

Step 1	*Determine the electrovalency of the ions that comprise the compound and write down their symbols. List the cation then the anion.* The symbol for the calcium ion is Ca^{2+} and the symbol for the phosphate ion is PO_4^{3-}.
Step 2	*Determine the ratio of ions required in order to achieve electrical neutrality. (Remember compounds have no overall charge.)* The ratio of negative to positive charges for calcium and phosphate ions is 3 : 2. That is, it takes two negatively charged phosphate ions to balance the charge of the three positively charged calcium ions.
Step 3	*Write the formula for the compound using the numbers in the ratios as subscripts. Use brackets to show the polyatomic component.* The formula for calcium phosphate is $Ca_3(PO_4)_2$.

Note the use of brackets in the formula to show that more than one molecular ion is needed to balance the electric charge. If the brackets were not used for $Ca_3(PO_4)_2$ then the formula would be an absurd Ca_3PO_{42}.

Swap and drop

Another way to determine the ionic formula is using the swap and drop rule.

1. Write the ions in the compound.
2. Swap the values of the ions.
3. Drop the number and make a subscript (to represent the number of atoms).

Notes:
- If part of the compound is polyatomic ensure brackets are used.
- If the numbers are identical, they cancel out.
- You do not need to write the number 1 for both charge or number of atoms, but it can help while doing the 'swap and drop'.

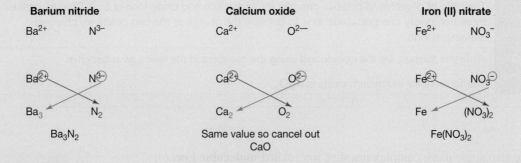

DISCUSSION

Imagine that no recognised system for naming elements and compounds existed. Describe some of the problems this would lead to.

ACTIVITY: The ionic compound formula game

Use the following to practise deriving the chemical formulae for ionic compounds.
- Create a set of playing cards with the name and valency of each of the cations and anions listed in tables 4.21 and 4.22. You need at least four identical cards for each ion.
- Organise a group of four students to play the card game. The aim of this game is to collect as many cards as possible by producing compounds with correct chemical formulae.
- Shuffle the cards and distribute them to the players.
- The dealer puts down one card.
- The rest of the players try to produce a chemical formula using their cards. The first person to come up with a correct chemical formula wins the hand and keeps the cards. They are put aside until the end of the game. The dealer will decide the winner of the hand.
- The person to the left of the dealer then puts down one of their cards.
- The other players in the game now try to produce a chemical formula using the cards they have in their hands. Again, the person to come up with a correct chemical formula wins that hand and the cards are put aside until the end of the game.
- The game continues moving to the next person until no one is able to produce a chemical formula. The game stops at this point.
- Each player then counts the number of cards they have produced formulae with. The winner is the person with the most cards.

Discuss and explain

1. Write a list of the formulae and the name of the compounds formed.
2. What is the best strategy to win the game?
3. Did you find the game useful in learning the formulae of compounds? Explain.

---

on Resources

📋 **eWorkbook** Chemical formulae (ewbk-13106)

4.7 Activities

learn on

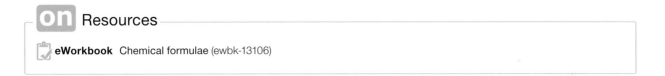

4.7 Quick quiz **on**	**4.7 Exercise**

Select your pathway

■ LEVEL 1	■ LEVEL 2	■ LEVEL 3
1, 2, 3, 7, 10	4, 8, 11, 12, 14	5, 6, 9, 13, 15

These questions are even better in jacPLUS!
- Receive immediate feedback
- Access sample responses
- Track results and progress

Find all this and MORE in jacPLUS ▶

Remember and understand

1. Identify two pieces of information that are available in a chemical formula.
2. Write the symbols for the following elements: sodium, hydrogen, potassium, lead, iodine and sulfur.
3. Write the names of the elements present in each of the following compounds.
   a. HNO_3
   b. $NaHCO_3$
   c. FeS

4. Define the following terms.
   a. The valency of an element
   b. The electrovalency of an element
5. Complete the following table comparing covalency and electrovalency.

TABLE Differences between valency and electrovalency

	Valency	Electrovalency
Type of bonding		
Is a sign for charge included?		

Apply and analyse

6. The chloride ion has the same valency as the sodium ion. However, it has a different electrovalency. Why?
7. Write down the valencies for the following elements: sodium, hydrogen, lead, iodine, magnesium and sulfur.
8. How many chloride (Cl^-) ions would be required to combine with each of the following ions to form an ionic compound?
   a. Calcium (Ca^{2+})
   b. Aluminium (Al^{3+})
   c. Silver (Ag^+)
9. What is the electrovalency of the cation in each of the following compounds?
   a. $FeCl_2$
   b. SnO_2
10. Consider the following formulae: H_2S, KBr, PF_3 and CCl_4. Write the names of these compounds.
11. Write a formula for each of the following.
   a. Oxygen gas
   b. Chlorine gas
   c. Lead
   d. Nitrogen monoxide
   e. Zinc oxide
   f. Potassium sulfate
   g. Calcium hydroxide
   h. Magnesium phosphate
12. Name the following compounds.
   a. NH_4Cl
   b. KI
   c. $Al(NO_3)_3$
   d. $Fe(OH)_3$
   e. $KHCO_3$
   f. $MgCO_3$
   g. HNO_3
   h. $(NH_4)_3PO_4$

Evaluate and create

13. The following ions can combine in different ways to form many compounds. Write the formulae and names of ten compounds using the ions listed. You must use each ion twice. *Hint:* iron and copper are transition metals.

   • Na^+ • Fe^{3+} • Li^+ • Cu^{2+} • Al^{3+}

   • Cl^- • OH^- • N^{3-} • O^{2-} • SO_4^{2-}

14. Using a clearly labelled diagram, explain why group 18 is not listed in table 4.19 showing valency of groups in the periodic table.
15. **SIS** Research the ingredients of five different chemicals (that you haven't heard of) listed in the ingredients for your shampoo. Find out the formula for each of these and identify if they have covalent or ionic bonding.

Fully worked solutions and sample responses are available in your digital formats.

LESSON
4.8 Thinking tools — Concept maps and mind maps

4.8.1 Tell me

What is a concept map?

A concept map is a flow chart that shows an understanding of interconnected ideas, linking the topic to main ideas and concepts within those ideas. An example of a concept map is shown in figure 4.47.

It is also known as a knowledge map or concept web. It allows you to focus on two main questions:

- What do I understand about this particular topic?
- How do ideas link in this topic?

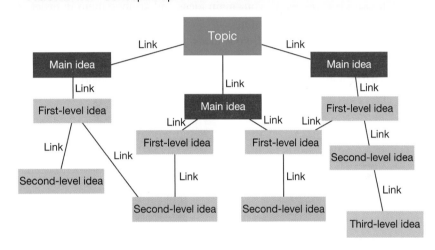

FIGURE 4.47 Concept map

What is a mind map?

A mind map is similar to a concept map, and allows you to show the hierarchy of ideas within a topic. An example of a mind map is shown in figure 4.48.

Mind maps are also referred to as model maps, memory maps, brain maps or learning maps, and not only show ideas, but also features of each of these.

A mind map allows for the summary of ideas and features in a topic.

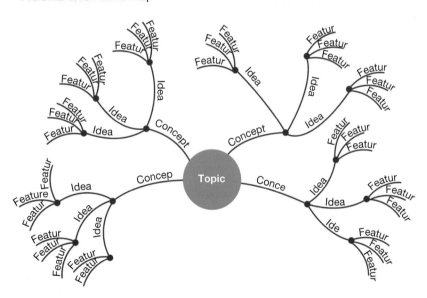

FIGURE 4.48 Mind map

How are concept maps and mind maps different?

Both mind maps and concept maps show the structure of a topic in a hierarchical way. While concept maps are usally organised horizontally, mind maps radiate from a central idea. Concept maps often show links and relationships between different ideas more clearly, whereas a mind map tends to separate ideas in groups.

4.8.2 Show me

Creating a concept map

To create a concept map, such as the one shown in figure 4.49, complete the following steps.

1. On small pieces of paper, write all the ideas you can think of about a particular topic (you may also do this digitally).
2. Start with your topic and place this at the top of your page.
3. Select the most important ideas and arrange them under your topic. Link these main ideas to your topic and write the relationship along the link.
4. Choose ideas related to your main ideas and arrange them in order of importance under your main ideas, adding links and relationships.
5. When you have placed all your ideas, try to add links between the branches and write in the relationships.

FIGURE 4.49   Sample concept map on chemical bonding

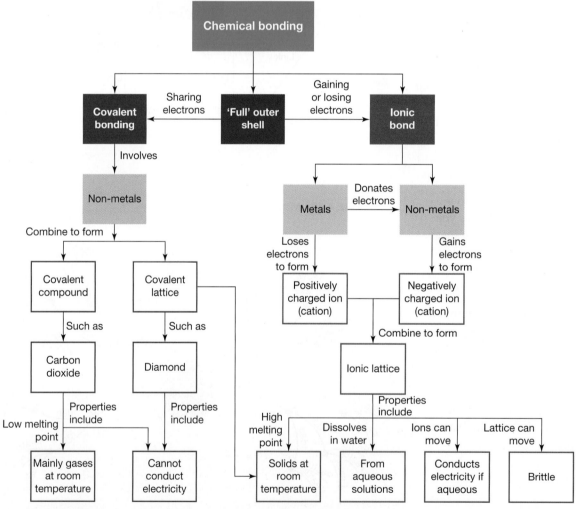

Creating a mind map

To create a mind map, such as the one shown in figure 4.50, complete these steps.

1. Start with your topic and place this in the centre and draw a number of lines branching from it.
2. On small pieces of paper, write all the ideas you can think of about a particular topic.
3. Select the most important ideas and write one on each branch.
4. Draw a number of lines branching from these main ideas.
5. Think of words or terms related to each of the key ideas, continuing to add branches until you run out of ideas.

FIGURE 4.50 Sample mind map on chemical bonding

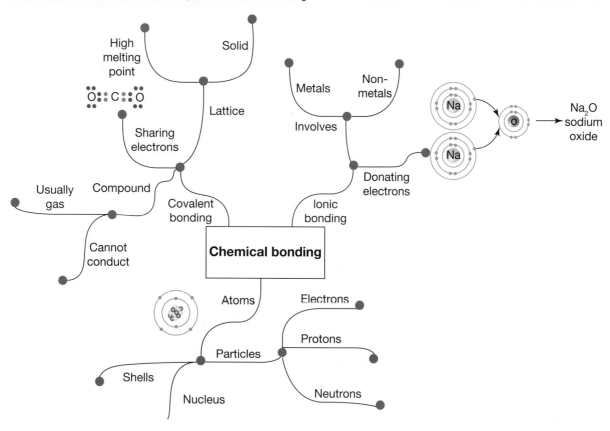

4.8.3 Let me do it

4.8 Activity

1. A concept map can be used to illustrate some of the important ideas associated with the atom and the links between them. Use the concept map shown below for the atom and complete it by adding links between the ideas.

These questions are even better in jacPLUS!
- Receive immediate feedback
- Access sample responses
- Track results and progress

Find all this and MORE in jacPLUS ▶

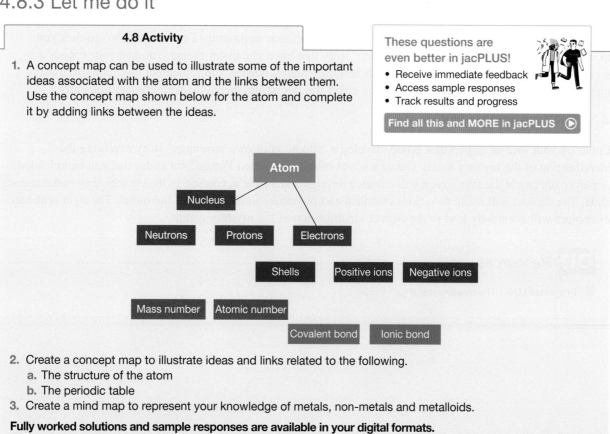

2. Create a concept map to illustrate ideas and links related to the following.
   a. The structure of the atom
   b. The periodic table
3. Create a mind map to represent your knowledge of metals, non-metals and metalloids.

Fully worked solutions and sample responses are available in your digital formats.

LESSON
4.9 Project — The mystery metal

4.9.1 Scenario

Your eccentric aunt loves combing through op-shops in search of overlooked treasures, and every time you spend a day with her she makes you go into one grubby shop after another, all smelling of mangy mink coats. One day during the school holidays, you are wandering idly in one of these old op-shops while your aunt haggles with the owner about the price for an old vase. You find a lump of metal in a drawer of an old dresser. The shopkeeper says that you can keep it and you put it in your pocket. Occasionally over the next few days you wonder what the metal is. Is it something valuable like platinum, or useful like aluminium? Or is it just an old lump of lead? By the end of the holidays, you've forgotten all about the lump of mystery metal.

FIGURE 4.51 What mysteries hide in a junk shop?

FIGURE 4.52 Can you identify a mystery metal?

When you get back to school, your science teacher announces that everyone in your class is to enter a competition that the Australian Chemistry Teachers Association is running. The competition requires you to write an online 'choose your own adventure' story that has a chemistry theme. You and your friends are scratching your heads trying to come up with an idea when, suddenly, you remember that lump of mystery metal you found in the junk shop. Maybe you could use that as the theme for your competition entry …

4.9.2 Your task

Either on your own or as part of a group, develop a 'choose your own adventure' story exploring the identification of the mystery metal. Create a series of interconnected PowerPoint slides that can be uploaded. A player starting at the first screen will advance through a storyline according to the choices they make at each slide. The choices will relate to various chemical and physical characteristics of the metal. The right sequence of choices will eventually lead to the correct identification of the mystery metal.

 Resources

💡 **ProjectsPLUS** The mystery metal (pro-0133)

LESSON
4.10 Review

Access your topic review eWorkbooks

 Resources

■ Topic review Level 1
ewbk-13108

■ Topic review Level 2
ewbk-13110

■ Topic review Level 3
ewbk-13112

4.10.1 Summary

The periodic table

- The periodic table shows all discovered elements in order of their atomic number or number of protons.

- Atoms are electrically neutral, the number of protons in an atom is the same as the number of electrons.

- The mass number of an atom is the sum of the number of protons and neutrons in the atom (the particles within the nucleus of the atom).

- The columns in the periodic table are known as groups. Elements in similar groups have similar physical and chemical properties.

- The periodic table can be split into metals, metalloids and non-metals.

- Metals are towards the left of the periodic table, and are mostly solid at room temperature, good conductors, malleable, lustrous and ductile.

- Non-metals are towards the right of the periodic table, and are dull, brittle and poor conductors. Many non-metals are gases.

- Metalloids are found between metals and non-metals on the periodic table and have properties of both metals and non-metals.

- The periodic table also displays a number of trends, both across a period and down a group. Examples of such trends are atomic number, atomic size, reactivity, melting points and metallic character.

- The elements after uranium have all been artificially produced. To date, 26 further such elements have been made and there are ongoing efforts to produce still more.

Electrons, electron configuration and shells

- Electrons are constantly moving and behave as a cloud of negative charge.

- An electron shell diagram is a simplified model of an atom, showing electrons arranged in a series of energy levels around the nucleus known as shells.

- The maximum number of electrons in each shell can be calculated using the rule $2n^2$, where n is the number of the shell.

- The arrangement of electrons for each element can be written as an electron configuration; for example, sodium is written as 2,8,1.

- Spectral evidence is the main experimental basis upon which the theory of electron shells is based.

- There are two main types of spectra — emission spectra and absorption spectra.

- The lines in each type of spectra are identical for a particular element and can be used as a means of identification.
- A spectroscope is an instrument used to analyse light and produce a spectrum.

Extension: Ionic bonding — Give and take

- Atoms want to gain or lose electrons to have a stable outer shell (eight for most shells, except the first shell, which has a maximum of two).
- Bonding involves outer shell electrons known as valence electrons.
- Ionic bonding occurs between metals and non-metals, in which metals give up electrons to non-metals.
- In ionic bonding, metals become positively charged cations (such as calcium becoming Ca^{2+}) and non-metals becoming negatively charged anions (such as oxygen becoming O^{2-}).
- Ionic compounds are usually solid at room temperature, with high melting points. They are brittle when a force is applied, can dissolve to form aqueous solutions and conduct electricity when aqueous.

Extension: Covalent bonding — When sharing works best

- Covalent bonding occurs between non-metals.
- As non-metals wish to gain electrons, they can share electrons to gain form stable outer shells.
- Electron dot diagrams can assist in illustrating covalent bonds by showing the sharing of the outer shell electrons.
- Electrons that are shared are known as bonding pairs. Those that are not shared are known as non-bonding pairs or lone pairs.

The reactivity of metals

- Different metals have different reactivities. This is shown using an activity (or reactivity) series.
- The activity series places the elements in decreasing order of reactivity:

$$Li \rightarrow K \rightarrow Na \rightarrow Ca \rightarrow Mg \rightarrow Al \rightarrow Mn \rightarrow Cr \rightarrow Zn \rightarrow Fe \rightarrow Ni \rightarrow$$
$$Sn \rightarrow Pb \rightarrow H \rightarrow Cu \rightarrow Hg \rightarrow Ag \rightarrow Au \rightarrow Pt.$$

Finding the right formula

- When naming covalent compounds, prefixes are used to name compounds, so CCl_4 is known as carbon tetrachloride.
- When determining the formula for covalent compounds, the valency of each atom needs to be considered.
- When naming ionic compounds, the metal component forms the first part of the name and the non-metal forms the second part of the name. The ending of the non-metal is changed to -ide.
- When determining the formula for ionic compounds, the electrovalencies need to be used to determine the number of each ion required to give an overall neutral charge.

4.10.2 Key terms

absorption spectrum the pattern of dark lines against a rainbow background obtained when white light is passed through a gas
activity series a classification of metals in decreasing order of reactivity
alkali metals very reactive metals in group 1 of the periodic table
alkaline earth metals reactive metals in group 2 of the periodic table
alloy a mixture of several metals or a mixture of a metal and a non-metal
anions atoms that have gained electrons to become negatively charged ions
aqueous solutions solutions in which water is the solvent
atoms very small particles that make up all things
atomic number the number of protons in the nucleus of an atom
bonding electrons shared electrons holding two atoms together
brittle breaks easily into many pieces
cations atoms that have lost electrons to become positively charged ions
conductors materials that allow electric charge to flow through them
covalent bond a shared pair of electrons holding two atoms together

covalent compounds compounds in which the atoms are held together by covalent bonds

displacement reactions chemical reactions involving the transfer of electrons from the atoms of a more reactive metal to the ions of a less reactive metal

double covalent bond a covalent bond involving the sharing of two pairs of electrons

ductile capable of being drawn into wires or threads; a property of most metals

duet rule the tendency of elements that only contain electrons in shell 1 to gain electrons to have two

electron very light, negatively charged particle inside an atom that moves around the central nucleus

electron configuration an ordered list of the number of electrons in each electron shell, from inner (low energy) to outer (higher energy) shells

electron dot diagrams diagrams using dots to represent the electrons in the outer shell of atoms and to show the bonds between atoms in molecules

electron shell diagram a diagram showing electrons in their shells around the nucleus of an atom

electrostatic attraction the attraction between two oppositely charged bodies

electrovalency the number of positive or negative charges on an ion

elements pure substances made up of only one type of atom

emission spectra the pattern of coloured lines obtained when emitted light from a sample is observed through a spectroscope

excited state the highest energy state of an atom, in which electrons jump to a higher shell

flame test a type of test in which an element is heated so the atom moves from a ground state to an excited state, and releases coloured light as the electrons fall back down

frequency the number of waves that pass a particular point each second

ground state the lowest energy state of an atom

groups columns of the periodic table containing elements with similar properties

halogens non-metal elements in group 17 of the periodic table

ionic bond an attractive force between ions with opposite electrical charge

ionic compounds compounds containing positive and negative ions held together by the electrostatic force

ions atoms that have lost or gained electrons and carry an electrical charge

isotopes atoms of the same element that differ in the number of neutrons in the nucleus

lattice a structure in which particles are bonded to each other to form crystal structures

lustre the shiny appearance of many metals

malleable able to be beaten, bent or flattened into shape

mass number the number of protons and neutrons in the nucleus of an atom

metals elements that conduct heat and electricity and are shiny solids that can be made into thin wires and sheets

metalloids elements that have features of both metals and non-metals

mineral ores rocks mined to obtain a metal or other chemical within them

molecules small particles of a chemical substance consisting of one or more types of atoms

molecular formula a shorthand statement of the elements in a molecule showing the relative number of atoms of each kind of element

molecular ions groups of atoms that have an overall charge and are treated as an entity; e.g. OH^-, SO_4^{2-}, NH_4^+

network lattice a lattice in which each atom is covalently bonded to adjacent atoms

neutron uncharged particle found in the nucleus of an atom

noble gases elements in the last column of the periodic table that are unreactive

non-bonding pairs electron pairs not involved in covalent bonding (also known as lone pairs)

non-metals elements that do not conduct electricity or heat and usually have a lower boiling and melting point than metals

nucleus the central part of an atom, made up of protons and neutrons

octet rule the tendency of atoms to gain or lose electrons to have eight in their outer shell

periods rows of elements on the periodic table

periodic law elements with similar properties occur at regular intervals when all elements are listed in order of atomic weight

periodic table the table listing all known elements, grouped according to their properties and in order of atomic number (number of protons in their nucleus)

polyatomic ions another term for a molecular ions, in which a group of atoms have an overall charge

proton a positively charged particle found in the nucleus of an atom

relative atomic mass mass of the naturally occurring mixture of the isotopes of an element

shells energy levels surrounding the nucleus of an atom into which electrons are arranged

single covalent bond a covalent bond involving the sharing of one pair of electrons

smelting the production of a metal in its molten state

spectroscope an instrument that breaks up the light it receives into distinct frequencies, colours or wavelengths

structural formula a diagram showing the arrangement of atoms in a substance with covalent bonds drawn as dashes

symbols one- or two-letter code(s) used for the elements

transition metal block a block of metallic elements in the middle of the periodic table

triple covalent bond a covalent bond involving the sharing of three pairs of electrons

valence electrons outer shell electrons involved in chemical reactions

valency the number of electrons that each atom needs to gain, lose or share to 'fill' its outer shell

wavelength the distance between the crest of one wave and the next

weighted mean the average mass of an element that is calculated from the percentage of each isotope in nature

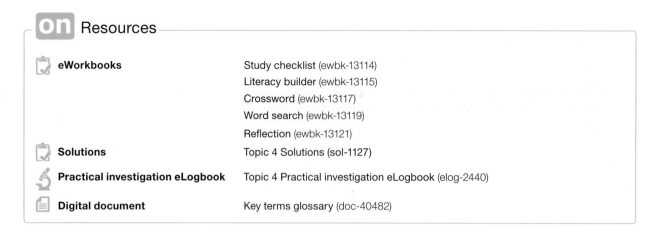
4.10 Activities

learn on

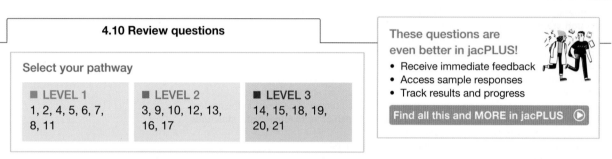

4.10 Review questions

Select your pathway

■ LEVEL 1	■ LEVEL 2	■ LEVEL 3
1, 2, 4, 5, 6, 7, 8, 11	3, 9, 10, 12, 13, 16, 17	14, 15, 18, 19, 20, 21

These questions are even better in jacPLUS!
- Receive immediate feedback
- Access sample responses
- Track results and progress

Find all this and MORE in jacPLUS ▶

Remember and understand

1. Explain why it is more useful to display the elements as a periodic table than as a list.
2. The periodic table is an arrangement of all the known elements. What information is given by the group and period numbers on the periodic table?
3. **SIS** Explain how the periodic table has helped chemists of both the past and present when they are searching for new elements.
4. Explain why water does not appear in the periodic table.
5. **MC** To which group of elements in the periodic table does the neon used in lighting belong?
   A. Group 1 (alkali metals)
   B. Group 2 (alkaline earth metals)
   C. Group 17 (halogens)
   D. Group 18 (noble gases)
6. List three properties that all (or almost all) metals have in common.
7. List three properties that most solid non-metals have in common.
8. As you move down the groups in the periodic table, state the reactivity change for the following.
   a. Metals
   b. Non-metals

Apply and analyse

9. Complete the following table for atoms of each element.

TABLE The sub-atomic particles and nuclide symbols of different elements

Nuclide symbol	Atomic number	Mass number	Protons	Neutrons	Electrons
a. $^{28}_{14}$Si	14	28		14	
b.	25	55			25
c.	79			118	
d.		207			82
e.				150	94

10. As you move across the periodic table, describe the changes that occur in the following. Explain the reason for each of these trends.
    a. Atomic number
    b. Mass number
    c. Melting points
    d. Metallic character

11. Although they look very different from each other and have very different uses, arsenic, germanium and silicon belong to the group of elements known as metalloids. Describe how metalloids are different from all of the other elements in the periodic table, using clear examples.

12. Write the formula for the following substances.
    a. Lead
    b. Carbon
    c. Oxygen gas
    d. Aluminium oxide
    e. Sodium fluoride
    f. Calcium carbonate
    g. Zinc chloride
    h. Iron(II) sulfide
    i. Sulfur dioxide

13. Complete the following table for the electron configuration of different atoms.

TABLE Electron configuration of different atoms

Name	Symbol	Atomic number	Electron configuration
a. Lithium		3	2,1
b.	C	6	
c.			2,6
d. Neon			
e.	Na		
f.		13	
g.			2,8,5
h. Chlorine			
i.	K		2,8,8,1
j.	Ca	20	

14. Complete the following table for the electron configuration of different ions.

TABLE Electron configuration of different ions

Ion name	Ion symbol	Atomic number	Electron configuration
a.		3	2
b.	Na^+		
c.		12	2,8
d.	N^{3-}		
e.		9	2,8
f. Sulfide ion			

15. The electron shell diagram shown has its first two shells filled. It could represent a neutral atom, a positive ion or a negative ion. Identify the names and symbols of the atom or ion if it represents the following.

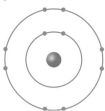

a. **MC** A neutral atom (identify one)
   A. Oxygen
   B. Neon
   C. Fluorine
   D. Argon
b. **MC** A positive ion (identify three)
   A. Al^+
   B. Al^{3+}
   C. Mg^{2+}
   D. Na^+
   E. Mg^{3+}
c. **MC** A negative ion (identify two)
   A. Cl^-
   B. F^-
   C. F^{2-}
   D. O^{2-}
   E. O^-
   F. Br^-

16. Complete the following table.

TABLE Names and formulae of different ionic compounds

Name	Formula
a. Lithium fluoride	
b. Potassium nitride	
c.	MgO
d. Copper(II) chloride	
e.	$Ca(NO_3)_2$
f.	Na_2S
g. Ammonium chloride	
h.	Fe_2O_3

Evaluate and create

17. Name and draw electron dot diagrams for the following covalent compounds.
    a. HCl   **b.** H_2O **c.** NH_3 **d.** CH_4 **e.** CO_2

18. State and explain the different melting points of ionic and covalent molecular compounds. Use clearly labelled diagrams to show this.

19. a. Explain why you are more likely to find pure gold than pure copper in the ground.
    b. Research and outline the locations of at least three copper mines and three gold mines in Australia.

20. Displacement reactions can be used to determine the reactivity of different metals.
    a. Use the activity series to predict whether a reaction will occur for each of the combinations shown in the following table. In each case, a piece of metal is added to an ionic solution. A shaded cell indicates that no experiment was performed for this combination.
       • Use a ✓ to indicate that a reaction will occur.
       • Use a ✗ to indicate that no reaction will occur.

TABLE Reactions between metals and metal salts

	Cu	Ag	Sn	Ca
$Cu(NO_3)_2$				
$AgNO_3$				
$Sn(NO_3)_2$				
$Ca(NO_3)_2$				

    b. Which of the metals listed could be added to a solution of silver nitrate to produce silver crystals?
    c. Write a word equation for the reaction between any of the metals from part **b.** and silver nitrate. States, symbols and Roman numerals are not required.

21. All atoms of the element magnesium have 12 protons. Of those atoms, 80 per cent have 12 neutrons. The relative atomic mass of magnesium is 24.3.
    a. State the atomic number of magnesium.
    b. What is the mass number of most magnesium atoms?
    c. Explain why all magnesium atoms don't have the same mass number.
    d. Predict whether there are likely to be more magnesium atoms with a mass number of 23 or 25 on Earth. Explain your answer.

Fully worked solutions and sample responses are available in your digital formats.

Hey teachers! Create custom assignments for this topic

Create and assign unique tests and exams

Access quarantined tests and assessments

Track your students' results

Find all this and MORE in jacPLUS

Online Resources

Below is a full list of **rich resources** available online for this topic. These resources are designed to bring ideas to life, to promote deep and lasting learning and to support the different learning needs of each individual.

4.1 Overview

eWorkbooks
- Topic 4 eWorkbook (ewbk-13083)
- Starter activity (ewbk-13085)
- Student learning matrix (ewbk-13087)

Solutions
- Topic 4 Solutions (sol-1127)

Practical investigation eLogbooks
- Topic 4 Practical investigation eLogbook (elog-2440)
- Investigation 4.1: Modelling an atom (elog-2442)

Video eLesson
- The atom (eles-1775)

Interactivity
- Model of the atom (int-5797)

4.2 The periodic table

eWorkbooks
- The structure of the atom (ewbk-13088)
- Periodic table (ewbk-13090)
- Elements and atomic numbers (ewbk-13092)
- Trends in the periodic table (ewbk-13094)

Practical investigation eLogbooks
- Investigation 4.2: Chemical properties of metals and non-metals (elog-2444)
- Investigation 4.3: Comparing properties of two metal families (elog-2446)

Video eLessons
- The unusual properties of mercury (eles-4184)
- Malleability (eles-2033)

Interactivities
- Periodic table (int-0758)

Weblink
- Periodic table trends

Digital document
- Blank periodic table (doc-34956)

4.3 Electrons, electron configuration and shells

eWorkbooks
- Electron shells (ewbk-13096)
- Electron configuration (ewbk-13098)

Practical investigation eLogbook
- Investigation 4.4: Flame tests (elog-2448)
- Investigation 4.5: Observing emission spectra (elog-2450)

Video eLesson
- Flame tests (eles-2712)

Interactivities
- Shell-shocked (int-0676)
- Electron shells (int-5798)

4.4 Extension: Ionic bonding — Give and take

eWorkbook
- Ionic bonding (ewbk-13100)

Interactivities
- Pass the salt (int-0675)
- Forming ions (int-5799)
- Forming ionic compounds (int-5800)
- Ionic models (int-6351)

4.5 Extension: Covalent bonding — When sharing works best

eWorkbook
- Covalent molecules (ewbk-13102)

Interactivity
- Making molecules (int-0228)

4.6 The reactivity of metals

eWorkbook
- Metals and the activity series (ewbk-13104)

Practical investigation eLogbooks
- Investigation 4.6: Investigating reactivity (elog-2452)
- Investigation 4.7: Quantified reactivity (elog-2454)

Video eLesson
- Davey and potassium (eles-1773)

Teacher-led video
- Investigation 4.7: Quantified reactivity (tlvd-10816)

4.7 Finding the right formula

eWorkbook
- Chemical formulae (ewbk-13106)

4.9 Project — The mystery metal

ProjectsPLUS
- The mystery metal (pro-0133)

4.10 Review

eWorkbooks
- Topic review Level 1 (ewbk-13108)
- Topic review Level 2 (ewbk-13110)
- Topic review Level 3 (ewbk-13112)
- Study checklist (ewbk-13114)
- Literacy builder (ewbk-13115)
- Crossword (ewbk-13117)
- Word search (ewbk-13119)
- Reflection (ewbk-13121)

Digital document
- Key terms glossary (doc-40482)

To access these online resources, log on to **www.jacplus.com.au**

5 Chemical reactions

CONTENT DESCRIPTION

Identify patterns in synthesis, decomposition and displacement reactions and investigate the factors that affect reaction rates (AC9S10U07)

Source: F–10 Australian Curriculum 9.0 (2024–2029) extracts © Australian Curriculum, Assessment and Reporting Authority; reproduced by permission.

LESSON SEQUENCE

SCIENCE INQUIRY AND INVESTIGATIONS

Science inquiry is a central component of Science curriculum. Investigations, supported by a **Practical investigation eLogbook** and **teacher-led videos**, are included in this topic to provide opportunities to build Science inquiry skills through undertaking investigations and communicating findings.

LESSON
5.1 Overview

Hey students! Bring these pages to life online

 Watch videos

 Engage with interactivities

 Answer questions and check results

Find all this and MORE in jacPLUS

5.1.1 Introduction

Engineers developing materials for spacesuits or the next generation of passenger aircraft need knowledge and understanding of chemical reactions to produce new materials that are strong, light and capable of resisting high temperatures. Other useful substances and materials such as fuels, metals and pharmaceuticals are also products of chemical reactions.

Chemical reactions are everywhere! They can be subtle, such as a minor change in temperature in an ice pack or fruit ripening. They can be spectacular, such as the reaction between sulfuric acid and icing sugar shown in the Video eLesson. They can be slow, like the rusting of a car, or fast, like the explosion of

FIGURE 5.1 When developing materials for spacesuits, engineers produce new materials that are light and capable of resisting high temperatures.

fireworks on New Year's Eve. Chemical reactions can be controlled, as in a laboratory, or out of control, as in a bushfire. Complex chemical reactions occur constantly around us and in us. Every second of the day food is being digested to provide energy and resources for the growth and repair of cells. Fortunately, the breakdown of food is a speedy process. The speed, or rate, of a chemical reaction is important not just in our bodies, but also in the manufacture of materials.

A **chemical reaction** occurs when the bonds between one or more reactant particles are broken, and new bonds are formed in the products. This can occur in one or more steps and this process is called a 'chemical change'. The language of chemistry, formulae and equations are used to summarise reactions. Models are used to show where atoms begin as a reactant and finish as a product. This ensures that the law of conservation of mass is maintained; that is, all the atoms that are present at the beginning are rearranged and present at the end. There are many ways of classifying chemical reactions and some reactions can be put into more than one category. This topic will examine the following types of reactions: synthesis, decomposition, and single and double displacement reactions.

chemical reaction the breaking of chemical bonds between reactant molecules and the forming of new bonds in product molecules

5.1.2 Think about chemical reactions

1. Why shouldn't you add water to concentrated acid?
2. Which two gases react explosively to form water?
3. Why can you find metallic gold in the ground but other metals such as aluminium are only found as compounds?
4. How are metals separated from their compounds?
5. Why is it that when you mix an acid and a base, which are two dangerous chemicals, you make a compound that is safe to sprinkle on your chips?
6. How can we make chemical reactions go faster or slower?

5.1.3 Science inquiry

Chemical reactions

You will already be quite familiar with common chemical reactions. Answer the questions below to review your knowledge.

Remember and explain

1. Each of the photos in figure 5.2 a. to c. depicts an example of a chemical reaction.
   a. What observation suggests that a chemical change has occurred in each of the photos?
   b. Discuss what further information would assist your responses to part **a**.
   c. In one test tube, a metal is displacing hydrogen, from an acid, forming the hydrogen gas (dihydrogen). Suggest with a reason or reasons which reaction fits this description.
   d. In another test tube, a solid is formed from the combination of two solutions in a precipitation reaction. Suggest with a reason or reasons which reaction fits this description.
   e. Finally, a reactive metal is dissolving in a solution of less reactive metal and the less reactive metal is changing from being in solution to being in metal form.
   f. Describe some other observations that could be made when a chemical reaction takes place.

FIGURE 5.2 Signs of chemical change in chemical reactions **a.** to **c.**

a.

b.
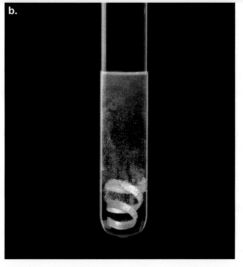

c.

2. Chemical reactions take place in all living things to keep them alive.
   a. Which chemical reaction takes place in all green plants in the presence of sunlight?
   b. Name the process by which food is broken down from large molecules into small molecules so that they can be absorbed by the body.
   c. Name the chemical reaction that takes place in every cell of all animals to transform stored energy from food into other forms of energy.

Inside chemical reactions

Chemical reactions take place when the bonds between atoms are broken and new bonds are formed. This creates a new arrangement of atoms and therefore at least one new substance.

3. Explain what happens to the chemical bonds during the chemical reaction between oxygen and hydrogen, as illustrated in figure 5.3.

FIGURE 5.3 The chemical reaction between oxygen and hydrogen

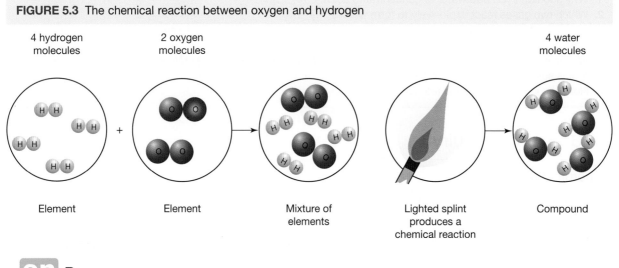

| 4 hydrogen molecules | 2 oxygen molecules | | | 4 water molecules |

Element   Element   Mixture of elements   Lighted splint produces a chemical reaction   Compound

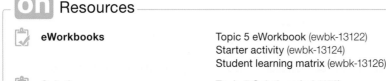 Resources

eWorkbooks	Topic 5 eWorkbook (ewbk-13122) Starter activity (ewbk-13124) Student learning matrix (ewbk-13126)
Solution	Topic 5 Solutions (sol-1128)
Practical investigation elogbook	Topic 5 Practical investigation eLogbook (elog-2456)

LESSON
5.2 The language of chemistry

LEARNING INTENTION

At the end of this lesson you will be able to write balanced chemical equations using correct chemical formulae.

5.2.1 Chemical formulae

To communicate with each other easily about chemical reactions, scientists all over the world need to use the same language. The language used by scientists in chemistry involves chemical symbols, **formulae** and **equations**

Chemists use chemical formulae as a shorthand way of representing molecules and compounds. Covalent compounds consist of non-metals chemically bonded together,

formulae set of chemical symbols showing the elements present in a compound and their relative proportions

equation statement describing a chemical reaction, with the reactants on the left and the products on the right separated by an arrow

usually in small groups of atoms called molecules; for example, water (H_2O) and carbon dioxide (CO_2). Often the name of a covalent compound can be helpful — carbon monoxide means that there is only one oxygen attached to the carbon atom whereas carbon dioxide has two oxygen atoms attached to the carbon atom. Remembering the seven molecules that are **diatomic** is useful when writing equations. These elements are always written as diatomic molecules and the formulae are H_2, N_2, O_2, F_2, Cl_2, Br_2, and I_2.

Compounds consisting of non-metallic elements are covalent compounds.

Compounds consisting of non-metallic elements and metallic elements are ionic compounds.

Ionic compounds consist of cations and anions in giant lattices, so their formulae show the ratio of cations to anions. Cations are positive ions because they have lost electrons, and anions are negative ions because they have gained electrons. Negative ions often end in -ide, for example, sulfide (S^{2-}), nitride (N^{3-}) and oxide (O^{2-}). Examples of ionic formulae include sodium chloride (NaCl), sodium hydroxide (NaOH) and zinc nitrate ($Zn(NO_3)$). Examples of common formulae for different compounds are listed in tables 5.1 and 5.2.

FIGURE 5.4 Seven elements (hydrogen, nitrogen, oxygen, fluorine, chlorine, bromine and iodine) that form diatomic molecules.

TABLE 5.1 The formulae of some common ionic compounds

Compound	Formula
Sodium hydroxide	NaOH
Sodium chloride	NaCl
Sodium sulfate	Na_2SO_4
Sodium citrate	$C_6H_5O_7Na_3$
Sodium hydrogen carbonate	$NaHCO_3$
Copper(II) hydroxide	$Cu(OH)_2$
Copper(II) sulfate	$CuSO_4$
Magnesium chloride	$MgCl_2$
Mercury(II) oxide	HgO

TABLE 5.2 The formulae of some common covalent compounds

Compound	Formula
Water	H_2O
Citric acid	$C_6H_8O_7$
Carbon dioxide	CO_2
Oxygen	O_2
Hydrochloric acid	HCl
Carbon monoxide	CO
Hydrogen	H_2
Methane	CH_4
Ammonia	NH_3

Writing formulae for ionic compounds

As discussed in topic 4, most ionic formulae commence with the cation first, which is generally a metal (or the ammonium ion), followed by the anion, which is usually a non-metal. The following steps are used for writing formulae for ionic compounds. For example, if the formula of tin nitrate is required:

diatomic molecule containing two atoms

Step	Example
1. Write the name of the compound.	tin nitrate
2. Write the formulae of the individual ions (from memory or using the table).	Sn^{2+} NO_3^-
3. Balance the charges so there are the same number of positive and negative charges. Two nitrate atoms will be required to accept the two electrons that tin has lost.	Sn^{2+} NO_3^- NO_3^-
4. Use a subscript to show that two nitrate ions will be in the formula. Note that it is important not to change the subscript in the nitrate ion. The final formula should not have any charges present.	$Sn(NO_3)_2$

5.2.2 Chemical equations

Word equations provide a simple way to describe chemical reactions by stating the reactants and products. Chemical equations that use formulae provide more information. They show how the atoms in the reactants combine to form the products.

Writing chemical equations involves some simple mathematics and a knowledge of chemical formulae. Chemical equations are set out in the same way as word equations, with the reactants to the left of the arrow and products to the right. However, they are different from word equations in three ways:

- Formulae are used to represent the chemicals involved.
- The physical states of the chemicals are often included.
- Numbers are written in front of the formulae in order to balance the numbers of atoms on each side of the equation.

The steps used in balancing equations are outlined below. For a detailed, step-by-step guide, see Science Quest 9, topic 7.

Balancing equations

1. Determine the reactants and products and write a word equation.
2. Determine the chemical formulae.
3. Write down the equation.
4. Balance the numbers of atoms.
5. Include the states.

CASE STUDY: Another example of balancing a chemical equation

During respiration carbon dioxide gas and water are produced from glucose and oxygen gas.

Step 1: Determine the reactants and products and write a word equation

The reactants are glucose and oxygen and the products are carbon dioxide and water.

As a word equation, this is:

$$glucose + oxygen\ gas \rightarrow carbon\ dioxide\ gas + water$$

Step 2: Chemical formulae

Determine the formulae for each reactant and product:

- glucose = $C_6H_{12}O_6$
- oxygen gas = O_2
- carbon dioxide = CO_2
- water = H_2O.

Step 3: Write down the equation

Replace the words in the word equation with the formulae:

$$C_6H_{12}O_6 + O_2 \rightarrow CO_2 + H_2O$$

Step 4: Balance the number of atoms

Count the number of atoms of each element for the reactants and products.

Element	Reactants	Products
C	6	1
H	12	2
O	8	3

If the number of atoms of each element is the same on both sides of the equation, the equation is already balanced. If not, numbers need to be placed in front of one or more of the formulae to balance the equation. These numbers are called coefficients and they multiply all of the atoms in the formula.

To balance the carbon atoms, put a 6 in front of CO_2:

$$C_6H_{12}O_6 + O_2 \rightarrow 6CO_2 + H_2O$$

Recount the number of atoms:

Element	Reactants	Products
C	6	6
H	12	2
O	8	13

To balance the hydrogen atoms, put a 6 in front of H_2O:

$$C_6H_{12}O_6 + O_2 \rightarrow 6CO_2 + 6H_2O$$

Recount the number of atoms:

Element	Reactants	Products
C	6	6
H	12	12
O	8	18

Finally, the oxygen atoms can be balanced by putting a 6 in front of the O_2 on the left:

$$C_6H_{12}O_6 + 6O_2 \rightarrow 6CO_2 + 6H_2O$$

The equation is now balanced. It can be checked by counting the number of atoms of each element on both sides of the new equation.

Element	Reactants	Products
C	6	6
H	12	12
O	18	18

Step 5: Include the states

Include the states for each. Check the given information for clues. Ensure that you include glucose as a solid, because the question specifies that solid glucose is used. Carbon dioxide and oxygen are stated as gases. Water will be in liquid state as respiration occurs below the boiling point of water.

$$C_6H_{12}O_6(g) + 6O_2(g) \rightarrow 6CO_2(g) + 6H_2O(l)$$

ACTIVITY: Modelling equations

Use a molecular modelling kit to model the equation showing the reaction between oxygen and hydrogen. Start with two oxygen atoms joined to make an oxygen molecule and two lots of two hydrogen atoms. Rearrange these to make two water molecules.

Resources

eWorkbooks   Chemical equations (ewbk-13127)
Balancing chemical equations (ewbk-13129)
A world of reactions (ewbk-13131)

Interactivity   Balancing chemical equations (int-0677)

5.2 Activities

learn on

| 5.2 Quick quiz | on | 5.2 Exercise |

Select your pathway

| ■ LEVEL 1 | ■ LEVEL 2 | ■ LEVEL 3 |
| 1, 2, 3, 5, 6 | 7, 11, 12 | 4, 8, 9, 10 |

These questions are even better in jacPLUS!
• Receive immediate feedback
• Access sample responses
• Track results and progress

Find all this and MORE in jacPLUS ▶

Remember and understand

1. **MC** What is the formula for aluminium chloride?
   A. Al_3Cl
   B. $AlCl_3$
   C. $AlCl$
   D. Al_3Cl_3

2. **MC** In what way/s are word equations different from equations in which chemical formulae are used?
   A. Word equations do not have the formulae of the chemicals involved.
   B. Word equations always include the states of the reactants and products.
   C. Numbers are used in word equations, so it is possible to know the numbers of atoms involved.
   D. All of the above.

3. **MC** How are the states (solid, liquid and gas) indicated in a chemical equation?
   A. The states are not indicated in a chemical equation.
   B. The symbols (s) for solid, (l) for liquid and (g) for gas are placed after each reactant and product.
   C. The symbols (1) for solid, (2) for liquid and (3) for gas are placed before each reactant and product.
   D. The symbols (s) for solid, (l) for liquid and (g) for gas are placed before each reactant and product.

4. Explain why it is necessary to balance chemical equations.

Apply and analyse

5. Write chemical formulae for the following compounds.
   a. Magnesium carbonate
   b. Nitrogen dioxide
   c. Sodium nitride

6. Match each metal given in the following table with its chemical symbol.

Metals	Symbols
a. Sodium	**A.** Cu
b. Mercury	**B.** Na
c. Magnesium	**C.** Mg
d. Copper	**D.** Hg

7. Write a word equation and a balanced chemical equation with the states for the following.
   a. When carbon monoxide gas and oxygen gas react to form carbon dioxide gas.
   b. When sodium hydroxide solution and hydrochloric acid solution react to form sodium chloride solution and water.
8. Write a balanced equation using formulae for the reaction that occurs when you eat a sherbet lolly. These sweets commonly contain citric acid $C_6H_8O_7$(aq) and sodium hydrogen carbonate $NaHCO_3$(aq). In the mouth, these chemicals dissolve in your saliva and react together to form sodium citrate solution, $C_6H_5O_7Na_3$(aq), carbon dioxide gas and water.

Evaluate and create

9. **a.** The formula for titanium nitride is TiN. What is the charge on a titanium ion?
   b. The formula for tungsten chloride is WCl_6. What is the charge on a tungsten ion?
10. Write a chemical equation with the states for the following.
    a. When mercury metal and oxygen gas react to form solid mercury(II) oxide.
    b. When magnesium metal and hydrochloric acid solution react to form hydrogen gas and magnesium chloride solution.
    c. When sodium metal and water react to form hydrogen gas and sodium hydroxide solution.
    d. When copper(II) sulfate solution and sodium hydroxide solution react to form solid copper(II) hydroxide and sodium sulfate solution.
    e. When iron metal reacts with oxygen gas to produce solid iron(II) oxide.
11. Create a mnemonic to remember the seven diatomic gases.
12. Create an instructional flowchart explaining how to balance a chemical equation. Use an example that has not already been covered in this lesson to show this process.

Fully worked solutions and sample responses are available in your digital formats.

LESSON
5.3 Safety with chemicals

LEARNING INTENTION

At the end of this lesson you will be able to explain the difference between dangerous goods and hazardous substances and be able to create and understand risk assessments and safety data sheets.

5.3.1 Chemicals can be a health hazard

Chemistry is an exciting subject where unlimited varieties of fascinating new materials are constantly being developed to make life better for society.

Many of the chemicals used in industry, medicine, schools, universities and homes can, however, be hazardous to your health. The hazards come about because these chemicals can react with parts of your body — inside or out. Apart from the dangers to your own health, chemicals can react with common substances such as water and air or have properties that cause great damage to property and the environment.

Laws exist, at both the national and state levels, to ensure that people who use harmful chemicals are informed about how to handle them safely. For this purpose, harmful chemicals are placed within one or both of the **dangerous goods** or **hazardous substances** groups.

5.3.2 Dangerous goods

Chemicals in the dangerous goods group are those that could be dangerous to people, property or the environment. Most dangerous goods are grouped into one of nine classes according to the greatest immediate risk they present. The nine classes are often represented using GHS (Globally Harmonized System) hazard pictograms. These hazard pictograms are shown in figure 5.5. Dangerous goods must be identified with the appropriate dangerous goods sign on their labels.

FIGURE 5.5 The nine GHS pictograms

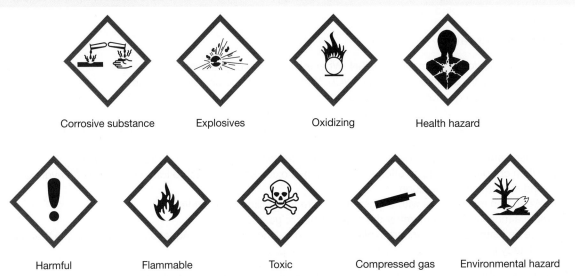

Corrosive substance   Explosives   Oxidizing   Health hazard

Harmful   Flammable   Toxic   Compressed gas   Environmental hazard

5.3.3 Hazardous substances

Chemicals in the hazardous substances group are those that have an effect on human health. The effect may be immediate, such as poisoning and burning, or long term, such as causing liver disease or cancer. Hazardous substances can enter the body in a number of ways. They can be inhaled, absorbed through the skin, ingested (swallowed) or injected.

Hazardous substances are identified on their labels by a signal word (or words) providing a warning about the substance, or the word 'Hazardous' printed in red. Signal words include 'dangerous poison', 'poison', 'warning' and 'caution'. Labels of hazardous substances also include:

- information about the risks of the substance
- directions for use
- safety information
- first aid instructions and emergency procedures.

FIGURE 5.6 A hazardous substance label

If the substance is also in the dangerous goods group, the label will include the appropriate diamond sign showing its class.

5.3.4 Safety data sheets

All employers are required by law to make sure that their employees are fully informed about chemicals in the workplace that are classified as dangerous goods and/or hazardous substances. A list of such chemicals stored or used in the workplace must be kept, along with a copy of each chemical's **safety data sheet (SDS)**. Chemical suppliers are required to provide an SDS for each of the hazardous substances or dangerous goods they supply. In turn, employers are required to make the SDS accessible to employees who are exposed to the chemicals.

An SDS is likely to consist of several A4 pages. Many of them can be downloaded directly from the internet or through a manufacturer. The information on an SDS should include:
- the ingredients of the product
- the date of issue — an up-to-date SDS should be no more than five years old
- information about health hazards and first aid instructions
- precautions that need to be taken when using the product
- information about safe storage and handling of the product.

The concentration of chemicals must be considered as well. For example, experiments carried out at school use only dilute acids. Concentrated acids are extremely dangerous and require extra care. If a concentrated acid needs to be diluted, then the acid must be added to water and not the reverse. If water is added to acid, the reaction releases so much heat that the acid can splatter and cause acid burns and equipment damage. When the acid is added to water, the solution is very dilute and there is less heat produced.

Assessing risk

A **risk assessment** identifies the potential hazards of an experiment and gives protective measures to minimise the risk. Before any experiment involving chemicals is conducted in your school laboratory, a risk assessment is carried out. The form of a risk assessment varies from school to school, but will always contain:
- a summary of the experiment
- a list of the risks and safety precautions for each chemical
- information about whether the chemical is classified as a hazardous substance or dangerous good
- a list of protective measures to be taken. These might include the use of a fume hood and/or the wearing of safety glasses or other protective items.
- first aid information.

Most of the information used in a risk assessment is obtained from the SDS for each chemical used. The date on the SDS used for each chemical must be stated to ensure that the risk assessment is up to date.

Part of a risk assessment sheet is shown in figure 5.7. Risk assessment sheets in schools are usually completed and signed by a qualified science teacher or laboratory technician. Your science teacher is required to carefully read the risk assessment sheet before allowing an experiment involving chemicals to commence.

Protective measures to minimise risk include wearing safety glasses, gloves and protective clothing, working in a well ventilated area and not disposing of dangerous chemicals in the sink after use. Corrosive chemicals should be stored in chemical resistant containers.

safety data sheet (SDS) document containing important information about hazardous chemicals

risk assessment document outlining potential hazards of an experiment and providing protective measures to minimise risk

FIGURE 5.7 Part of a risk assessment sheet for Investigation 4.6

RISK ASSESSMENT SHEET

ACTIVITY	Investigating reactivity	
REFERENCE	Science Quest 10	
	TOPIC 4: Chemical patterns	INVESTIGATION 4.6

SUMMARY OF EXPERIMENT
INVESTIGATING REACTIVITY

1. Place pieces of magnesium, copper, zinc, aluminium and iron in test tubes.
2. Add 1 M hydrochloric acid to the test tubes and observe the reaction.

PROTECTIVE MEASURES

GLASSES	GLOVES	DUST MASK	LAB COAT	FUME HOOD
x			x	x

SAFETY INFORMATION

REACTANT
Hydrochloric acid 1 mol L^{-1}

Hazards	Safety precautions
• **Low toxicity; not classified as a hazardous chemical**	• **Do not breathe gas/fumes/vapour/spray.** • **Wear suitable protective clothing.** • **Avoid contact with skin.**

FIRST AID	
SWALLOWED	Contact doctor or poisons centre. Give glass of water.
EYE	Wash with running water for 15 minutes. Seek medical attention.
SKIN	Remove contaminated clothing. Wash with soap and water.
INHALED	Fresh air. Rest. Keep warm.

REACTANT
Magnesium

Hazards	Safety precautions
• **Highly flammable in contact with water** • **Can cause violent reactions with ethanol, salts, sulfur, phosphorus and silica** • **Burns with a bright white light which can damage eyes**	• **Wear suitable clothing and eye protection.** • **Do not breathe dust.** • **Never add water to this product.** • **Keep locked up.** • **Avoid contact with skin.**

FIRST AID	
SWALLOWED	Rinse mouth with water.
EYE	Wash with running water.
SKIN	Wash with soap and water. For burns: immerse in cold running water. Bandage lightly. Seek medical attention.
INHALED	Blow nose. Rinse mouth with water.

on Resources

	eWorkbook	Understanding chemical hazards (ewbk-13133)
	Video eLesson	Saving acid wetlands (eles-1072)
	Interactivities	Hazardous substances (int-5891) The components of a risk assessment (int-8099)
	Weblinks	RiskAssess for Schools WorkSafe

5.3 Activities

5.3 Quick quiz on	5.3 Exercise

Select your pathway

■ LEVEL 1 1, 2, 3, 6	■ LEVEL 2 5, 7, 9, 10	■ LEVEL 3 4, 8, 11, 12

These questions are
even better in jacPLUS!
- Receive immediate feedback
- Access sample responses
- Track results and progress

Find all this and MORE in jacPLUS ⊙

Remember and understand

1. **MC** Dangerous goods are chemicals that could be dangerous to what?
   A. People
   B. Property
   C. The environment
   D. All of the above
2. **MC** If a chemical in the dangerous goods group is explosive, toxic and corrosive, how is the decision made about which class it is placed in?
   A. It is placed in the class based on the greatest immediate threat that the chemical presents.
   B. It is placed in all three classes.
   C. It will be placed in the flammable class.
   D. It will be placed in the toxic class.
3. Describe what the chemicals listed as hazardous substances have in common.
4. Concentrated hydrochloric acid can cause severe skin burns and eye damage and may cause harmful respiratory irritation. Which two pictograms would be used for this chemical?
5. **SIS** Explain the difference between the purposes of an SDS and a risk assessment.

Apply and analyse

6. Explain the purpose of safety data sheet in a laboratory.
7. Explain how you will know if a person in a laboratory has read the SDSs for hazardous substances and dangerous goods.
8. Identify four individuals whose responsibility it is to make sure that people have access to an SDS for each of the hazardous chemicals and dangerous goods that they store or use.
9. Describe how hazardous gases and liquids may enter the body.

Evaluate and create

10. **SIS** Explain why water may be considered a health hazard in a laboratory.
11. Many chemical suppliers provide access to SDSs online. Use the internet to search for an SDS on hydrochloric acid and use it to answer the following questions.
    a. What are some alternative names for hydrochloric acid?
    b. What are the health hazards of hydrochloric acid?
    c. What first aid treatment is recommended if hydrochloric acid:
       i. is ingested (swallowed)?
       ii. makes contact with an eye?
    d. What recommendations are made for the storage of hydrochloric acid?
12. **SIS** A student wishes to investigate the changes in colour a Bunsen burner flame turns in the presence of barium chloride, copper chloride and sodium chloride. Write a clear risk assessment for this investigation, exploring the chemicals and equipment used.

Fully worked solutions and sample responses are available in your digital formats.

LESSON
5.4 Synthesis reactions

LEARNING INTENTION

At the end of this lesson you will be able to list the different types of chemical reactions such as synthesis, decomposition, single displacement and double displacement. You will also be able to describe and give examples of synthesis reactions.

5.4.1 Types of chemical reactions

In a world where countless chemical reactions take place, it is helpful to classify the reactions. A chemical reaction occurs when one or more substances are converted into one or more new substances. They can be classified according to whether they release or absorb energy. They can also be grouped according to the nature of the reactants, the nature of the products, the way in which charged particles in atoms rearrange themselves, or even the number of reactants. Any one reaction can fall into several different groups. One method of classifying chemical reactions is according to these categories: synthesis, decomposition, single displacement and double displacement.

FIGURE 5.8 Types of chemical reaction

1. Synthesis or combination reaction

2. Decomposition reaction

3. Single-displacement reaction

4. Double-displacement reaction

5.4.2 Synthesis reactions

When two elements or compounds combine to form a more complex product these reactions are called **synthesis reactions**. Such reactions are also called **combination reactions**. The general form of a synthesis reaction is:

$$A + B \rightarrow AB$$

Usually, synthesis reactions are **exothermic**, where energy is released into the atmosphere in the form of light, heat or electricity. The reaction container may feel warm due to an increase in temperature.

Common synthesis reactions include:
- metals and oxygen combining
- two elements combining
- metallic oxides and carbon dioxide combining to form a metal carbonate
- non-metallic oxides and water combining to form an acid.

combination reaction chemical reaction in which two substances, usually elements, combine to form a compound, also **synthesis reactions**

exothermic reaction that releases energy

Example 1: Synthesis reactions of metals and oxygen

When a metal reacts with oxygen a metal oxide is formed.

$$\text{Metal} + \text{oxygen} \rightarrow \text{metal oxide}$$

The reaction of magnesium with oxygen is a spectacular example. Magnesium burns in air, producing a brilliant flash of white light, as shown in figure 5.9.

The chemical equation describing the combination of magnesium metal and oxygen gas to form magnesium oxide (a white powder) is:

$$2Mg(s) + O_2(g) \rightarrow 2MgO(s)$$

This reaction can also be classified in the following ways:

- As an exothermic reaction because energy is released.
- As an **oxidation** reaction because it is a reaction with oxygen. More information about oxidation reactions can be found in lesson 5.9.
- As a **combustion reaction** because it is a reaction with oxygen releasing energy in the form of heat and light. Combustion reactions are also oxidation reactions.

FIGURE 5.9 Magnesium ribbon burning in air with a brilliant white light

> **oxidation** chemical reaction involving the loss of electrons by a substance
>
> **combustion reaction** chemical reaction in which a substance reacts with oxygen and heat is released
>
> **corrosion** chemical reaction that wears away a metal
>
> **galvanised** metal coated with a more reactive metal that will corrode first, preventing metal from corroding

on Resources

▶ **Video eLesson** Magnesium burning (eles-2716)

Corrosion

The reaction of iron and oxygen can cause structural problems in vehicles and construction so methods of protection must be used to prevent this corrosion happening.

Corrosion is a chemical reaction in which a metal is 'eaten away' by substances in the air or water. The tarnishing of silver jewellery and cutlery, rust (see figure 5.10), and the green coating that appears on copper are all examples of corrosion. Corrosion is an oxidation reaction usually involving oxygen.

Rust protection

If you look at a sheet of **galvanised** iron, you will notice that it does not have a shiny metallic surface. Galvanised iron has been coated with a layer of zinc metal. The zinc, being more reactive than iron, prevents the iron underneath from reacting with oxygen and water in the air and rusting. Instead, the zinc corrodes first.

FIGURE 5.10 The rusting of these old car wrecks is an example of corrosion.

Corrosion of zinc

1. In the corrosion of the layer of zinc used to protect iron from rusting, a transfer of electrons from the zinc atoms to the oxygen molecules occurs. This causes the formation of positive zinc ions and negative oxide ions. This is the redox process, which happens in the presence of water. Redox reactions are discussed in lesson 5.9.

$$2Zn(s) + O_2(g) \rightarrow 2Zn^{2+}(aq) + O^{2-}(aq)$$

Zinc	Oxygen	Zinc	Oxygen
atoms	molecules	ions	ions

In this reaction, zinc atoms lose electrons, so zinc is oxidised. Oxygen molecules gain electrons, so oxygen is reduced. Remember that oxidation and reduction always occur together.

2. These oppositely charged ions attract and bond together to form the ionic compound zinc oxide. In this reaction, zinc atoms lose electrons, so zinc is oxidised. Oxygen molecules gain electrons, so oxygen is reduced. Remember that oxidation and reduction always occur together.

$$2Zn^{2+}(aq) + 2O^{2-}(aq) \rightarrow 2ZnO(s)$$

Zinc	Oxygen	Zinc oxide
ions	ions	Compound

Example 2: Synthesis of table salt

The reaction between sodium metal and chlorine gas to produce sodium chloride can demonstrate how electrons are transferred. Sodium metal atoms are oxidised because they lose electrons to the chlorine atoms to form sodium ions and chloride ions which are attracted due to their opposite charges.

FIGURE 5.11 Synthesis of sodium chloride

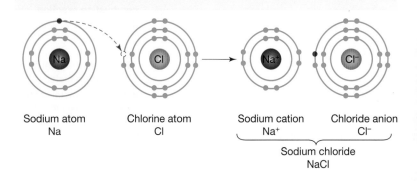

Sodium atom
Na

Chlorine atom
Cl

Sodium cation
Na⁺

Chloride anion
Cl⁻

Sodium chloride
NaCl

FIGURE 5.12 Sodium burns in chlorine to form sodium chloride

An extremely reactive metal reacts with a poisonous gas to produce a white solid, table salt, which fortunately is harmless, to sprinkle on food. The equation for this reaction is as follows:

$$2Na(s) + Cl_2(g) \rightarrow 2NaCl(s)$$

Sodium chloride is well known as table salt but there are other salts that may or may not be safe to eat. A salt is an ionic compound formed by the reaction of an acid and a base.

Example 3: Synthesis of water

The chemical reaction between hydrogen and oxygen in figure 5.3 is another example of a synthesis reaction.

$$2H_2(g) + O_2(g) \rightarrow 2H_2O(g)$$

When hydrogen burns in oxygen only water is formed. Hydrogen is considered a fuel of the future.

This is an important reaction that reduces dependence on fossil fuels as the product of this reaction is water and not carbon dioxide, which is a significant contributor to global warming.

FIGURE 5.13 A hydrogen fuel cell tram on a city street

---

on Resources

📋 **eWorkbook** Synthesis reactions (ewbk-13163)

5.4 Activities

learn on

5.4 Quick quiz on	**5.4 Exercise**

These questions are even better in jacPLUS!
- Receive immediate feedback
- Access sample responses
- Track results and progress

Find all this and MORE in jacPLUS ⊙

Select your pathway

■ LEVEL 1 1, 2, 3, 5	■ LEVEL 2 4, 6, 8, 9	■ LEVEL 3 7, 10, 11

Remember and understand

1. **MC** A synthesis reaction is one in which:
   A. a new compound is formed from a different compound.
   B. there is a reaction with oxygen to form a compound.
   C. a compound forms two elements or compounds.
   D. two or more substances combine to form a more complex compound.
2. **MC** Which of the following is a synthesis reaction?
   A. $2Pb(NO_3)_2(s) \rightarrow 2PbO(s) + 4NO_2(g) + O_2(g)$
   B. $3Mg(s) + N_2(g) \rightarrow Mg_3N_2(s)$
   C. $2H_2O(l) \rightarrow 2H_2(g) + O_2(g)$
   D. $2AgCl(s) \rightarrow 2Ag(s) + Cl_2(g)$
3. Explain why this reaction is a synthesis reaction: $2CaO(s) + 2H_2O(l) \rightarrow 2Ca(OH)_2$ (aq)

Apply and analyse

4. **a.** Consider the following reaction:

   > potassium metal + oxygen gas produces solid potassium oxide.

   Write a balanced equation using formulae for the reaction.
   b. Consider the following synthesis reaction:

   > carbon monoxide gas + oxygen gas forms carbon dioxide gas.

   Write a balanced equation using formulae for the reaction.
5. **MC** What are the coefficients of the reactants and product in the following equation?
   __P(s) + __O_2(g) → __ P_2O_5(s)
   A. 2, 5, 1   **B.** 5, 2, 1   **C.** 2, 5, 4   **D.** 4, 5, 2

6. Complete the following synthesis reactions by filling in the missing products.
   a. $2Li + S \rightarrow$
   b. $3Pb + N_2 \rightarrow$
   c. $2Al + 3Cl_2 \rightarrow$
7. Complete the following synthesis reactions by filling in the missing reactants.
   a. _____ + _____ $\rightarrow 2NH_3$
   b. _____ + _____ $\rightarrow 2HCl$
   c. _____ + _____ $\rightarrow 2FeCl_3$

Evaluate and create

8. Sodium metal is stored in oil. Find out and state why this is the case. Write out a word and symbol equation to show what would happen if it was not stored in oil.
9. **SIS** Iron is widely used for buildings, bridges, bicycle chains and containers, but it is likely to corrode when in contact with air and water. What is corrosion? Find out four examples of how iron is protected from corrosion.
10. **SIS** Another useful synthesis reaction is the reaction of lime with carbon dioxide to form calcium carbonate. Find out what lime is, write the equation for the reaction and research what calcium carbonate is used for.
11. Design an experiment to find out the conditions (water, air, salt) required for rusting. State your hypothesis first and state which is the control example.

Fully worked solutions and sample responses are available in your digital formats.

LESSON
5.5 Decomposition reactions

LEARNING INTENTION

At the end of this lesson you will be able to describe and give examples of decomposition reactions.

5.5.1 Decomposition reactions

A **decomposition reaction** is a breaking down of a substance. It usually occurs when a compound breaks down into two simpler substances. The general form of a decomposition reaction is:

$$AB \rightarrow A + B$$

Usually, decomposition reactions are **endothermic**, which means they require the input of energy in the form of light, heat or electricity to break the bonds in the reactants. The reaction container may feel cool as a result. If the reaction involves electricity then the process is called **electrolysis**.

Common decomposition reactions include:
- heating metal hydroxides
- heating metal carbonates
- breaking down compounds using electrolysis.

decomposition reaction chemical reaction in which one single compound breaks down into two or more simpler chemicals

endothermic reaction that absorbs energy

electrolysis the decomposition of a chemical substance by the application of electrical energy

Example 1: Heating metal hydroxides

Metal hydroxides decompose on heating to form metal oxides and water:

$$\text{Metal hydroxide} \xrightarrow{\text{heat}} \text{metal oxide} + \text{water}$$

If solid sodium hydroxide is heated, water vapour is released, leaving a white powder.

$$2NaOH(s) \rightarrow Na_2O(s) + H_2O(g)$$

Example 2: Heating metal carbonates

Metal carbonates decompose on heating to form metal oxides and carbon dioxide.

An example is heating a zinc carbonate to give a zinc oxide and a carbon dioxide gas:

$$ZnCO_3(s) \rightarrow ZnO(s) + CO_2(g)$$

The presence of carbon dioxide can be confirmed by directing the gas down through a tube into some limewater $(Ca(OH)_2)$ in a test tube — carbon dioxide is heavier than air — and observing if the solution becomes cloudy. The cloudiness indicates the presence of carbon dioxide gas.

$$CO_2(g) + Ca(OH)_2(aq) \rightarrow CaCO_3(s) + H_2O(l)$$

Carbon dioxide does not support combustion which is why it is used in some fire extinguishers. It will also put out a lit match.

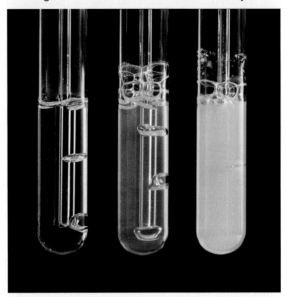

FIGURE 5.14 Bubbling carbon dioxide gas through lime water turns the solution cloudy

elog-2458

INVESTIGATION 5.1

Decomposing zinc carbonate

Aim
To observe a decomposition reaction

Materials

- laboratory coat and safety glasses
- zinc carbonate powder
- spatula
- Bunsen burner, heatproof mat and matches
- large Pyrex test tube and test tube rack

- test tube holder
- electronic balance
- marking pen
- stereo microscope
- Petri dish

CAUTION

Wear safety glasses and a laboratory coat. Make sure the test tube is not pointing at anyone.

Method

1. Place two spatulas of zinc carbonate powder in the test tube. Weigh the test tube and record the mass.
2. Mark the level of the powder in the test tube with the marking pen.
3. Heat the test tube gently in a blue Bunsen burner flame for 5 to 10 minutes.
4. While heating the test tube, hold a lit match at the mouth of the tube. Record your observations.
5. Allow the test tube to cool down. Note any change in the level of powder and then reweigh the test tube. Record the mass.
6. Place small amounts of zinc carbonate and the powder from the test tube in the Petri dish. Examine them using a stereo microscope. Record your observations.

Results

Record your observations, and ensure you note any recorded masses.

▶

5.5.2 Extraction of metals

Metal extraction is the method used to extract metals from their ores. This always involves chemical reactions and often smelting, in which the mineral ore is melted. Smelting is used in the extraction of iron, copper and zinc ores using decomposition reactions. Oxides of the less reactive metals undergo thermal decomposition reactions to produce the metal and oxygen.

The reactivity of a metal also influences how easily it is extracted. Silver and gold, for example, are unreactive and so are found naturally in their elemental state, which is why they have been known and used since ancient times. More reactive metals must be extracted from their ores. Ores are rocks that are mixtures of different compounds such as oxygen and carbon, and from which it is economically viable to separate out the metal.

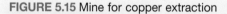

FIGURE 5.15 Mine for copper extraction

TABLE 5.3 Reactivity of metals and their extraction methods

Reactivity	Metal	Extraction
Unreactive	Silver, gold platinum	Found as free elements
Less reactive	Copper, mercury	Heating the metal ore in air
Reactive	Zinc, iron, tin, lead	Heating the metal ore with carbon
Very reactive	Aluminium, calcium, sodium, magnesium, potassium	Electrolysis of molten compound

Extraction of very reactive metals is achieved by electrolysis of the molten ores. This involves passing an electric current through the molten ore. This is a very expensive process because of the amount of electrical energy needed. Aluminium, for example, is extracted by electrolysis. The high cost of aluminium extraction is one of the reasons that recycling of aluminium cans is so important.

Extraction of iron

Iron is the most mined metal; it is used mainly to produce steel for manufacturing and construction. Chromium is added to produce stainless steel, which is resistant to corrosion and is used for many items in the kitchen.

Iron is obtained from the ores haematite and magnetite by **smelting** in a blast furnace. Smelting is a process of extracting metals from ores by heating them to remove other elements present. Iron is obtained from the iron oxide in the ore by making it react with carbon monoxide to remove the oxygen in a series of steps.

1. Making the carbon monoxide:

$$C + O_2 \rightarrow CO_2$$

$$CO_2 + C \rightarrow 2CO$$

2. Producing the iron metal:

$$Fe_2O_3 + 3CO \rightarrow 2Fe + 3CO_2$$

3. Removing impurities:
   The calcium carbonate in the limestone decomposes to form **calcium oxide**.

$$CaCO_3(s) \rightarrow CaO(s) + CO_2(g)$$

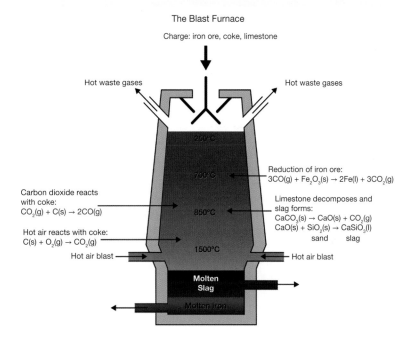

FIGURE 5.16 A blast furnace used to produce iron

The Blast Furnace

Charge: iron ore, coke, limestone

Hot waste gases

Hot waste gases

250°C

700°C

Reduction of iron ore:
$3CO(g) + Fe_2O_3(s) \rightarrow 2Fe(l) + 3CO_2(g)$

Carbon dioxide reacts with coke:
$CO_2(g) + C(s) \rightarrow 2CO(g)$

Limestone decomposes and slag forms:
$CaCO_3(s) \rightarrow CaO(s) + CO_2(g)$
$CaO(s) + SiO_2(s) \rightarrow CaSiO_3(l)$
sand   slag

Hot air reacts with coke:
$C(s) + O_2(g) \rightarrow CO_2(g)$

850°C

1500°C

Hot air blast

Hot air blast

Molten Slag

Molten iron

4. The calcium oxide then reacts with **silica** (sand) impurities in the iron ore, to produce slag, which is **calcium silicate**.

$$CaO(s) + SiO_2(s) \rightarrow CaSiO_3(l)$$

5. The iron has been separated from the iron oxide but is impure — it still contains some carbon and is quite brittle so it requires further treatment depending on what it is going to be used for.

smelting process to obtain a metal from its ore

EXTENSION: World's ten most precious metals

Rhodium (symbol Rh) is one of the rarest and most valuable precious metals. It is silvery white in colour and its major use is in the catalytic convertors in vehicles. In order, the world's most precious metals are:

1. rhodium
2. platinum
3. gold
4. ruthenium
5. iridium

6. osmium
7. palladium
8. rhenium
9. silver
10. indium.

  Resources

eWorkbook Decomposition (ewbk-13135)

5.5 Activities

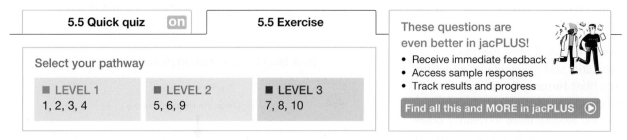

Remember and understand

1. **MC** Which of the following is an example of a decomposition reaction? Choose one or more options.
   - **A.** $2Ca(s) + O_2(g) \rightarrow 2CaO(s)$
   - **B.** $H_2O_2(aq) \rightarrow H_2O(l) + O_2(g)$
   - **C.** $PbCO_3(s) \rightarrow PbO(s) + CO_2(g)$
   - **D.** $2Al(s) + 3Cl_2(g) \rightarrow 2AlCl_3(s)$

2. Refer to the description of the extraction of iron in this lesson
   - **a.** Write an equation for a decomposition reaction.
   - **b.** Write an equation for a synthesis reaction.

3. Balance the following decomposition reactions:
   - **a.** $Fe_2O_3(s) \rightarrow Fe(s) + O_2(g)$
   - **b.** $NaCl(s) \rightarrow Na(s) + Cl_2(g)$
   - **c.** $NH_3(g) \rightarrow N_2(g) + H_2(g)$
   - **d.** $CaF(s) \rightarrow Ca(s) + F_2(g)$

4. **MC** A decomposition reaction is usually:
   - **A.** endothermic and forms more complex products.
   - **B.** endothermic and forms less complex products.
   - **C.** exothermic and forms more complex products.
   - **D.** exothermic and forms less complex products.

Apply and analyse

5. **SIS** **MC** Sam added some magnesium carbonate to a test tube and found it weighed 48 g. The test tube and contents were heated over a Bunsen burner for 5 minutes and allowed to cool.
   What would the final mass after heating be?
   - **A.** More than 48 g
   - **B.** Less than 48 g
   - **C.** Equal to 48 g
   - **D.** Unable to be determined

6. **SIS** Chris heated a blue powder in a test tube for 3 minutes. The solid became black and vapour condensed on the side of the tube.
   - **a.** Explain why this can be described as a chemical reaction.
   - **b.** Discuss whether this is an example of a synthesis reaction or a decomposition reaction.

7. **SIS** List three safety precautions that Chris needs to be aware of when doing the experiment described in question **6**.

Evaluate and create

8. The unbalanced equation for the reaction to produce iron is:
$$Fe_2O_3(s) + CO(g) \rightarrow Fe(l) + CO_2(g)$$
   - **a.** Balance the equation.
   - **b.** Explain why this could be considered to be a decomposition reaction.
   - **c.** Explain the symbol of state for the iron formed.

5.5 Quick quiz · on   5.5 Exercise

Select your pathway

■ LEVEL 1	■ LEVEL 2	■ LEVEL 3
1, 2, 3, 4	5, 6, 9	7, 8, 10

9. **SIS** Design an experiment to decide which of two metal carbonates produces the greatest percentage of carbon dioxide gas by mass on heating. State the independent and dependent variables and the factors that will be controlled.

10. Aluminium is a reactive and versatile element and is used to manufacture products from drink cans to window frames. The more reactive a metal, the more difficult it is to extract from its ore. Considerable energy in the form of electricity is required to isolate aluminium from its ore, bauxite. Research and summarise the life cycle of an aluminium can.

Fully worked solutions and sample responses are available in your digital formats.

LESSON
5.6 Single displacement reactions

LEARNING INTENTION

At the end of this lesson you will be able to describe and give examples of single displacement reactions, and be able to describe the features of acids.

A **single displacement reaction** is when one element replaces another in a compound. The general form of a single displacement reaction is:

$$A + BC(aq) \rightarrow AC(aq) + B$$

Common single displacement reactions include:
- reactive metals displacing less reactive metals from solution
- reactive metals displacing hydrogen from acids.

5.6.1 Displacement of metals from metal solutions

In the laboratory, waste solutions containing silver ions are never poured down the sink. They are collected and sent to commercial laboratories where the silver metal is recovered from the nitrate solution simply by adding a piece of copper wire.

Reactions of this type, where an element displaces another element from a compound, are called **displacement reactions**. In this example, copper has displaced the silver from the silver nitrate solution.

single displacement reaction reaction in which one element is substituted for another element in a compound

displacement reaction reaction in which an atom or a set of atoms is displaced by another atom in a molecule

---

The chemical equation for the displacement of the silver ion from silver nitrate by copper is:

copper	+	silver nitrate solution	$\rightarrow$	silver	+	copper nitrate solution
$Cu(s)$	+	$2\,AgNO_3(aq)$	$\rightarrow$	$2Ag(s)$	+	$Cu(NO_3)_2(aq)$

The reactivity series of metals can be used to determine the products of some single displacement reactions. This reactivity series of metals can be seen in table 5.4.

TABLE 5.4 Reactivity series of metals compared to hydrogen

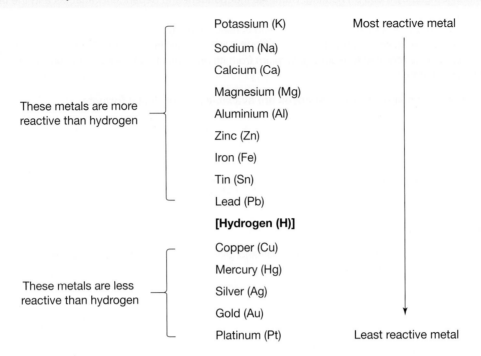

	Potassium (K)	Most reactive metal
	Sodium (Na)	
	Calcium (Ca)	
	Magnesium (Mg)	
These metals are more reactive than hydrogen	Aluminium (Al)	
	Zinc (Zn)	
	Iron (Fe)	
	Tin (Sn)	
	Lead (Pb)	
	[Hydrogen (H)]	
	Copper (Cu)	
	Mercury (Hg)	
These metals are less reactive than hydrogen	Silver (Ag)	
	Gold (Au)	
	Platinum (Pt)	Least reactive metal

A more reactive metal will displace a less reactive metal from a solution of the less reactive metal; for example, if a piece of copper is placed in a silver nitrate solution. Copper is more reactive than silver so, after a period, the copper displaces the silver in the solution. The copper metal atoms become copper ions, as observed in figure 5.17 with the blue colour. The electrons are accepted by the silver ions which form silver metal, which can also be seen in figure 5.17.

If a piece of silver is placed in a copper nitrate solution, there would be no reaction because silver is less reactive than copper. Silver will not displace the copper.

FIGURE 5.17 Copper displacing silver from silver nitrate

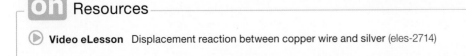

Resources

▶ **Video eLesson** Displacement reaction between copper wire and silver (eles-2714)

5.6.2 Displacement of hydrogen from acids

Acids

Acids are found in the home and in the laboratory. Acids can be **corrosive** substances. That means they react with solid substances, 'eating' them away. Acids have a sour taste and some acids, such as the sulfuric acid used in car batteries, are dangerously corrosive. The acids in ant stings and bee stings cause pain. Others, such as the acids in fruits and vinegar, are safe — even pleasant — to taste.

Acids can be strong or weak. Strong acids react more vigorously than weak acids. Usually, acids are diluted for use in school experiments; this means water is added to make them less hazardous.

- Strong acids include hydrochloric acid (HCl), sulfuric acid (H_2SO_4) and nitric acid (HNO_3).
- Weak acids include ethanoic acid (CH_3COOH), carbonic acid (H_2CO_3) and phosphoric acid (H_3PO_4).

WHAT DOES IT MEAN?

The word acid comes from the Latin word *acidus*, meaning 'sour'.

Acids and metals

The reactions of metals with acids are also examples of displacement reactions. When an acid reacts with some metals, the products are a salt and hydrogen gas. The activity series can be used to identify which metals will react with acids. The metals above hydrogen in table 5.5 will react, and the more reactive the metal, the more vigorous will be the reaction.

To confirm the production of hydrogen, the 'pop' test can be used. The hydrogen is collected using an inverted test tube because hydrogen is lighter than air. Then a lit match is placed at the mouth of the tube. If enough gas is collected, there will be a distinctive pop sound when the hydrogen reacts with the oxygen in the air, as seen in the following equation.

acid chemical that reacts with a base to produce a salt and water; edible acids taste sour

corrosive chemical that wears away the surface of substances, especially metals

$$2H_2(g) + O_2(g) \rightarrow 2H_2O(g)$$

Sodium is so reactive it can displace hydrogen from water.

 Resources

 Video eLesson Reaction between zinc powder and concentrated hydrochloric acid (eles-2589)
Sodium metal reacts with liquid (eles-2717)

---

elog-2460

INVESTIGATION 5.2

Reaction of acids with metals

Aim

To investigate the chemical reactions of an acid with a range of metals

Materials

- safety glasses
- bench mat
- test tubes and test tube rack
- pieces of metal such as copper, iron, zinc, magnesium, aluminium
- dropping bottle of 2 M hydrochloric acid solution
- rubber stopper
- matches

▶

Method

When an acid reacts with a metal, a salt is formed and hydrogen gas is given off. You can test for hydrogen gas by holding a lighted match at the mouth of the test tube. If the gas is hydrogen, it will explode and make a 'pop' sound.

CAUTION

Do not push the stopper into the test tube firmly. Just hold it in the top of the test tube for a few seconds.

---

1. Place a small piece of one of the metals in a test tube.
2. Add the acid to the test tube to a depth of 1 cm.
3. Observe the chemical reaction.
4. Test for hydrogen gas by holding a rubber stopper over the end of the test tube for a few seconds and then placing a lighted match at the mouth of the test tube.
5. Repeat the test with other metals.

Results

Construct a table to record your observations.

Discussion

1. When zinc metal reacts with hydrochloric acid, zinc chloride and hydrogen gas are formed. Write a word equation for this reaction.
2. When the lighted match produces a 'pop', the hydrogen gas is reacting with the oxygen in the air to form water. You may have noticed the water form at the top of the test tube after you performed the match test. Write a word equation for this chemical reaction.
3. Extension:
   a. Write chemical equations for the reactions between hydrochloric acid and each of the metals you tested.
   b. Write a chemical equation for the hydrogen gas 'pop' test.

Conclusion

Summarise the findings for this investigation about the reaction of acids with metals.

5.6 Activities

learn on

Remember and understand

1. **MC** Which of the following metals will react with acids?
   A. Copper
   B. Silver
   C. Zinc
   D. Mercury
2. **MC** What are the products of the reaction of iron and hydrochloric acid?
   A. Iron sulfate and chlorine
   B. Iron chloride and hydrogen
   C. Iron chloride and water
   D. Iron sulfate and hydrogen

3. MC A salt + _____ gas are formed in a chemical reaction between an acid and a reactive metal.
   A. oxygen
   B. carbon dioxide
   C. carbon monoxide
   D. hydrogen
4. Balance the following single replacement reactions and include states:
   a. $Al + CuCl_2 \rightarrow AlCl_3 + Cu$
   b. $Pb + HCl \rightarrow PbCl_2(s) + H_2$
   c. $Mg + H_2SO_4 \rightarrow MgSO_4 + H_2$

Apply and analyse

5. MC What is the name of the element that is present in all acid?
   A. Oxygen
   B. Carbon
   C. Chlorine
   D. Hydrogen
6. For each of the combinations of reactants, decide if a reaction will occur and write equations for those reactions. Write NR if no reaction will occur.
   a. $Ag + KNO_3 \rightarrow$
   b. $Fe + Pb(NO_3)_2 \rightarrow$
   c. $Zn + AgNO_3 \rightarrow$
   d. $Al + H_2SO_4 \rightarrow$
7. Complete the following equations by adding the reactants and balancing:
   a. ____ + ____ $\rightarrow ZnSO_4 + H_2$
   b. ____ + ____ $\rightarrow FeCl_2 + Cu$
   c. ____ + ____ $\rightarrow Sn(NO_3)_2 + Ag$
8. SIS Write word and symbol equations for the following reaction. Include states.
   A piece of magnesium is placed in a copper sulfate solution and brown metal is formed and the blue colour fades.
9. SIS Sophie places strips of three unknown metals in dilute hydrochloric acid and records the results in a table.

Metals	Results
X	Slow bubbling
Y	Very fast bubbling, test tube feels warm
Z	No observable change

   a. The metals are calcium, zinc and silver. Identify each metal and explain your answer.
   b. Name the gas produced and describe how Sophie could test for this gas.
   c. Which of the three metals would be reactive enough to react with water?

Evaluate and create

10. MC The most reactive metals in the list shown in table 5.5 would be found in which part of the periodic table?
    A. Top left
    B. Lower right
    C. Top right
    D. Centre
11. SIS An iron nail is placed in a solution of copper sulfate. Explain any observations that could be made over time and write an equation for the reaction.

Iron nail

Copper(II) sulfate solution

12. Describe the similarities and differences between metal displacement reactions and acid and reactive metal reactions.
13. Hydrochloric acid is present in the stomach. Acids are corrosive, so why is the stomach not broken down? Find out why and what the reason is for the presence of the acid.

Fully worked solutions and sample responses are available in your digital formats.

LESSON
5.7 Double displacement — neutralisation

LEARNING INTENTION

At the end of this lesson you will be able to describe the features of bases, and describe and give examples of neutralisation double displacement reactions.

In double displacement reactions partners are exchanged and the general form is:

$$AB + CD \rightarrow AD + BC$$

Common double displacement reactions include:
- neutralisation
- precipitation.

5.7.1 Neutralisation

As well as single replacement reactions, acids can take part in double displacement reactions. These reactions are also called **neutralisation reactions** and occur when an acid reacts with a **base**. To neutralise means to stop something from having an effect. To stop the properties of acids from having an effect, a base can be added. Similarly, to stop a base from having an effect, an acid can be added. But what are bases?

Bases

Bases have a bitter taste and feel slippery or soapy to the touch. Some bases are very corrosive, especially caustic soda (sodium hydroxide). Caustic soda will break down fat, hair and vegetable matter and is the main ingredient in drain cleaners. Other bases are used in soap, shampoo, toothpaste, dishwashing liquid and cloudy ammonia as cleaning agents. Bases that can be dissolved in water are called **alkalis**.

Like acids, bases can be strong or weak. Strong bases react more vigorously with substances than weak bases. The strength of an acid or base is measured by the pH scale.

- Strong bases include potassium hydroxide (KOH), sodium hydroxide (NaOH)and barium hydroxide ($Ba(OH)_2$).
- Weak bases include ammonia (NH_3), calcium carbonate ($CaCO_3$) and sodium carbonate (Na_2CO_3).

neutralisation reaction between an acid and a base

base chemical substance that will react with an acid to produce a salt and water; edible bases taste bitter

alkali base that dissolves in water

TABLE 5.5 Common acids and bases and their uses

Acid	Uses
Hydrochloric acid (HCl)	• To clean the surface of iron during its manufacture • Food processing • The manufacture of other chemicals • Oil recovery
Nitric acid (HNO_3)	• The manufacture of fertilisers, dyes, drugs and explosives
Sulfuric acid (H_2SO_4)	• The manufacture of fertilisers, plastics, paints, drugs, detergents and paper • Petroleum refining and metallurgy

Acid	Uses
Citric acid ($C_6H_8O_7$)	• Present in citrus fruits such as oranges and lemons • Used in the food industry and the manufacture of some pharmaceuticals
Carbonic acid (H_2CO_3)	• Formed when carbon dioxide gas dissolves in water; present in fizzy drinks
Ethanoic acid (CH_3COOH)	• Found in vinegar • The production of other chemicals, including aspirin

Base	Uses
Sodium hydroxide (NaOH) (caustic soda)	• The manufacture of soap • As a cleaning agent
Ammonia (NH_3)	• The manufacture of fertilisers and in cleaning agents
Sodium bicarbonate ($NaHCO_3$)	• To make cakes rise when they cook

Acids and bases

When a neutralisation reaction occurs between an acid and a base the products are a **salt** and water. Many neutralisation reactions occur in water. These reactions are said to occur 'in solution'.

salt one of the products of the reaction between an acid and a base

Neutralisation reactions

metal hydroxide + acid → salt + water

metal oxide + acid → salt + water

metal carbonate + acid → salt + carbon dioxide + water

metal hydrogen carbonate + acid → salt + carbon dioxide + water

Your stomach contains hydrochloric acid, which helps to break up food for digestion. Too much acid, however, can be a problem. If your stomach produces too much acid, you may need to take an antacid such as milk of magnesia. This medicine has the solid base magnesium oxide (MgO) suspended in it. This base reacts with the hydrochloric acid in your stomach according to the equation:

$$\underset{\text{base}}{MgO(s)} + \underset{\text{acid}}{2HCl(aq)} \rightarrow \underset{\text{salt}}{MgCl_2(aq)} + \underset{\text{water}}{H_2O(l)}$$

The products are the salt magnesium chloride and water. The salt contains the positive metal ion from the base and the negative non-metal ion from the acid.

The base sodium hydrogen carbonate, commonly known as bicarb, is a component of baking powder. It has the formula $NaHCO_3$ and contains the hydrogen carbonate ion HCO_3^-. When bases containing this ion react with acids, carbon dioxide gas is produced as well as salt and water. When hydrochloric acid and bicarb are mixed, the following reaction takes place:

$$\underset{\text{base}}{NaHCO_3(s)} + \underset{\text{acid}}{HCl(aq)} \rightarrow \underset{\text{salt}}{NaCl(aq)} + \underset{\substack{\text{carbon}\\\text{dioxide}}}{CO_2(g)} + \underset{\text{water}}{H_2O(l)}$$

In both of the reactions mentioned, the salts formed were metal chlorides, because they contained the chloride ion (Cl^-) from the hydrochloric acid. Neutralisation reactions between many different acids and bases are possible; therefore, it is possible to produce many different salts. Some of these reactions are summarised in table 5.7.

TABLE 5.6 Some examples of neutralisation reactions

Base	Acid	Negative ion present in salt	Salt	Formula of salt
Sodium hydroxide	**Sulfuric** acid	Sulfate SO_4^{2-}	Sodium **sulfate**	Na_2SO_4
Magnesium oxide	Hydro**chloric** acid	Chloride Cl^-	Magnesium **chloride**	$MgCl_2$
Sodium oxide	**Ethanoic** acid	Ethanoate CH_3COO^-	Sodium **ethanoate**	CH_3COONa
Copper(II) oxide	**Nitric** acid	Nitrate NO_3^-	Copper(II) **nitrate**	$Cu(NO_3)_2$

CASE STUDY: Different types of salts

Often when we hear the term salt, we automatically think of table *salt* (sodium chloride). However, sodium chloride is just one type of salt! Many salts are brightly coloured and can be highly poisonous.

Salts containing copper ions are usually blue, those containing nickel are pale green, those containing chromium can be green or orange, cobalt salts are pink, and manganese salts are usually black.

FIGURE 5.18 Many salts are brightly coloured.

SCIENCE INQUIRY SKILLS: Spills in the laboratory

How are acid or base spills cleaned up in the science laboratory? Find out from your teacher or laboratory technician how spills are dealt with in your school.

It is important to deal with spills quickly and safely. The first thing to do is to let your teacher know if you have spilled any substance. Methods of dealing with spills may include acid or base neutralisers, absorption pads, mops or granules and inactivators. One thing you would not do is to try to neutralise a strong acid with a base as the reaction could be violent and cause further problems.

INVESTIGATION 5.3

elog-2462

tlvd-10817

Exploring neutralisation reactions

Aim

To identify the products of a reaction between an acid and a base

Materials

- safety glasses, gloves and laboratory coat
- 50 mL burette
- retort stand, bosshead and clamp
- tripod and gauze mat
- Bunsen burner, heatproof mat and matches
- 20 mL pipette
- 100 mL conical flask
- pipette bulb
- white tile

- dropping bottle of phenolphthalein indicator
- wire shaped into a loop with a handle
- small funnel
- 1 M hydrochloric acid solution
- 1 M sodium hydroxide solution
- evaporating dish
- silver nitrate solution in a dropping bottle
- sample of sodium chloride
- test tube

CAUTION

Wear safety glasses, gloves and a laboratory coat.

Method

1. Rinse the burette with the hydrochloric acid solution and then, using the funnel, fill the burette with the hydrochloric acid solution.
2. Rinse the pipette with sodium hydroxide solution using the pipette bulb.
3. Set up the equipment as shown in the diagram. Use the pipette and bulb to transfer 20 mL of the sodium hydroxide solution into the conical flask.
4. Add a few drops of phenolphthalein indicator to the sodium hydroxide.
5. Add the acid from the burette carefully until the pink colour of the indicator disappears. The colour change indicates that the neutralisation reaction is complete.
6. Pour the contents of the flask into an evaporating dish. Heat the dish with the Bunsen burner and gently evaporate the water. Be careful — spattering may occur.
7. When the water has nearly evaporated, turn off the Bunsen burner and allow the dish to cool and the remaining water to evaporate without further heating.
8. Test the white crystals in the dish for the presence of sodium ions by placing a few crystals on a wire loop and heating in a Bunsen burner flame. Compare this flame colour with a known sample of sodium chloride.
9. Test for the presence of chloride ions by dissolving a few crystals in half a test tube of water and adding a few drops of silver nitrate solution. A white cloudiness indicates that chloride ions are present.

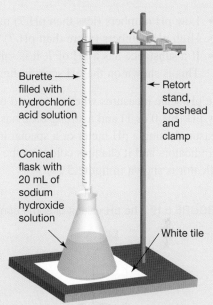

Burette filled with hydrochloric acid solution

Conical flask with 20 mL of sodium hydroxide solution

Retort stand, bosshead and clamp

White tile

Results

Record your observations for this investigation.

Discussion

1. Write a word equation for the neutralisation reaction.
2. Write a balanced equation, using formulae, for the neutralisation reaction.
3. Comment on the information that the flame and silver nitrate tests provided.
4. Which substance is the silver nitrate testing for?

Conclusion

Summarise the findings for this investigation about neutralisation reactions.

Acids and bases are usually colourless and so it is sometimes difficult to decide if enough acid and base have been combined to form a salt. A substance called an indicator is useful when working with acids and bases.

Indicators

Acid–base **indicators** are substances that can be used to tell whether a substance is an acid or a base. The indicators react with acids and bases, producing different colours in each. Some indicators are made from natural food dyes and some vegetables; for example, red cabbage can be used as a pH indicator. Two commonly used indicators are Universal indicator (figure 5.20), which turns red in an acid and blue in a base, and bromothymol blue, which turns yellow when added to an acid and a bluish-purple when added to a base.

indicator substance that changes colour when it reacts with acids or bases; the colour shows how acidic or basic a substance is

The pH scale

You can describe how acidic or basic a substance is by using the numbers on the **pH scale**. The pH scale ranges from 0 to 14. The pH scale is based on the amount of hydrogen ions present in the solution. Due to the way it is measured, the higher the concentration, the lower the pH and the stronger the acid.

- Low pH numbers (less than pH 7) mean that substances are acidic.
- High pH numbers (more than pH 7) mean that substances are basic.
- If a substance has a pH of 7, it is said to be neutral — neither acidic nor basic. This is shown on the pH scale in figure 5.19.

> **pH scale** the scale from 1 (acidic) to 14 (basic) that measures how acidic or basic a substance is
>
> **universal indicator** mixture of indicators that changes colour as the strength of an acid or base changes, indicating the pH of the substance

The pH scale measures whether an acid or base is strong or weak. For example, a strong acid has a very low pH (pH 0 or 1) and a strong base has a very high pH (pH 13 or 14). The pH of a substance can be measured using a pH meter or a special indicator called **universal indicator**. Universal indicator is a mixture of indicators and it changes colour as the strength of an acid or base changes. The colour range of universal indicator is shown in figure 5.21.

FIGURE 5.19 The pH values of some common substances

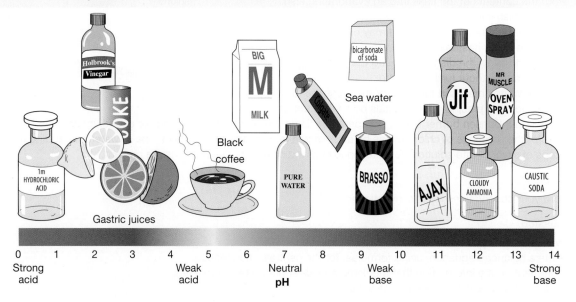

FIGURE 5.20 Universal Indicator paper turns blue when dipped into a basic solution.

FIGURE 5.21 The colour range of universal indicator. It is pink in strong acid (pH 1), blue in strong base (pH 14) and green in neutral solutions (pH 7).

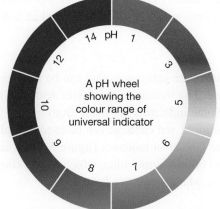

EXTENSION: Strength versus concentration

We have seen that acids and bases can be strong or weak. Strong acids or bases completely separate into their ions in water. This means they can react to their full extent in water.

For example, hydrochloric acid (HCl) is a strong acid, and sodium hydroxide (NaOH) is a strong base.

$$HCl \rightarrow H^+ + Cl^-$$
$$NaOH \rightarrow Na^+ + OH^-$$

Weak acids and bases only partially separate into ions in water.
- A strong acid or base completely separates in water.
- A weak acid or base only partially separates in water.

The concentration of an acid or base is different to its strength. Concentration is how much of the actual active substance is in the solution; this determines the extent of reaction.
- A concentrated acid or base will have a large number of active particles in a given volume.
- A dilute acid or base will have a small number of active particles in a given volume.

The difference between strong/weak and concentrated/dilute is shown in figure 5.22.

FIGURE 5.22 Concentrated and dilute weak and strong acids

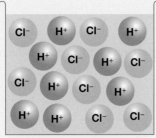
concentrated strong acid

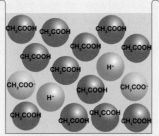
concentrated weak acid

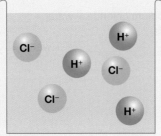
dilute solution of strong acid

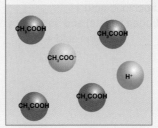
dilute solution of weak acid

5.7 Activities

learnon

| 5.7 Quick quiz on | 5.7 Exercise |

Select your pathway

■ LEVEL 1	■ LEVEL 2	■ LEVEL 3
1, 2, 3, 5, 8	4, 6, 7, 9, 10	11, 12, 13

These questions are
even better in jacPLUS!
- Receive immediate feedback
- Access sample responses
- Track results and progress

Find all this and MORE in jacPLUS ▶

Remember and understand

1. Copy and complete the table by describing the properties of acids and bases.

	Acids	**Bases**
pH		
Taste		
Properties		

2. **MC** What common property do some acids and bases have when they come into contact with solid substances?
   A. High pH
   B. Low pH
   C. They produce carbon dioxide
   D. They are corrosive
3. **MC** What is a salt?
   A. An ionic compound
   B. The result of a reaction between an acid and a base
   C. The result of a reaction between a metal and an acid
   D. All of the above
4. Explain the main difference between the reaction of a metal oxide with an acid and a metal carbonate with acid.

Apply and analyse

5. Write word equations for the reactions between:
   a. Hydrochloric acid and sodium hydroxide
      hydrochloric acid + _____ → _____ + water
   b. Hydrochloric acid and sodium bicarbonate
      hydrochloric acid + _____ → sodium chloride + water + _____
   c. Sulfuric acid (hydrogen sulfate) and sodium hydroxide
      sulfuric acid + _____ → _____ + water

6. **SIS** A pH meter is used to measure the pH of five different substances. The results are shown in the table.

TABLE pH of different substances

Substance	pH value
A	6.0
B	12.0
C	3.0
D	7.0
E	8.0

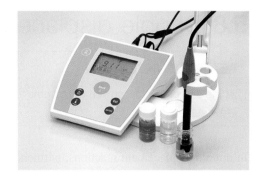

a. Which substance is most likely to be:
   i. oven cleaner
   ii. toothpaste?
b. Which substance could be:
   i. a weak acid
   ii. pure water
   iii. a strong acid?
c. Which two of the substances would you expect to be the most corrosive?

7. Write balanced chemical equations with the states for the following reactions. (Refer to lesson 5.2 for chemical formulae.)
   a. Solid sodium hydrogen carbonate and sulfuric acid react to form a sodium sulfate solution, carbon dioxide and water.
   b. Solid potassium hydroxide and hydrochloric acid react to form a solution of potassium chloride and water.
   c. Solid copper oxide reacts with sulfuric acid to form a solution of copper sulfate and water.

8. Identify the name of the salt that forms from the following reactions. (Refer to table 5.7 to assist.)
   a. Magnesium hydroxide and hydrochloric acid
   b. Potassium hydroxide and acetic acid
   c. Sodium carbonate and sulfuric acid

9. Explain why an antacid is used for heartburn.

Evaluate and create

10. When you add buttermilk (an acid) to baking soda (a base) in a mixing bowl the pH increases. True or false? Explain your response.

11. **SIS** Find the websites of two antacid products such as Gaviscon®, Mylanta®, Eno® or Alka-Seltzer®.
    a. Research and report on:
       i. the ingredients of the product or products
       ii. the claims made about each antacid product or products
       iii. advice and warnings
       iv. side effects.
    b. Find a medical site that provides information about antacids, including side effects, and report on your findings.

12. **SIS** You may have noticed some bare patches of soil under pine trees. This is due to the acidity in the soil. Research and explain how the acidity in the soil can be controlled.

13. **SIS** Universal indicator can be used to measure the pH of substances. Design an experiment that shows the change in pH during a neutralisation reaction. You may wish to research universal indicator and pH before you start.

Fully worked solutions and sample responses are available in your digital formats.

LESSON
5.8 Double displacement — precipitation

LEARNING INTENTION
At the end of this lesson you will be able to describe and give examples of precipitation double displacement reactions.

In double displacement reactions, partners are exchanged, and the general form is:

$$AB + CD \rightarrow AD + CB$$

5.8.1 Aqueous solutions

When table salt (sodium chloride) is dissolved in water to form an **aqueous solution**, it seems to disappear. The ions in the salt no longer bond together as a large array of positive and negative ions like they do as a solid. The sodium ions and the chloride ions separate when they dissolve (see figure 5.23).

FIGURE 5.23 Sodium chloride dissolving in water. In water, oxygen has a slight negative charge and hydrogen has a slight positive charge. Therefore, the chloride ions are attracted to the positively charged hydrogen and the sodium ions are attached to the negatively charged oxygen.

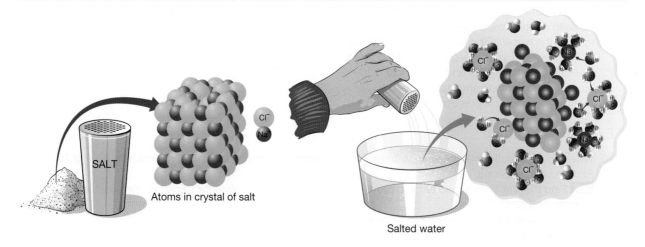

SALT

Atoms in crystal of salt

Salted water

Sodium chloride dissolving in water can be represented by the following equation:

$$NaCl(s) \xrightarrow{H_2O} Na^+(aq) + Cl^-(aq)$$

Ions in aqueous solutions are, therefore, separate entities and are able to react independently.

Ionic compounds dissolve in water to varying degrees. Some are soluble, others slightly soluble and others insoluble. Substances that are soluble will dissolve. Substances that are not soluble will form a precipitate.

aqueous solution solution in which water is the solvent

ionic compound compound containing positive and negative ions held together by the electrostatic force

Rules of solubility

1. All compounds containing either the Na^+, NO_3^-, NH_4^+ or K^+ ion will dissolve in water. Compounds containing these ions never form precipitates. (This is sometimes known as the SNAP rule: an acronym for Sodium ion, Nitrate ion, Ammonium ion and Potassium ion.)	*Example:* $NaCl$, NH_4Cl, K_2SO_4 and $AgNO_3$ are all soluble in water and therefore do not form precipitates.
2. Compounds containing the Cl^-, Br^- and I^- ions are soluble, except when they contain the Ag^+, Pb^{2+} or Hg^{2+} ions.	*Example:* $FeCl_3$, $ZnBr_2$ and AlI_3 are soluble, but $AgCl$, $HgBr_2$ and PbI_2 are not soluble and will form a precipitate.
3. Compounds containing the SO_4^{2-} ion are soluble, except for $BaSO_4$, $PbSO_4$ and $CaSO_4$.	*Example:* $ZnSO_4$ is soluble, but $BaSO_4$ is not soluble and will form a precipitate.
4. Compounds containing OH^-, CO_3^{2-} and PO_4^{3-} are insoluble except when they contain the ions Na^+, NH_4^+ or K^+.	*Example:* $BaCO_3$ and $Zn(OH)_2$ will form insoluble precipitates, but Na_2CO_3 and KOH will not.
5. Some compounds are slightly soluble. These include $Ca(OH)_2$, $PbCl_2$, $PbBr_2$, $CaSO_4$ and Ag_2SO_4.	

5.8.2 Precipitates

When two solutions containing dissolved ions are mixed together, the ions are able to come into contact with each other. Oppositely charged ions attract. In some cases, the attraction is strong enough to form ionic bonds and hence a new ionic compound. Some of these compounds are insoluble (unable to dissolve in water) and so a solid called a **precipitate** forms. Chemical reactions in which precipitates form are called **precipitation reactions**. When colourless lead nitrate solution and colourless potassium iodide solution are mixed together, a brilliant yellow precipitate is formed, as shown in figure 5.24.

FIGURE 5.24 The formation of the brilliant yellow precipitate, lead iodide, from the colourless solutions of lead nitrate and potassium iodide

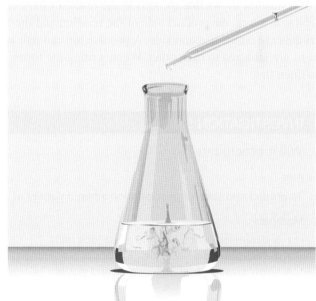

 **Resources**

▶ **Video eLesson** Precipitation (eles-2058)

precipitate solid product of a chemical reaction that is insoluble in water

precipitation reaction chemical reaction in which there is a water insoluble product

Changing partners

Another example of a precipitation reaction occurs between silver nitrate solution and sodium chloride solution (see figure 5.25). When these two colourless solutions are added together in a test tube, the contents become cloudy, indicating that a precipitate has formed. If the test tube is allowed to stand for a while, the solid settles to the bottom and we can see that a clear solution is also present. The products of the reaction are insoluble solid silver chloride (the precipitate) and sodium nitrate (not visible because it is soluble in water).

Silver nitrate, sodium chloride and sodium nitrate all dissolve in water. Therefore, they have the symbol (aq). Silver chloride does not dissolve in water, so it has the symbol (s) to indicate that it is solid.

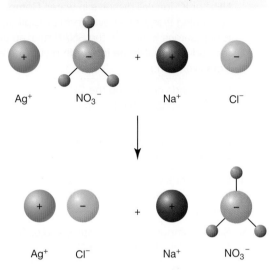

FIGURE 5.25 Ions can change partners when a chemical reaction takes place.

Ag^+ NO_3^- Na^+ Cl^-

Ag^+ Cl^- Na^+ NO_3^-

This precipitation reaction can be represented by the following equation:

silver nitrate   +   sodium chloride   →   silver chloride   +   sodium nitrate
$AgNO_3(aq)$ + $NaCl(aq)$ → $AgCl(s)$ + $NaNO_3(aq)$

The equation shows that the ions in the reactants have changed partners. The silver ion is paired with the chloride ion on the product side of the reaction and the sodium ion is paired with the nitrate ion. The opposite is the case on the reactant side. A positive ion can pair up only with a negative ion because oppositely charged ions are attracted to each other. When writing the formula of any new compound, the positive ion is always written first.

INVESTIGATION 5.4

Will it precipitate?

Aim

To predict and test for precipitation when a variety of solutions are added to each other

Materials

- 5 semi-micro test tubes and a test tube rack
- a white tile and a black tile
- safety glasses, a laboratory coat and gloves
- dropping bottles of the following solutions: copper sulfate, sodium chloride, silver nitrate, cobalt chloride, sodium hydroxide, potassium iodide

CAUTION

Wear safety glasses, a laboratory coat and gloves.

Method

1. For each combination of solution, predict if a precipitate will form using the rules in section 5.8.1.
2. Place 10 drops of copper sulfate solution in each test tube.
3. Predict which of the tubes will form a precipitate, record your predictions.
4. Add 10 drops of sodium chloride to the first test tube, 10 drops of silver nitrate to the second, and so on until each tube contains copper sulfate solution and one other solution.
5. Hold a black or white tile behind the test tube if necessary, to detect the presence of a precipitate. Record your results.
6. Tip the residues into a waste bottle. Wash out the test tubes thoroughly and this time place 10 drops of sodium chloride in each of the test tubes. Again, add one of the other solutions to each test tube (but not copper sulfate, which has already been tested). Record your results.
7. Repeat until all possible pairs of solutions have been tested.

Results

Record your observations in a table similar to the following. For each combination, place a cross if no precipitate appears or a tick if a precipitate appears. If a precipitate appears, note its colour.

TABLE The precipitation of combinations of aqueous solutions

	Copper sulfate	Sodium chloride	Silver nitrate	Cobalt chloride	Sodium hydroxide	Potassium iodide
Copper sulfate						
Sodium chloride						
Silver nitrate						
Cobalt chloride						
Sodium hydroxide						
Potassium iodide						

Discussion

1. Write word equations for each of the pairs that reacted to form a precipitate.
2. Use formulae to write chemical equations for each of the pairs that reacted to form a precipitate. You will need to balance some of the equations.
3. Check to see if the rules for solubility match your results.

Conclusion

Summarise the findings for this investigation about predicting and testing for a precipitation reaction.

Conservation of mass

Figure 5.26 shows the reaction between sodium sulfate and calcium chloride. Not only does a precipitate form during the reaction but, as the diagram shows, the mass of the products also equals the mass of the reactants, so mass is conserved.

FIGURE 5.26 The reaction between sodium sulfate and calcium chloride follows the Law of Conservation of Mass.

Regardless of how substances within a closed system
are changed, the total mass remains the same.

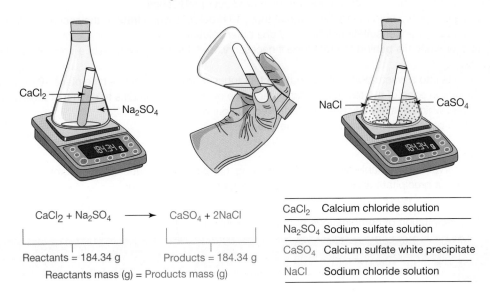

$$CaCl_2 + Na_2SO_4 \longrightarrow CaSO_4 + 2NaCl$$

Reactants = 184.34 g   Products = 184.34 g

Reactants mass (g) = Products mass (g)

$CaCl_2$	Calcium chloride solution
Na_2SO_4	Sodium sulfate solution
$CaSO_4$	Calcium sulfate white precipitate
NaCl	Sodium chloride solution

Resources

5.8 Activities

learn on

5.8 Quick quiz on	**5.8 Exercise**

These questions are
even better in jacPLUS!
- Receive immediate feedback
- Access sample responses
- Track results and progress

Find all this and MORE in jacPLUS ▶

Select your pathway

■ LEVEL 1 1, 3, 4	■ LEVEL 2 2, 5, 6	■ LEVEL 3 7, 8

Remember and understand

1. **MC** What is a precipitate?
   A. A substance that is soluble in water.
   B. An insoluble solid that can form when two aqueous ionic solutions react together.
   C. A solution in which water is the solvent.
   D. The product of a chemical reaction.

2. **MC** Which of the following is the correct equation for the reaction that occurs when the salt copper sulfate dissolves in water?
   A. $Cu^{2+}(s) + SO_4^{2-}(s) \rightarrow CuSO_4(aq)$
   B. $CuSO_4(s) \rightarrow Cu^{2+}(aq) + SO_4^{2-}(aq)$
   C. $CuSO_4(s) \rightarrow Cu^{2+}(l) + SO_4^{2-}(l)$
   D. $CuSO_4(s) \rightarrow Cu^{2+}(s) + SO_4^{2-}(s)$

3. **MC** Which of the following compounds will be insoluble in water? Select all possible answers.
   - A. $CuCO_3$
   - B. AgI
   - C. $NaCl$
   - D. $Mg(OH)_2$

4. **MC** What is an aqueous solution?
   - A. A substance that has melted and formed a liquid
   - B. A substance that has been dissolved in hydrochloric acid
   - C. A substance that has been dissolved in water
   - D. A substance in the form of gas

Apply and analyse

5. Describe the expected observations if potassium iodide (KI) and sodium carbonate (Na_2CO_3) are mixed together.

6. **SIS** Use the rules of solubility in section 5.3.1 to complete the following table.

TABLE Showing solubilities of compounds

Compound	Soluble	Slightly soluble	Insoluble
a. Sodium hydroxide			
b. Calcium sulfate			
c. Iron(II) carbonate			
d. Silver chloride			
e. Lead chloride			

Evaluate and create

7. **SIS** The following image shows the results of three precipitation reactions.

Test tube A contains a precipitate of copper(II) hydroxide, the precipitate in test tube B is iron(III) hydroxide and the precipitate in test tube C is iron(II) hydroxide.
For each of the precipitates, write the equation for a reaction that could have formed them. (*Hint:* use the rules of solubility given in section 5.3.1.)

8. **SIS** A student wishes to confirm the solubility rules for compounds containing the OH^- group. Design an investigation a student would conduct to explore this. Ensure you outline any health and safety requirements. You may wish to research different substances to investigate.

Fully worked solutions and sample responses are available in your digital formats.

LESSON
5.9 Redox reactions

LEARNING INTENTION

At the end of this lesson you will be able to describe and give examples of redox reactions.

Synthesis reactions, most decomposition reactions and single displacement reactions can also be categorised as redox reactions. Redox reactions are electron transfer reactions and are extremely important in industry and in our everyday lives.

A redox reaction is really two reactions occurring simultaneously. In the electron transfer process, one reactant loses electrons and another gains electrons. Loss of electrons is known as oxidation. The gain of electrons is called **reduction**. Oxidation and reduction always occur together, thus the two words are combined to form the word **redox**, which is used to describe reactions where electrons are transferred.

Consider the *synthesis* reaction of magnesium and oxygen to form magnesium oxide.

$$2Mg(s) + O_2(g) \rightarrow 2MgO(s)$$

The magnesium atoms have lost two electrons to form magnesium ions. The oxide atoms (in the oxygen molecule) have gained electrons and the oxygen is said to have been reduced. If one substance donates electrons, another substances must accept them. Oxidation and reduction always occur together.

FIGURE 5.27 Movement of electrons in a redox reaction

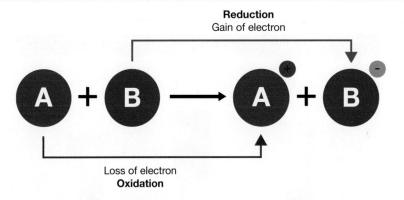

The mnemonic OIL RIG may help you to remember these processes: **o**xidation **i**s the **l**oss of electrons, and **r**eduction **i**s the **g**ain of electrons.

Some *decomposition* reactions are redox reactions and others are not.

When silver bromide is exposed to light it decomposes to silver metal and bromine vapour:

$$2AgBr(s) \rightarrow 2Ag(s) + Br_2(g)$$

reduction a chemical reaction involving the gain of electrons by a substance

redox oxidation and reduction considered together

Pale yellow silver bromide changes to grey silver metal and bromine gas.

Silver bromide is an ionic compound and so is composed of silver and chloride ions, but the products are neutral elements. This means that there must have been a transfer of electrons; that is, from the bromide ions to the silver ions. A transfer of electrons is a redox reaction. However, if calcium carbonate is heated, calcium oxide and carbon dioxide are produced:

$$CaCO_3(s) \rightarrow CaO(s) + CO_2(g)$$

Both the reactant, $CaCO_3$, and product, CaO, contain Ca^{2+} ions so it appears that no electron transfer has occurred and so this is not a redox reaction.

In the *displacement* reaction where copper displaces the silver, electrons are transferred from the copper atoms to the silver ions. Silver ions (Ag^+) in the solution gain electrons to form atoms of solid silver. Thus, silver ions are reduced. Copper atoms (Cu) lose electrons, forming copper ions (Cu^{2+}), which dissolve into a solution. The formation of copper ions changes the colour of the solution from colourless to blue. The copper atoms are oxidised. The nitrate ions are not involved in the electron transfer.

$$Cu(s) + 2AgNO_3(aq) \rightarrow 2Ag(s) + Cu(NO_3)_2(aq)$$

In general, if an element has become a compound or an element has been formed from a compound in a reaction, then a redox reaction has occurred.

EXTENSION: The early days of oxidation

In the early days of chemistry, oxidation was defined as the combination of a chemical with oxygen or as the removal of hydrogen from a compound. Reduction was defined as the opposite of oxidation; that is, the removal of oxygen or the combination of hydrogen with a chemical. Today, we know that oxygen and hydrogen are not necessarily involved at all in a redox reaction. An example is when sodium metal and chlorine gas are produced by passing an electric current through molten sodium chloride.

FIGURE 5.28 Sodium metal exploding on water

Sodium ions	+	chloride ions	→	sodium metal	+	chloride gas
$2Na^+(l)$	+	$2Cl^-(l)$	→	$2Na(s)$	+	$Cl_2(g)$

In this redox reaction, sodium is reduced because it gains electrons, and chlorine is oxidised.

Oxidation is now defined as the transfer of electrons from a reactant. In the redox reaction included here, chlorine is oxidised and sodium is reduced. However, when sodium metal is produced, it is extremely reactive and will explode on contact with water as shown in figure 5.28.

 Resources

eWorkbook Redox reactions (ewbk-13145)

5.9 Activities

5.9 Quick quiz on	5.9 Exercise

These questions are
even better in jacPLUS!
- Receive immediate feedback
- Access sample responses
- Track results and progress

Find all this and MORE in jacPLUS ▶

Select your pathway

■ LEVEL 1 1, 2, 3, 4	■ LEVEL 2 5, 6, 7, 10	■ LEVEL 3 8, 9, 11

Remember and understand

1. Complete the following table to summarise each of the groups of reactions discussed in this lesson

TABLE Descriptions and examples of different reaction types

Reaction type	Description	Example
a. Corrosion		
b. Displacement	When an element displaces another element to form a compound	
c.		Reaction between petrol and oxygen in a car engine
d.	When a compound breaks down into two or more simpler substances	
e. Synthesis		
f.	Reactions that involve the transfer of elements	Darkening

2. **MC** A redox reaction is one that has:
   A. at least one reactant and one product.
   B. a transfer of electrons.
   C. a transfer of ions.
   D. an element and a compound.
3. **MC** When does oxidation occur?
   A. Electrons are 'gained', or transferred to an atom or ion.
   B. Electrons are 'lost', or transferred away from an atom or ion.
   C. Electrons are 'gained' or 'lost' from an atom or ion.
   D. One single compound breaks down into two or more simpler chemicals.
4. **MC** What is reduction?
   A. Loss of electrons
   B. Gain of oxygen
   C. Gain of electrons
   D. Loss of hydrogen

Apply and analyse

5. Outline the difference between a redox reaction and a decomposition reaction.
6. Explain the meaning of the word *redox*.
7. Consider the reaction $Zn(s) + Cl_2(g) \rightarrow ZnCl_2(s)$
   Explain where the electrons are moving from and to in this reaction.
8. **MC** Magnesium displaces iron ions from a solution of iron sulphate. Which statement about this reaction is true?
   A. Iron ions are oxidised.
   B. Magnesium atoms are oxidised.
   C. Iron ions lose electrons.
   D. Magnesium atoms gain electrons.

Evaluate and create

9. Consider the following equations:

$$2Zn(s) + O_2(g) \rightarrow 2ZnO(s)$$
$$2Li + S \rightarrow Li_2S$$
$$3Mg(s) + 2N_2(g) \rightarrow Mg_3O_2(s)$$
$$Ca(s) + Cl_2(g) \rightarrow CaCl_2(s)$$

   a. State the elements in each equation that have lost electrons and so are oxidised.
   b. State the elements in each equation that have gained electrons and so are reduced.
   c. Explain if metals are more likely to lose or gain electrons.
   d. Explain if non-metals are more likely to lose or gain electrons.
10. **SIS** Practical work partners Sophie and Oliver heated solid magnesium carbonate carefully in a test tube. After a few minutes of heating there appeared to be no reaction.

$$MgCO_3 \rightarrow MgO + CO_2$$

   a. How could you prove to them that a reaction had occurred?
   b. Oliver said that this was an example of a redox reaction, but Sophie said it was not. Explain whether it was or was not a redox reaction.
11. **SIS** When a piece of zinc is placed in a copper sulfate solution the blue colour fades and a brown deposit forms.
   a. Explain the observation using the terms *oxidised* and *reduced*.
   b. Write an equation for the reaction.

Fully worked solutions and sample responses are available in your digital formats.

LESSON
5.10 Rates of reaction

LEARNING INTENTION

At the end of this lesson you will be able to explain the effect of a range of factors, such as temperature, concentration, surface area and catalysts, on the rate of chemical reactions.

5.10.1 Rates of reaction

The rates at which chemical reactions occur vary. Some reactions occur within a fraction of a second, while others may take days or even years. Sometimes it is necessary or convenient to speed up a chemical reaction. The rate of a reaction refers to how quickly the reactants are converted into products. This is particularly important in industry because the amount of product and the speed of production have significant financial implications.

Closer to home, we need to know how long food will take to cook or how long it will take for a medicine such as an antacid to make us feel better. Controlling the rate at which reactions occur is therefore of great interest to scientists. Sometimes it is necessary to slow reactions down; for example, corrosion of metals and decay of food.

The rate of a reaction can be increased by:
- increasing the temperature
- increasing the surface area of solid reactants
- increasing the concentration of the reactants
- adding a catalyst.

5.10.2 Collision theory

These factors can be explained by collision theory, which looks at the interaction of the particles involved in the reaction.

For a reaction to occur the particles (atoms, ions or molecules) must:
- collide with each other
- have sufficient energy to break bonds in the reactants — this is called the **activation energy** — otherwise the particles will just bounce off each other
- collide with the correct orientation.

FIGURE 5.29 The nitrogen monoxide (NO) and ozone (O_3) molecules must collide with the correct orientation for a reaction to occur.

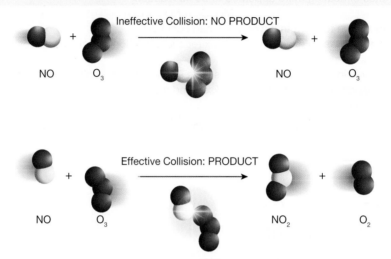

Ineffective Collision: NO PRODUCT

NO O_3 →   NO   + O_3

Effective Collision: PRODUCT

NO   + O_3 → NO_2 + O_2

5.10.3 Temperature and rate

Particles are in constant motion and have a range of energies. When the temperature of a reaction is increased, the average speed of the particles increases and so does the average kinetic energy. Not only will the particles collide more frequently, but more importantly, more particles will have enough energy to effectively collide.

FIGURE 5.30 Temperature is one of the factors affecting the reaction rate.

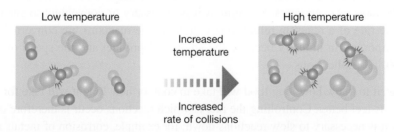

Low temperature

Increased temperature

Increased rate of collisions

High temperature

For example, if hydrogen gas and oxygen gas are mixed at room temperature, there is no reaction — the molecules do not have enough energy to break bonds in the reactants. If, however, a spark is introduced to the mixture, some of the molecules will react vigorously and the energy produced will rapidly cause other molecules to react.

$$2H_2(g) + O_2(g) \rightarrow 2H_2O(g)$$

Lowering the temperature slows down the particles and the reaction. Many types of organisms are found in food, including microbes. Chemical reactions that spoil food take place in microbes, but refrigeration cools the food and the microbes. This makes the chemical reactions inside the microbes slow down and so the food keeps for longer.

activation energy minimum energy required to start the reaction

5.10.4 Surface area and rate

Bath bombs are sold as solid balls. When they are added to water, the chemicals inside them begin to dissolve. The ball slowly disappears. But what if the same bath bomb was crushed into smaller pieces? A much larger surface area is exposed to the water, and the bath bomb dissolves much more quickly. If the solid reactants are crushed into a powder, then there is more surface accessible where the reactant particles to collide. Larger lumps will only have the outside surface available for collisions.

FIGURE 5.31 Crushing a bath bomb increases the surface area. This would result in a much fast reaction rate, so it would dissolve much quicker.

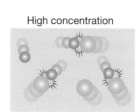

5.10.5 Concentration and rate

Particles in gaseous reactions or reactants in solution will have a greater opportunity to collide and react if there are more of them in a particular volume. That is, an increase in the concentration of reactants will increase the rate of collisions and, so, increase the rate of reaction.

In a reaction of magnesium and hydrochloric acid, the reaction is far more vigorous if concentrated hydrochloric acid is used.

FIGURE 5.32 Concentration is one of the factors affecting the reaction rate.

Low concentration

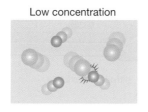

Increased concentration

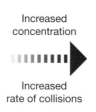

Increased rate of collisions

High concentration

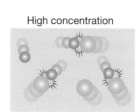

on Resources

 **Interactivity** Reaction rates (int-0230)

5.10.6 Catalyst and rate

Another way to increase the rate of a reaction is to use a **catalyst**. Catalysts are not changed by the reaction. There is always as much catalyst present at the end of a reaction as there was at the start. Catalysts work by helping bonds to break more easily; therefore, the reactants need less energy to react and the reaction is faster. A catalyst can be recovered and used again and again. We all make use of catalysts every day. Cars have catalytic converters; contact lenses are cleaned using a catalysed chemical reaction; and there are catalysts in the food you eat every day. There are also thousands of catalysts in your body without which you could not live. These biological catalysts are called **enzymes**.

catalyst chemical that speeds up reactions but is not consumed in the reaction

enzyme biological catalyst that speeds up reactions

Catalysts in industry

Industry makes use of many catalysts. For example:

- Iron and iron oxide are used to catalyse the production of ammonia gas. Ammonia is used to make fertilisers and explosives.
- Vanadium oxide (V_2O_5) is used in the production of sulfuric acid. One important reaction in this process, between sulfur dioxide gas and oxygen, has a very slow rate at room temperature. However, it proceeds rapidly at 450 °C in the presence of a vanadium oxide catalyst according to the equation:

$$2\,SO_2(g) \;+\; O_2(g) \xrightarrow[\;450\,°C\;]{V_2O_5} 2\,SO_3(g)$$

Note that the catalyst is written above the arrow and not on the side of the reactants. It is not changed as the reaction takes place.

- Crystalline substances made of aluminium, silicon and oxygen called **zeolites** are used to 'crack' (break up) the large molecules in crude oil to form the smaller molecules, such as octane, found in petrol.

> **zeolite** crystalline substance consisting of aluminium, silicon and oxygen that is used as a catalyst to break up large molecules in crude oil

Everyday catalysts

In the confined space of the internal combustion engine, the fuel does not completely react with oxygen. As a result, carbon monoxide (CO), a highly poisonous gas, is produced. Nitrogen oxides are other harmful gases produced by car engines. In order to reduce the amount of pollution from these gases, cars are fitted with catalytic converters as part of the exhaust system (see figure 5.33). These converters have a honeycombed surface that is coated with the metals platinum and rhodium, and with aluminium oxide (see figure 5.34). At the catalyst surface, the nitrogen oxides are converted to less harmful gases and the carbon monoxide is reacted with more oxygen to form carbon dioxide according to the equation:

$$2\,CO(g) + O_2(g) \longrightarrow 2\,CO_2(g)$$

Catalysts can also help clean your contact lenses. One cleaning product makes use of a platinum catalyst. A solution of hydrogen peroxide (H_2O_2) is poured into a small container that contains a platinum-coated disc. The platinum causes the peroxide to decompose according to the reaction:

$$2\,H_2O_2(aq) \xrightarrow{\text{platinum}} 2\,H_2O(l) + O_2(g)$$

Any microbes not tolerant to oxygen on the contact lenses are killed by the oxygen released.

FIGURE 5.33 Catalytic converters have a large surface area and are coated with a metal catalyst.

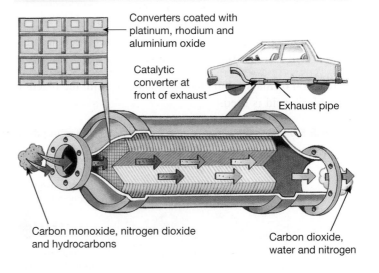

Converters coated with platinum, rhodium and aluminium oxide

Catalytic converter at front of exhaust

Exhaust pipe

Carbon monoxide, nitrogen dioxide and hydrocarbons

Carbon dioxide, water and nitrogen

FIGURE 5.34 The honeycombed surface of catalytic converters helps to maximise their surface area and help convert harmful gases into harmless ones.

Catalysts in living things

Almost every one of the chemical reactions that take place in your body is controlled by an enzyme. Enzymes are large protein molecules essential for digesting food, breaking down toxic waste products, and numerous other chemical processes that keep you alive and healthy. Enzymes are also used to make bread, cheese, vinegar and many other food products. Generally the reactants are called **substrates**. The enzyme **amylase**, which is present in your saliva, is involved in the breakdown of starch into sugar.

Your liver contains an enzyme called **catalase**. Catalase speeds up the breakdown of hydrogen peroxide, a toxic waste product produced in your cells.

FIGURE 5.35 An example of an enzyme catalysed reaction

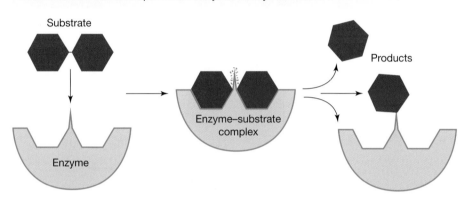

substrate a molecule acted upon by an enzyme

amylase enzyme in saliva that breaks down starch into sugar

catalase enzyme in the liver involved in the breakdown of hydrogen peroxide, a toxic waste product from cells in the body

INVESTIGATION 5.5

elog-2466

A liver catalyst

Aim

To observe the effect of a catalyst on a decomposition reaction

Materials

- heatproof mat
- 2 test tubes and test tube rack
- 20% hydrogen peroxide solution
- spatula
- fresh liver
- safety glasses
- mortar and pestle

Method

1. Pour hydrogen peroxide to a depth of 3 cm into the test tubes. Label the test tubes 1 and 2.
2. Grind a small piece of liver in the mortar and pestle. Add liver to test tube 1 only.

Results

Design a table and outline your observations for test tube 1 and test tube 2.

Discussion

1. What effect did the liver have on the breakdown of hydrogen peroxide?
2. What evidence suggests that a chemical reaction has taken place?
3. Suggest a reason the liver was ground up before it was placed in the hydrogen peroxide solution.
4. What is the function of test tube 2?

Conclusion

Summarise the findings for this investigation about the effect of a catalyst on a decomposition reaction.

5.10.7 Measuring rate

The reaction rate is measured as change over time. It can be determined experimentally by:
1. measuring the decrease in a reactant over time
2. measuring the increase in a product over time.

Methods include measuring changes in the mass of the reaction vessel, changes in pH, changes in colour or changes in the volume of gas evolved. For example, the reaction of an acid and a metal carbonate could be carried out in a flask on a balance, such as in figure 5.36. The mass could be recorded over time and plotted, and a graph like the one in the figure would be obtained.

The steeper the gradient of the graph the faster the rate of reaction.

FIGURE 5.36 Experiment and graph showing the change in mass of the reaction of a metal carbonate and acid over time

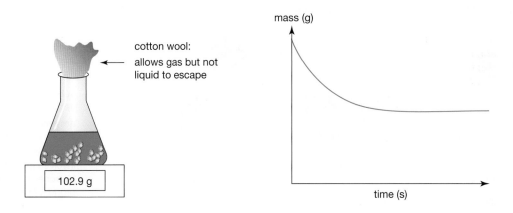

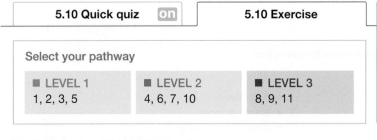

cotton wool:
allows gas but not
liquid to escape

102.9 g

mass (g)

time (s)

5.10 Activities

learnon

| 5.10 Quick quiz on | 5.10 Exercise |

These questions are even better in jacPLUS!
- Receive immediate feedback
- Access sample responses
- Track results and progress

Find all this and MORE in jacPLUS ▶

Select your pathway

| ■ LEVEL 1 | ■ LEVEL 2 | ■ LEVEL 3 |
| 1, 2, 3, 5 | 4, 6, 7, 10 | 8, 9, 11 |

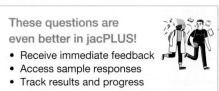

Remember and understand

1. **MC** Why does increased concentration result in a rate increase?
   A. The particles move faster so there are more collisions.
   B. There are more particles and so there are more collisions.
   C. The particles have more energy so there are more collisions.
   D. The surface area is greater so there are more collisions.

2. **MC** Increasing the temperature of a reaction increases the rate by:
   i. increasing the frequency of collisions.
   ii. increasing the activation energy.
   iii. increasing the average energy of the particles.
   A. i only
   B. ii and iii only
   C. i and iii only
   D. i, ii and iii

3. **MC** What is a catalyst?
   A. A catalyst is a chemical that helps speed up a reaction but is not changed by the reaction.
   B. A catalyst is a chemical that slows down a reaction but is not changed by the reaction.
   C. A catalyst is a chemical that speeds up a reaction and will be changed by the reaction.
   D. A catalyst is a chemical that slows down a reaction and will be changed by the reaction.

Apply and analyse

4. **MC** The equation for the reaction of sodium carbonate and hydrochloric acid is:
$$Na_2CO_3(s) + 2HCl(aq) \rightarrow 2NaCl(aq) + H_2O(l) + CO_2(g)$$
   Which of the following could be used to measure the rate?
   i. The mass of water produced
   ii. The pH of the reaction mixture
   iii. The mass of the flask and contents
   A. i only
   B. ii and iii only
   C. i and iii only
   D. i, ii and iii

5. **SIS** **MC** To make a rate of reaction experiment a fair test you must:
   A. repeat the experiment more than once.
   B. change all the variables except the one you are studying.
   C. use different concentrations for each test.
   D. change one variable and keep the others the same.

6. Explain why catalysts are important to industry. Provide at least three examples.

7. **SIS** A basic homemade bread recipe suggests this step:
   Dissolve yeast and 1/2 teaspoon sugar in lukewarm water and let stand until bubbles form on surface.
   Explain why the yeast is added to lukewarm water.

Evaluate and create

8. **SIS** Refer to the graph in figure 5.36.
   a. When is the rate of the reaction the fastest: at the beginning, middle or end of the reaction? Explain why this is the case.
   b. The Law of Conservation of Mass says that 'Mass cannot be created or destroyed'. Why is the mass decreasing as the reaction progresses?
   c. Explain why the graph becomes level.

9. **SIS** Describe the materials and method that could be used to produce a graph showing the rate of a reaction of calcium carbonate and hydrochloric acid experiment over time.

10. **SIS** Cooking potatoes in the oven is an example of a chemical reaction. Design an experiment that outlines at least two ways of changing the speed of the reaction involved in cooking potatoes.

11. **SIS** The graph shown represents the progress of three different experiments using zinc powder in excess hydrochloric acid.
    a. Explain which graph line shows the reaction with dilute hydrochloric acid.
    b. Explain which graph line shows the reaction that had twice the mass of zinc powder as the other two experiments.
    c. Explain which graph line shows the reaction of concentrated hydrochloric acid.

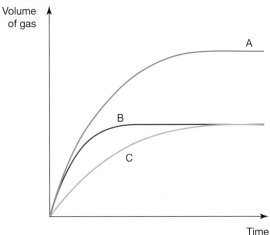

Fully worked solutions and sample responses are available in your digital formats.

LESSON
5.11 Making useful products

LEARNING INTENTION

At the end of this lesson you will be able to describe how First Nations Australians used chemical methods and reactions to prepare edible food products and produce ethanol.

You will also be able to describe photochromic reactions and types of polymers.

5.11.1 Useful products made by First Nations Australians

The First Peoples of Australia, through observation and experimentation across their vast history, have transformed natural materials into many useful products. For example, First Nations Australians have developed complex treatment processes to remove bitter and toxic compounds from tree nuts, which are important food sources. These tree nuts provide important dietary requirements including carbohydrates and protein. One such food is seeds from the black bean tree, which in their raw state are highly poisonous, causing serious symptoms including stomach pain, dizziness, vomiting and diarrhoea. Half a seed would be enough to send a person to hospital!

FIGURE 5.37 Seeds and seed case of black bean (*Castanospermum australe*), Wooroonooran National Park, Cairns

The process used to make the seeds edible begins with cooking them in a specially prepared fire pit lined with ginger leaves. They are cooked for many hours to make the starch more digestible. The cooked beans are cracked open, thinly sliced and pounded to break cell membranes. The last step involves a **leaching** process where the smashed seeds are placed in a finely woven basket and placed in a secure position in a fast-flowing creek for many days and the soluble toxins are washed away. The seeds are then dried and ground into flour for bread making or cooking.

Another type of reaction called **fermentation** was used by First Nations Australian to produce alcoholic beverages before the arrival of outside contact. Fermentation is a chemical change brought about by microorganisms — typically, a starch or sugar is converted into an alcohol. The alcohol produced from the fermentation of sugar is ethyl alcohol, commonly called ethanol. Scientists are investigating several different strains of yeast and bacteria found in the cider gum, a eucalyptus tree found in the Tasmanian highlands, that has been used in this fermentation process. The microorganisms involved are quite different from those used in winemaking.

leaching process of using water or another liquid to remove soluble matter from a solid

fermentation a chemical process in which the glucose is broken down to ethanol by the action of yeast

Fermentation reaction

The fermentation of sugar is as follows:

$$\text{Sugar} \xrightarrow{\text{yeast}} \text{ethanol} + \text{carbon dioxide}$$

$$C_6H_{12}O_6(aq) \xrightarrow{\text{yeast}} 2\,C_2H_5OH(aq) + 2\,CO_2(aq)$$

A drink called *way-a linah* was made from the sugar-rich sap of the cider gum tree. The process began by drilling a hole in the trunk of the tree with a sharp stone, and then a hole was dug at the base to collect the sap. It was then left to be fermentated by the local yeast. The cider-like beverage generated is a product of yeast metabolism, and carbon dioxide is formed as a by-product.

Alternatively, the First Nations Australians would prepare a trench near a swamp, and line it with a bark container which was filled with banksia cones or other nectar-laden flowers. They would leave this to soak in a sugar solution for many days while fermentation occurs. This drink is called *mangaitch*.

5.11.2 Glasses that darken outdoors

People who wear glasses often don't want to swap over to sunglasses when they go outside. **Photochromic** lenses solve the problem by darkening as the wearer moves from indoors out into bright sunshine (see figure 5.38). They lighten again when the wearer moves back into an area of low light. Plastic photochromic glasses use organic material that darkens the lenses when exposed to ultraviolet light.

FIGURE 5.38 Photochromic lenses darken in bright sunshine and lighten in areas of low light.

Glass photochromic glasses work due to the presence of silver chloride (AgCl) crystals in the glass. When in sunshine, ultraviolet light is absorbed by the silver chloride crystals and a redox reaction occurs. Electrons are transferred from the chloride ions to the silver ions according to the equation:

$$Ag^+ + Cl^- \rightarrow Ag + Cl$$

Silver particles then form in the glass, darkening the lens so that visible light is absorbed and reflected.

The fading of the dark glass is more complicated. The chlorine atoms are very reactive. To stop them reacting with the silver atoms and reversing the process too quickly, singly charged copper ions are dissolved in the molten glass during the manufacturing process. These ions react with the chlorine atoms to form chloride ions and doubly charged copper ions in the reaction:

$$Cu^+ + Cl \rightarrow Cu^{2+} + Cl^-$$

When the glasses are no longer in the sunlight, the doubly charged copper ions accept an electron from the silver atom. The silver ion re-forms and the dark lens becomes light again.

$$Cu^{2+} + Ag \rightarrow Cu^+ + Ag^+$$

5.11.3 Producing plastics

Plastics

Which material is strong, light in weight and cheap to make, comes in a huge range of colours and can be moulded into any shape?

It could only be one of the **synthetic** (manufactured) materials we know as **plastics**. All plastics are products of chemical reactions. They are used to manufacture food containers and packaging, ballpoint pens, plumbing materials, car parts, rubbish bins, cling films such as Glad® Wrap and a multitude of other items.

photochromic describes lenses made from glass that darkens in bright light

synthetic manufactured by people

plastic synthetic substance capable of being moulded

Monomers and polymers

All of the synthetic materials we call plastics are **polymers**. However, not all polymers are synthetic. Cotton, wool, leather and rubber are examples of natural polymers.

Polymers are very large molecules that consist of many repeating units called **monomers**. Monomers are small molecules and most contain the element carbon. Polymer molecules may, therefore, contain thousands of carbon atoms. The other elements commonly found in monomers and polymers include hydrogen, oxygen, chlorine, fluorine and nitrogen.

The chemical reactions that occur when polymers form can be modelled using plastic beads or blocks that click together to form a long chain. Each plastic bead or block represents a single monomer molecule. The long chain, which may contain thousands of monomers, represents a polymer molecule. The chemical reactions that join monomers together are known as **polymerisation** (polymer-forming) reactions.

The prefix *poly* (meaning 'many') is often used when naming polymers. For example, polyvinyl acetate (PVA) is a polymer made from the monomer known as vinyl acetate.

FIGURE 5.39 A polymer is made up of smaller units called monomers bonded together.

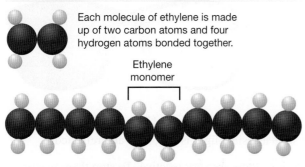

Each molecule of ethylene is made up of two carbon atoms and four hydrogen atoms bonded together.

Ethylene monomer

The polymer polyethylene is made of thousands of ethylene monomers all bonded together in long chains.

FIGURE 5.40 Polymer materials: PVC pipes, polyethylene bottle and nylon rope

Synthetic polymers

Polyvinyl chloride, also known as PVC, is formed from the monomer vinyl chloride. PVC is light, rigid and doesn't corrode, making it ideal for use in drainage and sewerage pipes.

Polyethylene is formed from the monomer ethylene. It is light, tough and resistant to most acids and bases. Polyethylene is used to make plastic bags, soft-drink bottles, buckets, cling wrap and many other household products.

Nylon is one of a group of polymers known as polyamides, which are formed from monomers joined by amide bonds. It is used to make fabrics for clothing, ropes, guitar strings, machine parts and much more.

polymer molecule made up of repeating subunits of monomers joined together in long chains

monomer small repeating unit that makes up a polymer, usually containing carbon and hydrogen, and sometimes other elements.

polymerisation chemical reaction joining monomers in long chains to form a polymer

polyvinyl chloride polymer formed from the monomer vinyl chloride; also known as PVC

polyethylene material formed from the monomer ethylene

nylon a group of polymers called polyamides used in clothing

Resources

⬥ **Interactivity** Natural polymers (int-5747)

CASE STUDY: Making nylon

A co-polymer is formed when two different monomers join together or polymerise. The two different monomers are shown in the figure 5.41; one begins and ends with a blue oxygen atom and the other begins and ends with a green nitrogen atom. When these two monomers join, a hydrogen ion is released from one and a hydroxide ion is released from the other. These form a water molecule, so this type of reaction is called a *condensation reaction*.

One of the many uses for nylon is in eyelash-lengthening mascara, where tiny threads of nylon are applied with a small brush that attach to the natural eyelash to make it appear longer.

FIGURE 5.41 Nylon is a co-polymer.

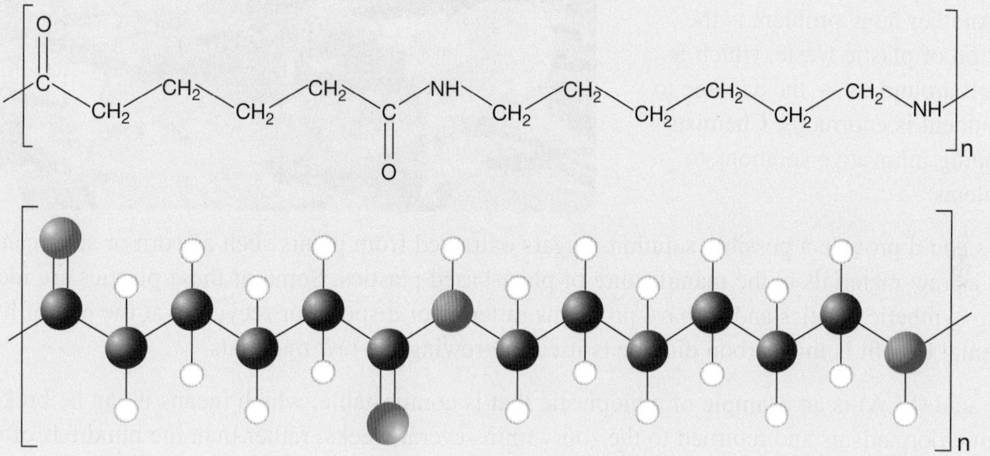

DISCUSSION

Australia was the first country in the world to use only plastic banknotes.
a. Research advantages and disadvantages of using plastic notes instead of paper notes.
b. Are any new types of money being designed? How do you think a new design could improve on plastic notes?

FIGURE 5.42 Australia's plastic notes

on Resources

eWorkbook	Making useful products (ewbk-13147)
Video eLessons	Road worker painting road markings (eles- 4174)
	A 3-D printer (eles-2720)
Interactivity	Making polymers (int-3849)

Using plastics

Plastics are versatile and useful materials, but chemists have been looking for alternatives. Plastics have been manufactured in the past from fossil fuels but the world's reserves of fossil fuels are limited, and eventually they will run out. Fossil fuels include coal, natural gas and oil, all of which were formed from decaying plants or animals over tens or hundreds of millions of years. Another huge problem is the accumulation of plastic waste, which is in us and all around us — the damage to the environment is enormous. Chemists are developing innovative solutions to these problems.

FIGURE 5.43 Plastic waste on the beach

Bioplastics could provide a possible solution. Sugars extracted from plants such as corn or sugarcane have been used as raw materials in the manufacture of plant-based plastics. Some of these plastics are identical in structure to synthetic plastics and so have problems in terms of disposal or recycling at the end of life. But one environmental benefit is that carbon dioxide is used in growing the raw materials.

Polylactic acid (PLA) is an example of a bioplastic that is compostable, which means it can be broken down by microorganisms and returned to the soil within several weeks, rather than the hundreds of years taken by conventional plastics. The composting, to be effective, must be done under controlled conditions in industrial facilities. Less energy is required to manufacture this bioplastic compared to synthetic plastics and it has a wide variety of uses from packaging to utensils, 3D printing and medical devices. A disadvantage of bioplastics is the large use of land and water, which could be used for food crops, so it is preferable to use non-edible parts of plants and other waste material from plants that would already be grown.

FIGURE 5.44 a. A prosthetic limb for amputees made from using a 3D-printer and polylactic acid; **b.** a bioware plastic tumbler made from polylactic acid that will compost within six days

SCIENCE AS A HUMAN ENDEAVOUR: Graphene

Carbon must be the champion of all elements. Incredibly, it can bond with itself and other elements to make millions of compounds. Many of these compounds form the basis of all living cells; it is present in hair, blood, bone, muscles, leaves, flowers and fruits. It is everywhere, from foods to fuels, pencils to plastics, diamonds and drugs.

Carbon is critical to life, and carbon compounds involved in living systems are known as organic chemicals. Carbon the element exists in different and diverse forms known as **allotropes.** Diamond is an allotrope of carbon; it is the hardest substance known and it has great reflective properties. Amazingly, that same element is also present in 'lead' pencils, but the atoms are arranged differently, in layers rather than in the 3D structure of diamond. These layers of material, known as graphite, are left on the paper when we write something with a pencil. Graphite has an unusual property for a non-metal — it can conduct electricity — and has interested scientists for many years.

FIGURE 5.45 Allotropes of carbon: **a.** diamond and **b.** graphite

allotrope different forms of a chemical element, for instance different arrangement of atoms in crystalline solids

graphene an allotrope of carbon consisting of a single layer of atoms arranged in a two-dimensional honeycomb lattice nanostructure

It wasn't until 2004 that scientist Andre Geim, while celebrating 'free' Friday afternoon with random experiments on graphite, peeled off layer after layer of graphite using sticky tape until he had layers a few atoms thick. He decided that this substance consisting of hexagonal rings of carbon atoms — which had been named **graphene** by earlier scientists — was worthy of serious study. He found that it conducted electricity better than silver or copper and was flexible, elastic and 200 times stronger than the equivalent thickness of steel. Graphene is also almost transparent, light, and a good conductor of heat. It is so thin that it takes 3 million layers to make a piece 1 mm thick. Geim was awarded the Nobel Prize in Physics for his discovery in 2010.

Considerable research is being carried out on the many possible applications of graphene including:

- light bulbs that use less energy, last longer and are less costly to manufacture
- water filters using graphene membranes
- electronics, because electrons can travel freely
- biomedical sensors to detect and target cancer cells
- removing CO_2 from emissions
- solar cells that would be thinner and lighter and less expensive than those based on silicon
- protective clothing, due to its strength
- transistors in computer chips that control the flow of electricity — using graphene could reduce their size and increase the speed of electron flow
- in TV and mobile phone screens, where graphene could improve conductivity, flexibility and transparency
- recharging batteries
- sporting equipment requiring strength and lightness
- corrosion prevention of metals.

FIGURE 5.46 Professor Andre Geim, who is responsible for the discovery of graphene, holding a model of graphene made of hexagonal rings of carbon atoms.

A team of researchers in Hong Kong has even developed graphene masks with an anti-bacterial efficiency of 85 per cent, which improves to almost 100 per cent after exposure to sunlight for 10 minutes. These low-cost masks are showing good results against two strains of coronavirus. Other graphene products are now produced commercially, and many more functions of this remarkable material are being developed by innovative scientists around the world.

5.11 Activities

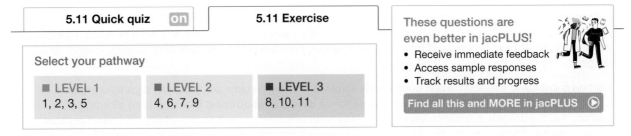

5.11 Quick quiz on	5.11 Exercise

Select your pathway

■ LEVEL 1	■ LEVEL 2	■ LEVEL 3
1, 2, 3, 5	4, 6, 7, 9	8, 10, 11

These questions are
even better in jacPLUS!
- Receive immediate feedback
- Access sample responses
- Track results and progress

Find all this and MORE in jacPLUS ▶

Remember and understand

1. **MC** First Nations Australians cook and then leach black bean seeds because it:
   A. improves the taste.
   B. makes them softer.
   C. increases nutrients.
   D. removes poisons.
2. **MC** Which of the following is NOT a polymer?
   A. PVC
   B. Nylon
   C. Polyethylene
   D. Ethanol
3. a. What are polymers?
   b. Explain with examples the difference between synthetic and natural polymers.
4. Explain two environmental advantages of using polylactic acid to make plastic drinking cups.

Apply and analyse

5. Complete the following table.

Monomer	Polymer
Ethene	
Styrene	
	Polylactic acid
	Polypropene
Tetrafluoroethene	

6. Photochromic glasses work due to the presence of silver chloride crystals in the glass. Explain how the glasses darken in bright sunshine and lighten in areas of low light.
7. Explain if the reaction occurring in photochromic glasses is a redox reaction.
8. Explain what is leached out of the black beans in the flowing creek.

Evaluate and create

9. **SIS** Provide two reasons that the black beans are pounded by First Nations Australians before the leaching process.
10. **SIS** Design an experiment to compare the strength of plastic supermarket bags with the strength of other types of bags. Ensure your tests are fair.
11. **SIS** Choose three types of biodegradable plastic bags and design an experiment to find out which degrades the fastest.

Fully worked solutions and sample responses are available in your digital formats.

LESSON
5.12 Thinking tools — Target maps

5.12.1 Tell me

What is a target map?

A target map is a very useful thinking tool that can assist you to identify (target) which concepts are relevant to a topic and which concepts are not relevant to a topic. They are sometime called circle maps.

Why use a target map?

Target maps identify and describe the lessons and concepts that relate to the topic of content. Compared to other maps (such as single bubble maps or mind maps), target maps separate ideas into relevant and non-relevant material, providing a point of focus at a glance.

FIGURE 5.47 Setting up a target map

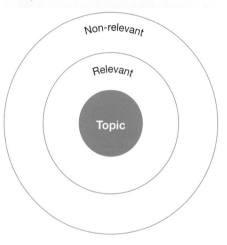

5.12.2 Show me

To create a target map:
1. Draw three concentric circles on a sheet of paper.
2. Write the topic in the centre circle.
3. In the next circle, write words and phrases that are relevant to the topic.
4. In the outer circle, write words and phrases that are not relevant to the topic.

FIGURE 5.48 A target map of acid–base reactions

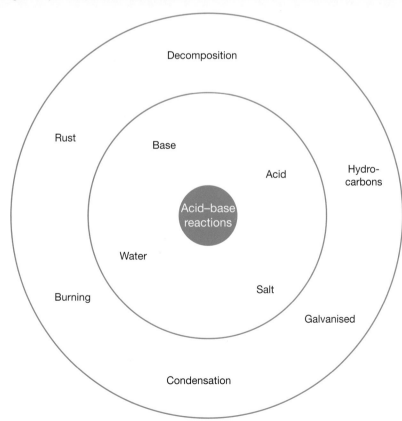

5.12.3 Let me do it

5.12 Activity

1. Create a target map of different factors that can change the rate of a chemical reaction. Try to include at least eight terms.
2. Use the following terms to create target maps that are relevant to the following:
   a. All chemical reactions
   b. Precipitation reactions
   c. Corrosion reactions
   Terms:
   - Synthesis
   - Double displacement
   - Soluble
   - Solution
   - Ionic
   - Aqueous
   - Decomposition
   - Galvanised
   - Burning
   - Base
   - Salt
   - Precipitate
   - Melting
   - Reactants
   - Evaporation
   - Single displacement
   - Neutralisation
   - Insoluble
   - Solid
   - Changing partners
   - Solubility rules
   - Hydrocarbons
   - Condensation
   - Rust
   - Acid
   - Water
   - Oxygen
   - Energy absorbed
   - Products
   - Broken bonds

3. Use the following terms to create target maps that are relevant to the following:
   a. Salts
   b. Redox reactions
   c. Catalysts
   Terms:
   - Double displacement
   - Rate
   - Decomposition
   - Lose electrons
   - Polymer
   - Platinum
   - Magnesium chloride
   - Zeolites
   - Sodium acetate
   - Copper nitrate
   - Sodium sulfate
   - Neutralisation
   - Single displacement
   - Gain electrons
   - Synthesis
   - Sodium chloride
   - Enzymes
   - Coal
   - Propane
   - Ethanol
   - Vanadium oxide

Fully worked solutions and sample responses are available in your digital formats.

LESSON
5.13 Project — Flavour fountain

5.13.1 Scenario

The Sparky Cola Corporation has a series of advertisements for which they are famous that always involve a Mentos lolly being dropped by various means into a plastic bottle of Sparky Cola, causing a huge foaming jet to burst out of the bottle. As a good science student, you know that the jet is the result of the carbon dioxide dissolved in the cola being able to form sizable gas bubbles very quickly on the rough surface of the lolly. You may well have even done this trick yourself.

Sparky Cola have decided that they want bigger jets than ever before and they pride themselves on using real video footage rather than CGI. They are providing a special prize at the next Science Fair for the project that determines how the biggest jet can be produced from a single Mentos lolly and 600 mL of cola. You are to provide not only a scientific report but also video footage of your highest fountain that they can use in their next ad. You and your friends are determined to win the cash prize at the Science Fair and the TV fame for your Flavour Fountain footage!

FIGURE 5.49 Carbon dioxide dissolved in cola forms sizable gas bubbles very quickly.

5.13.2 Your task

You will design and carry out an investigation that will test a number of different factors (for example, regular cola or diet cola) to determine which will produce the highest cola fountain from a 600 mL plastic bottle of cola and a Mentos lolly. Your findings will be presented in the form of a scientific report. You will also produce video footage of the highest fountain that you can make using what you have discovered.

 **Resources**

💡 **ProjectsPLUS** Flavour fountain (pro-0114)

LESSON
5.14 Review

Access your topic review eWorkbooks

 Resources

■ Topic review Level 1
ewbk-13149

■ Topic review Level 2
ewbk-13151

■ Topic review Level 3
ewbk-13153

5.14.1 Summary

The language of chemistry

- Chemical formulae show the number and type of atoms in one molecular unit.
- Chemical equations use formulae to show how atoms in a reaction rearrange to form new products.
- Chemical equations should always be balanced with the same number of atoms in the reactants and the products.

Safety with chemicals

- Many chemicals can be health hazards. Dangerous goods and hazardous substances must have appropriate hazard warnings to alert people to the risks.
- Safety data sheets and risk assessments are vital in any experiment to understand the chemicals in use, and to assess and minimise any risks.

Synthesis reactions

- A synthesis reaction is when two elements or compounds combine to form a more complex product.
- For example, the synthesis of water from hydrogen and oxygen.

Decomposition reactions

- A decomposition reaction is when a compound breaks down into two simpler substances.
- For example, the decomposition of zinc carbonate into zinc oxide and carbon dioxide.

Single displacement reactions

- A single displacement reaction is when one element replaces another in a compound.
- For example, the displacement of hydrogen from an acid by a metal.

Double displacement - neutralisation

- A double displacement reaction is when two reactants exchange a component.
- For example, neutralisation involves the reaction of an acid and base to form a salt and water.
  - metal hydroxide + acid → salt + water
  - metal oxide + acid → salt + water
  - metal carbonate + acid → salt + carbon dioxide + water
  - metal hydrogen carbonate + acid → salt + carbon dioxide + water

Double displacement - precipitation

- In a precipitation reaction, an insoluble compound, known as a precipitate, forms.
- The rules of solubility will determine whether substances are soluble or insoluble.

Redox reactions

- Redox reactions involve the transfer of electrons between reactants through the processes of reduction and oxidation.
- Reduction is gain of electrons.
- Oxidation is loss of electrons.
- Corrosion is a type of oxidation reaction in which a metal rusts and is eaten away.
- Combustion is a redox reaction.

Rates of reaction

- The rate or speed of a reaction can be changed by changing the amount of substance, the temperature or the surface area.
- Catalysts (or enzymes in the body) are chemicals that speed up the rate of the reaction but are not used up.

Making useful products

- First Nations Australians used chemical skills and knowledge of reactions to make toxic foods edible.
- First Nations Australians used fermentation reactions to make alcoholic beverages from bush plants.
- Photochromic reactions can be used to make sunglasses darker in sunlight.
- It is better for the environment if plastics can be made from plants instead of from fossil fuels.

5.14.2 Key terms

acid chemical that reacts with a base to produce a salt and water; edible acids taste sour

activation energy minimum energy required to start the reaction

alkali base that dissolves in water

allotrope different forms of a chemical element, for instance different arrangement of atoms in crystalline solids

amylase enzyme in saliva that breaks down starch into sugar

aqueous solution solution in which water is the solvent

base chemical substance that will react with an acid to produce a salt and water; edible bases taste bitter

catalase enzyme in the liver involved in the breakdown of hydrogen peroxide, a toxic waste product from cells in the body

catalyst chemical that speeds up reactions but is not consumed in the reaction

chemical reaction the breaking of chemical bonds between reactant molecules and the forming of new bonds in product molecules

combination reaction chemical reaction in which two substances, usually elements, combine to form a compound, also **synthesis reactions**

combustion reaction chemical reaction in which a substance reacts with oxygen and heat is released

corrosion chemical reaction that wears away a metal

corrosive chemical that wears away the surface of substances, especially metals

dangerous good chemical that could be dangerous to people, property or the environment

decomposition reaction chemical reaction in which one single compound breaks down into two or more simpler chemicals

diatomic molecule containing two atoms

displacement reaction reaction in which an atom or a set of atoms is displaced by another atom in a molecule

electrolysis the decomposition of a chemical substance by the application of electrical energy

endothermic reaction that absorbs energy

enzyme biological catalyst that speeds up reactions

equation statement describing a chemical reaction, with the reactants on the left and the products on the right separated by an arrow

exothermic reaction that releases energy

fermentation a chemical process in which the glucose is broken down to ethanol by the action of yeast

formulae set of chemical symbols showing the elements present in a compound and their relative proportions

galvanised metal coated with a more reactive metal that will corrode first, preventing metal from corroding

graphene an allotrope of carbon consisting of a single layer of atoms arranged in a two-dimensional honeycomb lattice nanostructure

hazardous substance chemical that has an effect on human health

indicator substance that changes colour when it reacts with acids or bases; the colour shows how acidic or basic a substance is

ionic compound compound containing positive and negative ions held together by the electrostatic force

leaching process of using water or another liquid to remove soluble matter from a solid

monomer small repeating unit that makes up a polymer, usually containing carbon and hydrogen, and sometimes other elements.

neutralisation reaction between an acid and a base

nylon a group of polymers called polyamides used in clothing

oxidation chemical reaction involving the loss of electrons by a substance

pH scale the scale from 1 (acidic) to 14 (basic) that measures how acidic or basic a substance is

photochromic describes lenses made from glass that darkens in bright light

plastic synthetic substance capable of being moulded

polyatomic ion ion containing more than one type of atom

polyethylene material formed from the monomer ethylene

polymerisation chemical reaction joining monomers in long chains to form a polymer

polymer molecule made up of repeating subunits of monomers joined together in long chains

polyvinyl chloride polymer formed from the monomer vinyl chloride; also known as PVC

precipitate solid product of a chemical reaction that is insoluble in water

precipitation reaction chemical reaction in which there is a water insoluble product

redox oxidation and reduction considered together

reduction a chemical reaction involving the gain of electrons by a substance

risk assessment document outlining potential hazards of an experiment and providing protective measures to minimise risk

safety data sheet (SDS) document containing important information about hazardous chemicals

salt one of the products of the reaction between an acid and a base

single displacement reaction reaction in which one element is substituted for another element in a compound

smelting process to obtain a metal from its ore

substrate a molecule acted upon by an enzyme

synthetic manufactured by people

universal indicator mixture of indicators that changes colour as the strength of an acid or base changes, indicating the pH of the substance

zeolite crystalline substance consisting of aluminium, silicon and oxygen that is used as a catalyst to break up large molecules in crude oil

 Resources

eWorkbooks	Study checklist (ewbk-13155)
	Literacy builder (ewbk-13156)
	Crossword (ewbk-13158)
	Word search (ewbk-13160)
	Reflection (ewbk-13162)
Solution	Topic 5 Solutions (sol-1128)
Practical investigation eLogbook	Topic 5 Practical investigation eLogbook (elog-2456)
Digital document	Key terms glossary (doc-40483)

5.14 Activities

5.14 Review questions

Select your pathway

■ LEVEL 1	■ LEVEL 2	■ LEVEL 3
1, 2, 5, 7, 8, 15	3, 4, 11, 14, 18	10, 12, 13, 17, 19, 20

Remember and understand

1. **MC** What is the only reliable evidence indicating that a chemical reaction has taken place?

   A. A change in temperature
   B. A change in state
   C. Formation of a new substance
   D. Disappearance of one or more reactants

2. Outline what a polymer is and outline why plastics are often considered polymers.

3. **MC** Refer to the reactivity series of metals and state which of the following reactions will take place:
   A. $PbO + Cu \rightarrow CuO + Pb$
   B. $2Ag + 2HCl \rightarrow 2AgCl + H_2$
   C. $Cu + FeSO_4 \rightarrow CuSO_4 + Fe$
   D. $Zn + PbCl_2 \rightarrow ZnCl_2 + Pb$

4. When an aqueous solution of barium hydroxide reacts with an aqueous solution of ammonium hydroxide, the temperature of the products becomes low enough to freeze water.
   a. What is an aqueous solution?
   b. This is an example of an exothermic chemical reaction. True or false?
   c. Where does the energy transferred to or from the reactants go?

5. **MC** Which of the following is a balanced equation?
   A. $Na + 2Cl \rightarrow 2NaCl$
   B. $MgO + 2HCl \rightarrow MgCl + H_2$
   C. $MgO + 2HCl \rightarrow MgCl_2 + H_2O$
   D. $2Na + Cl \rightarrow 2NaCl$

6. Using the following table, write the three steps and the final balanced equation and show physical state symbols.

TABLE The process for balancing a chemical equation

Balancing a chemical equation	Example: Ethene gas will burn in air in a combustion reaction. This type of reaction produces CO_2 and H_2O.
Step 1: Start with the word equation and name all of the reactants and products.	_____ gas + _____ gas → carbon dioxide + water
Step 2: Replace the words in the word equation with formulae and rewrite the equation.	Ethene gas = C_2H_4 Oxygen gas = _____ (reactants) _____ = CO_2 Water vapour = _____ (products)
Step 3: Count the number of atoms of each element (represented by the formulae of the reactants and products). Balance the equation.	(see sub-table below)
Step 4: Rewrite the equation ensuring it is balanced. Include the states.	_____ → _____

Element	Reactants	Products
C		
H		
O		

7. **MC** Which of the following are products of the reaction between silver nitrate and sodium chloride?
   A. Silver nitrate and sodium chloride
   B. Nitrogen chloride and silver sodium
   C. They do not react so there will be no products.
   D. Silver chloride and sodium nitrate

Apply and analyse

8. The two reactants in the chemical reaction taking place in the test tube shown are aqueous solutions. Identify and describe the type of chemical reaction taking place.

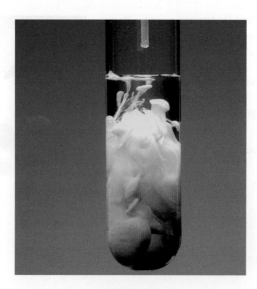

9. **SIS** Many chemicals are classified as dangerous goods and/or hazardous substances. Explain these two terms and what can be done to reduce accidents while undertaking experiments in a lab.

10. Write a balanced equation with states for the following reactions.
   a. Aluminium metal + oxygen gas → solid aluminium oxide
   b. Solid carbon + oxygen gas → carbon dioxide gas
   c. Copper(II) sulfate solution + sodium hydroxide solution → solid copper(II) hydroxide + sodium sulfate solution
11. Identify the reaction type for each of the reactions listed in the following table.

TABLE Different types of chemical reactions

	Reaction	Reaction type
A.	Aluminium metal + oxygen gas → solid aluminium oxide	
B.	Potassium metal + oxygen gas → solid potassium oxide	
C.	Solid copper carbonate → solid copper oxide + carbon dioxide gas	
D.	Iron metal + sulfur powder (S_8) → solid iron sulfide (FeS_2)	
E.	Copper sulfate solution + zinc metal → copper metal + zinc sulfate solution	
F.	Copper(II) sulfate solution + sodium hydroxide solution → solid copper(II) hydroxide + sodium sulfate solution	
G.	Solid magnesium hydroxide + hydrochloric acid → magnesium chloride + water	

12. Identify which of the reactions listed in the following table are redox reactions. Justify your response.

TABLE Examples of redox reactions

	Reactions	Are they redox reactions?
A.	Aluminium metal + oxygen gas → solid aluminium oxide	
B.	Potassium metal + oxygen gas → solid potassium oxide	
C.	Solid carbon + oxygen gas → carbon dioxide gas	
D.	Solid copper carbonate → solid copper oxide + carbon dioxide gas	
E.	Iron metal + sulfur powder (S_8) → solid iron sulfide (FeS_2)	
F.	Copper sulfate solution + zinc metal → copper metal + zinc sulfate solution	
G.	Copper(II) sulfate solution + sodium hydroxide solution → solid copper(II) hydroxide + sodium sulfate solution	
H.	Solid magnesium hydroxide + hydrochloric acid → magnesium chloride + water	

13. **SIS** a. During a reaction of dilute sulfuric acid and solid sodium carbonate the rate of reaction decreases as the reaction proceeds. Use collision theory to explain why this occurs.
   b. Write an equation with states for this reaction.
   c. Describe two methods of measuring the rate of this reaction over time.
   d. Describe two ways of increasing the rate of this reaction.
14. One of the chemical reactions used during the production of sulfuric acid makes use of the catalyst vanadium oxide (V_2O_5). The chemical equation for this reaction is:
$$2H_2 + O_2 \rightarrow 2H_2O$$
Why doesn't V_2O_5 appear as one of the reactants?

15. Predict the salts that would result from the neutralisation reaction between the following.
    a. Magnesium oxide and hydrochloric acid
    b. Copper(II) oxide and sulfuric acid
    c. Sodium hydroxide and acetic acid
    d. Sodium oxide and nitric acid

Evaluate and create

16. **SIS** In an experiment to test the effect of the amount of liver on the breakdown of hydrogen peroxide, the following results were obtained.

Mass of liver (g)	Volume of oxygen released (cm^3)
0.5	2.5
1.0	5.1
2.0	9.8
2.5	11.5

   a. **MC** Which of the following is the correct word equation for the reaction occurring in the experiment?
      A. Water + oxygen → hydrogen peroxide
      B. Hydrogen peroxide → water + oxygen
      C. Hydrogen peroxide → water and carbon dioxide
      D. Water + carbon dioxide → hydrogen peroxide
   b. Graph the results of the experiment.
   c. What does the graph show about the effect of the liver on the rate of this reaction?
   d. Why does the liver affect this reaction?
17. Sophie is suffering from indigestion. Explain whether it is advisable to give her some orange juice to help her feel better.
18. Collision theory states that for a reaction to occur the reactant particles must collide.
    a. Describe two reasons why a collision may not result in a reaction.
    b. What is activation energy and what is its significance in relation to chemical reactions?
19. Explain how temperature, surface area and concentration could affect the rate of reactions. Summarise an experiment that would allow you to show this.
20. **SIS** You are asked to prepare solid zinc sulfate using zinc carbonate and any other chemicals that you require.
    a. State the chemical(s) that you require.
    b. Name the type of reaction.
    c. Write a balanced equation.
    d. Identify any hazards and safety measures that are necessary.
    e. List the materials and method for the experiment.

Fully worked solutions and sample responses are available in your digital formats.

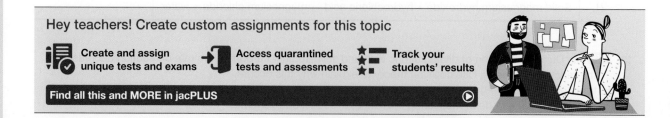

Hey teachers! Create custom assignments for this topic

Create and assign unique tests and exams

Access quarantined tests and assessments

Track your students' results

Find all this and MORE in jacPLUS

Online Resources

 **on** Resources

Below is a full list of **rich resources** available online for this topic. These resources are designed to bring ideas to life, to promote deep and lasting learning and to support the different learning needs of each individual.

5.1 Overview

 eWorkbooks
- Topic 5 eWorkbook (ewbk-13122)
- Starter activity (ewbk-13124)
- Student learning matrix (ewbk-13126)

Solutions
- Topic 5 Solutions (sol-1128)

Practical investigation eLogbook
- Topic 5 Practical investigation eLogbook (elog-2456)

Video eLesson
- Reaction of magnesium ribbons in acid (eles-2571)

5.2 The language of chemistry

 eWorkbooks
- Chemical equations (ewbk-13127)
- Balancing chemical equations (ewbk-13129)
- A world of reactions (ewbk-13131)

Interactivity
- Balancing chemical equations (int-0677)

5.3 Safety with chemicals

eWorkbook
- Understanding chemical hazards (ewbk-13133)

Video eLessons
- Saving acid wetlands (eles-1072)

Interactivities
- Hazardous substances (int-5891)
- The components of a risk assessment (int-8099)

 Weblinks
- RiskAssess for Schools
- WorkSafe

5.4 Synthesis reactions

eWorkbook
- Synthesis reactions (ewbk-13163)

Video eLessons
- Magnesium burning (eles-2716)

5.5 Decomposition reactions

eWorkbook
- Decomposition (ewbk-13135)

Practical investigation eLogbook
- Investigation 5.1: Decomposing zinc carbonate (elog-2458)

5.6 Single displacement reactions

 Video eLessons
- Displacement reaction between copper wire and silver (eles-2714)
- Reaction between zinc powder and concentrated hydrochloric acid (eles-2589)
- Sodium metal reacts with liquid (eles-2717)

Practical investigation eLogbook
- Investigation 5.2 Reaction of acids with metals (elog-2460)

5.7 Double displacement — neutralisation

eWorkbooks
- Acids and bases (ewbk-13137)
- Indicators and neutralisation (ewbk-13139)
- Neutralisation and salts (ewbk-13141)

Practical investigation eLogbook
- Investigation 5.3: Exploring neutralisation reactions (elog-2462)

Video eLesson
- Universal indicator solution (eles-2304)

Interactivity
- pH rainbow (int-0101)

Teacher-led video
- Investigation 5.3: Exploring neutralisation reactions (tlvd-10817)

5.8 Double displacement — precipitation

eWorkbook
- Precipitation (ewbk-13143)

Practical investigation eLogbook
- Investigation 5.4: Will it precipitate? (elog-2464)

Video eLessons
- Precipitation (eles-2058)
- Priestley and the Law of Conservation of Mass (eles-1767)

5.9 Redox reactions

eWorkbook
- Redox reactions (ewbk-13145)

5.10 Rates of reaction

Practical investigation eLogbook
- Investigation 5.5: A liver catalyst (elog-2466)

Video eLesson
- A platinum disc catalyst (eles-2719)

Interactivity
- Reaction rates (int-0230)

5.11 Making useful products

eWorkbook
- Making useful products (ewbk-13147)

Video eLessons
- Road worker painting road markings (eles-4174)
- A 3-D printer (eles-2720)

Interactivities
- Natural polymers (int-5747)
- Making polymers (int-3849)

5.13 Project — Flavour fountain

 ProjectsPLUS
- Flavour fountain (pro-0114)

5.14 Review

eWorkbooks
- Topic review Level 1 (ewbk-13149)
- Topic review Level 2 (ewbk-13151)
- Topic review Level 3 (ewbk-13153)
- Study checklist (ewbk-13155)
- Literacy builder (ewbk-13156)
- Crossword (ewbk-13158)
- Word search (ewbk-13160)
- Reflection (ewbk-13162)

Digital document
- Key terms glossary (doc-40483)

To access these online resources, log on to **www.jacplus.com.au**

6 The mysterious Universe

CONTENT DESCRIPTION

Describe how the big bang theory models the origin and evolution of the universe and analyse the supporting evidence for the theory (AC9S10U03)

Source: F–10 Australian Curriculum 9.0 (2024–2029) extracts © Australian Curriculum, Assessment and Reporting Authority; reproduced by permission.

LESSON SEQUENCE

SCIENCE INQUIRY AND INVESTIGATIONS

Science inquiry is a central component of Science curriculum. Investigations, supported by a **Practical investigation eLogbook** and **teacher-led videos**, are included in this topic to provide opportunities to build Science inquiry skills through undertaking investigations and communicating findings.

LESSON
6.1 Overview

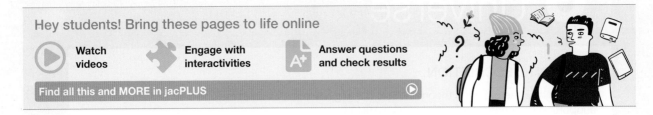
6.1.1 Introduction

On any cloudless night, a pattern of stars, galaxies and clouds of gas appears to glide across the sky above our heads. Yet against this backdrop, changes are taking place — often hard to see and sometimes spectacular, but always raising questions in our minds about the past and the future. How and when did the Universe all begin?

Until almost 400 years ago, most astronomers believed that Earth was at the centre of the Universe and was surrounded by a 'celestial sphere' on which the stars were attached. The Moon, the Sun and planets all were believed to orbit Earth. Over time, the idea that the Sun was the centre of the Universe became accepted. We now know that Earth is just a tiny part of the Solar System, which is a tiny speck in a galaxy known as the Milky Way. The Sun is one of up to 400 billion stars in the Milky Way, and the Milky Way galaxy is one of more than 100 billion galaxies in the Universe.

FIGURE 6.1 A beautiful clear night sky with clouds of gas and dense stars

on Resources

▶ **Video eLesson** Hubble and the expansion of the Universe (eles-1766)

Watch this video to learn about the great discovery of billions of galaxies outside our galaxy by Edwin Hubble. Hubble used the largest telescope of his time, the 250 centimetre Hooker telescope on Mount Wilson in California, to resolve one of the great debates in early twentieth century astronomy.

6.1.2 Think about the Universe

1. Where are stars formed?
2. Why do stars appear to be different colours?
3. How old is the Universe?
4. How do stars change across their lifetime?
5. What can we actually see from space?
6. The Universe started with a 'big bang', but how will it end?

6.1.3 Science inquiry

Our knowledge of the Universe

Throughout human history, we have looked up at the sky and asked questions about what we saw. What were the bright lights in the sky? What were they made of? How big are those objects? How far away were they? Are other objects above Earth that we can't see?

As technology has evolved, the type of information available to us has changed and increased. We still, however, struggle to answer some of these questions and find our place in the Universe.

To understand how our knowledge of the Universe has evolved:
- Choose an ancient civilisation and research their theories of the Universe.
- Consider the technologies available and the information each civilisation had available to them.
- Critique their theories of the Universe by considering whether or not the theory explained all the available evidence and presented a coherent explanation of astronomical events.

Use the following questions to guide your research:
- What technology was available to observe the sky?
- What types of measurements could be made? Is this data qualitative or quantitative?
- Where were observations made from? Which part of Earth? Multiple locations or just one?
- Where and how were observations recorded? Were they available for future generations as primary or secondary sources?
- How does the theory account for each observation? Does the theory make sense in the context of everyday life for these people?

You can research any ancient civilisation for this task, but you may like to consider the traditions of First Nations Australians, the Incas of Peru, the ancient Egyptians, ancient Chinese societies or the ancient Greeks.

on Resources

 eWorkbooks
Topic 6 eWorkbook (ewbk-12156)
Starter activity (ewbk-12159)
Student learning matrix (ewbk-12158)

 Solutions
Topic 6 Solutions (sol-1129)

 Practical investigation eLogbook Topic 6 Practical investigation eLogbook (elog-2224)

LESSON
6.2 What can we observe in the changing night sky?

LEARNING INTENTION

At the end of this lesson you will be able to describe the relationship between the movement of Earth and the changing position of stars and constellations in the sky and how centuries of observations of the night sky have changed our ideas of the structure of the Universe.

6.2.1 Looking upwards

When you look up into the sky on a clear night, you will see countless specks of light stretching from horizon to horizon. Have you seen the fixed patterns move to a different position in the sky over time? It is as if these patterns, or **constellations**, of stars appear to be dotted on a crystal sphere that is rotating around Earth. Have you noticed that the constellations appearance coincides with the seasons? Or that some stars' brightness subtly change over days and years? This would take some considerable observations! Western science has known about these variable stars for about 200 years, but the oral traditions of First Nations Australians' astronomers going back thousands of years describe the relative dimming of different stars, indicating that they had already made these observations.

It is surprising how much we now understand about the Universe by looking upwards towards the heavens with curiosity and wonderment.

ACTIVITY: First Nations Australians' astronomy

1. Research the stories of the Ancestor's Camp and the Fishing Line constellation. This was one of the first recorded reports of the Magellanic Clouds, which are dwarf satellite galaxies orbiting the Milky Way. What are the stories? How do they compare to the Western stories of the night sky?
2. Multiple cultures around the world saw Orion as a great hunter. One theory is that this story originated before the continents broke apart and each culture took this story with them. The Yolngu peoples of north-east Arnhem land in the Northern Territory refer to Orion as Djulpan, a King Fish. Research the story and explain what the three stars of Orion's Belt relate to.

6.2.2 Constellations

Ancient astronomers grouped stars according to the shapes they seemed to form. The shapes were usually of gods, animals or familiar objects. The most well-known constellations are the twelve groups we know as the signs of the zodiac. These constellations follow the **ecliptic** and their names include Taurus (the bull), Leo (the lion) and Sagittarius (the archer).

The Southern Cross (Crux Australia) is easily visible just after sunset in the Southern Hemisphere. Just below it is the Emu in the Sky (figure 6.2). The dark patches of the Milky Way — the space between the stars — forms the emu. The dark patch next to the Southern Cross forms the head of the emu. This patch is the Coalsack Nebulae. The legs of the emu extend into the constellation of Sagittarius. First Nations Australians use the stars to help identify seasonal change. For example, the shape that Emu in the Sky takes in the stars reflects what is happening on earth. It indicates when it's time to gather emu eggs or to leave the emus alone because they are breeding and laying.

Today, astronomers recognise 88 constellations. When observed from Earth, the stars in each constellation appear to be very close to each other. But the stars that make up constellations can be located at very different distances from Earth. For example, the star Betelgeuse in the Orion constellation is approximately 650 light-years from Earth, whereas the star Bellatrix in the same constellation is about 240 light-years from us.

constellations groups of stars that were given a particular name because of the shape the stars seem to form in the sky when viewed from Earth

ecliptic the path that the Sun traces in the sky during the year

FIGURE 6.2 The great Emu in the Sky.

FIGURE 6.3 a. Sagittarius, one of the constellations of the zodiac, is based on a centaur (half-man half-horse). **b.** Can you identify which parts of the constellation represents the centaur and the bow and arrow?

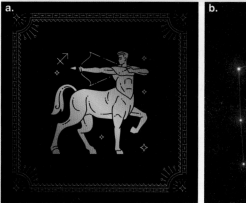

6.2.3 Constellations 'on the move'

We know now that the celestial sphere proposed by the Greek astronomer Ptolemy in 150 CE was wrong (figure 6.4). The apparent motion of the fixed pattern of stars at night, shown in time-lapse photography, is due to the rotation of Earth.

The apparent change in position of the constellations is due to Earth's orbit around the Sun. Sky charts, sometimes called sky maps, star maps or star charts, show the position of constellations, stars and the planets from different locations for each month of the year.

FIGURE 6.4 In 150 CE, the Greek astronomer Ptolemy suggested that the stars were attached to a 'celestial sphere' that rotated above our heads.

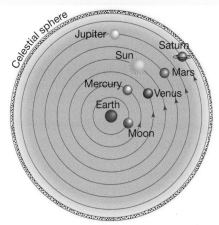

6.2.4 Looking more carefully

The development of the telescope in the sixteenth century allowed Earth-bound astronomers to see objects in the sky with much greater precision than ever before. Observations using telescopes showed that many different types of objects in the sky could be identified. These included single or double stars, groups of stars called **galaxies**, clusters of galaxies, and clouds of gas and dust called **nebulae**.

In 1718, English astronomer Edmond Halley, who is well known for identifying the comet named after him, used his telescope to check three particularly bright stars: Sirius, Procyon and Arcturus. He found that the position of each one relative to surrounding stars was noticeably different from the positions recorded by ancient Greek astronomers centuries before. There were even slight differences between Halley's observations and those of Danish astronomer Tycho Brahe about 150 years earlier. Never again could the stars be described as 'fixed in the heavens'.

FIGURE 6.5 The Horsehead Nebula in the constellation of Orion. A nebula is a cloud of dust and gas, visible as a glowing or dark shape in the sky against a background of stars.

galaxies very large groups of stars and dust held together by gravity

nebulae clouds of dust and gas that may be pulled together by gravity and heat up to form a star

---

ACTIVITY: Twinkling stars

Fill a glass dish with water. Take a sheet of aluminium foil large enough to cover the base of your dish. Crinkle this into a ball and open it out again. Put this crinkled foil under the glass dish.

Darken the room and shine a line down at an angle into the dish. Stir the water and observe the reflected image.

What observations did you make?

DISCUSSION

The people of Mabuiag Island in the Torres Strait used stellar twinkling as a sign of heavy winds, temperature changes and approaching rain. The Wardaman people of the Northern Territory use the twinkling of these stars to predict the approach of the wet season.

Research why stars twinkle due to atmospheric changes that come with the wet season and rain. Could these observations have allowed the First Peoples of Australia to make these predictions reliably?

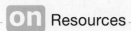 Resources

 Weblinks Star maps
The night sky time-lapse

6.2 Activities

6.2 Quick quiz on	6.2 Exercise

Select your pathway

■ LEVEL 1	■ LEVEL 2	■ LEVEL 3
1, 4, 8, 10	2, 3, 5, 6	7, 9, 11

These questions are even better in jacPLUS!
- Receive immediate feedback
- Access sample responses
- Track results and progress

Find all this and MORE in jacPLUS ▶

Remember and understand

1. **MC** Why do stars appear to rotate during the night?
   A. The Sun is rotating.
   B. Earth is rotating.
   C. The stars are rotating.
   D. The Moon is rotating.
2. **MC** Why do stars gradually change their position in the night sky throughout the year?
   A. Earth's position in its orbit around the Sun changes.
   B. Stars are constantly orbiting the Sun.
   C. Stars are getting closer to Earth.
   D. Stars are getting further away from Earth.
3. **MC** What do the constellations of the zodiac have in common?
   A. They are all named after ancient astronomers.
   B. They lie along the ecliptic.
   C. The stars in the constellations of the zodiac are very close to Earth.
   D. All of the above.
4. Explain constellations and give two examples of these.
5. Explain how the invention of the telescope changed our view of the night sky from Earth.

Apply and analyse

6. Explain why early civilisations found observations about the night sky useful.
7. Explain why the constellations have changed since the earliest descriptions, and why this changed our ideas about the Universe.
8. Explain why it is easier to observe the night sky in rural areas than in the city.
9. All of the stars in the constellation Orion appear to be the same distance from Earth. Explain how observers on Earth know that they are not.

Evaluate and create

10. Create a story board of the creation stories of two civilisations. Note the similarities. Explain why they could have these similarities.
11. **SIS** The ancient Greeks observed the wandering stars. The First Peoples of Australia observed these too. They describe the celestial road that the ancestor spirits walk across and would occasionally turn backwards to chat. How are both these descriptions similar and what is the current explanation for what was observed?

Fully worked solutions and sample responses are available in your digital formats.

LESSON
6.3 How can we measure cosmic distances?

LEARNING INTENTION

At the end of this lesson you will be able to describe the major components of the Universe and use the astronomical units (AU, parsec and light-year) including the scientific notation for large distances.

If you were to look out at the night sky you would observe that some stars are brighter than others. And you could observe that the position of some stars changes during the night. Is it simply that some stars are closer than others and so appear brighter and are moving faster? Are all the stars the same size? Objects further away appear smaller so to judge the size of the objects you do need to know how distant they are.

6.3.1 Cosmic distances

To get an idea of how big the Universe is, the first measurement needed is the distance from Earth to the Sun which is called the **astronomical unit** (AU). Halley devised the method to measure this: by knowing the distance between two observers on Earth and by timing the transit of Venus as the planet moves across the face of the Sun, the distance can be calculated using geometry. This relies upon the principle of **parallax**, the apparent movement of an object due to actual movement of the observer. Recording the transit of Venus was the main purpose of Captain James Cook's voyage to Tahiti in June 1769. The astronomical unit (AU) was calculated to be 150 million kilometres.

The distance to stars outside our Solar System can also be determined by using parallax. This led to the next astronomical unit, the **parsec**. This is very useful to astronomers as it is calculated directly from observed angles. One parsec, by definition, is the distance to an object whose parallax angle is one arcsecond ($\frac{1}{3600}$ of a degree) (figure 6.6). The radius of Earth's orbit equals one astronomical unit, so an object that is one parsec distant is 206 265 AU.

FIGURE 6.6 Relationship between the AU and a parsec

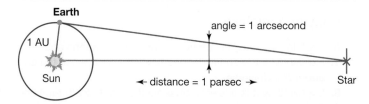

In 1838 Friedrich Bessel used a more sophisticated telescope to observe a shift in position of stars due to parallax, as Earth moved around the Sun. This was the first piece of evidence that the Sun not the centre of the Universe. In the early 1900s it was thought that there was only one galaxy in the Universe with the Sun effectively at the centre of the galaxy. We now know that the Sun is in a small galaxy which is just one in 2 trillion galaxies.

In 1905 Einstein's theory of special relativity posited that light always travels at the same speed. Once the speed of light was accepted as a constant, the distance light travels in one year (a light-year) was accepted as another way of measuring cosmic distances. A light year is 0.306 601 parsecs or 63 241 AU.

astronomical unit unit of length equal to the mean distance between Earth and the Sun, defined as approximately 150 million kilometres

parallax apparent movement of close stars against the background of distant stars when viewed from different positions around Earth's orbit

parsec unit for expressing distances to stars and galaxies, it represents the distance at which the radius of Earth's orbit subtends an angle of one second of arc

What is a light-year?

Distances in space are significantly larger than distances on Earth and, therefore, we need new units to measure distance. A **light-year** is not a measure of time! It's a measure of distance. In one year, light travels a distance of 9 500 000 000 000 or 9.5×10^{12} kilometres. This distance is called a light-year.

Very large numbers are often written using **scientific notation**. This allows us to avoid writing lots of zeros and also makes the numbers easier to read, because the reader does not have to count the zeros. For example, the distance between Earth and the Sun averages 150 million kilometres. This could be written as 150 000 000 km or, in scientific notation, as 1.5×10^8 km.

One billion is equal to one thousand million; that is, 1 000 000 000 or 10^9.

Some other examples are:
- $45\,000\,000\,000 = 4.5 \times 10^{10}$
- $700\,000\,000\,000\,000\,000 = 7.0 \times 10^{17}$.

elog-2227

INVESTIGATION 6.1

The effect of parallax

Aim

To observe the effect of parallax

Materials

- a number of traffic cones (witch's hats)
- pencil and paper

Method

1. Mark a circle on the largest open area in your school to represent Earth's orbit around the Sun.
2. Place a series of traffic cones at different distances from the circle to represent stars nearby and far away.
3. Take a walk around Earth's 'orbit' and, at several different points, sketch the appearance of the 'stars' relative to one another and to even more distant objects such as trees and fence posts.

Results

Your results are your sketches of the 'stars' appearance relative to one another.

Discussion

1. Looking at your sketches, did the positions of the stars relative to one another appear to change as you moved around the orbit?
2. Can you see any difference between the relative movements of the nearby stars and those of the more distant stars?

Conclusion

Summarise the findings of this investigation about the effects of parallax.

 Resources

 eWorkbook Observing stars (ewbk-12183)

Video eLessons Twinkle, twinkle (eles-0071)

Star trails (eles-2724)

light-year the distance travelled by light in one year
scientific notation a way of writing very large or small numbers, using power notation; e.g. 1.5×10^8 km

6.3.2 Scaling the cosmic distance ladder

How do you measure the vast distances across the cosmos? The first challenge was to measure the distance of Earth to the Sun (AU: astronomical unit). The international effort to measure the transit of Venus across the Sun accurately determined it to be 150 million kilometres, but this is small in terms of the Universe. How can you measure the size of the Universe?

SCIENCE AS A HUMAN ENDEAVOUR: Scaling the cosmic distance ladder

In the early 1900s the optical observatories at Mount Wilson in California, along with the discovery of photography, allowed a group of female astronomers, who became known as the 'Harvard computers', to make significant discoveries in astronomy. Cecilia Payne worked out the chemical composition of the Sun. Annie Jump Cannon helped classify the stars. Antonia Maury published the first catalogue of stellar spectra and discovered the evidence of binary stars. The work of Harlow Shapley and Henrietta Swan Leavitt in 1918 displaced the Sun to the outer regions of the galaxy. And Leavitt discovered a way to measure the Universe, which allowed astronomers to measure distances up to 20 million light-years and allowed Hubble and others to prove that the Milky Way was just one galaxy of many.

FIGURE 6.7 Mt Wilson observatory in California, USA.

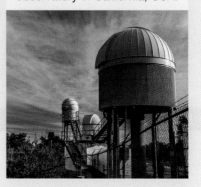

ACTIVITY: Cosmic distance ladder

The challenge to measure the distances to the stars relied upon advances in technology. Observatories, including the one at Mount Wilson, as well as work by individuals such as Henrietta Leavitt, led to a way to measure the Universe: the development of a cosmic distance ladder. Each rung of the ladder made the Universe larger and larger, and was the result of international scientific projects and sometimes international rivalry.

Research the different methods to determine the size of the Universe: stellar parallax, standard candles (Cepheid variable, type 1a supernova) and red shift.
1. What distances does each method reach to?
2. What technology was relied on to undertake the method?
3. What was the main scientific breakthrough that was the result of each development?

6.3 Activities

learn[on]

| 6.3 Quick quiz [on] | 6.3 Exercise |

These questions are even better in jacPLUS!
- Receive immediate feedback
- Access sample responses
- Track results and progress

Find all this and MORE in jacPLUS ⊙

Learning pathways

| ■ LEVEL 1 | ■ LEVEL 2 | ■ LEVEL 3 |
| 1, 2, 5 | 3, 4, 6, 9 | 7, 8, 10 |

Remember and understand

1. **MC** Which of the following has the astronomical units in order from smallest to largest?
   A. Astronomical unit, light-year, parsec
   B. Astronomical unit, parsec, light-year
   C. Parsec, light-year, astronomical unit
   D. Light-year, parsec, astronomical unit
2. **MC** What is the definition of the AU?
   A. the mean distance Earth–Moon
   B. the mean distance Earth–Sun
   C. the diameter of Earth's orbit
   D. the radius of the Sun's orbit

3. Define a parsec and describe two significant times where it was used.
4. **MC** The closest star to our Sun, Proxima Centauri, is 40 208 000 000 000 km away. This is closest to
   A. 2.7×10^5 AU
   B. 2.7×10^4 AU
   C. 4.0×10^{10} km
   D. 4.25 light-years
5. **MC** The average distance between Mars and the Sun is 228 million kilometers. This is closest to
   A. 1.52 AU
   B. 15.2 AU
   C. 152 AU
   D. 0.152 AU

Apply and analyse

6. Explain what we mean by the term *parallax*.
7. Explain how observations of a stellar parallax effect change our ideas about the Universe.
8. Choose a constellation such as Orion and, using the known distances to each star, explain why the constellation shape is only 'true' in one direction.

Evaluate and create

9. The estimated distances from Earth to some stars and galaxies are listed in the following table. How long would it take to reach each of them, travelling at the speed of light (about 300 000 km/s)?

TABLE The estimated distances from Earth to some stars and galaxies

Name of star/galaxy	Description	Distance from Earth
Sun	Our own star	1.5×10^8 km
Proxima Centauri	The closest star after the Sun	4.0×10^{13} km
Centre of Milky Way	Our own galaxy	2.5×10^{17} km
Magellanic Clouds	One of the closest galaxies	1.5×10^{18} km
Andromeda galaxy	One of the closest galaxies	1.4×10^{19} km
Quasars	Very distant objects	1.4×10^{23} km

10. **SIS** The following figures show two photos of the night sky. Both photos were taken at midnight from the same place in rural Victoria. The left image was taken in January and the right image was taken in July.
    a. Describe the differences between the two images. Are the same stars and constellations visible in both images?
    b. Two stars have been labelled A and B in each photo. Which star is closer to Earth? Explain your answer.

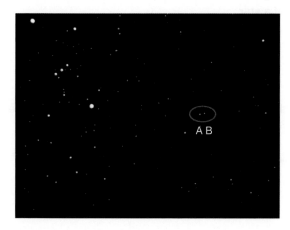 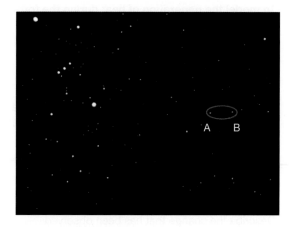

Fully worked solutions and sample responses are available in your digital formats.

LESSON
6.4 What does starlight tell us?

LEARNING INTENTION

At the end of this lesson you will be able to describe how stars' light spectra and brightness is used to identify what the stars are made of, their movements and their distances from Earth.

If you think about how far away the stars are, it is surprising that we know so much about what they are made of, how far away they are and even how they live and die.

So what does the light or spectra of a star tell us?

6.4.1 Star shine

Stars shine because of the reactions that are occurring inside them. Every star is born of a nebula made of gas and dust. The overall mass and the elements that make it up determine what type of star will result. Nebulae are really like star nurseries (figure 6.8). The Great Orion Nebula is large enough to be seen with the naked eye.

In dense areas of dust and gas, gravity takes hold and the gas and dust begin to collapse, forming a cloud. The collapse continues under the influence of gravity, forming visible globules in the nebula cloud. As the globules collapse further, the formation of any original gas cloud is accelerated. The now dense cloud is known as a protostar, where the temperature is not high enough for nuclear fusion to occur. At the same time, the increasing pressure causes the temperature to rise. This effect is modelled in investigation 6.2.

FIGURE 6.8 New stars are born in giant, dense clouds of hydrogen gas, like the 'Pillars of Creation' in the Eagle Nebula.

elog-2229

INVESTIGATION 6.2

Heat produced by compressing a gas

Aim

To model the generation of heat during the formation of a star

Materials

- a bicycle pump
- a tyre with inner tube

Method

1. Using an energetic pumping action, inflate a tyre with the bicycle pump. Alternatively, just pump the bicycle pump with your finger partially covering the open end so the air does not escape.
2. Now feel the body of the pump.

Results

Make notes about what you observe.

Discussion

Describe the change that has been observed.

Conclusion

How does an increase in air pressure affect the temperature of the surroundings?

Nuclear fusion

The temperature and pressure of the gases in a protostar eventually rise high enough for atomic nuclei to become joined together by a process called **nuclear fusion**. As a result of fusion, two isotopes of hydrogen, deuterium (hydrogen-2) and tritium (hydrogen-3), combine to form helium nuclei, neutrons and vast amounts of energy. This fusion reaction is the source of energy in suns. While our Sun burns hydrogen to make helium, larger suns fuse larger atomic nuclei forming heavier elements. These larger suns are hotter as the gravity force is greater, increasing the pressure within the sun.

FIGURE 6.9 This nuclear fusion reaction in stars releases vast amounts of energy.

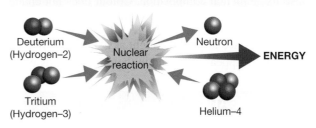

Deuterium (Hydrogen–2)
Tritium (Hydrogen–3)
Nuclear reaction
Neutron
Helium–4
ENERGY

6.4.2 Star brightness

You must have noticed that a torch is brighter when it is closer to you. This is true for stars as well. How bright a star appears to us (its **apparent magnitude**) depends on its actual brightness (its **absolute magnitude**) and its distance from Earth. A dim star close to us may appear brighter than a really bright star a long way away. To calculate the absolute magnitude of a star, astronomers must know how far away it is. The colour of a star depends on its surface temperature: red stars are cool, and white and blue stars are hot.

FIGURE 6.10 The Great Orion Nebula

How bright a star or planet appears as viewed from Earth is measured on a scale of apparent magnitude. This scale currently extends from approximately –30 (the brightest objects in the sky) to around +30 (the dimmest objects in the sky). On this scale, brighter objects have the lowest apparent magnitudes. For example, the Sun has an apparent magnitude of –27. A full Moon has an apparent magnitude of approximately –13. The brightest stars have apparent magnitudes between –1 and 1. The weakest objects visible with the naked eye have an apparent magnitude of approximately 6.

The absolute magnitude is a measure of how much light an object emits. The Sun is much smaller than Rigel in Orion and it emits a lot less light. However, it appears brighter to us because it is much closer than Rigel. The Moon emits no light of its own.

nuclear fusion joining together of the nuclei of lighter elements to form another element, with the release of energy

apparent magnitude brightness of a star as seen from Earth

absolute magnitude actual brightness of a star

Table 6.1 shows some typical values of apparent and absolute magnitudes.

TABLE 6.1 Apparent and absolute magnitudes of stars and constellations

Star and constellation	Apparent magnitude	Absolute magnitude
Sun	–27	+4.7
Sirius (Canis Major)	–1.5	+8.7
Canopus (Carina)	–0.73	–4.6
Alpha Centauri (Centaurus)	–0.33	+4.7
Rigel (Orion)	+0.11	–7.5
Beta Centauri (Centaurus)	+0.60	–5.0
Betelgeuse (Orion)	+0.80	–5.0
Aldebaran (Taurus)	+0.85	–0.3
Alpha Crucis (Southern Cross)	+0.90	–3.9

6.4.3 Star colour

A star's colour forms the basis for classifying stars. From a star's colour we can determine its temperature and also its elemental composition. Colour does not change with distance but with temperature. As stars heat up the colour goes from a red glow to bright blue. Most stars are either red or orange.

A quick glance around the night sky shows us that stars differ quite noticeably from one another, both in how bright they appear to us and in their colour. This is modeled in investigation 6.3.

elog-2231

INVESTIGATION 6.3

Seeing the colours of stars

Aim

To investigate the relationship between star colour and brightness

Background information

The constellation Orion is visible from every inhabited place on Earth. It is most easily recognised from the line of three stars that represent the hunter's belt. Remember, the constellations were named by observers in the Northern Hemisphere, so to Southern Hemisphere observers the constellations appear upside down. This is why Orion's sword points upwards from the belt when viewed from Australia. This group of stars making up the sword and the belt is often known as the Saucepan.

Orion's sword, pointing upwards from the belt, contains a misty patch visible to the naked eye. This is the Orion Nebula, labelled M42 by the astronomer Messier, who prepared a catalogue of such objects in an attempt to prevent observers being distracted by them. Through binoculars, stars can be seen embedded in the gas and dust of the Orion Nebula, and new stars have been seen as they begin to emit light. The Orion Nebula and other similar formations are the birthplaces of the stars.

Materials

- star atlas (optional)
- pair of binoculars (optional)

Method

1. Use the background information provided, a star atlas or an astronomy computer program to help you to find the constellation Orion (the Hunter). Alternatively, find a colour photograph of the constellation Orion. The star α-Orionis (also known as Betelgeuse) is a red giant that has a diameter bigger than Earth's orbit. It appears quite visibly red to the naked eye and this distinctive colour shows up even more clearly through binoculars. The star β-Orionis (also known as Rigel) is 60 000 times as bright as the Sun.
2. Compare the brightness and colours of Betelgeuse and Rigel.

Results

Construct a table to record your results.

Discussion

1. Explain how the colours of Betelgeuse and Rigel compare.
2. Which of Betelgeuse and Rigel appears to be brighter? Relate your observation to table 6.1.

Conclusion

Summarise the findings of this investigation about the relationship between star colour and brightness.

6.4.4 Star spectra

To classify the stars the Harvard Computers team looked at the spectrum of light and photographs of the stars. When the **spectrum** of light from a star is analysed, some dark lines are observed (see figure 6.11). These correspond to colours of light that have been absorbed by substances in the star. Different substances absorb different colours of light. By identifying the **wavelengths** of the colours missing from the spectrum, astronomers can find out which elements are present in the star.

spectrum light from a source separated out into the sequence of colours, showing the different frequencies

wavelengths invisible radiation similar to light but with a slightly higher frequency and more energy

FIGURE 6.11 The spectrum of white light from a nearby star. The black lines show which colours have been absorbed by elements in the star. The numbers indicate the wavelength of the light in nanometres.

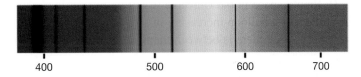

In many cases, missing colours in the spectra of stars are shifted from their expected positions. A shift to lower or 'redder' frequencies is called a **red shift** and results from a star's movement away from Earth. A shift to higher or 'bluer' frequencies is called a **blue shift** and is caused by a star's movement towards Earth.

FIGURE 6.12 The spectrum of an object moving towards you (blue shift), or away from you (red shift).

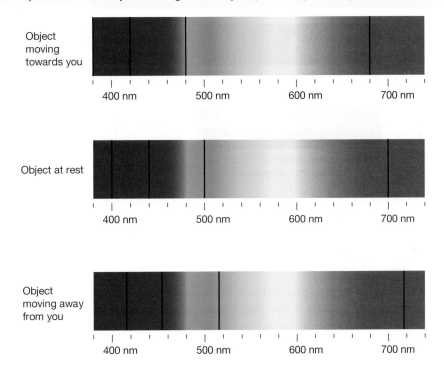

Nearby objects show a range of shifts. Some stars, like Sirius (the Dog star), are moving away from us and others are moving slowly towards us. Some even show alternate red and blue shifts in step with changes in brightness, suggesting that these stars have an invisible dark companion (star) orbiting them. The brightness is reduced as the circling star passes between us and the main star, while the shift is caused by the main star moving in response to the gravitational pull of its dark companion.

6.4.5 Retreating galaxies

A relatively small number of galaxies, including the nearby Andromeda galaxy, are moving towards Earth, but the majority of galaxies are moving away from us at a considerable speed. Even more extraordinary is the relationship between the size of the red shift and the distance from Earth. This was first investigated by the astronomer Edwin Hubble and is now referred to as Hubble's law. The law was proposed in 1929 and it states that the further a galaxy is, the greater its red shift and so the faster it is moving away from us.

While this finding appears to put Earth in a very special position at the centre of a rapidly expanding Universe, it is in fact an illusion. Observers anywhere in the Universe will see the surrounding galaxies moving away from them at a speed that is consistent with Hubble's law.

red shift shift of lines of a spectral pattern towards the red end when the source is moving away from the observer

blue shift shift of lines of a spectral pattern towards the blue end when a light source is moving rapidly towards the observer

6.4.6 The expanding Universe

Scientists explain the movement of most objects away from others in the Universe by using the expansion of space (see figure 6.13). Empty space in the Universe is getting bigger, increasing the distance between objects. This is similar to the surface of a balloon increasing in size as it is being blown up. The further apart two objects are, the more empty space there is between them, so the faster they are moving apart.

FIGURE 6.13 Expansion of space explains red shift. Any light waves passing through the expanding space increase their wavelengths and become more 'red'.

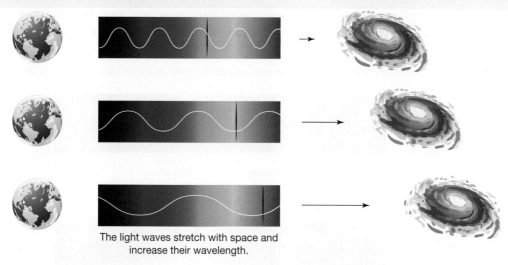

The light waves stretch with space and increase their wavelength.

This expansion can be prevented by the force of gravity. The gravitational pull of Earth prevents it getting larger. Similarly, the gravitational pull of the Sun prevents the Solar System from getting larger. At the edge of the Solar System, our galaxy's gravitational forces prevent the galaxy from getting larger. The expansion can only happen in outer space, far away from objects.

Light from stars travels through empty space to reach Earth. Light waves do not have mass and cannot prevent the expansion of space. Instead, the light waves stretch with space and increase their wavelength. This causes red shift: longer wavelengths are closer to the red end of the electromagnetic spectrum.

Light from some parts of the Universe will never reach Earth due to the expansion of the Universe, so we may never know how large the Universe is.

DISCUSSION

When we view the Universe, we are looking back in time because it takes years for the light from nearby stars to reach Earth. How useful is this historical information? Can scientists make predictions about the future of the Universe using information from the past?

INVESTIGATION 6.4

elog-2233

tlvd-10818

Red shift of light

Aim

To observe the effect of the expansion of space on light waves

Materials

- a large, thick elastic band
- a ruler a texta or marker
- a pair of scissors
- a partner

Method

1. Cut the elastic band to create one long piece of rubber.
2. Draw a light wave on the elastic band, use the whole length of the elastic band (as shown in the figure).
3. Choose two points on the light wave, and label them A and B.
4. Label the start of the wave, 'X' and the end of the wave, 'Y'.
5. Measure the distance between X and A, X and B, and X and Y.
6. Stretch the elastic band a short distance and repeat your measurements.
7. Stretch the elastic band as far as you can and repeat your measurements.

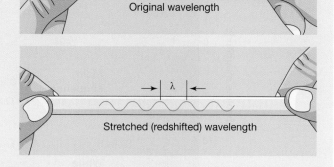

Original wavelength

Stretched (redshifted) wavelength

Results

Construct a table to record your measurements.

Discussion

1. Explain which distance, XA, XB or XY, increased the most when you stretched the elastic band a short distance. Was it the same when you stretched the elastic band as far as you could?
2. Draw a labelled and to scale diagram showing the results of your experiment. Use the diagram to explain why light waves are redshifted.
3. The same experiment can be done on the surface of a balloon. Describe the steps you would need to take to ensure you obtained reliable results on a curved surface.
4. If you had drawn a shorter wave on your elastic band at the start of the experiment, would you have obtained the same trend in your results? Explain.

Conclusion

Summarise the findings from this investigation and how the expansion of space affects light waves.

on Resources

eWorkbook	The expanding Universe (ewbk-12185)	
Video eLesson	The expanding Universe (eles-0038)	
Interactivities	Shifting spectral lines (int-0678)	
	Expansion of the Universe (int-0057)	

6.4.7 Hertzsprung–Russell diagram

An interesting way of displaying the data collected about stars was developed independently by two astronomers, Ejnar Hertzsprung from Denmark and Henry Norris Russell from the USA. The Hertzsprung–Russell method has been named after both of them. In the Hertzsprung–Russell diagram shown in figure 6.14, the absolute brightness of a star is plotted against its surface temperature, which is deduced from its colour. Blue stars have hotter surface temperatures than red stars.

When data for many stars are plotted, most of them, including our Sun, fall into what is known as the **main sequence**. The main sequence, or central diagonal line of the diagram, is where stars spend most of their lives. In a stable main sequence star, hydrogen is steadily turned into helium by the process of fusion.

> **main sequence** area on the Hertzsprung–Russell where the majority of stars are plotted that produce energy by fusing hydrogen to form helium

FIGURE 6.14 The Hertzsprung–Russell diagram sorts stars according to their absolute magnitude (or luminosity) and surface temperature.

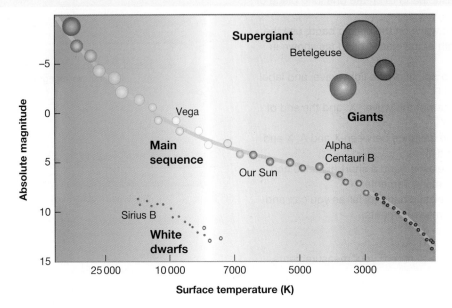

Astronomers suggest that all stars begin their existence in the main sequence and spend the largest part of their life there. This explains why most of the stars observed at a particular time are found in the main sequence. Exactly where a star is found along the main sequence is determined by its mass. Low-mass stars tend to be cooler and less bright than high-mass stars.

Other types of stars show up very clearly on the Hertzsprung–Russell diagram but in much smaller numbers than in the main sequence. The names of these stars — white dwarfs, red giants, blue giants and supergiants — clearly describe their characteristics. These are normally older stars that are no longer using hydrogen as their main fuel source. The rarer types are stars passing relatively quickly through later stages of development on the way to extinction as their nuclear fuel runs out.

6.4.8 The death of a star

During the life of a star there is a balance between the gravitational pull of the mass of the star and the energy released, with nuclei fusion pushing the star outward. Once the fuel is used up, gravity is said to win. The path the star takes depends upon the size of the star, as illustrated in figure 6.15.

Medium size main sequence stars, such as our Sun, consume hydrogen in their core and form helium for billions of years. As helium builds up in the core of the star, the region where energy is produced by the fusion of hydrogen becomes a shell around the core. As the hydrogen is consumed, the star starts to fuse oxygen and carbon in the core. With not enough mass to fuse any heavier elements it gets hotter and hotter. The shell gradually expands and the star swells to 200 or 300 times its original size, cooling as it does so, to become a **red giant**. A Sun-light star might shine 100 times more brightly than it did in its stable period as part of the main sequence.

The brightness of many red giants varies greatly because they have become unstable after many millions of years of light. The red giant Mira in the constellation Cetus (the Whale) was the first variable red giant discovered. The brightness of Mira increases and decreases over a huge range in a cycle that averages 320 days. Not surprisingly, it is known as a **pulsating star**. The shorter cycles of some pulsating stars are so predictable that they can be used as reference points to measure vast interstellar distances in the Universe.

> **red giant** star in the late stage of its life in which helium in its core is fused to form carbon and other heavier elements
>
> **pulsating star** star that periodically expands due to an increase in core temperature and then cools and contracts under gravity

FIGURE 6.15 Large stars follow a different evolutionary sequence than smaller stars like the Sun.

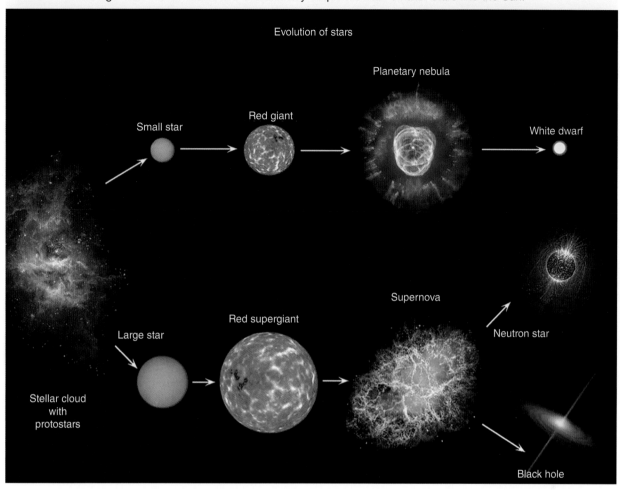

This will eventually happen to our Sun, which will grow large enough to swallow the inner planets, including Earth. This unstable phase will continue until the final death stages, during which it will turn into a **white dwarf** and a **planetary nebula**.

Massive stars die more violently. Stars several times the mass of our Sun swell into much larger red giants called **supergiants**. The huge size means that nuclei fuse to form the heavier elements up to iron. The increase in density and heat builds up until a **supernova** occurs. During this explosion the heaviest elements such as gold and lead are formed. The matter making up the star is hurled into space along with huge amounts of energy. A supernova can emit as much energy in a month as the Sun radiates in a million years. Supernova events are very rare, being seen only every 200 to 300 years on average and fading within a few years.

What remains of a supernova is extremely dense; the pull of gravity becomes so great that even the protons and electrons in atoms are forced together. They combine to form neutrons, and the resulting solid core is known as a **neutron star**. If the remaining core is more massive, the force of gravity is great enough to suck in everything — even light. Such a core becomes a **black hole**.

white dwarf the core remaining after a red giant has shed layers of gases

planetary nebula ring of expanding gas caused by the outer layers of a star less than eight times the mass of our sun being thrown off into space

supergiant very large star that is expanding while running out of fuel, and will eventually explode

supernova huge explosion that happens at the end of the life cycle of supergiant stars

neutron star extremely dense remnants of a supernova in which protons and electrons in atoms are fused to form neutrons

black hole the remains of a star, which forms when the force of gravity of a large neutron star is so great that not even light can escape

The elements we know are produced in these cosmic events as shown in figure 6.16.

FIGURE 6.16 The cosmic source of elements in the Solar System

- Big Bang fusion
- Cosmic ray fusion
- Exploding massive stars
- Exploding white dwarfs
- Merging neutron stars
- Dying low-mass stars
- Very radioactive isotopes; nothing left from stars

| 1 H |
3 Li	4 Be
11 Na	12 Mg
19 K	20 Ca	21 Sc	22 Ti	23 V	24 Cr	25 Mn	26 Fe	27 Co	28 Ni	29 Cu	30 Zn
37 Rb	38 Sr	39 Y	40 Zr	41 Nb	42 Mo	43 Tc	44 Ru	45 Rh	46 Pd	47 Ag	48 Cd
55 Cs	56 Ba		72 Hf	73 Ta	74 W	75 Re	76 Os	77 Ir	78 Pt	79 Au	80 Hg
87 Fr	88 Ra

| 2 He |
5 B	6 C	7 N	8 O	9 F	10 Ne
13 Al	14 Si	15 P	16 S	17 Cl	18 Ar
31 Ga	32 Ge	33 As	34 Se	35 Br	36 Kr
49 In	50 Sn	51 Sb	52 Te	53 I	54 Xe
81 Ti	82 Pb	83 Bi	84 Po	85 At	86 Rn

| 57 La | 58 Ce | 59 Pr | 60 Nd | 61 Pm | 62 Sm | 63 Eu | 64 Gd | 65 Tb | 66 Dy | 67 Ho | 68 Er | 69 Tm | 70 Yb | 71 Lu |
| 89 Ac | 90 Th | 91 Pa | 92 U | 93 Np | 94 Pu |

ACTIVITY: Exploring supernovas

Find out more about the formation and destruction of a supernova. For example, when was the last supernova seen? Can we predict when the next one will be seen?

EXTENSION: Black holes

A black hole is the remnants of large neutron stars, with a force of gravity so large that not even light can escape from it. But then, how do we know they exist if we can't see them? Scientists are able to study the effect they have on matter that is nearby. For example, if a star travels too close to a black hole, it will be pulled in by its gravity and the black hole will tear the star apart. This causes the star's matter to accelerate and heat up, giving off X-rays that can be detected.

Given the Milky Way galaxy has a black hole at the center, does this mean we'll be torn apart soon?

FIGURE 6.17 An illustration of what a black hole looks like

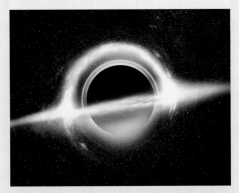

6.4 Activities

learn on

| 6.4 Quick quiz on | 6.4 Exercise |

Select your pathway

■ LEVEL 1	■ LEVEL 2	■ LEVEL 3
2, 3, 4, 5, 6, 8	1, 7, 9, 10, 13	11, 12, 14, 15

These questions are even better in jacPLUS!
- Receive immediate feedback
- Access sample responses
- Track results and progress

Find all this and MORE in jacPLUS ▶

Remember and understand

1. Explain why there are black lines in the spectra of the light emitted by stars.
2. State which colour of light has the higher frequency — red or blue.
3. What is a red shift?
4. **MC** Are most stars found in the main sequence of the Hertzsprung–Russell diagram?
   A. Yes, most stars are found in the main sequence and stay there for their lifetime.
   B. Yes, most stars are found there, but do move from one sequence to another during their lifetime.
   C. No, most stars are not usually found in the main sequence but may stay there for a small part of their lifetime.
   D. No, most stars are not found there. Most stars spend the majority of their lives outside the main sequence.
5. **MC** To which group of stars shown on the Hertzsprung–Russell diagram does the Sun belong?
   A. Giants
   B. Supergiants
   C. White dwarfs
   D. The main sequence
6. **MC** In which galaxy can Earth and the rest of our Solar System be found?
   A. Milky Way
   B. Centaurus
   C. Orion
   D. Moro

Apply and analyse

7. Complete the following passage.
   The light from a star is often analysed by its _____ instead of its _____. Long _____ correspond to low _____ and short _____ correspond to high _____. The spectrum of colours emitted by excited atoms of _____ on Earth contains the wavelength _____. This same wavelength is observed in the spectrum of light from the bright star _____ Vega is moving _____ from Earth.

8. The most distant objects in the Universe are estimated to be about 14 billion light-years from Earth. Explain why the age of the Universe is thought to be at least 14 billion years.
9. Explain why nebulae are often referred to as star nurseries.
10. Is it likely that our own star, the Sun, will become a supernova? Explain your answer.

Evaluate and create

11. **SIS** Observe the diagram of the Milky Way. Would light from the stars in the Sagittarius arm of the Milky Way display red shift or blue shift? Explain your answer.

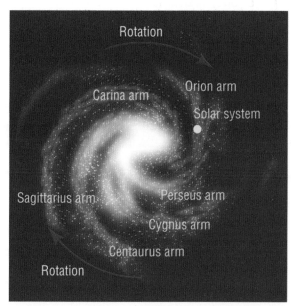

12. **SIS** To investigate the expansion of space, you conduct an experiment. You draw some stars on the surface of a balloon and measure the distance between them and Earth before you blow up the balloon and after blowing up the balloon. The results from your experiment are shown in the following table.

TABLE The modelled distance between the stars and Earth

Star	Initial distance (cm)	Final distance (cm)	Difference in distance (cm)
A	0.92	2.3	1.38
B	1.44	3.6	2.16
C	1.56	3.9	2.34
D	1.80	4.5	2.70
E	2.04	5.1	3.06
F	3.12	7.8	4.68
G	3.28	8.2	4.92

a. Use this information to draw a graph of your results. Put the 'Final distance' of the star from Earth on the x-axis and the difference in distance on the y-axis of your graph.
b. What shape is your graph? Draw in a line of best fit.
c. Describe the trend in your graph.
d. If an extra star had been drawn on the balloon, star H, which was located 6 cm from Earth after the balloon was inflated, how far would you expect this star to have travelled from its initial position?

13. The following table lists information about three bright stars.
a. Which star has the greatest actual brightness?
b. Which star is the faintest as seen from Earth?

TABLE The apparent and absolute magnitudes of three stars

Star	Apparent magnitude	Absolute magnitude
Rigel	0.11	−7.5
Aldebaran	0.86	−0.3
Canopus	−0.73	−4.6

14. **SIS** Using the Hertzsprung-Russell diagram shown in figure 6.14, answer the following questions.
a. Estimate the surface temperature of Alpha Centauri B.
b. Give the absolute magnitude of Vega.
c. Can this diagram be used to determine whether Vega is brighter than Alpha Centauri B when viewed from the surface of Earth? Explain your answer.

15. **SIS** The following table shows the surface temperature and absolute magnitudes for some stars. Plot these on a Hertzsprung-Russell diagram to determine whether or not they are located on the main sequence.

TABLE The surface temperature and absolute magnitudes for three stars

Star	Surface temperature (K)	Absolute magnitude
Procyon A	6600	2.6
Barnard's Star	2800	13.2
Eridani B	10000	11.1

Fully worked solutions and sample responses are available in your digital formats.

LESSON
6.5 How did it all begin?

LEARNING INTENTION

At the end of this lesson you should be able to describe current theories of the Universe and explain the evidence that supports these theories.

6.5.1 Was there a beginning?

When and how did the Universe begin? Was there a beginning? Perhaps it was always there. If there was a beginning, will there be an end? The study of the answers to these questions is called **cosmology**.

Following Edwin Hubble's discoveries about the expanding Universe, two major theories about the beginning of the Universe became popular — the **Big Bang theory** and the **Steady State theory**.

ACTIVITY: The Big Bang or Steady State

Create a poster, PowerPoint presentation, animation or comic strip to compare the Big Bang theory and Steady State theory.

You should ensure that you include the following:
- the history of both theories
- the evidence of both theories
- information about why the Big Bang theory is more commonly accepted.

6.5.2 The Big Bang

According to the most commonly accepted theory among cosmologists, the Universe began about 15 billion years ago with a 'big bang'. This is shown in figure 6.18.

FIGURE 6.18 From the 'big bang' to one billion years old: the Universe's formation according to the Big Bang theory

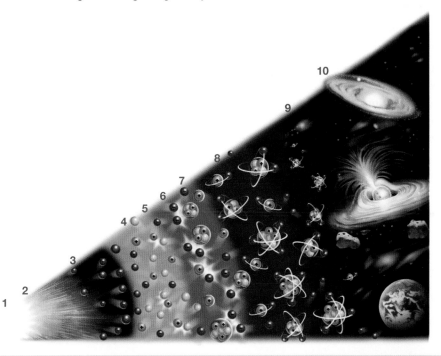

cosmology the study of the beginning and end of the Universe

Big Bang theory a theory that states that the Universe began about 15 billion years ago with the explosive expansion of a singularity

Steady State theory a theory that states that there was no beginning to the Universe and that the Universe does not change in appearance

The stages in the formation of the Universe through the Big Bang theory

The stages of the Big Bang (as shown in figure 6.18) can be summarised as follows.

1. **The Big Bang ($t = 0$)**

   It's hard to imagine, but at this moment there was no space and no time. All that existed was energy, which was concentrated into a single point called a **singularity**.

2. **One ten million trillion trillion trillionths of a second later** $t = +\dfrac{1}{1^{43}}s$

   Time and space had begun. Space was expanding quickly and the temperature was about 100 million trillion trillion degrees Celsius. (The current core temperature of the Sun is 15 million degrees Celsius.)

3. **One ten billion trillion trillionths of a second after the Big Bang** $t = +\dfrac{1}{1^{34}}s$

   The Universe had expanded to about the size of a pea. Matter in the form of tiny particles such as electrons and **positrons** (positively charged electrons) had formed. Particles collided with each other, releasing huge amounts of energy in the form of light. Until this moment there was no light.

4. **One ten thousandth of a second after the Big Bang** $t = +\dfrac{1}{10^{4}}s$

   Protons and neutrons had formed as a result of collisions between smaller particles. The Universe was very bright because light was trapped as it was continually being reflected by particles.

5. **One hundredth of a second after the Big Bang** $t = +\dfrac{1}{100}s$

   The Universe was still expanding and cooling rapidly. It had grown to the same size as our Solar System, but there was still no such thing as an atom.

6. **One second after the Big Bang ($t = +1\ s$)**

   The Universe was probably more than a trillion trillion kilometres across. It had cooled to about ten billion degrees Celsius.

7. **Five minutes after the Big Bang ($t = +5\ min$)**

   The nuclei of hydrogen, helium and lithium had formed among a sea of electrons.

8. **Three hundred thousand years after the Big Bang ($t = +300\ 000$ years)**

   The Universe was about one thousandth of its current size. It had cooled to about 3000 °C. Electrons had slowed down enough to be captured by the nuclei of hydrogen, helium and lithium, forming the first atoms. There was now enough empty space in the Universe to allow light to escape to the outer edges. For the first time, the Universe was dark.

9. **Two hundred million years after the Big Bang ($t = +200\ 000\ 000$ years)**

   The first stars had appeared as gravity pulled atoms of hydrogen, helium and lithium together. **Nuclear reactions** took place inside the stars, causing the nuclei of the atoms to fuse together to form heavier nuclei. Around some of the newly forming stars, swirling clouds of matter cooled and formed clumps. This is how planets began to form.

10. **One billion years after the Big Bang ($t = +1\ 000\ 000\ 000$ years)**

    The Universe was beginning to become a little 'lumpy'. The force of gravity was stronger in 'lumpier' regions pulling even more matter towards these areas, causing the first galaxies to form.

singularity a single point of immense energy present at the time of the Big Bang

positrons positively charged electrons

nuclear reactions reactions involving the breaking of bonds between the particles (protons and neutrons) inside the nuclei of atoms

---

ACTIVITY: The first second

In your own words, write an account (about 200 words) of the first second after the Big Bang.

You may wish to include diagrams to show each stage.

The Einstein connection

The Big Bang theory would not make any sense at all if it were not for Albert Einstein's famous equation. How could matter be created from nothing? Well, the singularity before the Big Bang was not 'nothing'. It was a huge amount of energy (with no mass) concentrated into a tiny, tiny point.

Einstein proposed that energy could be changed into matter. His equation $E = mc^2$ describes the change. Einstein's equation also describes how matter can be changed into energy. That is what happens in nuclear power stations and nuclear weapons. This equation also explains the energy of stars because the nuclear fusion reaction converts mass into energy.

6.5.3 Galaxies evolve

According to the Steady State theory, which was proposed in 1948, there was no beginning of the Universe. It was always there. The galaxies are continually moving away from each other. In the extra space left between the galaxies, new stars and galaxies are created. These new stars and galaxies replace those that move away, so that the Universe always looks the same. The images of closer, younger galaxies and older, more distant galaxies also support the Big Bang model. The early, distant galaxies are more disturbed because of collisions with other galaxies, they are smaller and they contain fewer heavy elements ('heavy element' is an astronomer's expression for all elements other than hydrogen and helium). Physically and chemically, they are different.

A huge debate between those who supported the Steady State theory and those who supported the Big Bang theory raged from 1948 until 1965. During that period, the evidence supporting the Big Bang theory grew.

The red shift

The red shift provides evidence for an expanding Universe. This evidence supports the Big Bang theory and causes problems for those supporting the Steady State theory. A steady-state Universe could expand only if new stars and galaxies replaced those that moved away. There is no way to explain how these new stars and galaxies could be created from nothing. Apart from that, these young stars and galaxies have not been found by astronomers.

The elements

The amounts of hydrogen and helium in the Universe support the Big Bang theory. According to the Steady State theory, the only way that helium is produced is by the nuclear reactions taking place in stars. About 8.7 per cent of the atoms in the Universe are helium. This is far more than could be produced by the stars alone. The percentage of helium atoms can, however, be explained by their creation as a result of the Big Bang.

Cosmic microwave background radiation

When George Gamow and Ralph Alpher proposed their version of the Big Bang theory in 1948, they calculated that the Universe would now, about 15 billion years after creation, have a temperature of 2.7 °C above **absolute zero**. That's approximately −270 °C. Anything with a temperature above absolute zero emits radiation. The nature of the radiation depends on the temperature. Gamow predicted that, because of its temperature, the Universe would be emitting an 'afterglow' of radiation. This afterglow became known as cosmic microwave background radiation.

This radiation was discovered by accident in 1965. Engineers trying to track communications satellites picked up a consistent radio noise that they couldn't get rid of. The noise wasn't coming from anywhere on Earth, because it was coming from all directions. It was the cosmic microwave background radiation predicted by Gamow. Its discovery put an end to the Steady State theory, leaving the Big Bang theory as the only theory supported by evidence currently available. Even Fred Hoyle, who had ridiculed the idea of a 'big bang', admitted that the evidence seemed to favour the Big Bang theory.

absolute zero temperature at which the particles that make up an object or substance have no kinetic energy, approximately −273.15 °C

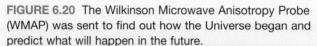
6.5.4 Measuring background radiation

In 1989, a satellite named COBE (COsmic Background Explorer) was put into orbit around Earth to accurately measure the background radiation and temperature of the Universe (figure 6.20). COBE could detect variations as small as 0.00003 °C. As predicted by Gamow, it detected an average temperature of −270 °C.

FIGURE 6.20 The Wilkinson Microwave Anisotropy Probe (WMAP) was sent to find out how the Universe began and predict what will happen in the future.

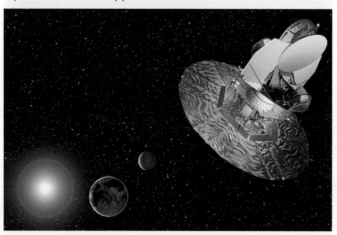

In 2001, a probe called WMAP (Wilkinson Microwave Anisotropy Probe) was sent into orbit around Earth at a much greater distance to gather even more accurate data, detecting temperatures within a millionth of a degree. WMAP's first images were released by NASA in February 2003.

The computer-enhanced image of cosmic microwave background radiation shown in figure 6.21 was produced by the WMAP mission. The background radiation detected was released only 380 000 years after the Big Bang — the first radiation to escape. It shows how the temperature varied across the Universe as it was 380 000 years after the Big Bang. The blue parts of the map are the cooler regions. These regions were cool enough for atoms, and eventually galaxies, to form. The red parts are warmer regions. The map shows that galaxies are not evenly spread throughout the Universe. They support the theory of an expanding Universe that began with a big bang.

FIGURE 6.21 WMAP image of cosmic microwave background radiation

6.5.5 Will it ever end?

Will the expansion of the Universe continue forever? If the Universe does stop expanding, what will happen to it? Several competing theories offer answers to these questions. One theory suggests not enough mass exists in the Universe for gravity to be able to pull it all back, so it will continue to expand forever. Other theories suggest that the Universe will eventually end. According to these theories, the end will come when:

- the Universe snaps back onto itself in a 'big crunch' (the **Big Crunch theory**). If this happens, the end result will be a single point — singularity. Some cosmologists believe that the big crunch will be followed by another big bang.
- the expansion of the Universe continues and stars use up their fuel and burn out, causing planets to freeze (the **Big Chill theory**). The Universe would then consist of scattered particles that never meet again.
- the Universe rips itself apart violently as a result of expanding at an increasing speed (the **Big Rip theory**). According to this theory, the end of the Universe will also be the end of time itself.

Big Crunch theory a theory that proposes that the Universe will snap back on itself resulting in another singularity

Big Chill theory a theory that proposes that the expansion of the Universe will continue indefinitely until stars use up their fuel and burn out

Big Rip theory a theory that proposes that the Universe will rip itself apart due to accelerating expansion

DISCUSSION

Which of the three theories about the end of the Universe do you think is the most likely to be correct? Give reasons for your answer.

on Resources

	eWorkbook	The Big Bang (ewbk-12165)
	Video eLesson	Entropy: Is the end of the Universe nearer than we thought? (eles-1073)
	Weblink	How will the Universe end, and could anything survive?

6.5 Activities

| 6.5 Quick quiz   on | 6.5 Exercise |

Select your pathway

■ LEVEL 1	■ LEVEL 2	■ LEVEL 3
1, 2, 5, 7, 11	3, 6, 8, 10, 12	4, 9, 13, 14, 15

These questions are even better in jacPLUS!
- Receive immediate feedback
- Access sample responses
- Track results and progress

Find all this and MORE in jacPLUS ▶

Remember and understand

1. **MC** What is the science of cosmology?
   A. A theory that states that the Universe began about 15 billion years ago with the explosive expansion of a singularity
   B. The study of the way that the Universe changes
   C. The study of a single point of immense energy present at the time of the Big Bang
   D. A theory that proposes that the expansion of the Universe will continue indefinitely until stars use up their fuel and burn out

2. **MC** How old is the Universe believed to be?
   A. Approximately 15 billion years
   B. Approximately 15 million years
   C. Approximately 150 billion years
   D. Approximately 150 million years

3. **MC** According to the Big Bang theory, what was present at the time the Universe began?
   A. Matter in the form of tiny particles such as electrons and positrons about the size of a pea
   B. A single atom
   C. Nothing
   D. A single point called a singularity

4. Complete the following statement: WMAP is able to provide a picture of the Universe as it was _____ years after the Big Bang. Before that time, _____ (and other _____ radiation) was not escaping from the Universe _____ from before that time could not have been detected.

Apply and analyse

5. Explain when galaxies began to form.
6. Explain the link between Einstein's famous equation $E = mc^2$ and the Big Bang theory.
7. Describe the two theories about the 'beginning' of the Universe.
8. How did the Steady State theory explain that the Universe was expanding, yet remained the same?
9. What evidence put an end to the Steady State theory?
10. Describe three major pieces of evidence that supported the Big Bang theory.
11. Name and describe three theories about how the Universe might end.

Evaluate and create

12. Predict what would have happened to the Universe if, one million years after the Big Bang, the matter in it was perfectly evenly distributed and not moving.
13. Explain why neutral atoms were not likely to form during the first five minutes after the Big Bang. Justify your response.
14. Explain why scientists go through the expense of measuring background radiation with a satellite or space probe when it could be done from Earth.
15. Draw flowcharts to describe:
    a. the Big Bang theory
    b. how cycles in the Big Bang and Big Crunch theories might work together.

Fully worked solutions and sample responses are available in your digital formats.

LESSON
6.6 Advancing science, technology and engineering

LEARNING INTENTION

At the end of this lesson you should be able to describe how advances in technology have enabled scientists to improve their theories of the Universe.

6.6.1 Looking with light (the electromagnetic spectra)

For hundreds of years, optical telescopes have been used to observe what lies beyond the Solar System. To find out what's in deep space, in the most distant parts of the Universe, observing visible light is not enough. We rely on other parts of the electromagnetic spectrum (figure 6.22).

FIGURE 6.22 The electromagnetic spectrum

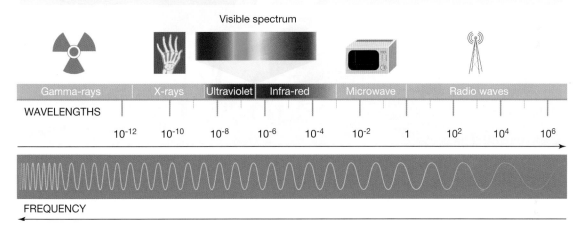

Until the accidental discovery in 1931 that stars emitted radio waves as well as light, the only way to observe distant stars and galaxies was with light telescopes. Like light and other forms of **electromagnetic radiation**, radio waves travel through space at a speed of 300 000 kilometres per second. Radio waves from deep in space are collected by huge dishes and reflected towards a central antenna. The waves are then analysed by a computer, which produces an image that we can see.

Radio telescopes can detect tiny amounts of energy. In fact, the total amount of energy detected in ten years by even the largest radio telescopes would light a torch globe for only a fraction of a second. They can detect signals from much further away than light telescopes. Figures 6.23 and 6.24 show two radio telescopes in China and New Mexico respectively.

Unlike light waves, radio waves can travel through clouds in Earth's atmosphere, and can be viewed in daylight as well as at night. Radio waves also pass through clouds of dust and gas in deep space.

FIGURE 6.23 The Five-hundred-meter Aperture Spherical radio Telescope (FAST), in China, is the largest single radio telescope in the world.

electromagnetic radiation waves produced by the acceleration of an electric charge and that have an electric field and a magnetic field at right angles to each other

radio telescope a telescope that can detect radio waves from distant objects

FIGURE 6.24 The Very Large Array in New Mexico consists of 27 dishes, each with a diameter of 25 metres, arranged in a Y shape.

6.6.2 Satellites

More than 2500 satellites are currently orbiting Earth, many of them constantly watching Earth's surface and atmosphere. Others provide views of the Universe that could never be seen from Earth's surface.

The James Webb Space Telescope (JWST) (see figure 6.25a) was launched on 25 December 2021. It is NASA's successor to the Hubble telescope. It was the cumulation of twenty years of work, and cost more than $10 billion dollars. The James Webb is solely an infrared observatory orbiting the Sun around 1.5 million kilometres above Earth. That is further than the Moon and it cannot be serviced by astronauts. It is so powerful that it may be able to indirectly 'observe' evidence about dark matter as it also uncovers the history of the Universe, and galaxy and planet formation.

FIGURE 6.25 **a.** James Webb Space Telescope **b.** The primary mirror of the JWST is 6.5 m in diameter **c.** A JWST test image of a distance star

The JWST test image of a distance star (see figure 6.25c) was photobombed by thousands of ancient galaxies. This star, which is 1000 fainter than the human eye can see, is 2000 light years away. The galaxies are several billions of years old. The first images from the JWST were released by NASA on 12 July 2022. They revealed sharper and clearer views of cosmic features of the distance Universe than had ever been captured before.

Looking in, looking out

Artificial satellites can be used to look at Earth or to look into space. An inward-looking satellite can sweep the surface of Earth every day, using cameras and remote sensors to observe and measure events on the surface hundreds or thousands of kilometres below. An outward-looking satellite can see directly into space, its view unobstructed by the atmosphere, pollution or dust. Light pollution, an increasing problem for Earth-bound observers as our cities grow, is not an issue for an observer in space.

Inward-looking satellites are used for:
- collecting weather and climate data, providing early warning of events (such as volcanic activity and changing ocean currents) and showing long-term trends
- collecting data used for mineral exploration, crop analysis, mapping, and identifying long-term erosion or degradation
- strategic defence ('spy-in-the-sky') systems
- communications for telephones, television, radio and computer data.

Outward-looking satellites are used for:
- observing the other planets and bodies circling the Sun
- observing stars, galaxies and other remote objects in space
- watching for comets and asteroids that may hit Earth
- listening for signs of extraterrestrial life.

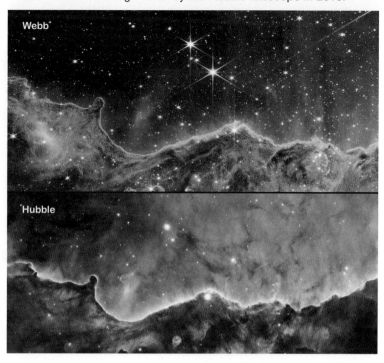

FIGURE 6.26 An image of the Carina Nebula taken by the JWST in 2022 versus the image taken by the Hubble telescope in 2010.

Webb

Hubble

FIGURE 6.27 A satellite image of smoke from a fire.

Resources

Video eLessons The radio telescope at Arecibo (eles-2726)
The Very Large Array satellite dishes at Twilight in New Mexico, USA (eles-2727)

The International Space Station

The International Space Station (ISS) was completed in 2011 with the support of a number of different space agencies. The primary purpose of the ISS was to provide laboratories in space for research into microgravity and fields such as medicine, geology and technology. The ISS also provides the opportunity to investigate the effect of a space environment on humans and prepare for the exploration of Mars and beyond.

The ISS is able to provide similar images to those from other inward-looking satellites, but also has the advantage of having a crew who can respond to unfolding events immediately, rather than waiting for further 'instructions' from the ground.

FIGURE 6.28 The International Space Station in orbit

This is especially helpful when natural disasters such as volcanic eruptions, earthquakes and tsunamis occur. In addition, the orbit of the ISS is different from most other satellites and is able to collect images at different and often more suitable times. The ISS orbits Earth once every 90 minutes and the crew can direct cameras to anywhere on Earth, making it extremely versatile.

 Resources

 Weblinks   Spot the ISS depending on your location
Live Space Station tracking map

ultraviolet radiation invisible radiation similar to light but with a slightly higher frequency and more energy

infra-red radiation invisible radiation emitted by all warm objects

Hubble Space Telescope

The Hubble Space Telescope is an example of an outward-looking satellite. It was carried into orbit about 600 kilometres above Earth's surface by the space shuttle *Discovery* in 1990. The Hubble Space Telescope (see figure 6.29) collects images by collecting and analysing data in the form of visible light, **ultraviolet radiation** and **infra-red radiation** from deep space. It produces spectacularly clear images that are relayed back to Earth by radio waves.

The Hubble Space Telescope was the first space telescope that could be serviced while in orbit, and its useful life has been dependent on transporting astronauts to and from Earth aboard space shuttles. NASA's space shuttle program has now ceased, so servicing is no longer possible. When

FIGURE 6.29 The Hubble Space Telescope is the only telescope designed to be serviced and improved in space by astronauts.

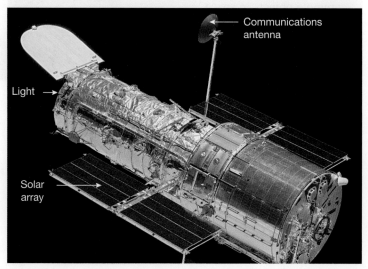

the orbiting telescope stops functioning it will be 'deorbited' by an uncrewed space mission so that it plunges harmlessly into the ocean. However, the recent success of the privately owned Space X mission taking astronauts to the ISS means these plans for the Hubble Space Telescope may change. The Hubble Space Telescope is still sending images back to Earth and is expected to remain operating until at least the mid 2030s.

Several other space telescopes are in orbit around Earth that collect radiation from parts of the electromagnetic spectrum. They include the Chandra X-ray Observatory, carried into orbit by the space shuttle *Columbia* in 1999. Most **X-rays** coming from neutron stars and black holes approaching Earth are absorbed by the atmosphere. The Chandra is able to gather data that could never be collected by X-ray telescopes on Earth's surface.

Data overload?

The unprecedented amount of data coming from telescopes of all types on the ground and in orbit requires processing by **supercomputers** with capabilities well beyond those of personal computers and even large computers used in most industries. IT specialists play a crucial role in developing new and faster computer systems to ensure that exploration of the Universe is not limited by data overload.

X-rays high energy electromagnetic waves that can be transmitted through solids and provide information about their structure

supercomputers computers with the fastest processing capacity available

dark energy a theoretical form of energy that may exist throughout the Universe

on Resources

▶ **Video eLessons**   The Hubble Space Telescope (eles-2725)
                        Expedition 34: International Space Station (eles-2728)
🧩 **Interactivity**   Hubble Space Telescope (int-5867)

6.6.3 Discovering dark energy

In 1997, a group of astrophysicists, including Australian National University's Professor Brian Schmidt (see figure 6.30), presented evidence that the expansion of the Universe was speeding up. Their research took several years to complete and required the analysis of images of the most distant supernovas that could be observed. The images were from light and infra-red telescopes in several locations, including Australia, and from the Hubble Space Telescope.

Using the red shift and brightness of Type 1a supernovae it was found that gravity was not slowing the expansion of the Universe but that the Universe's expansion was accelerating. This was due to the presence of **dark energy**, a repulsive force which is calculated to make up approximately 68 per cent of the Universe.

The successful outcome of the research was only possible with the use of advanced digital imaging sensors and powerful state-of-the-art computers. Brian Schmidt and two other members of the team of astrophysicists were awarded the Nobel Prize for Physics in 2011 for their research.

FIGURE 6.30 Professor Brian Paul Schmidt at ANU won the Nobel Prize for Physics in 2011.

Dark energy appears to be the dominant component of the physical Universe, yet there is no persuasive theoretical explanation for its existence or magnitude … the nature of dark energy ranks among the most compelling of all outstanding problems in physical science.

Source: [astro-ph/0609591] Report of the Dark Energy Task Force (arxiv.org)

Optical telescopes were used in the 1970s to develop the spectral class of stars, to see the red shift and what stars were made of. Vera Rubin (figure 6.31) observed the outer galaxies and measured the red shift of individual galaxies. She calculated the speed of the galaxies and found that they are rotating at such a speed that they should be tearing themselves apart. The gravity needed to keep the galaxy together requires a larger mass than we can see. She had discovered the first evidence of invisible matter called **dark matter**. Dark matter makes up most of the Universe, but we can only detect it from its gravitational effects.

FIGURE 6.31 Vera Rubin, who discovered the first evidence of dark matter.

ACTIVITY: Observatories across the electromagnetic spectrum

Mark a chart of the different parts of the electromagnetic spectrum, showing what each part detects. As a challenge, also list the name of a key telescope or observatory on the chart.

6.6.4 Telescope of the future

The Square Kilometre Array (SKA) is a new radio telescope currently being constructed with the cooperation of seven countries including Australia, the United Kingdom, South Africa (figure 6.32) and China.

The SKA will consist of thousands of dish-shaped antennas in South Africa and Australia, linked by optical fibre. It will be about 50 times as sensitive as the best of the current generation of radio telescopes and will be able to scan the sky up to 10 000 times faster. The SKA is scheduled to be in full operation in 2027. It will allow astronomers to investigate events that took place within the first second after the Big Bang and fill gaps in our knowledge of the period when the Universe became dark and the first galaxies formed.

The SKA is sensitive enough to detect weak extraterrestrial signals (if they exist) and planets in other galaxies capable of supporting life. The SKA should also provide answers to questions about distant galaxies, dark energy, gravity and magnetism.

> **dark matter** a component of the universe that can be detected from its gravitational attraction

FIGURE 6.32 An artist's impression of the dishes that will be installed in the Karoo desert in South Africa

 Resources

 eWorkbook Telescopes (ewbk-12167)

6.6.5 A new window on the Universe: gravitational wave astronomy

Einstein's thought experiments on gravity predicted that there was a fourth dimension called space-time. Space-time was dynamic and could be warped into a black hole. While a hundred years ago theoretical physicists could predict the events that could warp time and space, they had no means to detect the resultant wave.

That was until the Laser Interferometer Gravitational-Wave Observatory (LIGO). This was designed to detect the minute ripples in space-time caused by passing gravitational waves that formed when neutron stars or black holes collide or when a supernova occurs. These ripples are invisible as they are not any part of the EM spectra but they travel at the speed of light.

In 2015 the first gravitational waves from two black holes that collided 1.3 billion years ago were detected by two laser interferometers 3000 km apart on Earth. In 2016 OzGrav, part of the international LIGO-Virgo KAGRA collaboration, was funded. The OzGrav scientists are developing the instrumentation for the advanced LIGO, the Square Kilometre Array and the next generation of gravitational wave detectors.

NEMO (Neutron Star Extreme Matter Observatory) is an Australian proposal which hopes to develop by 2035 the means to test the hypothesis that when neutron stars collide, they form a hyper-massive neutron star. This massive neutron star would be two times the mass of our Sun and would last a few hundred milliseconds before collapsing into a black hole. The gamma ray that would be ejected could be detected by this observatory.

By using gravitational waves observations, we can further our understanding of dark matter and could measure the expansion of the Universe. This will allow us to observe the un-seeable wonders of the Universe.

6.6 Activities

| 6.6 Quick quiz on | 6.6 Exercise |

These questions are even better in jacPLUS!
- Receive immediate feedback
- Access sample responses
- Track results and progress

Find all this and MORE in jacPLUS ▶

Select your pathway

■ LEVEL 1	■ LEVEL 2	■ LEVEL 3
1, 3, 4	2, 6, 7	5, 8, 9

Remember and understand

1. **MC** What are two advantages of Earth-based radio telescopes over light telescopes?
   A. They are able to detect radio waves through daylight hours.
   B. Images produced by single radio telescopes are extremely sharp.
   C. They observe visible light to produce images.
   D. They are able to detect radio waves through cloud and atmosphere.

2. Complete the following description of single radio telescopes.
   Images produced by single radio telescopes are not very _____ . To solve this problem, signals from groups of telescopes pointed at the _____ are combined to produce _____ images.

3. Outline the advantages of space telescopes over telescopes on Earth's surface.

4. **MC** What information can be revealed by images from radio telescopes?
   A. Size
   B. Position
   C. Composition of stars
   D. All of the above

Apply and analyse

5. Which part or parts of the electromagnetic spectrum have been collected by the Hubble Space Telescope?

6. **MC** Select two reasons the International Space Station is useful as an inward-looking satellite.
   A. It has a crew who can respond to unfolding events immediately, unlike other satellites, which must receive new programming from the ground.
   B. The images it produces are sharper than any other telescope in the world.
   C. Its primary purpose is to provide laboratories in space for research into microgravity and fields such as medicine, geology and technology.
   D. It is able to collect images at different and often more suitable times.

Evaluate and create

7. Explain why fast computers are important to exploration of the Universe.

8. Explain why orbiting space telescopes have a limited lifetime.

9. **SIS** Outline at least four reasons there can be uncertainty about the launch dates of future space missions and projects like the Square Kilometre Array.

Fully worked solutions and sample responses are available in your digital formats.

LESSON
6.7 Thinking tools — Matrices

6.7.1 Tell me

What is a matrix?

A matrix is a very useful thinking tool that can assist you in identifying similarities and differences between topics. They are sometimes called tables, grids or decision charts.

Why use a matrix?

A matrix helps to identify common points between topics. If you have multiple topics of which you need to compare features, a matrix is an incredibly helpful way to do this.

6.7.2 Show me

To create a matrix:
1. Write the topics being compared in the left column of the matrix (the example shown in figure 6.34 lists different stars).
2. Write the characteristics to be compared along the top row of the matrix.
3. If a characteristic applies to a topic, put a tick in the appropriate cell of the matrix.
4. The matrix now shows how the various topics are related.

FIGURE 6.33 An example matrix or decision chart

Topic	Feature A	Feature B	Feature C	Feature D
1	✓		✓	✓
2		✓		
3		✓		✓

FIGURE 6.34 A matrix comparing different stars

	Type of star	Forms from small stars	Positive absolute magnitude
Red giant	✓	✓	
White dwarf	✓	✓	✓
Supergiants	✓		

6.7.3 Let me do it

6.7 Activity

1. Copy and complete the following matrix around the Big Bang and Steady State theories.

These questions are even better in jacPLUS! 
- Receive immediate feedback
- Access sample responses
- Track results and progress

Find all this and MORE in jacPLUS

Statement	Big Bang theory	Steady State theory
a. The Universe has no beginning.		✓
b. The Universe began with a single point called singularity.		
c. The Universe is expanding.		
d. The Universe always looks the same.		
e. The red shift in the spectrum of visible light coming from stars and galaxies provides evidence for the theory.		
f. New stars and galaxies are created to replace those that move away due to expansion of the Universe.		

Statement	Big Bang theory	Steady State theory
g. This theory explains the amount of the helium in the Universe.		
h. This theory is supported by the measurement of the current temperature of the Universe (about −270 °C).		

2. Use matrixes to compare the following.
   a. Radio telescopes and light telescopes
   b. Inward-looking satellites and outward-looking satellites
   c. Red giants and white dwarfs
   d. Living in space and living on Earth

Fully worked solutions and sample responses are available in your digital formats.

LESSON
6.8 Review

Access your topic review eWorkbooks

 Resources

■ Topic review Level 1
ewbk-12177

■ Topic review Level 2
ewbk-12179

■ Topic review Level 3
ewbk-12181

6.8.1 Summary

What can we observe in the changing night sky?

- The stars in the sky rotate in position, but remain in fixed patterns or constellations.
- Groups of stars are known as galaxies.
- Clusters of galaxies, clouds of gas and dust are referred to as nebulae.

How can we measure cosmic distances?

- Large distances can be expressed using the astronomical unit, the parsec or the light-year.

What does starlight tell us?

- Denser sections of dust and gas eventually reach such a high density that gravity takes hold and a nebula cloud forms.
- It is this denser cloud that eventually forms into stars.
- The brightness of a star depends on its actual brightness (its magnitude) and its distance from Earth.
- The relationship between absolute magnitude and temperature is observed in the Hertzsprung–Russell diagram.
- The colour of a star depends on its temperature.
- Smaller stars eventually expand and become red giants. Eventually these become planetary nebulae as outer layers are thrown off. The star that remains fades into a white dwarf and eventually dies and disappears.
- Larger stars come to a more violent end and form into supergiants. These blow up in explosions known as supernovas, which may form into neutron stars or black holes.
- The spectra of stars can be in certain frequencies. Stars that are moving away from Earth shift towards lower, 'redder', frequencies known as red shift. Stars moving closer to Earth experience blue shift.
- Due to the red shift effect, we can infer that most objects in the Universe are moving away from us, becoming more red in appearance.

How did it all begin?

- The two major theories on the formation of the Universe are the Big Bang theory and the Steady State theory.
- The Big Bang theory suggests that no space or time existed at the beginning of time. All that existed was energy, concentrated into a singularity. The Universe formed and expanded from this singularity.

Advancing science, technology and engineering

- Light telescopes have been used to observe what lies beyond. As technology becomes more powerful, we can see more and more of the Universe, including through the use of radio telescopes, which can detect tiny amounts of energy.
- Satellites constantly orbit Earth — some provide us information about Earth, while others provide views of the Universe, by collecting visible light, infra-red radiation, ultraviolet radiation and microwaves.

6.8.2 Key terms

absolute magnitude actual brightness of a star

absolute zero temperature at which the particles that make up an object or substance have no kinetic energy, approximately −273.15 °C

apparent magnitude brightness of a star as seen from Earth

astronomical unit unit of length equal to the mean distance between Earth and the Sun, defined as approximately 150 million kilometres

Big Bang theory a theory that states that the Universe began about 15 billion years ago with the explosive expansion of a singularity

Big Chill theory a theory that proposes that the expansion of the Universe will continue indefinitely until stars use up their fuel and burn out

Big Crunch theory a theory that proposes that the Universe will snap back on itself resulting in another singularity

Big Rip theory a theory that proposes that the Universe will rip itself apart due to accelerating expansion

black hole the remains of a star, which forms when the force of gravity of a large neutron star is so great that not even light can escape

blue shift shift of lines of a spectral pattern towards the blue end when a light source is moving rapidly towards the observer

constellations groups of stars that were given a particular name because of the shape the stars seem to form in the sky when viewed from Earth

cosmology the study of the beginning and end of the Universe

dark energy a theoretical form of energy that may exist throughout the Universe

dark matter a component of the universe that can be detected from its gravitational attraction

ecliptic the path that the Sun traces in the sky during the year

electromagnetic radiation waves produced by the acceleration of an electric charge and that have an electric field and a magnetic field at right angles to each other

galaxies very large groups of stars and dust held together by gravity

infra-red radiation invisible radiation emitted by all warm objects

light-year the distance travelled by light in one year

main sequence area on the Hertzsprung–Russell where the majority of stars are plotted that produce energy by fusing hydrogen to form helium

nebulae clouds of dust and gas that may be pulled together by gravity and heat up to form a star

neutron star extremely dense remnants of a supernova in which protons and electrons in atoms are fused to form neutrons

nuclear fusion joining together of the nuclei of lighter elements to form another element, with the release of energy

nuclear reactions reactions involving the breaking of bonds between the particles (protons and neutrons) inside the nuclei of atoms

parallax apparent movement of close stars against the background of distant stars when viewed from different positions around Earth's orbit

parsec unit for expressing distances to stars and galaxies, it represents the distance at which the radius of Earth's orbit subtends an angle of one second of arc

planetary nebula ring of expanding gas caused by the outer layers of a star less than eight times the mass of our sun being thrown off into space

positrons positively charged electrons

pulsating star star that periodically expands due to an increase in core temperature and then cools and contracts under gravity

radio telescope a telescope that can detect radio waves from distant objects

red giant star in the late stage of its life in which helium in its core is fused to form carbon and other heavier elements

red shift shift of lines of a spectral pattern towards the red end when the source is moving away from the observer

scientific notation a way of writing very large or small numbers, using power notation; e.g. 1.5×10^8 km

singularity a single point of immense energy present at the time of the Big Bang

spectrum light from a source separated out into the sequence of colours, showing the different frequencies

Steady State theory a theory that states that there was no beginning to the Universe and that the Universe does not change in appearance

supercomputers computers with the fastest processing capacity available

supergiant very large star that is expanding while running out of fuel, and will eventually explode

supernova huge explosion that happens at the end of the life cycle of supergiant stars

ultraviolet radiation invisible radiation similar to light but with a slightly higher frequency and more energy

wavelengths invisible radiation similar to light but with a slightly higher frequency and more energy

white dwarf the core remaining after a red giant has shed layers of gases

X-rays high energy electromagnetic waves that can be transmitted through solids and provide information about their structure

---

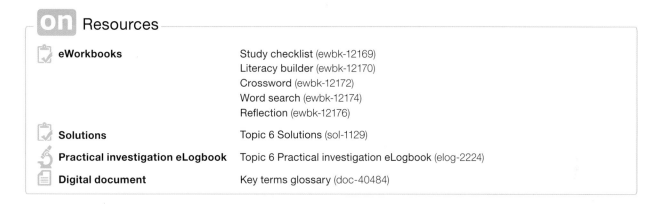

on Resources

eWorkbooks	Study checklist (ewbk-12169)
	Literacy builder (ewbk-12170)
	Crossword (ewbk-12172)
	Word search (ewbk-12174)
	Reflection (ewbk-12176)
Solutions	Topic 6 Solutions (sol-1129)
Practical investigation eLogbook	Topic 6 Practical investigation eLogbook (elog-2224)
Digital document	Key terms glossary (doc-40484)

6.8 Activities

learn on

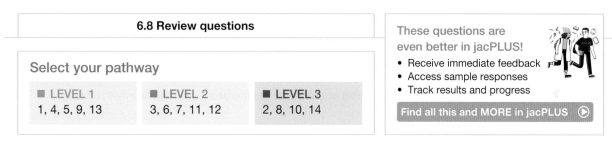

6.8 Review questions

These questions are even better in jacPLUS!
- Receive immediate feedback
- Access sample responses
- Track results and progress

Find all this and MORE in jacPLUS

Select your pathway

■ LEVEL 1	■ LEVEL 2	■ LEVEL 3
1, 4, 5, 9, 13	3, 6, 7, 11, 12	2, 8, 10, 14

Remember and understand

1. **MC** During which process is energy emitted by stars?
   A. Gas and dust collapsing and forming a cloud (nebulae)
   B. Nuclear fission
   C. Extinction
   D. Nuclear fusion

2. **MC** Explain the difference between the apparent magnitude of a star and its absolute magnitude.
   A. The apparent magnitude is a measure of how bright a star appears to be from Earth, while the absolute magnitude is a measure of how far away from Earth a star is.
   B. The absolute magnitude is a measure of how bright a star appears to be from Earth, while the apparent magnitude is a measure of how far away from Earth a star is.
   C. The apparent magnitude is a measure of how bright a star appears to be from Earth, while the absolute magnitude is a measure of how much light a star emits.
   D. The absolute magnitude is a measure of how bright a star appears to be from Earth, while the apparent magnitude is a measure of how much light a star emits.

3. **MC** What makes stars group together to form galaxies?
   A. Gravitational attraction   B. Nuclear fission
   C. Absolute magnitude   D. The ecliptic

4. What do the black lines in the spectra of light emitted by a star tell us?

5. What do each of the following abbreviations stand for?
   a. COBE   b. WMAP   c. ISS

6. Complete the following paragraph to explain the difference between a neutron star and a black hole.
   A _____ is formed from the dense remains of a _____. It has a solid core of _____, which is formed when gravity forces the protons and electrons together. If this neutron core is sufficiently large (about three times the mass of our Sun), the gravitational forces are so great that everything is 'sucked in' — even light. Such a core becomes a _____.

Apply and analyse

7. **SIS** Use the data in the provided table to answer the following questions.

TABLE Apparent and absolute magnitudes of stars and constellations

Star and constellation	Apparent magnitude	Absolute magnitude
Sun	−27	+4.7
Sirius (Canis Major)	−1.5	+8.7
Canopus (Carina)	−0.73	−4.6
Alpha Centauri (Centaurus)	−0.33	+4.7
Rigel (Orion)	+0.11	−7.5
Beta Centauri (Centaurus)	+0.60	−5.0
Betelgeuse (Orion)	+0.80	−5.0
Aldebaran (Taurus)	+0.85	−0.3
Alpha Crucis (Southern Cross)	+0.90	−3.9

Which of the stars Alpha Centauri, Betelguese and Rigel:
   a. is brightest when viewed from Earth on a clear night?
   b. has the greatest actual brightness?
   c. is faintest when viewed from Earth on a clear night?
8. **SIS** How have scientists gained their knowledge of the life and death of stars if the processes involved take millions of years to occur?
9. **SIS** Two different theories about the beginning of the Universe emerged during the twentieth century.
   a. Name the two theories.
   b. Describe which theory proposed that there was no beginning.
   c. Explain which lost favour in 1965. Why did it lose favour?
10. Explain what cosmic microwave background radiation is and why it exists.

Evaluate and create

11. Explain the difference between a ground and a space telescope, and provide one example for each.
12. **SIS** Outline two major advantages of using radio telescopes instead of light telescopes to view events in deep space from Earth's surface.
13. Explain the purpose of the International Space Station.
14. Explain why Brian Schmidt was awarded the Nobel Prize for Physics in 2011.

Fully worked solutions and sample responses are available in your digital formats.

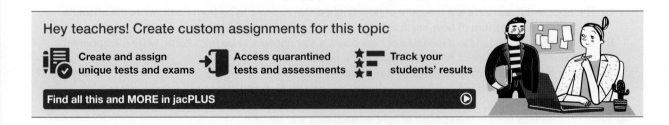

Hey teachers! Create custom assignments for this topic

Create and assign unique tests and exams

Access quarantined tests and assessments

Track your students' results

Find all this and MORE in jacPLUS

Online Resources

Below is a full list of **rich resources** available online for this topic. These resources are designed to bring ideas to life, to promote deep and lasting learning and to support the different learning needs of each individual.

6.1 Overview

eWorkbooks
- Topic 6 eWorkbook (ewbk-12156)
- Starter activity (ewbk-12159)
- Student learning matrix (ewbk-12158)

Solutions
- Topic 6 Solutions (sol-1129)

Practical investigation eLogbook
- Topic 6 Practical investigation eLogbook (elog-2224)

Video eLesson
- Hubble and the expansion of the Universe (eles-1766)

6.2 What can we observe in the changing night sky?

Weblinks
- Star maps
- The night sky time-lapse

6.3 How can we measure cosmic distances?

eWorkbook
- Observing stars (ewbk-12183)

Practical investigation eLogbook
- Investigation 6.1: The effect of parallax (elog-2227)

Video eLessons
- Twinkle, twinkle (eles-0071)
- Star trails (eles-2724)

6.4 What does starlight tell us?

eWorkbooks
- The expanding Universe (ewbk-12185)
- The brightness of stars (ewbk-12161)
- Star life cycles (ewbk-12163)

Practical investigation eLogbooks
- Investigation 6.2: Heat produced by compressing a gas (elog-2229)
- Investigation 6.3: Seeing the colours of stars (elog-2231)
- Investigation 6.4: Red shift of light (elog-2233)

Teacher-led video
- Investigation 6.4: Red shift of light (tlvd-10818)

Video eLessons
- The expanding Universe (eles-0038)
- Biggest bang (eles-1074)
- Supernova (eles-2032)

Interactivities

- Shifting spectral lines (int-0678)
- Expansion of the Universe (int-0057)
- Star cycle (int-0679)

6.5 How did it all begin?

eWorkbook
- The Big Bang (ewbk-12165)

Video eLesson
- Entropy: Is the end of the Universe nearer than we thought? (eles-1073)

Weblink
- How will the Universe end, and could anything survive?

6.6 Advancing science, technology and engineering

eWorkbook
- Telescopes (ewbk-12167)

Video eLessons
- The radio telescope at Arecibo (eles-2726)
- The Very Large Array satellite dishes at Twilight in New Mexico, USA (eles-2727)
- The Hubble Space Telescope (eles-2725)
- Expedition 34: International Space Station (eles-2728)

Interactivity
- Hubble Space Telescope (int-5867)

Weblinks
- Spot the ISS depending on your location
- Live Space Station tracking map

6.8 Review

eWorkbooks
- Topic review Level 1 (ewbk-12177)
- Topic review Level 2 (ewbk-12179)
- Topic review Level 3 (ewbk-12181)
- Study checklist (ewbk-12169)
- Literacy builder (ewbk-12170)
- Crossword (ewbk-12172)
- Word search (ewbk-12174)
- Reflection (ewbk-12176)

Digital document
- Key terms glossary (doc-40484)

To access these online resources, log on to **www.jacplus.com.au**

7 Patterns of global climate change

CONTENT DESCRIPTION

Use models of energy flow between the geosphere, biosphere, hydrosphere and atmosphere to explain patterns of global climate change (AC9S10U04)

Source: F–10 Australian Curriculum 9.0 (2024–2029) extracts © Australian Curriculum, Assessment and Reporting Authority; reproduced by permission.

LESSON SEQUENCE

SCIENCE INQUIRY AND INVESTIGATIONS

Science inquiry is a central component of Science curriculum. Investigations, supported by a **Practical investigation eLogbook** and **teacher-led videos**, are included in this topic to provide opportunities to build Science inquiry skills through undertaking investigations and communicating findings.

LESSON
7.1 Overview

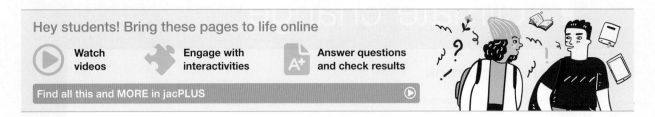
7.1.1 Introduction

We are living in the Anthropocene era — an age in which humans are dominating and disrupting many of our planet's natural systems. Advances in technology have increased our knowledge and understanding of life on our planet. These advances have also enabled the development and mass production of objects designed to make our lives easier. But the increasing demand for the latest fashion in mobile phones, appliances, cars, computers, electronic devices and accessories, whitegoods, kitchen gadgets, home appliances, toys and all of the associated plastic packaging, comes at a cost to our planet. The production of various items and the consumer lifestyle have led to a greater demand for fossil fuels. This, in turn, has resulted in an increased release of carbon dioxide and other greenhouse gases into the atmosphere. With a better understanding of how elevated carbon dioxide levels contribute to climate change, there's a growing emphasis on governments to take active measures in response including setting targets to reduce both fossil fuel consumption and carbon dioxide emissions, and instigating changes within the manufacturing sector to align with these goals.

Simultaneously, society is undergoing a transformation as individuals become more conscious of the far-reaching effects of climate change. People are seeking ways to minimize their carbon footprint by engaging in practices such as reusing or recycling goods, adopting electric vehicles, and even cultivating their own food. Demonstrations by students and advocacy groups have also become increasingly commonplace. These protests not only highlight the urgent need to address climate change but also urge governments to take significant actions before it becomes irreparable.

In the Anthropocene era, we also must consider the lingering impact of activities like the radioactive fallout from the atomic bombs used in 1945 and subsequent nuclear bomb testing. These historical actions have ramifications for our future lives and the environment. This consideration becomes paramount in our examination of humanity's influence on Earth. How can we apply our scientific knowledge to safeguard our planet and its inhabitants? If we remain inactive, how will the Earth transform over the course of the next century?

FIGURE 7.1 A coal-powered plant and student activists protesting in 2020

Video eLesson Climate change protests (eles-4175)

Watch this video to observe part of the Strike for Climate rally in 2019, as more individuals push for change in order to protect the environment.

7.1.2 Think about global systems

1. What affects are humans and their activities having on the Earth?
2. What has both a 'layer' and a 'zone' in it?
3. When is the 'laughing gas' nitrous oxide nothing to laugh about?
4. If global cooling did increase the size of the human brain, what effects might global warming have?
5. How are organisms responding to the changes in their environment?
6. Why do people support or not support the idea of climate change?

7.1.3 Science inquiry

Disappearing ice

When we think about the North or South Pole of Earth, we often picture penguins and polar bears, the extreme cold, and vast icy landscapes. Over the past two decades, scientists have been closely studying Antarctica and its enormous sheets of ice to better understand how the melting ice might affect life on our planet. They're using information and computer models to predict how the complex interactions between the atmosphere and changes in the oceans are causing the ice to melt. As our planet's climate changes and temperatures of both water and air rise, we're witnessing more and more of this ice melting.

This melting ice in Antarctica isn't just affecting the animals and plants that live there – it's also contributing to global warming. Ice is very reflective, almost like a mirror, and has a high albedo, which is the fraction of light or radiation that is reflected by a surface. It reflects back a lot of the sunlight that hits it. But as the ice melts, it uncovers the dirt, rocks, and vegetation below, which don't reflect as much sunlight and instead absorbs more heat. This creates a cycle where more melting leads to even more warming.

In places like Australia, scientists predict that sea levels could go up by 15 to 30 centimeters by the year 2050. The World Climate Research Programme (WCRP) estimates that if the Earth keeps getting warmer quickly, global sea levels might rise by 1.3 to 1.6 meters by the year 2100. This is a big deal because it could cause flooding in coastal areas and change the way our oceans and weather systems work.

FIGURE 7.2 The landscape of Antarctica is changing due to climate change.

Questioning and predicting

1. **a.** Compare how much ice in Antarctica has melted from the 1990s to 2020 using the weblink **Cool Antarctica**.
   b. Suggest why scientists are concerned about the ice loss in Antarctica.

Planning and conducting

2. Research and describe how scientists are able to determine the rate of ice loss in Antarctica and how these findings help shape legislation and future planning.

Analysing and evaluating

3. **a.** List examples of human activities that are contributing to climate change on Earth, resulting in ice loss.
   b. Rank the human activities in part **a** that you believe has the highest contribution to climate change, to the lowest contribution. Justify your response.
   c. Determine what is the threshold that can be reached before there is irreversible loss of all or part of the Antarctic ice sheet?

Communicating

4. **a.** Research the predicted rise in sea levels over the next 100 – 200 years and compare the different values scientists are coming up with. Explain why there is a difference in the predicted outcomes.
   b. Report how rising sea levels would affect coastal areas and islands and the future planning required for these areas.

LESSON
7.2 Global systems

> **LEARNING INTENTION**
>
> At the end of this lesson you will be able to describe how the biosphere supports all life on Earth, the importance of the various Earth cycles and the impact humans are having on these cycles.

7.2.1 Interacting spheres

Our planet is made up of a number of interconnected systems (figure 7.3). Four of these are considered to be global systems. These are the **biosphere**, the **atmosphere**, the **hydrosphere** and the **lithosphere** (also called the geosphere). Interactions occur between all of these spheres with each other and with the radiant energy of the sun.

The biosphere

Earth consists of a thin layer in which all living things (**biota**) live. This includes the atmosphere, the ocean depths, and the upper part of Earth's crust and its sediments. The layer on Earth in which all life exists is called the biosphere.

biosphere the layer on Earth in which all life exists. It includes all living things (biota) and all ecosystems

atmosphere the layer of gases around the Earth

hydrosphere Earth's water from the surface, underground and air

lithosphere the outermost layer of the Earth; includes the crust and uppermost part of the mantle

biota the living things within a region or geological period

Within the biosphere there are regions called **biomes**. As biomes are influenced by environmental factors (such as latitude, temperature and rainfall) that determine the type of vegetation that can survive in an area, they are named according to the dominant vegetation (figures 7.6 and 7.7). Within the biomes are **ecosystems** and within these, materials are cycled between the atmosphere, hydrosphere and lithosphere within the carbon, nitrogen and phosphorus cycles.

on Resources

🔗 **Weblink**　The geosphere and biosphere

The atmosphere

Earth's atmosphere (as shown in figure 7.4) is divided into five layers; the **troposphere**, the **stratosphere**, the mesosphere, the thermosphere and the exosphere. The ionosphere, a region, consisting of ionised atoms and electrons, overlaps the upper three layers and extends to the 'edge of space'.

FIGURE 7.3　How the different spheres interact

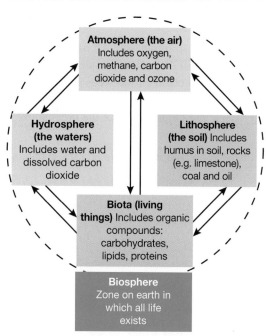

The troposphere is about 6 to 17 kilometres high, depending on how close you are to the Earth's equator, and contains 75% of the total mass of the atmosphere. It has, by far, the greatest influence on the weather. The stratosphere is approximately 35 to 40 kilometres thick and contains a region known as the ozone layer. While ozone allows visible and infrared radiation from the Sun through, it absorbs ultraviolet (UV) radiation. This reduces the amount of damaging UV radiation from the Sun reaching the Earth's surface. Excessive absorption leads to skin cancer in humans.

FIGURE 7.4　Layers in Earth's atmosphere

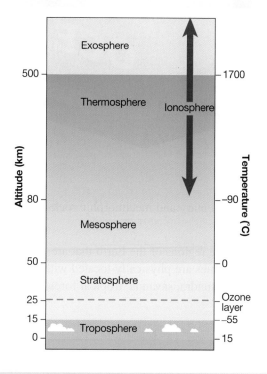

biomes　regions of the Earth divided according to dominant vegetation type

troposphere　the layer of the atmosphere closest to the Earth's surface, containing closely packed particles of air

stratosphere　the second layer of the atmosphere up to about 55 km above the Earth's surface, between the troposphere and the mesosphere

The hydrosphere

The waters of our planet make up the hydrosphere. Water comes from the surface of the planet, underground water supplies and from the air. The water cycle (seen in figure 7.5) demonstrates how water continually moves around in different states (ice, liquid and gas).

Why is water so important? All organisms on Earth require water to survive. For example, if plants do not have water, they will die and, consequently, consumers of these plants will also perish. This will have a detrimental effect on food chains, food webs and ecosystems. The evaporation of water from the sea surface is also important for the movement of heat as part of the climate system. This evaporation helps cool the surface of the ocean and helps to reduce the greenhouse effect.

FIGURE 7.5 The water cycle and the continual movement of water in the hydrosphere

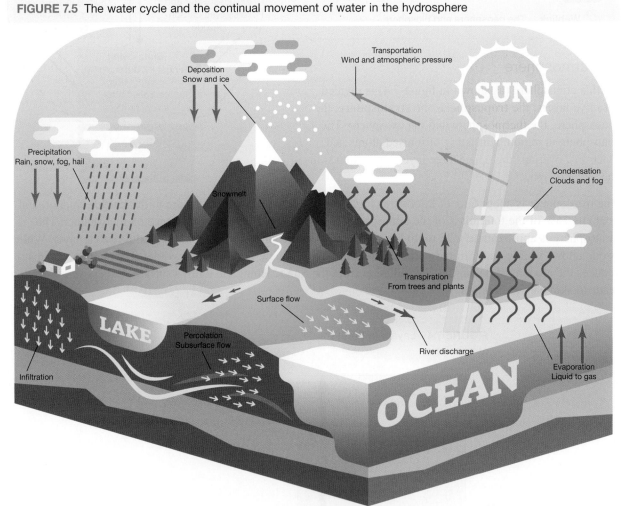

The lithosphere

The Earth's rocky crust (consisting of igneous, sedimentary and metamorphic rocks) and soil make up the lithosphere.

As mentioned earlier, biomes within the biosphere are regions of the Earth that are named according to their dominant type of vegetation. Some of these biomes are physically located within the lithosphere. Figures 7.6 and 7.7 show examples of these, such as tundra, savanna, tropical forest, desert and temperate grassland biomes.

FIGURE 7.6 The different biomes on Earth

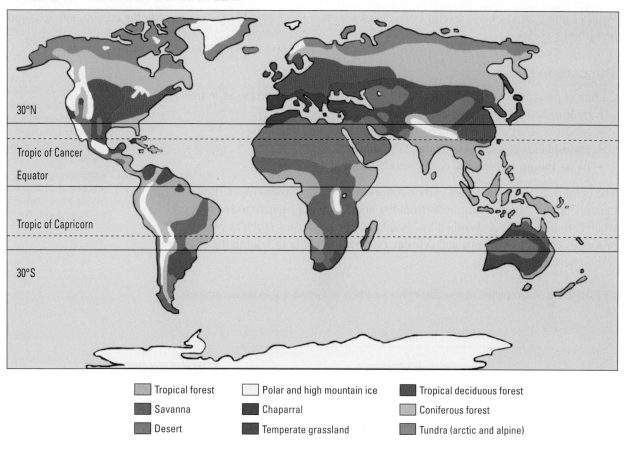

▢ Tropical forest	▢ Polar and high mountain ice	▢ Tropical deciduous forest
▢ Savanna	▢ Chaparral	▢ Coniferous forest
▢ Desert	▢ Temperate grassland	▢ Tundra (arctic and alpine)

FIGURE 7.7 Examples of different types of biomes: **a.** tropical forest **b.** savanna **c.** desert **d.** polar **e.** chaparral **f.** temperate grassland **g.** temperate deciduous forest **h.** coniferous forest **i.** tundra

7.2.2 Carbon in the biosphere

Within Earth's atmosphere, several cycles work together to support life. These include the carbon, nitrogen and phosphorus cycles.

The carbon cycle

Carbon is present in various forms within the biosphere. It can be found in the:
- hydrosphere, as dissolved carbon dioxide
- lithosphere, as coal or oil deposits and rocks such as limestone
- atmosphere, as methane or carbon dioxide
- living things, as proteins, carbohydrates or lipids.

The carbon cycle displays how carbon moves through the biosphere, as shown in figures 7.8 and 7.9. Carbon travels from the non-living atmosphere to living things when carbon dioxide is absorbed by photosynthetic organisms (such as plants). The processes of **photosynthesis** and **cellular respiration** are both vital in the cycling of carbon between organisms.

photosynthesis food-making process in plants that takes place in chloroplasts, and uses carbon dioxide, water and energy from the Sun

cellular respiration the chemical reaction involving oxygen that moves the energy in glucose into the compound ATP

FIGURE 7.8 A simplified illustration of how carbon is cycled within an ecosystem

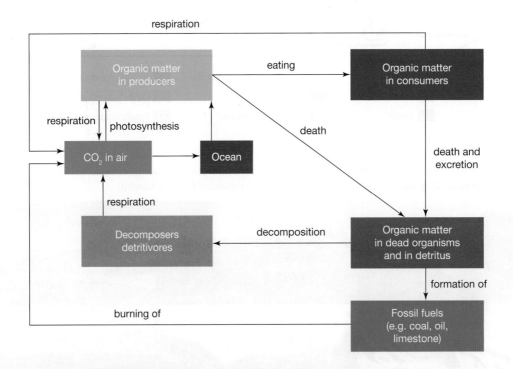

Each form of carbon in the carbon cycle can only hold a certain amount of carbon. Any changes in the cycle will result in carbon being shifted to other reservoirs, increasing or decreasing the amount in the reservoir. The release of carbon gases into the atmosphere has resulted in an increase in temperatures on Earth.

FIGURE 7.9 A simplified version of the carbon cycle

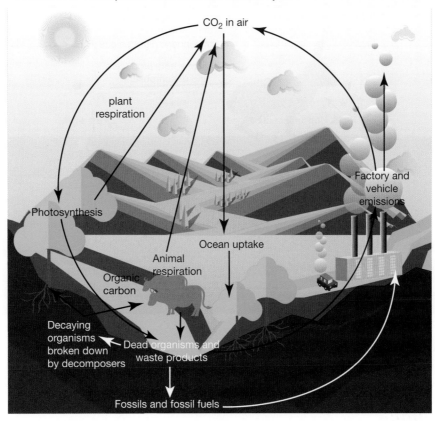

The production and use of carbon in the form of carbon dioxide (CO_2) by the ecosystem is summarised in table 7.1.

TABLE 7.1 Main ways CO_2 is produced and used

Production of CO_2 in ecosystem	Uptake of CO_2 by ecosystem
Burning of fossil fuels	Ocean uptake
Cellular respiration by various organisms (including decomposers, consumers, producers and detritivores)	Photosynthesis (mainly by plants)

The nitrogen cycle

The nitrogen cycle represents how nitrogen moves through the biosphere. Nitrogen is important because:
- our cells require it to make our DNA and proteins such as enzymes and hormones
- plants require it to make chlorophyll, which is used in the process of photosynthesis.

However, nitrous oxide (N_2O), which is produced as an intermediate during denitrification and nitrification (figure 7.10), absorbs radiation and traps heat in the atmosphere like carbon.

The phosphorus cycle

The phosphorus cycle represents how phosphorus moves from the lithosphere to the hydrosphere, through food chains and back. Phosphorus is an important component of fertilisers, which are used to provide nutrients to plants and improve their growth and crop yield. As there is no major gaseous forms of phosphorus, this has little impact on climate change.

FIGURE 7.10 A simplified version of the nitrogen cycle

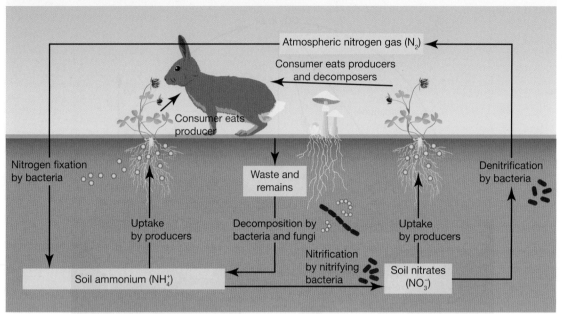

FIGURE 7.11 The phosphorus cycle

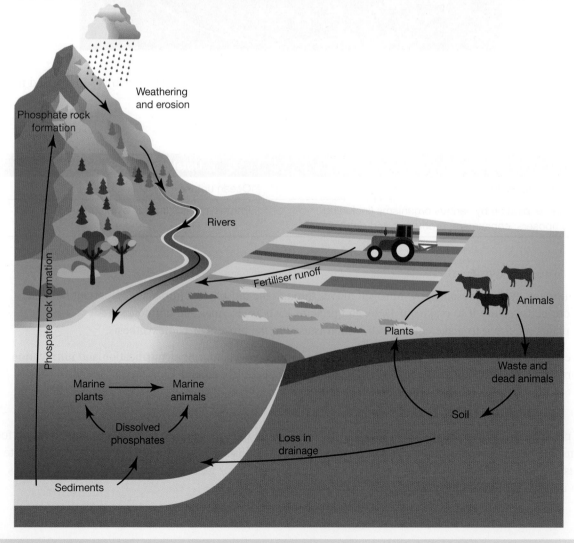

7.2.3 Human impact on Earth's cycles

Humans are becoming more aware of the detrimental effect of our actions on Earth's climate, biomes, oceans, ecosystems and living systems. People are finally beginning to understand and accept that changes need to be made about how we utilise our planet before it is too late.

So what are the main factors that have a detrimental effect on our planet and its cycles?

Deforestation

Deforestation has a major impact on Earth's cycles and ecosystems. Humans clear forests to build houses, cities and produce food and fuel. The adverse effects of deforestation include:

- the release of more carbon dioxide to the atmosphere as there are fewer green plants available for photosynthesis.
- a negative impact on the carbon, nitrogen and phosphorus cycles. The reduction in trees changes how these three elements are able to move through their cycles, which in turn affects ecosystems, plant life, habitats and so on.
- an increased probability of **desertification** occurring. This is when a flourishing, viable habitat disintegrates into a dry and arid region. As there is now less water on the surface for the Sun to evaporate, this means that there is more energy to warm the ground and the lower atmosphere.

FIGURE 7.12 Large sections of forests are cleared for mining, food, agriculture and housing.

Mining

Mining for precious metals used in items such as mobile phones or for non-renewable energy sources such as coal, requires land to be cleared, as seen in figure 7.13. Mining sites encompass enormous areas. The natural landscape is destroyed and a great deal of waste material is produced. Consequently, this results in more carbon byproducts being released into the atmosphere due to deforestation and the use of fossil fuels in machinery and vehicles.

FIGURE 7.13 Bird's eye view of an opencast mining quarry

CFCs

The ozone layer is a layer of O_3 gas high in the stratosphere. Its purpose is to prevent damaging UV rays from penetrating the stratosphere and reaching Earth. Many years ago, the greenhouse gas, chlorofluorocarbons (CFCs), were used extensively in refrigeration and aerosols, which resulted in the destruction of the ozone layer — specifically above

deforestation the process of clearing trees to convert the land for other uses

desertification the process in which fertile regions become more dry and arid

Australia. This gas is able to absorb infrared radiation and trap substantially more heat than carbon dioxide. Through government and media intervention, the use of CFCs for refrigeration and aerosols was banned, removing a greenhouse gas contributor.

Increases in human population

An increase in the human population creates a demand for more food, which results in the need for more livestock. This, in turn, leads to increased deforestation and a reduction of nutrients in the soil. In addition, an increase in livestock results in more methane being released into the atmosphere. More methane in the atmosphere will result in climate warming as methane is a greenhouse gas and traps more heat in the atmosphere than carbon dioxide.

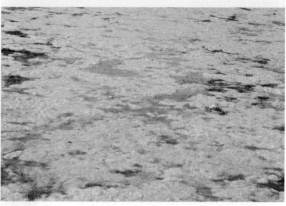

FIGURE 7.14   Algal bloom in freshwater due to eutrophication

More fertilisers are needed to enhance crop growth, and excess nitrogen and phosphorus can end up in our water system — this affects the natural cycling of water, nitrogen and phosphorus. In some instances, this has led to **eutrophication** (an excess of minerals and nutrients, which induces excessive growth of algae leading to oxygen depletion in the water) and the death of organisms within those ecosystems (figure 7.14). It is predicted that eutrophication will increase methane emissions into the atmosphere by about 30 to 90 per cent over the next 100 years.

Industrial wastes

Many factories across the world release wastes into the atmosphere, waterways and surrounding land. Australia has an extremely regulated industry and the release of wastes into our environment is closely monitored through the Environmental Protection Agency (EPA). Many environmental monitoring systems are in place throughout the world, however, not all monitoring systems are reliable or monitored appropriately.

FIGURE 7.15   The effects of acid rain on trees

Thus, industrial wastes that contain gaseous contaminants such as carbon dioxide, methane, nitrous oxide, hydrofluorocarbons, perfluorocarbons, sulphur hexafluoride and nitrogen trifluoride can be released into the atmosphere and have an impact on Earth's climate.

Sulfur dioxide and nitrous oxides can react with water vapour in the air to form acidic gases which accumulate in rain clouds. If the concentration of these acids are high enough, when rain falls from the cloud, these acids also fall. The result is **acid rain**. As seen in figure 7.15, acid rain has a detrimental effect on plants, which means more carbon dioxide remains in the atmosphere.

eutrophication a form of water pollution involving an excess of nutrients leaching from soils

acid rain rainwater, snow or fog that contains dissolved chemicals that make it acidic

Travel

Humans need to move around, whether internationally or locally, and this often involves the use of aeroplanes and cars. As most of these transport options burn petrol (a carbon-based product), carbon dioxide is released into the atmosphere.

There is no longer any doubt that the increased levels of carbon dioxide in our atmosphere have resulted in increased global temperatures. Evidence of this is seen in the unprecedented extent of bleaching of coral on the Great Barrier Reef, the melting and collapse of icecaps, the drowning of polar bears that cannot find icebergs to swim to, the rising sea levels and 'sinking' of islands under water, and many, many other examples. Climate change events will be further explored in lesson 7.4.

FIGURE 7.16 a. Cars release carbon dioxide when they burn petrol and diesel. b. Increased greenhouse gases have resulted in the melting of icecaps.

CASE STUDY: The water cycle and desertification

When you think of desertification, you likely think of sand dunes, infertile land, and dry, hot and inhospitable places. However, this is not to be mistaken for a desert biome.

Desertification usually occurs when the soil has been changed in such a way that water can no longer penetrate to the roots of the plants. This can be caused by the over-farming and over-grazing of land, or by fire or severe drought. If the land cannot recover from these changes (even after the problem has been removed), it is classified as desertified. In these situations, the water cycle can be permanently disrupted. This occurred significantly in the Sahel region in Africa, as shown in figure 7.17b, from the late 1990s, leading to the loss of fertile land.

FIGURE 7.17 a. The effects of desertification on land b. An image of Mali (in the Sahel region in West Africa), 9 June 2000

Scientists can identify possible desertification by looking at rainfall and plant growth over time (see figure 7.18). This allows them to observe times of not just severe drought, but the impact this has on vegetation.

FIGURE 7.18 Rainfall records and vegetation measurements for the Sahel from 1981 to 2000. This shows the fluctuations from an average expected reading and the drop in rainfall and vegetation in the late 1990s.

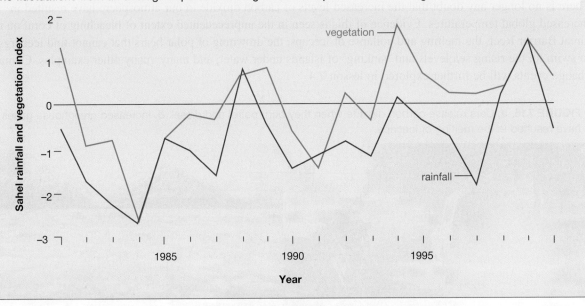

elog-2470

INVESTIGATION 7.1

Effect of deforestation

Aim

To explore how deforestation affects the water cycle

Materials

- plants (any small plant will work)
- glass jars (large enough that multiple plants can sit in soil)
- lid or cling wrap and rubber band to cover glass jar
- soil
- water

Method

1. Write a clear hypothesis for the investigation.
2. Fill two glass jars halfway with equal amounts of soil.
3. In one jar, plant around six to seven plants in the soil. In the other jar, plant only one or two plants.
4. Add equal amounts of water to each jar (this will be dependent on the size of your jar and ensure not to add too much).
5. Seal the jars with its lid or cling wrap and a rubber band and observe what happens.

Results

Construct a table and note your observations. You should note the observations every day for approximately two weeks.

Discussion

1. Identify variables you controlled between the two trials.
2. Describe any errors that occurred in your investigation and how these may be improved.
3. Write four sentences explaining how deforestation can affect an ecosystem.

Conclusion

Summarise the findings of this investigation about how deforestation affects the water cycle.

INVESTIGATION 7.2

Effect of acid rain on plants

Aim

To investigate how acid rain affects plants

Materials

- 2 x same plant in soil
- low concentration acid (0.01 mol L^{-1} HCl)
- water
- measuring tape

Method

1. Write a clear hypothesis for the investigation.
2. Measure and record the heights of the plants.
3. Regularly water one plant with just water and the second plant with water and low concentration acid, ensuring equal amounts of liquid is added to each plant.
4. Observe what happens and record your findings.

Results

Construct a table and note your observations. You should note observations every day for approximately two weeks.

Discussion

1. Identify the dependent and independent variables in this investigation.
2. Describe how acid rain forms normally.
3. Does acid affect the growth of a plant? Justify your response using data from your investigation.

Conclusion

Summarise the findings for this investigation about how acid rain affects plants.

Extension

Test the pH of the soil for both plants before watering, water as per the method, then test the pH every day before watering to collect data along with the observations.

EXTENSION: Will planting more trees solve our problems?

Visit the **Australian State of the Environment: Land** weblink and read through the information about land clearing (or research your own information about land clearing and deforestation in Australia) and answer the following questions.
1. Carefully examine the graph (Figure 34) showing annual areas of primary clearing, provided in the weblink.
   a. How has land clearing in Australia changed over the years? Describe any patterns observed.
   b. Determine the period that had the smallest amount of land clearing. Explain why.
2. Research what desertification is and determine if it possible that parts of Australia are experiencing desertification? What other information would you need in order to confirm if it is happening or not?
3. Would planting more trees solve the issue of deforestation and the amount of carbon dioxide in the atmosphere? Justify your reasoning.

 Resources

eWorkbooks	Recycling carbon (ewbk-13170)	
	Cycles in nature (ewbk-13172)	
Video eLessons	Removing algae from a waterway (eles-2731)	
	Excessive clearing and deforestation (eles-2905)	
Interactivity	Lake Urmia (int-5605)	
Weblinks	Australian State of the Environment: Land	
	WWF Footprint Calculator	

7.2 Activities

| 7.2 Quick quiz | on | 7.2 Exercise |

These questions are even better in jacPLUS!
- Receive immediate feedback
- Access sample responses
- Track results and progress

Find all this and MORE in jacPLUS ▶

Select your pathway

■ LEVEL 1	■ LEVEL 2	■ LEVEL 3
1, 3, 4, 7, 9	2, 5, 6, 10, 11	8, 12, 13, 14

Remember and understand

1. The biosphere includes all of the _____ on Earth, along with the _____, _____ and _____ cycles.
2. The stratosphere is uppermost in the atmosphere. It is approximately 35 to 40 kilometers thick and contains the ozone layer. How does the ozone layer assist life on Earth?
3. Match the type of biome to its picture. Biome types: chaparral, coniferous forest, desert, polar, savanna, temperate deciduous forest, temperate grassland, tropical forest, tundra.

4. Photosynthesis is a _____-_____ process in plants. It uses the gas, _____ _____, water and _____ from the Sun to produce glucose.
5. Describe the process of cellular respiration.
6. Identify if the following statement is true or false, providing reasons for your response.
   Depleted areas of the ozone layer can result in smaller amounts of damaging UV rays getting through, hence protecting living organisms.
7. List the criteria used to divide regions into biomes.

Apply and analyse

8. For each of the cycles, identify where and how the non-living parts of the biosphere (atmosphere, lithosphere and hydrosphere) and the living parts (biota) interact in the following.
   a. Carbon cycle
   b. Nitrogen cycle
   c. Phosphorus cycle
   d. Water cycle
9. Suggest why an increase in CFCs in the atmosphere would be of concern. Based on this, what recommendations would you have on the use of CFCs?
10. Provide examples of two environmental factors that contribute to the type of biome that exists in a particular area. Explain how each contributes to the type of biome.

Evaluate and create

11. Research and produce a report on the effects of human activity on the nitrogen and phosphorus cycles.
12. **SIS** Use the provided weblink for the **WWF Footprint Calculator** to complete the following.
    a. Hypothesise how many Earths would be required to sustain your current lifestyle.
    b. Answer each question on the calculator to the best of your ability, adding details to improve the accuracy of your footprint. Explain the outcome of your footprint after entering all your data. Do you think this is an accurate representation of your footprint?
    c. Identify ways you could reduce your footprint so that fewer Earths would be required.
    d. Is it possible for the answers in part **c.** to be implemented by the rest of society? Justify your reasoning.
    e. Scientists like to formulate questions to investigate scientifically in the hope of discovering new ways to reduce carbon and greenhouse emissions. Formulate two questions you would want to investigate as a scientist, and identify how you would collect the information required and what you would need to find out to help answer the question.
13. Refer back to figure 7.18 and the following figure showing rainfall data from the Sahel region in Africa.

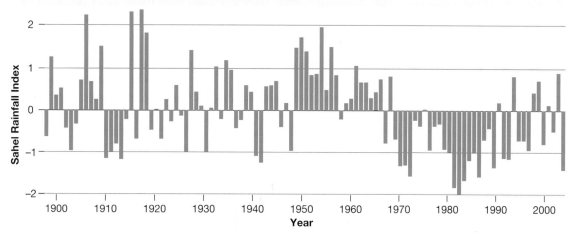

Fluctuations from an average expected reading in rainfall in the Sahel, Africa

a. Is the Sahel becoming desertified? Justify your reasoning.
b. Is it possible for other factors to be affecting the rainfall and vegetation growth? Explain.
14. **SIS** By recording rainfall data, scientists can analyse this data to identify changes to rainfall in various areas.
    a. Research climate data online and compare the rainfalls for one city and one country town and copy the graphs into a document to analyse.
    b. Identify any similarities, differences or trends in the graphs for the city and country town.
    c. Was data not recorded during any time frames? Why?
    d. Why would differences in the rainfall occur in these two areas?

Fully worked solutions and sample responses are available in your digital formats.

LESSON
7.3 Climate patterns

LEARNING INTENTION

At the end of this lesson you will be able to explain factors that influence Earth's climate, and describe how climate change will affect ocean currents and wind patterns.

7.3.1 Influences of the variation of Earth's climate

When scientists discuss climate change, they are referring to a significant difference in the average temperature of a region, as recorded over many, many years. We all know that we experience seasons and that Earth's climate changes in a cyclical fashion over the year. Therefore, Earth's climate is always changing — it always has and it always will. So why has climate change become the single most important issue for so many people in the twenty-first century? It is because climate change is occurring on a potentially catastrophic level, resulting in substantial changes to habitats, environments, waterways, oceans, land masses and atmospheric layers?

When scientists discuss climate *patterns,* they are not referring to the controversial discussion of global warming. Instead, they are talking about the regular, cyclical pattern of seasons, weather conditions, rainfall and temperatures in a region over time. Scientists have identified that the variation of climate over Earth's surface is largely the result of three major influences:

1. The tilt of Earth's axis and the amount of energy from the Sun reaching Earth's surface.
2. The differing abilities of land and water to absorb and emit radiant heat.
3. The features of the land.

The tilt of Earth's axis and the amount of energy from the Sun reaching the surface

The shape of Earth is almost spherical. So the energy from the Sun that reaches Earth's surface is spread over a larger area in the polar regions than near the equator. Polar regions will experience up to six months with little or no radiation from the Sun, while the region around the equator will always be warm because it experiences the same amount of radiation all year round. It is the difference in surface temperature between the poles and the tropics that causes the movement of air that we know as wind.

FIGURE 7.19   The shape and tilt of Earth affects the distribution of the Sun's radiation.

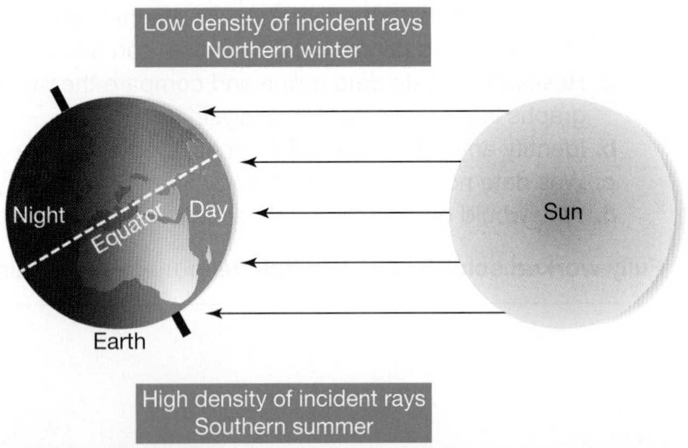

Low density of incident rays
Northern winter

Night   Equator   Day

Sun

Earth

High density of incident rays
Southern summer

INVESTIGATION 7.3

elog-2474

Exploring the axis tilt of Earth

Aim
To investigate how the tilt of Earth affects the Sun reaching the surface

Materials
- a ball
- ruler
- infrared thermometer

- protractor
- heat lamp

Method

1. Hold the ball and, using a heat lamp, shine the light directly onto the ball, holding the lamp approximately 10 cm away. Record the temperature of the surface of the ball.
2. Adjust the angle of the ball and measure the angle from the horizontal. Shine the light on the ball and record the temperature.
3. Continue adjusting the angle of the light shining on the ball.

Results

Copy and complete the following table (adding additional rows if required).

TABLE The surface temperature of a ball at different angles

Angle	Surface temperature

Discussion

1. Identify the independent and dependent variables in this investigation.
2. This is a model of how light from the Sun shines on Earth. Describe some limitations of this model.
3. Summarise your findings and relate this to the axis of Earth and climate.
4. Explain three improvements that could be made to improve the accuracy of the investigation.

Conclusion

Summarise the findings for this investigation about the tilt of the Earth and the Sun reaching the surface.

The differing abilities of land and water to absorb and emit radiant heat

During daylight hours the land absorbs **radiant heat** from the Sun more quickly than water. At night, heat is radiated from the land more quickly than from the water. As a result, the ocean temperature changes less on a daily basis than air and land temperatures, and coastal climates are protected from the high and low temperature extremes of inland climates. Weather stations and space satellites can collect data to allow for these differences to be measured (see figure 7.20).

radiant heat heat transferred by radiation, such as from the Sun to the Earth

FIGURE 7.20 a. Weather stations contain devices such as a thermometer to measure temperature, a barometer to measure atmospheric pressure, a hydrometer to measure humidity, an anemometer to measure wind speed and a wind vane to measure wind direction. b. Some of this information can also be collected from space, using satellites to monitor weather from Earth's orbit.

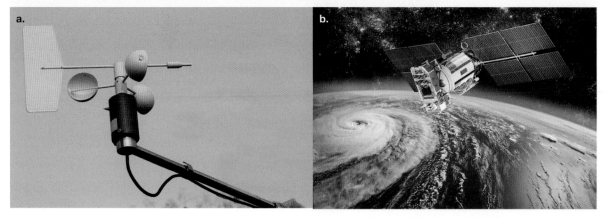

The features of the land

The temperature of the part of the atmosphere above Earth's land masses decreases with increased height above sea level. In addition, mountain ranges have a dramatic effect on the climate of nearby regions. They can block the path of wind blowing towards them, forcing the air to move quickly upwards to form almost permanent clouds, as water vapour in the air condenses quickly.

Other features of the land that affect climate:
- Sandy soils reflect more energy from the Sun than dark, fertile soils.
- Fresh snow reflects up to 90 per cent of the Sun's energy that reaches it.
- Heavily vegetated areas absorb much more of the Sun's radiation than bare land because plants use it to photosynthesise.

7.3.2 Climate patterns and climate change

Climate change also affects ocean currents and wind patterns, impacting the way that these circulate.

Ocean currents

The water in Earth's oceans is constantly moving in currents. Ocean currents are the result of Earth's rotation and the temperature difference between the tropics and poles.
- Warm surface water near the equator sinks and cools as it moves towards the poles, while the cold water in polar regions rises and warms as it moves towards the equator.
- Warm and cold ocean currents (shown in figure 7.21) move huge volumes of water past coastal regions and have a major influence on their climates. The Gulf Stream, for example, carries warm water from the equator into the North Atlantic Ocean, keeping Great Britain, Norway and Iceland warmer than other regions at similar latitudes. Cold water currents cool coastal regions that would otherwise be hot.

FIGURE 7.21 The movement of ocean currents

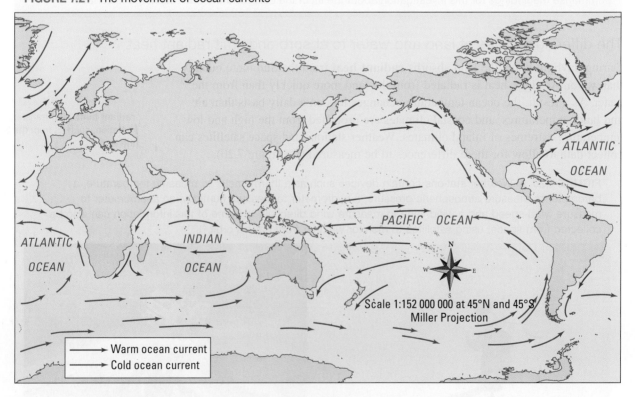

Carbon dioxide gas in the air traps heat in our atmosphere. This heat is absorbed by oceans, causing ocean temperatures to rise (see figure 7.22). This warmer body of water affects the movement of ocean currents, which counteracts the uneven distribution of solar radiation on Earth's surface.

FIGURE 7.22 Changes in the long-term average global ocean heat content, 1955 to 2020

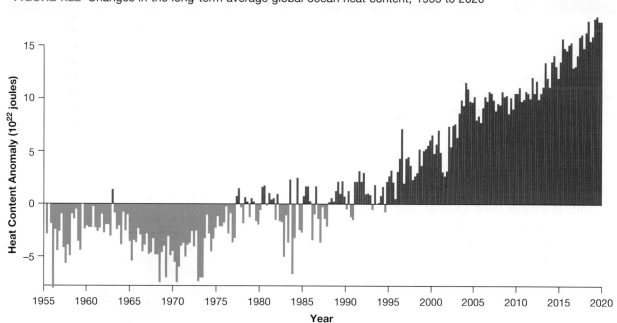

The currents around Australia

Around Australia, major currents circulate hot and cold water in the ocean (see figure 7.23).

FIGURE 7.23 Surface currents around Australia

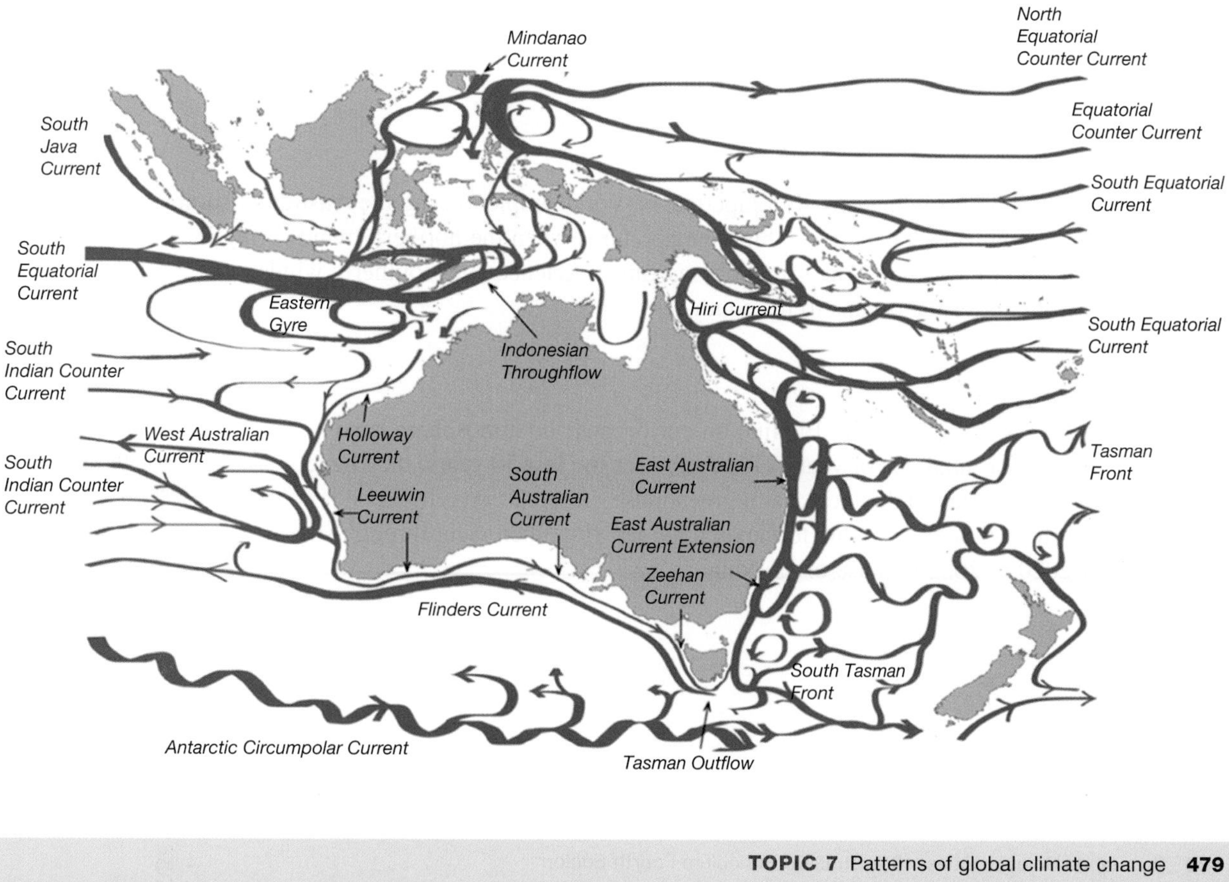

The three major currents that influence Australia's marine environment are:
- the East Australian Current
- the Indonesian Throughflow
- the Leeuwin Current.

These currents help:
- redistribute heat between the ocean and the atmosphere
- redistribute heat from warmer regions to the cooler regions of the globe.

Changes in these currents have led to marine heatwaves in the south-east Indian Ocean in 2010–11 and in eastern Tasmania during 2015–16. These changes have also affected the exchange of water between the open ocean and in-shore regions, altering the nutrients of the water.

Wind

The differences in surface temperature between the poles and the tropics cause the large-scale **convection currents** that create wind. Cold air near the poles sinks and moves towards the equator, and hot air near the equator rises and moves towards the poles.

Figure 7.24 shows the effects of these convection currents during March and September, when the Sun is directly over the equator.

The winds shown are called **prevailing winds** and are generally those most frequently observed in each region. The direction of prevailing winds is complicated by:
- latitude
- the rotation of Earth on its axis, the tilt of Earth's axis and Earth's orbit around the Sun
- friction caused by the land surface
- ocean currents
- local variations in air pressure and temperature
- variations in water and land temperature
- altitude.

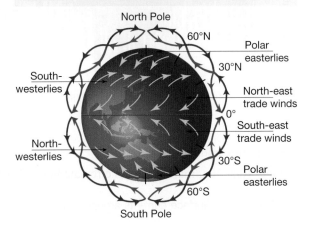

FIGURE 7.24 Convection currents of cool and warm air around the globe

The wind direction in turn influences air temperatures. For example, during the Australian summer, regions along most of the south coast experience high temperatures when the northerly winds bring in hot and dry air from above the land to the north. The same regions can experience cold southerly winds, which bring in cool and damp air from above the oceans to the south.

La Niña and El Niño

The climate patterns of La Niña and El Niño on our oceans and atmosphere have occurred naturally for hundreds of years. However, over the last 50 years, they have become more extreme due to the increase in greenhouse gas emissions. In 2015, Australia went through a drought due to the extreme El Niño, then went through severe flooding over 2020 to 2022 because of La Niña.

convection currents process of heat transfer in gases and liquids in which lighter and hotter materials rise, and cooler and denser materials sink

prevailing winds winds most frequently observed in each region of the Earth

FIGURE 7.25 Australia has experienced more extreme weather events including droughts and floodings.

So how do La Niña and El Niño cause these extreme weather patterns? It depends on the event, where you are and the season.

As seen in figure 7.26, cold water in the east is progressively warmed by heat from the sun as it travels west due to the trade winds, but during El Niño events, the trade winds weaken or reverse and send the warm water east. This results in more heat being released into the atmosphere from the warmer water, along with wetter and warmer weather to regions including the south of the United States and the Gulf of Mexico. Regions like Australia, central Africa and Asia tend to experience drier conditions.

During La Niña events, warmed water from the east travels further west than it normally would due to stronger trade winds. This can lead to drought in the southern United States and heavy rains further north. In Australia, South-East Asia and southeastern Africa there can be an increase in rainfall and can lead to catastrophic floods.

FIGURE 7.26 The difference between a normal year and an El Niño year.

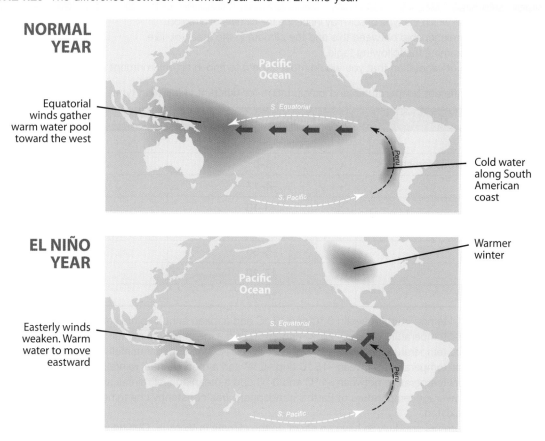

7.3 Activities

learn on

Remember and understand

1. List three major factors that influence the variation of climate over Earth's surface.
2. What causes the large-scale convection currents in the air that create prevailing winds?
3. Identify if the following statement is true or false, providing reasons for your response.
   The tilt of Earth's axis results in the polar regions receiving little or no solar radiation for six months of each year.
4. List five factors that determine the wind direction at any given time or place.
5. Outline the causes of warm and cold ocean currents.

Apply and analyse

6. Explain why Great Britain, Norway and Iceland experience warmer climates than other regions at similar latitudes.
7. **a.** Explain ways in which the type of soil may influence the climate of a region.
   b. Describe the relationship between this and the climate in the area you live.
8. **SIS** Carefully examine the following table.
   Identify the type of vegetation that you would most expect to find in an environment with the following.

TABLE The different temperatures and precipitation for different vegetation types

Vegetation type	Mean annual temperature (°C)	Mean annual precipitation (cm)
Tundra	−15–5	0–100
Northen coniferous forests	−5–0	50–150
Mediterranean	−4–17	0–60
Grassland	3–18	50–100
Temperate deciduous forest	3–19	50–300
Desert	−5–30	0–50
Savanna	17–30	50–200
Tropical forests	18–30	100–450

   a. A mean annual temperature between 0 °C and 15 °C and a mean annual rainfall around 50–100 cm
   b. A mean annual temperature between 20 °C and 28 °C and a mean annual rainfall around 250–400 cm
   c. A mean annual temperature between 20 °C and 28 °C and a mean annual rainfall around 20–30 cm
   d. A mean annual temperature between −15 °C and −5 °C and a mean annual rainfall around 80 cm
   e. A mean annual temperature between −4 °C and 0 °C and a mean annual rainfall around 100–150 cm
9. Explain why the average temperature of Earth's atmosphere was constantly changing for millions of years before humans existed.
10. Identify which currents are likely to have an effect on the climate in Victoria. Outline what impact these may have.

11. **SIS** Research, discuss and reflect on each of the following statements about climate change and state your own opinion.
    a. Australia has vast resources of coal, much of which is exported. The Australian coal industry provides employment and other benefits for the economy. If targets for the reduction of global emissions are high enough to damage the Australian coal industry, the government should not agree to them.
    b. Developing countries that have little or no industry have not contributed to global warming. These countries should be allowed to increase their carbon dioxide emissions so that they can develop industries and improve their living standards.
12. **SIS** a. Find out the mean annual temperature and mean annual rainfall of your local environment. Using the table from question 8, what type of vegetation would you expect to find there? Is this the case? If it is not, suggest possible reasons for the difference.
    b. Find out what climate change is predicted to occur in your local area due to global warming. Which vegetation would be best suited to this type of environment?

Fully worked solutions and sample responses are available in your digital formats.

LESSON
7.4 Climate change

> **LEARNING INTENTION**
>
> At the end of this lesson you will be able to describe climate change and the enhanced greenhouse effect, and explain the underlying causes of these changes.

7.4.1 Enhanced greenhouse effect

The Earth's atmosphere acts like a giant invisible blanket that keeps temperatures on our planet's surface within a range that supports life. Within the atmosphere, greenhouse gases trap some of the energy leaving Earth's surface to help maintain these warm temperatures. The maintenance of Earth's temperatures by these atmospheric gases is called the **greenhouse effect**. Without the greenhouse effect, the average temperature on Earth would be about −18 °C, so life would not exist as we know it. The main greenhouse gases are carbon dioxide, methane, nitrous oxide, water and ozone. In the past, other greenhouse gases that include chlorofluorocarbons and hydrofluorocarbons, were also known to be significant contributers to the greenhouse effect. However, due to a worldwide campaign to minimise the use of these gases in refrigerants and aerosols, today their effect on global warming is minimal. For many years, scientists have analysed the concentration of greenhouse gases in the atmosphere and recorded and publicised 'safe' and 'acceptable' levels of these gases that favour our atmosphere and environment, and that will maintain temperatures on land and sea. At these safe levels, the greenhouse effect supports our life on Earth.

What's the problem?

Global temperatures have been increasing at an accelerated rate since the Industrial Revolution and are expected to continue to do so. The observed rise in the average near-surface temperature of Earth is known as **global warming**. This has resulted in melting icecaps, rising sea levels, increased coastal flooding, unusual weather patterns and ocean currents, and consequent threats to the survival of many organisms and species.

> **greenhouse effect** a natural effect of the Earth's atmosphere trapping heat, which keeps the Earth's temperature stable
>
> **global warming** the observed rise in the average near-surface temperature of the Earth

What's the cause?

The Industrial Revolution of the nineteenth century resulted in the exponential extraction (mining) and use of fossil fuels in many industries, including for transport and the production of electricity. In the scientific world it is widely accepted that our increased dependence on fossil fuels is the major cause of global warming. Burning fossil fuels such as coal and oil, along with deforestation, travel, industrial waste and increase in the human population has resulted in increased levels of greenhouse gases (mainly nitrous oxide, carbon dioxide and water), which have increased the amount of heat trapped in our atmosphere, causing Earth to heat up. The

FIGURE 7.27 Greenhouse gases and the enhanced greenhouse effect

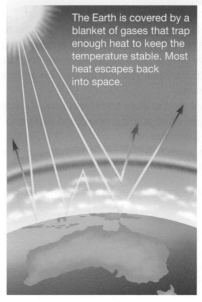

The Earth is covered by a blanket of gases that trap enough heat to keep the temperature stable. Most heat escapes back into space.

More carbon dioxide and other greenhouse gases in the air trap more heat from the Sun. The Earth's temperature will rise.

heating up of Earth is called the enhanced greenhouse effect, shown in figure 7.27. Some sources of these human-produced greenhouse gases are shown in figure 7.28.

int-5884

FIGURE 7.28 Some sources of greenhouse gases

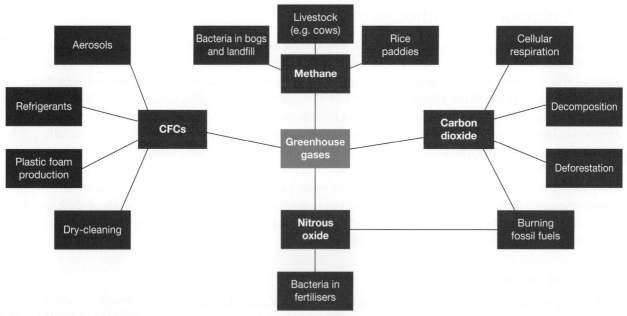

Carbon emissions

Photosynthesis and cellular respiration

Light energy, carbon dioxide and water are used by phototrophic organisms such as plants to make glucose and oxygen. This process is called photosynthesis.

FIGURE 7.29 The process of photosynthesis

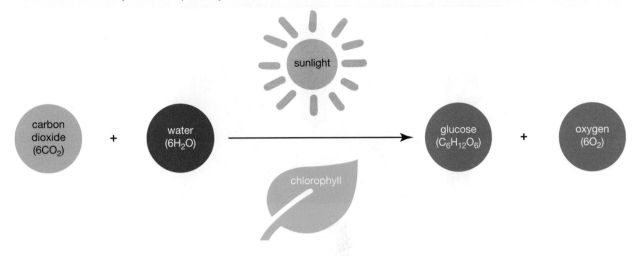

All living things use cellular respiration. During this process, glucose is converted into a form of energy that the cells can use. Carbon dioxide is one of the products of this reaction (figure 7.30).

FIGURE 7.30 The process of cellular respiration

In terms of the carbon cycle, carbon dioxide is taken from the atmosphere during photosynthesis and released back into the atmosphere during cellular respiration. This suggests that, if producers are reduced in number or removed from ecosystems, less carbon dioxide will be removed from the atmosphere, resulting in an overall increase in this gas.

As well as this, often the land in which deforestation occurs is used for other purposes, such as urban and residential buildings or farming. Often, the replacements produce carbon dioxide — for example, cattle produce carbon dioxide by cellular respiration. This explains why cutting down trees and replacing them with buildings or crops with lower photosynthetic rates can contribute to the enhanced greenhouse effect, because more carbon dioxide is being released into the atmosphere.

Fossil fuels

Over millions of years and under extreme heat and pressure, once-living animal and plant material was converted into fossil fuels such as oil. This oil was then refined by humans to create fuels for cars as well as different types of plastics. Unfortunately, due to the rise in the human population and reliance on various modes of transport to move people and products, the levels of carbon dioxide released through combustion has increased dramatically. The addition of more industrial buildings that burn fossils fuels has also added to this rise. This has resulted in an increase in Earth's average global temperature by over 1 °C since 1880, with two-thirds of warming occurring since 1975!

Evidence

For thousands of years, snow has fallen in Antarctica. The snow turns to ice, which builds up over time. Dust, gases and other substances from the air became trapped in the ice. These trapped substances provide information about what was in the air at the time the snow fell. Scientists have used **ice cores** to track the air temperature and concentration of carbon dioxide near Earth's surface in the past.

> **ice cores** samples of ice extracted from ice sheets containing a build-up of dust, gases and other substances trapped over time

FIGURE 7.31 Ice cores drilled from more than 3.7 kilometres below the surface. Parts of the ice are more than 150 000 years old.

The graphs in figure 7.32 indicate the results of testing the ice core samples for carbon dioxide and air temperature.

int-5885

FIGURE 7.32 **a.** Graphs of carbon dioxide concentration shown in parts per million (ppm) by volume, **b.** the temperature variation over the same time and **c.** a graph of Earth's surface temperature and atmospheric carbon dioxide 1880–2019

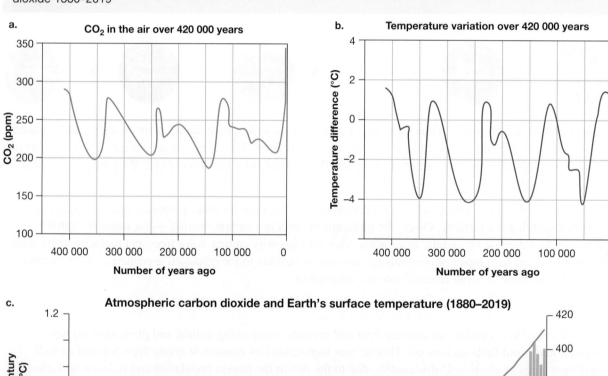

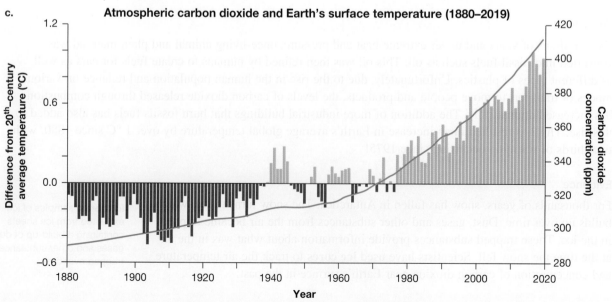

Methane

While carbon dioxide has generally been the main focus of climate change research and action, methane also contributes to the enhanced greenhouse effect and accounts for about 20 per cent of the greenhouse gases produced. The amount of methane being produced has consistently risen in recent years, as shown in figure 7.33.

FIGURE 7.33 Trends in global methane since 1983 show methane levels are increasing.

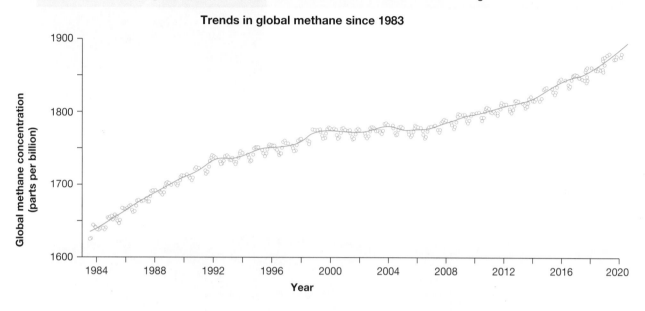

Trends in global methane since 1983

Methane (CH_4) is a natural gas that is colourless and odourless, and is produced through agriculture (cattle), wetlands, fossil fuel production and use, biomass burning and other natural emissions including melting **permafrost**, as shown in figure 7.34a. It is also highly flammable, as shown in figure 7.34b.

permafrost ground that is frozen for at least two continuous years

FIGURE 7.34 a. Methane bubbles in frozen lakes. As the Earth warms and the permafrost melts, this methane will be released into the atmosphere. **b.** Methane can easily flame when released into the atmosphere, as seen with the flames of burning methane on Mount Chimera, the largest abiogenic methane burning in the world.

SCIENCE AS A HUMAN ENDEAVOUR: Eyes in space

A number of satellites are gathering data on Earth's biosphere from a distance. This type of data collection is called **remote sensing**. For example, the satellite *Terra* (shown schematically in figure 7.35) has a number of instruments that gather different types of data on how Earth is changing in response to both natural changes and those caused by humans. This includes data of the land, oceans and atmosphere.

Scientists from different fields are also working together on collaborative projects that use data from remote-sensing observations to improve forecasting systems such as those that warn of future floods. They also explore sea surface temperature, using satellites and devices as shown in figure 7.36.

FIGURE 7.35 A diagram of Terra, the flagship satellite of the Earth Observing System

eles-2732

FIGURE 7.36 Colour-coded image of the sea surface temperature as revealed by an AVHRR (Advanced Very High Resolution Radiometer) carried on a satellite. Red represents the hottest and purple the coolest sea surface temperature.

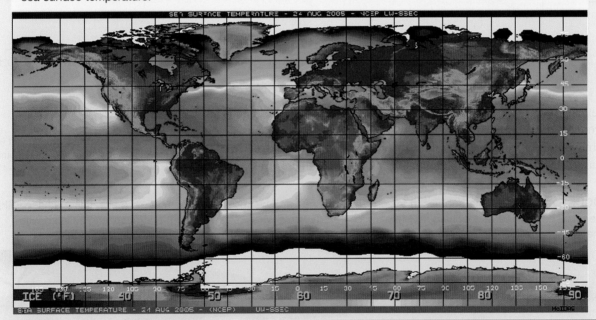

DISCUSSION

1. Which of the following actions would you be prepared to take so that you can contribute to the fight against global warming?
   - Walk, cycle or use public transport rather than relying on someone to drive you to school, work or leisure activities.
   - Change your diet so that you eat less meat and more fruit and vegetables.
   - Recycle paper, aluminium and steel cans, glass and plastics.
   - Stop using electric clothes dryers and use outdoor clothes lines in dry weather and indoor clothes airers in wet weather.
   - Buy fewer clothes or buy clothes second-hand.
2. Select one of the actions in question 1 that the government could enforce by passing new laws, and explain how it could be done.

remote sensing data collection about Earth's biosphere completed from space by devices such as satellites

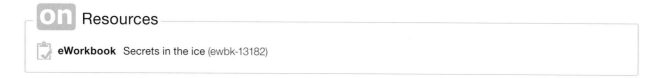
7.4.2 Climate models

Meteorologists and other scientists use computer modelling to make predictions about climate change and the possible consequences. The computer programs used to model climate change simulate the circulation of air in the atmosphere and water in the oceans. An immense amount of data collected from the atmosphere, ocean and land surface is used, together with mathematical equations that describe the circulation. The laws of Physics and Chemistry, including the Laws of Conservation of Energy and Newton's Laws of Motion, are an important part of the modelling process.

An example of this is looking at the change in global temperatures. It is predicted that by 2100, it could be 1.1 to 5.4 °C warmer than it is today!

FIGURE 7.37 Earth's temperature change prediction in 2100. The black line represents an average of a set of temperatures. The coloured lines represent average expected temperatures under three emissions scenarios that may occur in the future. These scenarios are as follows: humans worldwide make more sustainable choices and use renewable energy (purple), a balance is achieved between use of fossil fuels and renewable sources (green), and humans continue to accelerate carbon emissions (pink).

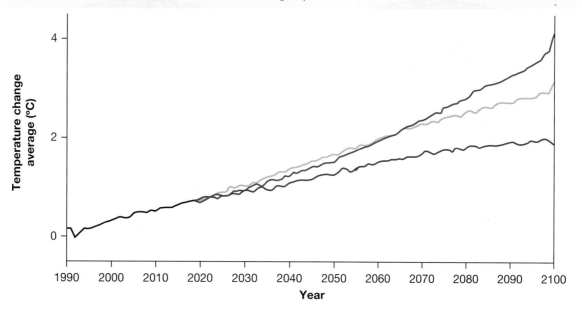

Although that increase doesn't sound like much, the consequences are very serious. Computer modelling suggests that the global temperature will not increase evenly across the continents. According to the CSIRO, in Australia temperatures could increase by up to 2 °C by 2030 and up to 6 °C by 2070.

As a consequence, there will be:
- more hot days and fewer cold days
- an increase in rainfall in the north-east and a decrease in the south
- more bushfires
- more destructive tropical cyclones.

Palaeoclimates

Palaeoclimates offer a unique perspective in that they can show the wide range of climates over various time scales, and transitions between them. It is the study of Earth's climate using geological and biological evidence to reconstruct past climates around the world. This information can be used to develop climate models for future climate studies. Figure 7.38 shows examples of various palaeoclimates throughout Earth's history.

palaeoclimate the climate at a specific point in geological history

FIGURE 7.38 The study of palaeoclimates throughout history may help us develop climate models to predict climates of the future. (Kya = thousand years ago; Mya = million years ago)

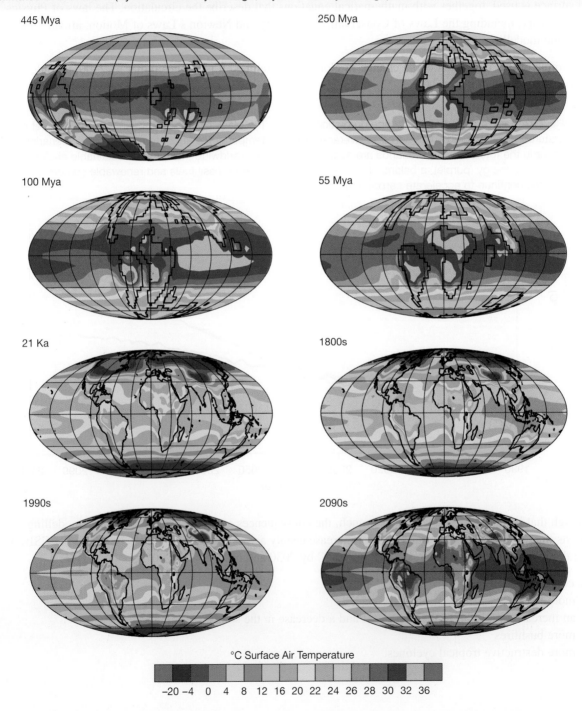

Scientists use proxies to help determine past climates. Proxies are biological, physical or chemical materials that are preserved as geological evidence. These are often preserved in palaeoclimate archives, which can be gathered and analysed by scientists.

These studies include looking at ice cores, tree rings, sediments, coral and other climate archives. Figure 7.39 shows examples of evidence palaeoclimate scientists gather.

FIGURE 7.39 The kind of geological evidence palaeoclimate scientists look for to help them determine what climates in the past were like

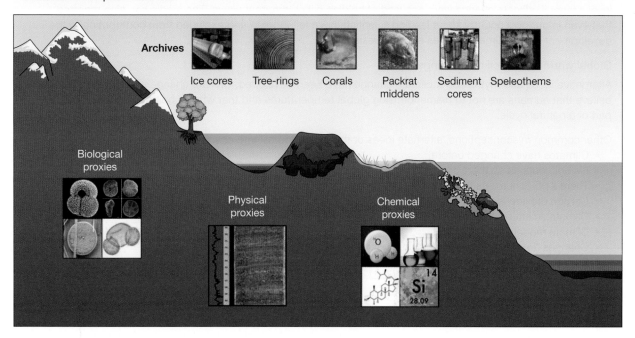

7.4.3 Global warming is a hot topic

As the physicist Niels Bohr reportedly said, 'Prediction is very difficult, especially of the future.'

While most scientists agree that an increase in the amount of carbon dioxide in the atmosphere is the main cause of global warming, they argue about the details of the cause and about the effects of global warming. The key arguments that scientists are involved in investigating and discussing can be divided into three categories:

1. Are humans responsible for global warming?
2. What will the effects of global warming be?
3. What can be done to stop global warming?

FIGURE 7.40 a. Climate protests in Melbourne in 2019 and **b.** a scientist measuring greenhouse gas emissions

Climate science and policy

Global warming is a thorny problem, and clashes occur over climate science and policy. While some refer to this as the climate debate, to those deeply immersed in it, it may feel more like an ugly war. It has included frontline battles between science and opinion, politics, media and human psychology. There has been scepticism, outright denial, disrespect and even name-calling!

An Australian newspaper reported that, in one country, scientists trying to present evidence for human involvement in climate change were accused of holding elitist, arrogant views. The media has also reported that even in our own country some leading scientists have felt ignored and excluded from contributing to the development of key climate policies and discussions.

Global warming and climate change myths

Alternative ideas and myths about climate change have been developed. Climate change sceptics, for example, believe that humans are not to blame for rising global temperatures and that what is being experienced is merely part of a natural cycle.

Other common misconceptions, alternate ideas or myths around climate change:
- Climates have changed before.
- It's cold outside so there can't be 'global warming'.
- There's no consensus the planet is actually warming.
- Carbon dioxide is not a pollutant.
- Climate scientists are conspiring to push 'global warming'.
- Climate models are unreliable.
- Changes are due to sunspots/galactic cosmic rays.
- Scientists manipulate all data tests to show a warming trend.

Climate science

Climate scientists are trying to find evidence against the hypothesis that global warming is caused mainly by humans releasing greenhouse gases into the atmosphere. That is, they are considering that the hypothesis may be wrong and are assessing other ways in which this warming may be occurring. Over the past 40 years, however, no evidence against the hypothesis has been found.

A difficulty for climate scientists is not just about predicting how the climate will change, but also in estimating the level of uncertainty within the prediction.

It is important to observe evidence over time — while trends across a year may seem insignificant, exploring the changes to the climate over a longer period of time is much more powerful, and allows for trends and patterns around the changing climate to be more easily observed. As will be explored in lesson 7.6, we are exploring many solutions that can help our changing climate.

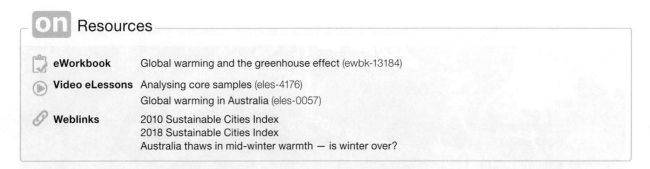

on Resources

eWorkbook	Global warming and the greenhouse effect (ewbk-13184)	
Video eLessons	Analysing core samples (eles-4176)	
	Global warming in Australia (eles-0057)	
Weblinks	2010 Sustainable Cities Index	
	2018 Sustainable Cities Index	
	Australia thaws in mid-winter warmth — is winter over?	

7.4 Activities

learn

7.4 Quick quiz on	7.4 Exercise

Select your pathway

■ LEVEL 1	■ LEVEL 2	■ LEVEL 3
1, 3, 4, 5, 7	2, 6, 8, 10	9, 11, 12

These questions are even better in jacPLUS!
- Receive immediate feedback
- Access sample responses
- Track results and progress

Find all this and MORE in jacPLUS ▶

Remember and understand

1. Match the following terms to their definition.

Term	Definition
a. The greenhouse effect	**A.** The maintenance of Earth's temperatures by the atmospheric gases
b. The enhanced greenhouse effect	**B.** The rising temperature of Earth
c. Global warming	**C.** Increase of greenhouse gases in the atmosphere that trap heat, causing the atmosphere to heat up

2. a. State whether the following statements are true or false.

Statements	True or false?
i. Earth's atmosphere consists of a blanket of gases that trap heat to keep the temperature of our planet stable.	
ii. Explain your response for each.	
iii. Scientists have used ice cores to track the air temperature and concentration of carbon dioxide near Earth's surface in the past.	
iv. Ozone (O_3) produced by photochemical reactions involving emissions from motor vehicles is a significant contributor to the enhanced greenhouse effect.	

b. Explain your response for each.
3. Suggest four consequences of global warming.
4. **MC** What is the greenhouse gas associated with sources such as rice paddies, livestock (for example, cows), bacteria in bogs and landfill?
   A. Carbon dioxide
   B. CFCs
   C. Methane
   D. Nitrous oxide
5. Identify the missing terms to complete the sentence, using the following terms.
   Word bank: Carbon, cellular respiration, photosynthesis, taking up, release, fossil
   Within the _____ cycle, _____ involves the _____ of carbon dioxide from the atmosphere, whereas _____ and burning _____ fuels _____ carbon dioxide as a waste product into the atmosphere.

Apply and analyse

6. Describe the links between photosynthesis, cellular respiration, decomposition, fossil fuels and global warming.
7. Describe and provide examples showing why ozone in Earth's stratosphere is important to humans and all other life on Earth.

▶

8. Explain how scientists are able to determine the air temperature and the amount of carbon dioxide in the atmosphere hundreds of thousands of years ago when such measurements were not recorded.
9. Using a clear diagram, explain how the thawing of permafrost could increase the rate of global warming.

Evaluate and create

10. **SIS** Many people do not believe that climate change and global warming are taking place. Others acknowledge they are taking place but do not believe they are serious problems or that they are caused by humans. Use the internet and other sources to list the arguments that these two groups of people use to support their beliefs and write an argumentative statement on your personal opinion on this topic.
11. **SIS** One of the difficulties of using models to predict future events such as carbon dioxide emissions is that scientists need to make assumptions about a series of possible future states based on known facts, rather than on accurate measurements of events from the past. This provides the opportunity for bias in selection. Find out more about the computer models used to predict these events and whether any bias may be present. Create a report to summarise your findings.
12. **SIS** Refer to the **Sustainable Cities Index** developed by the Australian Conservation Foundation (ACF). This index is based on a range of environmental, social and economic issues. It provides a snapshot of the performance of 20 of our largest cities and ranks them on their sustainability.
    a. Select the city closest to where you live. How did it rate in this index? Do you agree with the ACF's findings? Justify your response. Suggest ways in which your city's score could be improved.
    b. Select one of the criteria used and find out more about the method used to collect the data.
    c. Which of the 20 Australian cities scored as our most sustainable city? For which criterion did it score the highest? Suggest reasons for its high score.
    d. Which city scored the lowest? Suggest reasons for its low score and what it could do to improve this score.
    e. Use the weblink for the 2018 Sustainable Cities Index to describe how Australian cities ranked globally in the different categories, and explain how this compares to the 2010 index.

Fully worked solutions and sample responses are available in your digital formats.

LESSON
7.5 Biodiversity and climate change

LEARNING INTENTION

At the end of this lesson you will be able to explain how sensitive climates are to small changes and describe how this can affect the biodiversity of living things and lead to extinction.

7.5.1 Climate sensitivity and biological implications

Will parts of Earth get too hot for humans? Computer models are predicting that this could happen in some parts of the tropics in the future. Scientists have suggested that under these hot and humid conditions, even someone standing in the shade in front of a fan could die of heat stress.

Changes in Earth's climate due to global warming will affect the survival of organisms. The survival of every living thing on Earth is dependent on the characteristics and stability of its habitat. Climate change will affect some living things more than others (figure 7.41).

FIGURE 7.41 If Earth's temperature increases, what will this mean for polar bears in the Arctic?

Climate sensitivity

How hot things get will depend on how much more carbon dioxide is pumped into the atmosphere and how much warming it produces. This is defined as **climate sensitivity**. The Intergovernmental Panel on Climate Change (IPCC) suggests that temperatures may rise between 1.9 and 4.5 °C (around 3 °C) for every doubling of carbon dioxide concentration in the atmosphere. However, the IPCC's computer model is based only on fast feedback processes and excludes slower processes such as the release of methane from thawing permafrost.

With a climate sensitivity of around 1.9 °C, it may take centuries for our planet to warm by 7 °C. With a climate sensitivity of around 4.5 °C, however, the increase could reach 7 °C within a century if we continue at our current levels of carbon dioxide production. The effects of increased temperature can be seen in figure 7.43.

climate sensitivity the measure of temperature change in the climate, dependent on the amount of carbon dioxide released into the atmosphere

FIGURE 7.42 Increases and decreases are seen in various factors of our environment.

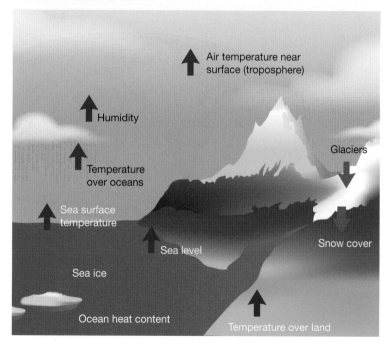

Will climate change shape human evolution?

Could Earth get too hot for humans? Does enough variation exist within our species that if things do get too hot to handle, at least some of us will survive?

To function normally we need to maintain a core body temperature of around 37 °C. If this core temperature rises above 42 °C, we die.

Some researchers have used climate computer models to predict the impact of different levels of global warming on populations. Their data suggest that an increase of around 7 °C in the environment may result in heat and humidity that makes many locations on Earth intolerable. They suggest that at increased temperatures of 12 °C, about half of the land inhabited today (including Australia) would be too hot to live in (refer to figure 7.43).

FIGURE 7.43 An increase in heat and humidity due to climate change could render half the world uninhabitable. In regions where the 'wet-bulb' temperature (the temperature to which objects can be cooled by evaporation) exceeds 35 °C (the human heat-stress limit), it would be impossible for people to survive without some kind of cooling system.

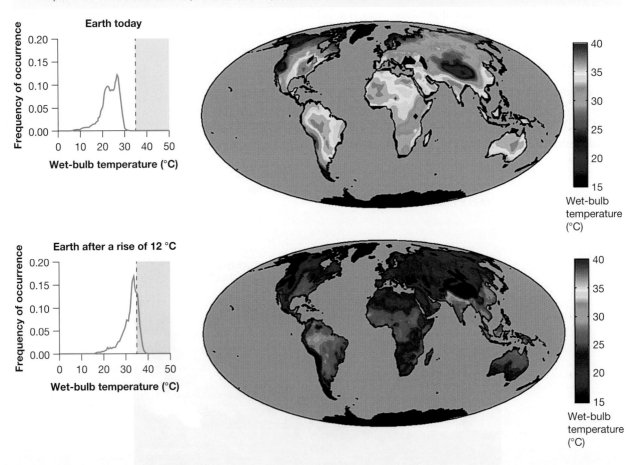

People living in the affected areas would need to wear 'cooling suits', live underground or stay in constantly air-conditioned environments. Organisms such as livestock or people who cannot afford these buffers may perish.

Adapting for the heat?

If Earth keeps warming up, over the long term will we see genetic shifts to select those variations that increase chances of survival? What will a human in a hot future world look like? Some evolutionary biologists have suggested slimmer and taller body shapes that radiate heat better, while at the same time carrying enough fat to be reproductively successful, would be selected for. Some palaeontologists, however, suggest that heat stress would be likely to favour the evolution of smaller mammals.

Dealing with disease

With warmer temperatures and global transport and global populations, scientists predict that humans may be more at risk of disease than at any other time in history. There may be an increased incidence of diseases such as food poisoning, skin cancers, eye cataracts and a new range of tropical diseases.

The presence of genes that provide quick resistance against the onslaught of future diseases is another factor that will determine who survives and who does not.

Are humans still evolving?

A hypothesis has suggested that global cooling was essential for the large brains of humans to evolve. If this hypothesis is supported, does this mean that global warming may lead to a reduction in the size of the human brain? Other scientists suggest that our modern brains have enabled us to develop culture and that as long as we have culture and technology, we will have a buffer against hot climates.

Research suggests that the human brain is still evolving. Scientists have identified two genes involved in regulating brain size that have been subject to recent natural selection.

 **Resources**

▶ **Video eLesson** Polar bears (eles-2908)

7.5.2 The impact of biodiversity

When the first traces of life appeared on Earth about 3500 million years ago, the climate was hostile. Lightning bolts blasted through a warm atmosphere of hydrogen, methane, ammonia, water vapour and carbon dioxide. There was no oxygen until the first living organisms produced it through photosynthesis. Since then, the composition of gases in Earth's atmosphere and its temperature have been constantly changing.

FIGURE 7.44 a. Earth was a hostile place 3500 million years ago. Fossils provide evidence of structures called stromatolites. They existed in warm sea water and consisted of cyanobacteria, one of the earliest forms of life. **b.** Stromatolites seen today in Shark Bay, Western Australia

Biodiversity

The evolution of life forms on Earth has occurred because some organisms are better suited to a particular environment than others. For some to be better suited than others, variation or diversity is needed.

In a global sense, biodiversity refers to the total variety of living things on Earth, their genes and the ecosystems in which they live. Biodiversity (or biological diversity) exists at the gene, species and ecosystem levels.

Genetic diversity

Genetic diversity can be considered in terms of variation within the genes, which are made up of DNA. Individuals within a species share the same genes that code (with an environmental influence) for a particular feature or characteristic. However, alternative forms of these genes (alleles) can exist within the individuals.

Genetic variation is important for the long-term survival of a species because it increases the chance that at least one of the variations will enable some of the population to survive to reproduce the next generation (refer to topics 2 and 3, which explore genetics and evolution respectively).

Species diversity

Species diversity can be considered in terms of diversity in populations. While the combination of alleles (different forms of a gene) for a trait within an individual is called a genotype, the combination of all the alleles within a group of individuals of the same species living in a particular place at a particular time (population) is called a **gene pool**.

All environments change over time. It is the diversity (or variation) of the alleles within the gene pool that contributes to the number of possible combinations that could be used to produce the next generation. Increased variety in the expression of these alleles as phenotypes (traits) of the offspring means an increased chance that some of these offspring will be able to survive in the environment in which they are born and will live — even if that environment changes.

If little variation exists in the gene pool, there is less chance of the offspring being able to survive possible changes in their environment such as climate and the availability of habitat, food, mates or other resources. The consequences of this limited diversity within the population may lead to the **extinction** of the species.

In Australia, an example of an animal with low genetic diversity is the Tasmanian devil. Due to a low population (either through hunting or weather events), more inbreeding occurred. This has made them more susceptible to diseases — in particular, the devil facial tumour disease (DFTD), as seen in figure 7.45. With at least 80 per cent of the wild population suffering from DFTD, will the Tasmanian devil survive if the climate changes?

> **gene pool** the total genetic information of a population, usually expressed in terms of allele frequency
>
> **extinction** complete loss of a species when the last organism of the species dies

FIGURE 7.45 a. The unique Tasmanian devil and **b.** the disease currently ravaging its population

BACKGROUND KNOWLEDGE: Passing genes on to the next generation

Genes code for proteins, which lead to a great number of different phenotypes or traits. We can predict how these genes are inherited.

Let's explore this with peppered moths, which can be black or white. This affects their ability to survive in certain conditions.

The allele (variation of a gene) for moth colour can be assigned as follows:
- B — black
- b — white.

If a black moth with the genotype Bb breeds with a white moth with genotype bb, we can predict the resulting offspring using a Punnett square, as shown in table 7.2. (Refer to topic 2 for information about Punnett squares.)

This shows a 50 per cent chance that their offspring will be black and a 50 per cent chance their offspring will be white.

Sexual reproduction allows for an increase in genetic diversity within a species.

FIGURE 7.46 Genetic variation in moths leads to different colours. The colour of the moths helps with their ability to camouflage and escape predators.

TABLE 7.2 A Punnett square

	B	b
b	Bb	bb
b	Bb	bb

DISCUSSION

The sex of turtles is affected by the temperature. Above 29 °C, mainly females are produced, while below 29 °C, more males are produced. Turtle eggs develop successfully in a temperature range of 25 to 35 °C. If Earth gets too warm or too cold in certain areas, what effect will this have on turtle populations?

Ecological diversity

Ecological diversity can be considered in terms of the diversity in ecosystems. The extinction of a particular species within an ecosystem may affect the survival of other species within that ecosystem. The extinct species' disappearance will have consequences for the food supplies of others within its food web. Unless other species can take its place without having a negative effect on others, the survival of other species may be threatened.

Increased biodiversity within ecosystems can reduce the consequences of losing a species to which the survival of others is linked. Likewise, reduced biodiversity in these ecosystems can lead to the extinction of other species.

In the early 1900s in Yellowstone National Park in the United States, wolves were eliminated because the park did not provide protection for predators and they were hunted through government-approved predator control programs. Removing this apex predator from this ecosystem, however, resulted in major changes in the park, as shown in table 7.3. The re-introduction of wolves to Yellowstone National Park enabled the ecosystem to be rebalanced.

TABLE 7.3 Ecosystems are sensitive to changes. A balancing act is required to ensure the survival of species.

Pre-1995	Post-1995 after wolves re-introduced
Elk populations increased because their main predator was no longer around.	Elk populations decreased due to the presence of the wolves (their main predator). This changed the elk behaviour, pushing them into less favourable habitats, raising their stress levels and lowering thier overall birth rate.
Coyote populations increased because wolves are their predators.	Coyote populations decreased dramatically; after about two years of the re-introduction of wolves, coyote populations had reduced by 50 per cent.
Local flora died because of overgrazing by the elk, including the aspen and cottonwool.	Flora and fauna were able to recover beause the elk now had to spread themselves out. This lead to an increase of berries, which benefitted the grizzly bear population.
No known beaver populations inhabited the park.	By 2011, nine beaver colonies had been formed. Because the elk now had to cover more land, they were not destroying the willow plant, which the beavers needed to survive the winter. Beaver dams help counter erosion, provide shaded water for fish and habitats for a variety of mammals.
Cougars and coyotes took over territories once belonging to the wolves.	Cougars and coyotes returned to their more traditional mountainside territory.

Mass extinctions

Many scientists believe that we are currently experiencing the sixth mass extinction. Five other mass extinctions have occurred as a result of global climate change. When we think about extinctions, this refers to the deaths of species that occur over hundreds or thousands of years, so a mass extinction is not often an instantaneous event.

Those with the view that humans are to blame divide this sixth extinction into two phases. The first phase began about 100 000 years ago when the first modern humans began to spread throughout the world. The second phase began when humans started to use agriculture around 10 000 years ago.

FIGURE 7.47 Five mass extinctions in the past

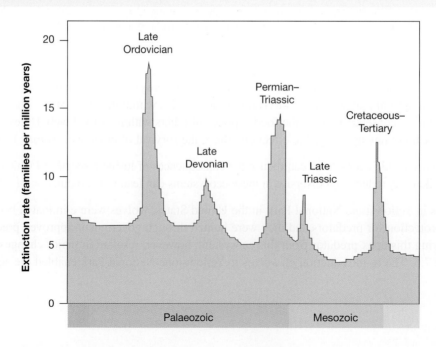

DISCUSSION

A 2008 study has shown that as the human population has increased, so has the extinction of flora and fauna. This is shown in the graph in figure 7.48.

FIGURE 7.48 Data relating to humans and species extinction

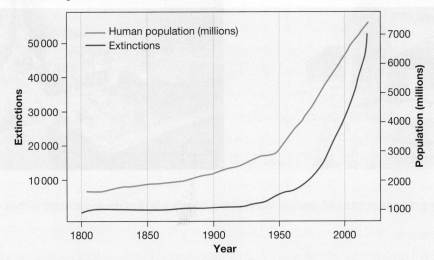

Data source: Scott, J.M. 2008. *Threats to Biological Diversity; Global, Continental, Local.* U.S. Geological Survey, Idaho Cooperative Fish and Wildlife, Research Unit, University of Idaho

Based on this data, do you think humans are the cause of a sixth main extinction event? What do you think the future might look like for animals and what might we need to do to combat this?

The extinction of amphibians

Since 1980, more than 150 species of amphibians have become extinct. This compares poorly with background extinctions of 1 every 250 years. Depletion of the ozone layer has been revived as an explanation for the extinction of amphibians after the discovery that increased ultraviolet-b radiation makes striped marsh frog tadpoles more vulnerable to predators.

FIGURE 7.49 Many amphibian species have become extinct.

on **Resources**

Weblinks   Ongoing loss of biodiversity in amphibians
Risk of extinction to amphibians

7.5.3 Impact on oceans

Some marine life will suffer and could even become extinct because of changes in water temperature. Changing temperatures and ocean currents could separate some marine species from their food source. Some marine animals depend on microscopic plankton that float along with the currents. Others depend on species from warmer or colder layers of water than the layer in which they live. It is also possible that some species will suffer from the reduction of oxygen dissolved in ocean water because of increases in temperature. The habitats of some species could be destroyed by rising sea levels.

Rising sea levels are likely to cause the flooding of low-lying islands and coastal regions, such as those shown in figure 7.50.

FIGURE 7.50 **a.** The low-lying Pacific nation of Kiribati is planning to relocate its population because of the threat of rising sea levels. **b.** Significant changes in distribution of about 30 per cent of coastal fish species in south-east Australia are being blamed on climate change.

FIGURE 7.51 The change in mass of glaciers found in the US show a disturbing decrease in their size.

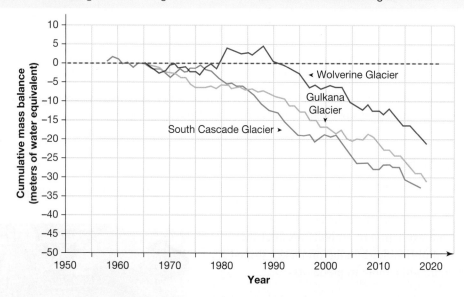

According to tide-gauge records, the average global sea level has increased by between 10 cm and 20 cm during the past 100 years (see figure 7.52). Sea levels are expected to rise further due to:

- the warming ocean water and its resulting thermal expansion.
- the melting of glaciers, the polar ice caps and the ice sheets of Greenland and Antarctica. According to NASA, sea ice in the Arctic is melting at the rate of 12.6 per cent every ten years. Of the world's 88 glaciers, 84 are receding due to melting ice.

Scientists from both the CSIRO Climate Adaptation Flagship and the Wealth from Oceans Flagship have identified shifts in the distribution of 43 coastal fish species in south-east Australia, likely due to these changing sea levels.

FIGURE 7.52 Projected sea level rise for a scenario where the greenhouse emission rate is high but is stabilised by 2100 using a range of technologies and strategies to reduce the emission rate

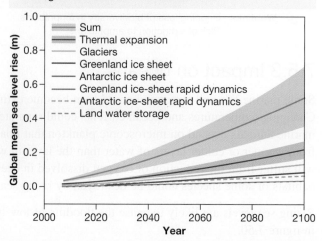

FIGURE 7.53 Photos taken of the Columbia Glacier, Alaska, in **a.** 2009 and **b.** 2015

a.

b.

elog-2476

tlvd-10819

INVESTIGATION 7.4

Investigating rising ocean levels

Aim

To model how temperature affects how ocean levels rise

Materials

- a tray of ice cubes (around 20 ice cubes)
- 5 × 250 mL beakers (larger sizes may be used, but it can be more difficult to see height change)
- water
- water baths set to varying temperatures
- ruler
- timer

Method

1. Write a hypothesis for your investigation before you commence.
2. Fill each beaker with 100 mL of water. Ensure the beakers are all the same shape so 100 mL is the same height.
3. Add 4 ice cubes in each beaker and record the water level.
4. Place each beaker in a water bath of varying temperatures (20 °C, 25 °C, 30 °C, 35 °C). Leave one at room temperature, ensuring you record the temperature.
5. Start a timer and start timing for 5 minutes.
6. After 5 minutes, measure the height of the water level in each beaker.
7. If all the ice hasn't melted yet, place each beaker in the water baths for another 5 minutes and repeat the measurement.

Results

Complete the following results table. You should add an appropriate heading for your table.

TABLE Changes in water levels at different temperatures

Beaker	Temperature of water bath	Initial water height (cm)	Final water height (cm)	Change in water level (cm)
1				
2				
3				
4				
5				

Discussion

1. Identify all the variables in this investigation, and determine if they are independent, dependent or controlled.
2. Why was it important to use beakers of the same size and shape?
3. How does temperature affect water levels?
4. What does this suggest about ocean levels if the temperature continues to rise globally?
5. How might this impact biodiversity and the survival of species?

Conclusion

Summarise the findings for this investigation about how temperature affects how ocean levels rise.

on Resources

📋 **eWorkbook**   How climate change impacts biodiversity (ewbk-13186)

7.5 Activities

learn on

7.5 Quick quiz	on	7.5 Exercise

These questions are even better in jacPLUS!
- Receive immediate feedback
- Access sample responses
- Track results and progress

Find all this and MORE in jacPLUS ▶

Select your pathway

■ **LEVEL 1**
1, 2, 4, 6

■ **LEVEL 2**
3, 7, 8, 10

■ **LEVEL 3**
5, 9, 11, 12

Remember and understand

1. Match the type of diversity with its most appropriate description.

Term	Description
a. Biodiversity	**A.** Diversity in ecosystems
b. Genetic diversity	**B.** Total variety of living things on Earth, their genes and the ecosystem in which they live
c. Species diversity	**C.** Diversity in populations of organisms
d. Ecosystem diversity	**D.** Variation within genes (alleles), which are made up of DNA

2. **a.** State the core body temperature that humans need to maintain.
    b. Describe what happens if the core body temperature of a human rises above 42 °C.
3. Suggest strategies that people living in areas affected by extreme heat and humidity might use to survive.
4. What is meant by the term *climate sensitivity*?
5. Identify the missing words to complete the sentence.
   Genetic variation is important for the long-term survival of a species because it _____ the chance that at least one of the variations will enable _____ of the population to survive to _____ the trait on to the next generation.

Apply and analyse

6. Suggest a connection between global warming and changed abiotic factors (non-living) within ecosystems.
7. Select a land-based living organism. Outline the likely effect on this organism caused by the following.
   a. Rising sea levels
   b. An increase in average temperatures

8. Do living organisms always have a negative effect on their environment? Justify your response and include a supporting example.
9. Suggest different ways in which Australia's biodiversity has changed due to climate change, providing at least two supporting examples.

Evaluate and create

10. While the yields of some types of crops, such as wheat and rice, may increase in conditions with higher carbon dioxide concentrations, increases in temperatures may be detrimental to other types of crops. Research and create a visual summary of the effects of global warming on at least three types of crops.
11. Some palaeontologists suggest that mammals get smaller as the climate gets warmer. Investigate this hypothesis and record your evidence for or against it with current examples.
12. **SIS** The biggest problem connected to the effects of climate in Kakadu's coastal floodplain is the rise in sea level. Salt water has already intruded in various parts of the park and has affected the local populations of *Melaleuca* (paperbark) trees and magpie geese. Research and create a report on the current and possible effects of rising sea levels in Kakadu.

Fully worked solutions and sample responses are available in your digital formats.

LESSON
7.6 Cool solutions

LEARNING INTENTION

At the end of this lesson you will be able to describe the various solutions being investigated to help reduce carbon and greenhouse gas emissions, and explain how countries are working together on these solutions.

7.6.1 Finding solutions

While climate change may seem daunting, we are in an incredibly powerful position to instigate change. What can we do to reduce climate change and global warming?

No-one can be certain about the actual consequences of climate change. So many variables influence climate that computer modelling cannot provide completely accurate predictions. However, plenty of evidence indicates that the levels of the greenhouse gases carbon dioxide, methane and nitrous oxide have been increasing over the past 100 years and will continue to increase.

It is clear that global warming must be slowed by reducing the emission of greenhouse gases. This is no easy task and requires the combination of a number of solutions.

FIGURE 7.54 Quantum computers are able to process data more quickly and more accurately.

Climate models

Climate models play a crucial role in helping scientists understand how changes in various factors impact Earth's climate. These models start with collecting and carefully analysing a vast amount of information gathered by satellites and specialised equipment. This information includes details about the temperature of the air and the ocean, patterns of rainfall, different gases released into the atmosphere, and even how forests are being cut down. Once this data is collected, scientists use advanced computer programs to run simulations that predict how these changes might affect the climate.

To make these predictions as accurate as possible, scientists are turning to quantum computers (figure 7.54). These computers are special because they can handle huge amounts of data and process it much faster than regular computers. This means that the predictions from these models become more precise, as they can consider many more factors that influence the climate. But it's important to understand that even with all this technology, climate modeling isn't perfect. Since different models and interpretations of data can lead to slightly different predictions, there's still some uncertainty involved.

The information that comes out of these models helps us figure out what actions should be taken to avoid extreme weather patterns. By understanding how changes in temperature, gas emissions, and other factors affect the climate, scientists can suggest targets that need to be achieved to reduce the effects of climate change.

Removing and storing carbon dioxide

Geosequestration is a process that involves separating carbon dioxide from other flue gases in fossil fuel power stations, compressing it and piping it to suitable sites. At least 65 suitable sites (for example, depleted oil and gas wells) have been identified in Australia that are capable of taking up to 115 million tonnes of carbon dioxide each year.

Research on this process dates back to the 1970s. Although considerable problems exist with the technology, there is renewed interest in further developing it. It is hoped that it may be used to remove carbon dioxide from the atmosphere and, hence, reduce global warming.

FIGURE 7.55 If carbon dioxide can be removed from the atmosphere and stored, its concentration in the atmosphere could be reduced.

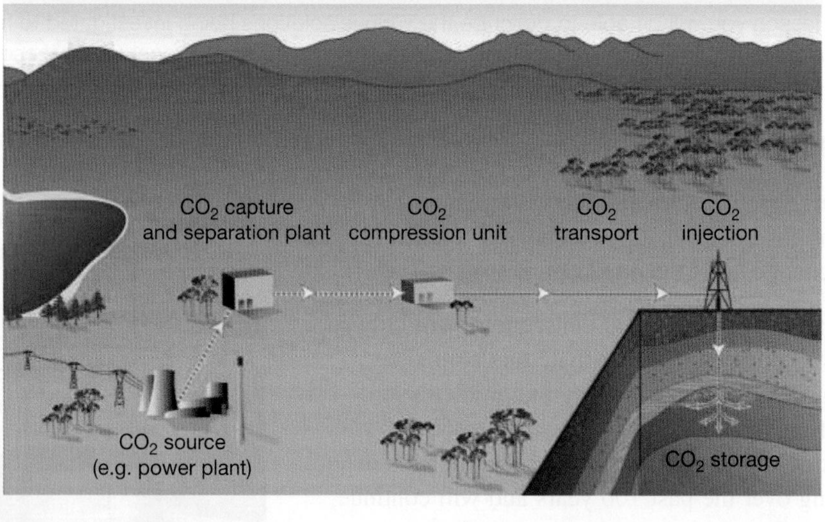

Sustainable forests

We live in a consumer society. The things we want and need often require large amounts of energy to manufacture and consequently result in the emission of carbon dioxide into the atmosphere. Scientists in the forestry and related industries have suggested one way to reduce carbon dioxide emissions is to produce and use wood products that have been grown under sustainable forest management strategies. Nick Roberts, Forests NSW chief executive, is passionate about the role that sustainably harvested native forests can play in combating climate change. The view that wood products produced under this sustainable management have the potential to maintain or increase forest carbon stocks is also supported by the IPCC.

In 2009, Fabiano Ximenes, a forest research scientist, and his colleagues from the NSW Department of Primary Industries (DPI) analysed the carbon content of paper and wood products in landfill and found that at least 82 per cent of the carbon originally in the sawn timber remained stored in the wood.

geosequestration the process that involves separating carbon dioxide from other flue gases, compressing it and piping it to a suitable site

This research suggested that wood products could act as a carbon 'sink', not only during use, but even after disposal (figure 7.56). Carbon sinks are natural sites that absorb high levels of carbon, and so remove it from the atmosphere.

FIGURE 7.56 How carbon stored in wood can be locked away, rather than being released into the atmosphere. Would burning wood be another way to store carbon?

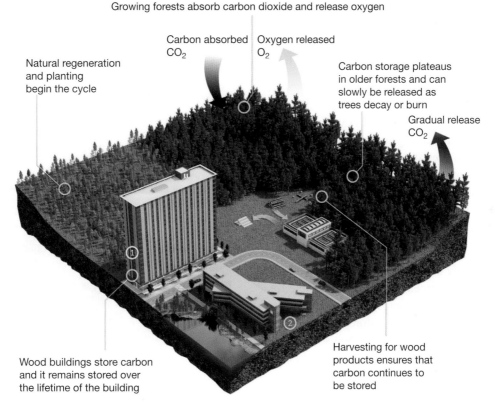

Growing forests absorb carbon dioxide and release oxygen

Carbon absorbed CO_2

Oxygen released O_2

Natural regeneration and planting begin the cycle

Carbon storage plateaus in older forests and can slowly be released as trees decay or burn

Gradual release CO_2

Wood buildings store carbon and it remains stored over the lifetime of the building

Harvesting for wood products ensures that carbon continues to be stored

Renewable resources

Renewable resources are becoming increasingly important as we recognize the harmful impact of relying on fossil fuels, which contribute to more than 75 percent of global greenhouse gas emissions. Fossil fuels are a finite resource, and their use contributes to environmental problems. To address this, there's a growing shift towards using renewable energy sources that are sustainable and have a lesser impact on the environment. This shift also offers companies the advantage of having a broader range of options for powering our world.

In Australia, some of the main sources of renewable energy include solar, wind, and hydro power.

Solar energy is derived from the Sun's rays and is converted into electricity using solar panels. As technology advances, solar panels have become more efficient, with an average efficiency of around 22 percent. The amount of electricity generated depends on factors like your location and the number of solar panels you have. However, solar panels do require a significant amount of space, whether it's on your rooftop or an open area. They can only generate electricity during daylight hours, and their efficiency is affected by factors like cloud cover.

FIGURE 7.57 A wind turbine farm in Taralga, New South Wales.

Wind energy is a highly efficient form of renewable energy, capable of converting a significant portion (between 60 to 90 percent) of the wind's energy into power. Wind turbines are used to harness this energy.

Unlike solar panels, wind turbines work whenever there's enough wind to turn the blades. However, they are dependent on favorable weather conditions and don't generate electricity without wind. The downside is that wind turbines are quite large and may require clearing land for their installation. They can also pose challenges to wildlife, including birds.

Hydro energy utilizes the energy from moving water to generate electricity. This is achieved by using turbines placed in the path of flowing water. Hydroelectric power stations can range from small to large, and they generate electricity as long as there is water flow. However, not all rivers or streams are suitable for hydro energy generation due to their flow and power potential. Larger hydroelectric power stations often involve damming water to create reservoirs. While this generates electricity, it also has negative consequences. Damming water alters ecosystems and disrupts fish migrations, river flow patterns, and water quality.

Afforestation and reforestation initiatives

Afforestation and **reforestation** initiatives is a strategy aimed at making our planet greener while also tackling climate change. Afforestation involves planting trees in areas that were not previously covered with forests while reforestation focuses on replanting trees in areas where forests used to exist but were lost due to activities like logging or wildfires.

FIGURE 7.58 Afforestation of a former sand mine

Trees play a critical role in fighting climate change. As trees grow, they absorb carbon dioxide from the atmosphere, a major greenhouse gas responsible for global warming, through photosynthesis. Trees convert carbon dioxide into oxygen and store carbon in their trunks, branches, and roots. This carbon storage helps reduce the overall amount of carbon dioxide in the atmosphere, contributing to efforts to slow down climate change.

Forests also have a cooling effect on the Earth. Their dense canopy provides shade, reducing the amount of sunlight that reaches the ground. This can help regulate temperatures, making the surrounding area cooler. Additionally, trees release water vapor through a process called transpiration. This water vapor rises into the atmosphere, eventually forming clouds and increasing the likelihood of rainfall.

afforestation planting trees in areas that had no recent cover

reforestation planting trees in forest areas that have been destroyed or damaged

metagenomics technology that combines DNA sequencing with molecular and computational biology

Investigating methane emissions by animals

Australian agriculture accounts for about 16 per cent of our national greenhouse emissions, and 67 per cent of this is methane emissions from livestock.

CSIRO Livestock Industries (CLI) is conducting research that aims to characterise the microbiome (assortment of microbes in the foregut) of Australian marsupials such as the Tammar wallaby (*Macropus eugenii*), shown in figure 7.59.

One project involves **metagenomics**, a technology that combines DNA sequencing with molecular and computational biology. This technology is being used by the scientists to study methanogens — bacteria that are involved in breaking down plant fibre in the wallaby's gut.

FIGURE 7.59 The Tammar wallaby

While these bacteria produce methane, the levels are a lot lower than those produced by cows and sheep. CSIRO's research may lead to discoveries about why marsupials produce far fewer greenhouse emissions than cows and sheep, and contribute to new biotechnologies that may help us to reduce agricultural greenhouse emissions.

7.6.2 Mitigation strategies

Infrastructure

In the face of more frequent and severe weather events, ensuring that our infrastructure is resilient and less vulnerable has become a pressing concern. Infrastructure refers to the essential systems and facilities that support our communities, like roads, bridges, water supply, electricity, and communication networks.

Floods isolate towns, making it hard for help and supplies to reach them, and cuts off vital routes for people to evacuate or access resources. During these times, having access to clean drinking water becomes critical. However, floods can contaminate water sources with sewage or other pollutants, making the water unsafe to consume. The disruption doesn't stop there; power lines and communication networks can be damaged, leaving people without electricity and a way to connect with others.

To make our infrastructure more resilient, we need to adapt it to the changing climate. For instance, roads and bridges could be designed to withstand stronger storms and flooding. Electrical and communication systems can be fortified to be more weather-resistant and efficient. Regular maintenance, updates, and repairs are key to reducing disruptions caused by weather events. When building new facilities, their location should be carefully considered, ensuring they are situated away from high-risk zones prone to flooding or other climate-related hazards.

While these adaptations are crucial, challenges exist. Costs, limited time, prioritising tasks, deciding who owns and manages these assets, and aligning with climate change policies can all slow down the process of improving infrastructure.

Any mitigation strategy implemented must be well thought out and considered. For example, sea walls, which are built along the coastline to protect houses from flooding. There are a number of potential impacts of these structures including:
- erosion of the shore
- safety risks from water coming over the wall
- disrupting natural habitats
- the cost to build and maintain the wall.

Governments also need to consider where people build their homes. Constructing houses in flood-prone areas, like floodplains, can lead to significant risks and damages during extreme weather events. Balancing development with environmental and climate considerations is vital for the safety and well-being of communities.

Relocation

As the effects of climate change become more pronounced, we're witnessing an increase in extreme weather events like floods and wildfires. Additionally, the melting polar ice caps are leading to rising sea levels, which is a major concern. This situation is prompting people to consider moving to safer regions, a process known as relocation.

One notable example of relocation due to climate change is occurring in the Solomon Islands. For instance, consider Nuatambu Island. In the past, it covered an area of around 30 080 square meters. However, measurements taken in 2014 revealed that it had shrunk to just 13 980 square meters, losing more than half of its livable space. As a result, 11 houses on the island were lost due to the encroaching sea.

The challenges posed by forced relocations due to climate change are complex and multifaceted. Governments must take into account various factors to ensure a smooth transition. Decisions need to be made regarding where these displaced communities will be relocated. Are there specific regions that are safer and more suitable for their needs? Then comes the issue of financing the entire process. Who will cover the costs of moving, building new infrastructure, and providing necessary services in the new location? Additionally, compensation might be necessary for those who have lost their homes and belongings.

Overall, the process of managing forced relocations due to climate change is a vital consideration for governments and organizations worldwide. It involves careful planning, addressing the needs of affected communities, and ensuring that the challenges brought about by climate change are met with appropriate responses and support.

Water management

Water management becomes increasingly important as temperatures climb due to climate change. Water is crucial for our hydration, considering that about 60% of our bodies are made up of water. However, only a tiny portion, around 0.5% of all the water on Earth, is available as freshwater that we can drink and use for various purposes. The impacts of climate change can affect this limited supply of freshwater.

FIGURE 7.60 Wetlands of the Barwon River and Lake Connewarre, Victoria.

Rising sea levels, a consequence of climate change, pose a threat to our freshwater sources. As the sea level increases, saltwater can intrude into underground freshwater reservoirs, contaminating them and reducing the amount of usable water. Moreover, climate change brings about more frequent and intense floods and droughts. These extreme weather events can lead to water pollution due to the introduction of pathogens and pesticides into water sources.

To address these potential problems, innovative solutions are being explored. One effective approach is the creation of wetlands. They play a dual role in absorbing and storing carbon dioxide, one of the main greenhouse gases responsible for global warming. Additionally, wetlands act as a defense against extreme weather events, such as cyclones and heavy rains, and they absorb excess water and rainfall, reducing the impact of floods.

As the global population continues to grow, the demand for food increases as well. Agriculture, a major source of food, requires a substantial amount of water. Implementing water-efficient technologies in agriculture can help reduce the strain on freshwater resources. These technologies aim to make the most of available water, ensuring that crops receive the right amount of water without unnecessary waste.

7.6.3 Paris agreement

On November 4, 2016, a significant step was taken in the global fight against climate change when 175 countries came together to sign the Paris Agreement. Since then, the number of participating countries has grown to 194. This agreement holds legal weight, and it reflects the collective commitment of these nations to address the urgent issue of climate change. Their shared goal is to reduce the emission of greenhouse gases in order to limit the rise in global temperatures to 1.5 degrees Celsius.

FIGURE 7.61 To meet the 1.5 °C global warming target set out in the Paris Agreement, net zero carbon emissions must be met by 2050.

The Paris Agreement recognizes that different countries have different levels of resources and capabilities to combat climate change. To ensure that everyone is on board and working towards the same goal, there's a focus on providing support to countries that need it the most.

Financial support: Transitioning to cleaner and more sustainable technologies, and building infrastructure that reduces greenhouse gas emissions, requires significant financial investment. The agreement acknowledges this and mandates that developed countries assist by providing financial resources. Additionally, other participating parties can also contribute on a voluntary basis. This financial assistance is crucial for countries to effectively implement their emission reduction plans.

Technology development and transfer: One of the key aspects of achieving the goals set by the Paris Agreement is the development and sharing of advanced technologies. These technologies can help countries reduce their greenhouse gas emissions and become more resilient to the effects of climate change. Technology transfer involves sharing these advancements with developing countries, ensuring that everyone has access to the tools needed for effective climate action.

Capacity-building: Tackling climate change can be a daunting task, especially for developing countries with limited resources. The Paris Agreement recognizes this challenge and calls on developed countries to lend their support in capacity-building efforts. This means providing assistance in terms of knowledge, skills, and training so that developing countries can effectively address the impacts of climate change and contribute to global efforts.

DISCUSSION

Earth's nine lives

Is it time to think about our relationship with our environment in a new way? Researchers at the Stockholm Environment Institute in Sweden have identified nine planetary life-support systems that provide planetary boundaries that they argue should be adhered to in order to live sustainably. These are:
- rate of biodiversity loss
- climate change
- nitrogen and phosphorus cycles
- stratospheric ozone depletion
- atmospheric aerosol loading
- chemical pollution
- ocean acidification
- fresh-water use
- change in land use.

What are some changes that need to occur in each of these areas in order for humans to live sustainably?

on Resources

🔗 **Weblink**   Paris Agreement

7.6.4 Biospherics

If humans are to colonise other planets such as Mars, we would need to develop viable biospheric systems that use physical and chemical techniques to recycle clean air and fresh water and remove accumulating wastes. As biospheric systems increase in size, however, the basic concepts of cycling of elements and the importance of biodiversity have direct implications on a number of different issues. These include global warming, the protection of endangered species, sufficient food supplies, effective waste removal and clean water requirements.

In the early 1980s, John Allen and a number of colleagues formed Space Biospheres Ventures. John and his team designed and built an artificial world — Biosphere 2 — to develop a closed ecological system for research and education. It was named Biosphere 2 not because it was the second artificial world created, but because the first biosphere, Biosphere 1, is Earth itself.

SCIENCE AS A HUMAN ENDEAVOUR: Biosphere 2

Biosphere 2 was designed as an eco-technological model for space exploration and colonisation. This bioengineered facility was intended to grow food, cleanse the air, and recirculate and purify water for its inhabitants. This was to be achieved without exchange of materials (including atmospheric gases) with the outside world. The purpose of this cyber-Earth was for scientists to gather information to assist in the development of strategies to solve some of Earth's environmental problems and the hurdles of developing human colonies in space.

Biosphere 2 covers 13 000 square metres, has 6500 windows and contains living quarters and greenhouses containing food crops. Five artificial environments that represent the different biomes are enclosed within the structure: a desert, a salt marsh, a tropical savanna, an ocean and a rainforest.

FIGURE 7.62 Biosphere 2, southern Arizona

FIGURE 7.63 Plan of Biosphere 2. The glass and structure components acted as a filter for incoming solar radiation so that almost all UV radiation was absorbed.

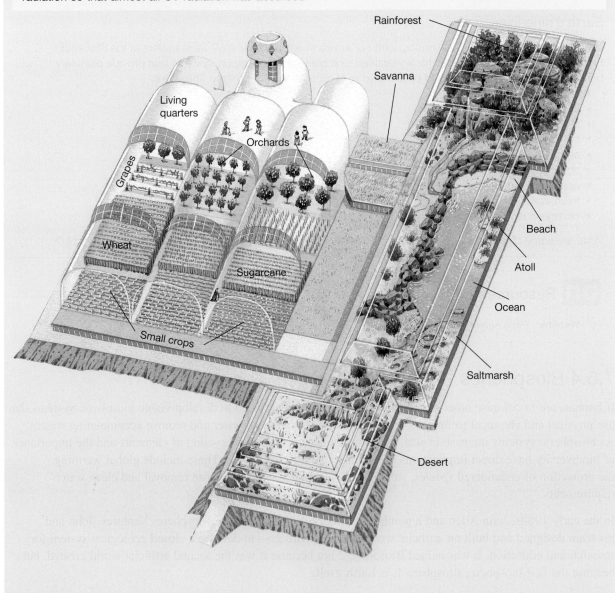

The experiment begins

Shortly after sunrise on 26 September 1991, eight people and 3800 species of plants and animals were locked inside this artificial world for two years. Worldwide, millions of television viewers watched. The crew had been prepared by years of training and working on developing systems for Biosphere 2. They had also had nine preliminary one-week semi-closed experiments over the previous five months.

FIGURE 7.64 a. Abigail Alling stopped her graduate work at Yale University on blue whales to enter Biosphere 2 as the manager of oceans and marshes. She created and operated the world's largest artificial ecological marine system, a mangrove marsh and ocean coral reef, for the Biosphere 2 project. **b.** Alling was one of the original eight people to live inside Biosphere 2.

Gasping for oxygen

By the end of the first year of their mission, the Biospherians reported deteriorating air and water quality. Oxygen concentrations in the air had fallen from 21 per cent to 14 per cent. This oxygen level was barely enough to keep them alive and functioning. At the same time, carbon dioxide concentrations were undergoing large daily and seasonal variations and nitrous oxide in the air had reached mind-numbing levels. In January 1993, fresh air was pumped in to replenish the dome's atmosphere and rescue the inhabitants. Investigations indicated that the missing oxygen was being consumed by microbes in the excessively rich food crop soil.

It was very fortunate that the fresh concrete used in the structure's construction absorbed carbon dioxide released by microbial metabolism. If this carbon dioxide sink hadn't been available, the air would have become unbreathable much earlier.

DISCUSSION

During the crew's time in Biosphere 2, a solar eclipse occurred in 1992. The crew were glued to monitors at this time, observing oxygen levels. Why were they so fixated on this?

More carbon cycling

One of Arizona's cloudiest seasons on record occurred between October 1991 and February 1992. The carbon dioxide concentration inside Biosphere 2 rose to about 3400 ppm (parts per million). The combination of this effect and an unusually dark cloudy period in the last week of December greatly reduced photosynthesis. During this period, the rise in carbon dioxide was kept below 4000 ppm by the operation of a recycler, which captured carbon dioxide and precipitated it into calcium carbonate (limestone). The calcium carbonate could later be released into the air by heating the limestone. This experience provided an insight into how to maximise photosynthesis and minimise soil respiration. Hence, Biosphere 2's goal to maintain its atmosphere was achieved despite the low light conditions.

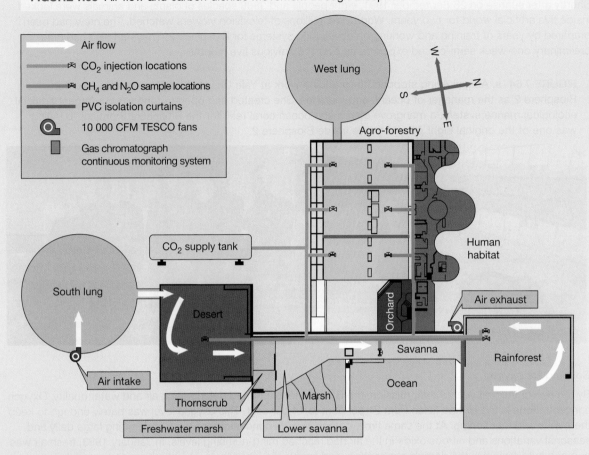

FIGURE 7.65 Air flow and carbon dioxide movement through Biosphere 2

Legend:
- Air flow
- CO_2 injection locations
- CH_4 and N_2O sample locations
- PVC isolation curtains
- 10 000 CFM TESCO fans
- Gas chromatograph continuous monitoring system

West lung
Agro-forestry
Human habitat
CO_2 supply tank
South lung
Desert
Orchard
Air exhaust
Air intake
Savanna
Rainforest
Thornscrub
Marsh
Ocean
Freshwater marsh
Lower savanna

Getting hungry

Ideally, the chemical-free agriculture system inside Biosphere 2 recycled all human and domestic animal waste products. It also initially included dozens of crop varieties to provide nutritional balance and allow for crop rotation. Biosphere 2, however, encountered considerable food production problems.

Due to unexpected crop failures, far less food was produced than had been projected. Only 60 per cent of the sunlight made it through the glass pane of Biosphere 2's space frame. Cloudy days also reduced the light available to plants for photosynthesis. A combination of unprecedented cloudy weather for the second straight year (20 per cent below the low rate of sunshine for 1992) and increased insect pest problems contributed to reduced food production.

Meat was a rare commodity that crew members ate only once a week. Coffee was only consumed once every fortnight because it was considered non-essential so not much space was allocated to it. When the mission ended, the average weight loss per person was around 13 kilograms and, interesting enough, the crew lost the enzymes needed to digest meat because they ate so little of it while locked away!

Survivors

Of the 25 introduced vertebrate species, 18 became extinct. All of the insect pollinators died, which prevented most plants from producing seeds. This led to food supplies falling to dangerous levels. Weedy vines flourished in the carbon-rich atmosphere and threatened to choke out more desirable plants. Although the majority of insects disappeared, ants and cockroaches thrived and overran everything, including workers.

int-5890

FIGURE 7.66 Water was conserved inside the Biosphere 2 wilderness environments. Condensation, artificial rain or irrigation (by sprinkler systems), evaporation, transpiration and sub-soil drainage were the major internal water cycling components.

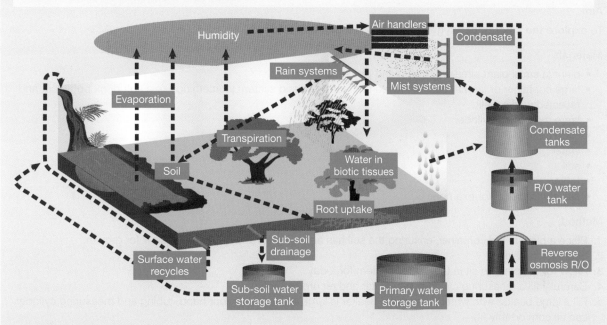

Other uses of Biosphere 2

Other experiments in Biosphere 2 include flushing it with carbon dioxide and using it to predict Earth's future.

Because carbon dioxide is a fundamental requirement for photosynthesis, scientists have long suspected that higher carbon dioxide levels will fuel extra plant growth. Some of them have even suggested that rising carbon dioxide levels may boost global harvests. Other scientists have suggested that trees and shrubs around the world will help alleviate the problems of global warming by soaking up some of the additional carbon dioxide. This brings some thought-provoking questions to mind:

- If extra plant growth does appear, will all crop plants be affected in the same way; if not, what are the implications?
- If extra carbon is taken up by the natural biosphere, how long will it stay there?
- What would happen if the carbon dioxide quickly went into the soil and was then returned to the atmosphere?
- Will the carbon dioxide be safely locked up in the forests?
- Are there carbon dioxide levels that may kill off trees and shrubs, resulting in release of their accumulated carbon in one catastrophic burst?
- How long (and with what effects) can a group of people live in an artificial closed system?
- Does the experience of Biosphere 2 bring us any closer to living on Mars?

The future of Biosphere 2

As of 2007, the research of Biosphere 2 was taken over by the University of Arizona after the site had been sold two years earlier. Research continues to occur around using the artificial ecosystem to support life. Biosphere 2 is also open for tours for individuals wishing to explore the world's largest closed system.

INVESTIGATION 7.5

How much oxygen is produced in a closed system?

Aim

To explore the amount of oxygen produced in a closed system

Materials

- plant (a small plant already growing and potted is best)
- large container or jar that can be closed to create a closed system (something like a soft drink bottle cut and resealed works well)
- large measuring cylinder
- large beaker
- water
- soil
- tubing
- sealant (silicone, for example)

Method

1. Place a plant in the container, ensuring the soil has ample water and nutrients to allow for growth.
2. Seal the container.
3. Allow for the plant to sit in the closed system for a day.
4. Carefully fill a measuring cylinder with water and record the volume.
5. Fill a large beaker with water (ensure the beaker is large enough that your hand, tubing and measuring cylinder can all comfortably fit).
6. Seal the measuring cylinder.
7. The next two stages must be done quickly, so make sure you are set up. Place the tubing in the container with the plant (ensure that oxygen can only escape through the tubing and that everything is sealed).
8. Place the sealed measuring cylinder of water upside-down in the beaker of water. Carefully pull the seal off and place the other end of the tubing in the measuring cylinder (only a centimetre or two in).
9. The oxygen produced will displace the water. Record the displacement of the water every 10 seconds, until the volume stays at a fixed point.

Results

Construct a table and note your observations.

Discussion

1. Why was an upside-down measuring cylinder used?
2. Is oxygen able to be produced in a closed system?
3. Why was this system closed?

Conclusion

Summarise the findings for this investigation about the amount of oxygen produced in a closed system.

---

on Resources

eWorkbooks	Slowing global warming (ewbk-13188)
	A dome away from home (ewbk-13190)
Video eLesson	Biosphere 2, southern Arizona (eles-2912)
Interactivity	The survival game (int-0217)
Weblink	Biosphere 2

7.6 Activities

7.6 Quick quiz on	7.6 Exercise

Select your pathway

■ LEVEL 1	■ LEVEL 2	■ LEVEL 3
1, 3, 4, 7, 9	2, 5, 8, 13, 14, 16	6, 10, 11, 12, 15, 17

Remember and understand

1. Suggest why no-one can be certain about the actual consequences of global warming.
2. If we can't be certain about the consequences of global warming, why bother about it?
3. Identify the following statement as true or false. Provide reasons for your response.
   Within Biosphere 2, water was cycled through condensation, artificial rain, evapotranspiration and sub-soil drainage.
4. What is geosequestration and why is it important?
5. What is metagenomics? What is it used for?
6. Suggest how manufacturing and using wooden products that have been produced using sustainable management may help fight global warming.
7. Explain why CSIRO scientists are studying Tammar wallabies in their research related to global warming.

Apply and analyse

8. Explain why it is necessary for the Australian government to create legislation to address the problem of global warming.
9. Explain why you would expect an increase in carbon dioxide levels and a decrease in oxygen levels if a large number of people entered Biosphere 2.
10. Consider Biosphere 2.
    a. Why was Biosphere 2 not called Biosphere 1?
    b. List the five artificial environments enclosed within Biosphere 2.
    c. Describe the purpose of Biosphere 2.
    d. State one way in which Biosphere 2 is similar to Earth.
    e. Why was it fortunate that the fresh concrete in the structure absorbed carbon dioxide?
    f. What improvements might be made to a hypothetical Biosphere 3?
11. As sea levels rise, forced relocations are going to occur. Discuss who should be responsible for the costs of relocating and compensating people who lose their homes.

Evaluate and create

12. Research why scientists have differing views on climate change and what they believe will happen if we do not reduce emissions.
13. **SIS** Wind energy is a source of renewable energy; however, this technology also has some negative aspects. Do the positives outweigh the negatives? Research two of these issues (using data where appropriate) and discuss how you would go about resolving them.
14. **SIS** Write a report on the experience, using findings of the Biosphere 2 project.
15. **SIS** Due to the moist, artificially generated climate, shrubs and grasses, rather than desert plants, overran the desert area.
    a. Suggest why this occurred.
    b. If you were the scientist assigned to solve this problem, suggest how you could increase the number of desert plants. Write a clear experimental method to test this.
16. **SIS** Research the Paris Agreement.
    a. Summarise the aims of this agreement.
    b. Do you believe all nations should sign this agreement? Justify your response.
    c. What are some ways that you can individually act to help meet the goals of the Paris Agreement?
17. Report what the effects of sea levels rising by 1.3 m in Australia would be and the impact it would have on the population.

Fully worked solutions and sample responses are available in your digital formats.

LESSON
7.7 Thinking tools — Fishbone diagrams

7.7.1 Tell me

What is a fishbone diagram?

A fishbone diagram is also known as a cause and effect diagram. It groups and categorises different causes under main cause groups, allowing the different factors leading to a specific effect to easily be seen.

FIGURE 7.67 The structure of a fishbone diagram

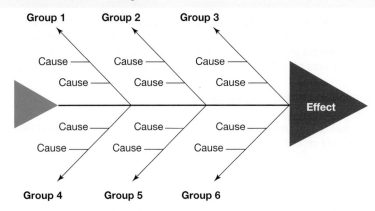

7.7.2 Show me

In order to create a fishbone diagram, you should:
1. Write as many causes as you can that can lead to a certain event.
2. Sort these causes into different groups, trying to have no more than three causes per group.
3. Determine the order of the cause groups (if any).
4. Draw the fishbone structure — try to use a large sheet of paper so you have lots of room to write on each 'bone'.
5. Fill in your information in your fishbone diagram. On each bone, place the name of the group at the end or along the bone and each of the causes coming off this.

Figure 7.68 shows an example of a fishbone diagram for global warming.

FIGURE 7.68 A fishbone diagram showing causes of global warming

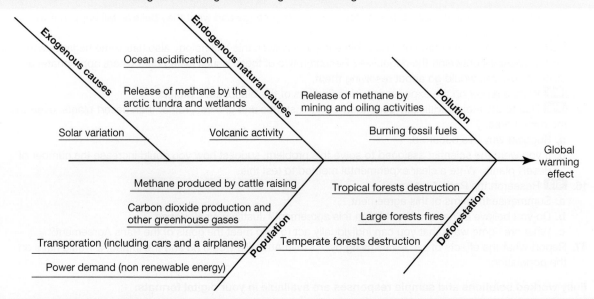

7.7.3 Let me do it

These questions are even better in jacPLUS!
- Receive immediate feedback
- Access sample responses
- Track results and progress

Find all this and MORE in jacPLUS ▶

7.7 Activity

1. Create fishbone diagrams for the following events.
   a. Influences of the variation of climate
   b. Changing biodiversity
   c. carbon dioxide and methane levels in the atmosphere
2. Read the following article. Using the information in the article, create a fishbone diagram showing the link between causes and the effect of developing the new type of washing machine.

Wash clothes with thin air

It could be a godsend for drought-stricken communities — a washing machine that needs no water or powder yet cleans clothes in a jiffy.

Scientists in Singapore have invented a revolutionary appliance called the Airwash and it has already caught the eye of one major manufacturer.

The machine works by blasting dirty clothes with jets of air primed with negative ions, which have the effect of clumping dust together, deactivating bacteria and neutralising odours.

The result, the inventors claim, is clean, fresh-smelling clothes that come out of the machine completely dry — meaning an end to clothes lines and perhaps even a death knell for the tumble dryer.

And since no water is involved, fabrics unsuitable for conventional machines — such as leather and suede — can be washed at home instead of having to be dry cleaned.

Negative ions are molecules that have gained an electric charge. Odourless, tasteless and invisible, they are created when molecules in the atmosphere break apart due to fast-moving air and sunlight. In nature, they are found in invigorating environments such as pine forests and where breaking waves pound the seashore.

The Airwash is inspired by the way clothes used to be beaten against river rocks near waterfalls, which are another of nature's negative-ion generators. A prototype has been built by Gabriel Tan and Wendy Chua of the National University of Singapore and Electrolux is watching closely.

The average householder spends nine months in a lifetime doing the washing and the Airwash designers believe any machine that makes the chore easier will be welcomed.

Mr Tan said: 'But as well as being boring, laundry uses up scarce water supplies and pollutes with chemical detergents.'

Source: UK white goods (www.ukwhitegoods.co.uk/appliance-industry-news/226-washing-machine-news/1565-washclotheswiththinair)

Fully worked solutions and sample responses are available in your digital formats.

LESSON
7.8 Project — Fifty years later

Scenario

Consider the following.

- **260 million years BCE:** A massive volcano in what is presently China erupts, causing atmospheric and oceanic changes and leading to the extinction of 95 per cent of life in the oceans and 70 per cent of land-based life.
- **95 million years BCE:** Undersea volcanic activity triggers a mass extinction of marine life and buries a thick mat of organic matter on the sea floor.
- **72 000 BCE:** The Lake Toba volcano in Indonesia ejects nearly 3000 cubic kilometres of material into the atmosphere, cutting off much of the sun's light to Earth's surface and for so long that 50 per cent of humanity dies out.
- **2000 CE:** The UK science program **Horizon** uses the term **supervolcano** to describe volcanoes capable of massive eruptions covering huge areas with lava and ash and causing long-term weather effects and mass extinctions.
- **2030 CE:** The supervolcano under Yellowstone National Park erupts cataclysmically, destroying half of the USA and changing Earth's atmosphere and surface conditions for centuries to come.

FIGURE 7.69 Grand Prismatic Spring in Yellowstone National Park

The year is now 2080. Fifty years after the eruption, the gases and ash that the eruption produced, as well as the destruction of large sections of land, have affected the critical environmental cycles of Earth's environments, and human civilisation has had to change its ways in order to survive. Some things remain the same, though — we still have radio and television of a sort. Not surprisingly, with the 50th anniversary of the Yellowstone eruption (or 'Y-day', as it is known) coming up, lots of TV programs will be focusing on the critical event that changed our world forever.

Your task

As part of a small documentary film company, you will produce a 5-minute segment for a special edition of a TV science show that will be aired on the 50th anniversary of Y-day. In this segment, a science journalist will interview a variety of experts in a retrospective of what happened on Y-day, how the environment has changed over the 50 years since the eruption, and what humanity can expect to happen in the next 50 years.

 Resources

 ProjectsPLUS Fifty years later (pro-0248)

LESSON
7.9 Review

Access your topic review eWorkbooks

on Resources

■ Topic review Level 1	■ Topic review Level 2	■ Topic review Level 3
ewbk-13192	ewbk-13194	ewbk-13196

7.9.1 Summary

Global systems

- The biosphere, atmosphere, hydrosphere and lithosphere (or geosphere) are four of Earth's global systems.
- The ozone layer, found in the stratosphere, absorbs ultraviolet radiation.
- Interactions occur between Earth's global spheres with each other and the radiant energy of the Sun.
- The biosphere is the layer on Earth in which all life exists. It includes all living things (biota) and all ecosystems.
- Within ecosystems, materials are cycled between the atmosphere, hydrosphere and lithosphere within the carbon, nitrogen, phosphorus and water cycles.
- Human activities are having an effect on Earth's cycles, including through clearing forests, releasing harmful chemicals from the burning of fossil fuels and propellants, an increasing population and mining.

Climate patterns

- The variation of climate over Earth's surface is influenced by the tilt of Earth's axis, the differing abilities of land and water to absorb and emit radiant heat, and the features of the land.
- Ocean currents are determined by the temperature differences between the tropics and poles and Earth's rotation.
- The direction of prevailing winds is affected by several factors, including latitude, the rotation of Earth, ocean currents and altitude.
- Due to the levels of carbon dioxide in the atmosphere, the ocean currents and wind patterns are changing.

Climate change

- Greenhouse gases help keep Earth at the right temperature to support life. However, global warming is threatening the survival of all life due to the enhanced greenhouse effect.
- Rises in carbon emissions and methane are increasing Earth's temperature.

Biodiversity and climate change

- The climate is sensitive to small changes in the concentration of carbon dioxide. Changes in the climate also affect other organisms, which can have an impact on genetic, species and ecological diversity.
- Increases in Earth's temperature will lead to a rise in ocean levels as ice and glaciers melt.

Cool solutions

- Climate models are created using quantum computers to help predict what may happen in the future.
- By separating carbon dioxide from other gases (geosequestration), it can be stored underground, reducing the amount of carbon dioxide in the atmosphere.
- By growing sustainable forests, less carbon dioxide is released into the atmosphere.
- Renewable resources such as solar, hydro and wind can be used to generate energy instead of using fossil fuels.
- Afforestation and reforestation will help reduce the amount of carbon in the atmosphere as trees take in carbon dioxide for photosynthesis.
- Infrastructure needs to be maintained, repaired and updated so that it can still operate during weather events. Any new structures should be situated away from high-risk zones.
- As sea levels rise, people in low lying areas around the world will need to relocate.
- Precious drinking water needs to be protected from climate change as rising sea levels can pose a threat to freshwater sources.
- In 2016, the Paris Agreement came into effect to limit the rise in global temperatures to 1.5° C.
- Viable biospheric systems are required if humans wish to colonise other planets in the future. Biosphere 2 was a bioengineered facility created to test if inhabitants could survive in the facility without exchanging any materials with the outside world.

7.9.2 Key terms

acid rain rainwater, snow or fog that contains dissolved chemicals that make it acidic

afforestation planting trees in areas that had no recent cover

atmosphere the layer of gases around the Earth

biodiversity total variety of living things on Earth

biomes regions of the Earth divided according to dominant vegetation type

biosphere the layer on Earth in which all life exists. It includes all living things (biota) and all ecosystems

biota the living things within a region or geological period

cellular respiration the chemical reaction involving oxygen that moves the energy in glucose into the compound ATP

climate sensitivity the measure of temperature change in the climate, dependent on the amount of carbon dioxide released into the atmosphere

convection currents process of heat transfer in gases and liquids in which lighter and hotter materials rise, and cooler and denser materials sink

deforestation the process of clearing trees to convert the land for other uses

desertification the process in which fertile regions become more dry and arid

eutrophication a form of water pollution involving an excess of nutrients leaching from soils

extinction complete loss of a species when the last organism of the species dies

gene pool the total genetic information of a population, usually expressed in terms of allele frequency

geosequestration the process that involves separating carbon dioxide from other flue gases, compressing it and piping it to a suitable site

global warming the observed rise in the average near-surface temperature of the Earth

greenhouse effect a natural effect of the Earth's atmosphere trapping heat, which keeps the Earth's temperature stable

hydrosphere Earth's water from the surface, underground and air

ice cores samples of ice extracted from ice sheets containing a build-up of dust, gases and other substances trapped over time

landfills areas set aside for the dumping of rubbish

lithosphere the outermost layer of the Earth; includes the crust and uppermost part of the mantle

metagenomics technology that combines DNA sequencing with molecular and computational biology

palaeoclimate the climate at a specific point in geological history

permafrost ground that is frozen for at least two continuous years

photosynthesis food-making process in plants that takes place in chloroplasts, and uses carbon dioxide, water and energy from the Sun

prevailing winds winds most frequently observed in each region of the Earth

radiant heat heat transferred by radiation, such as from the Sun to the Earth

reforestation planting trees in forest areas that have been destroyed or damaged

remote sensing data collection about Earth's biosphere completed from space by devices such as satellites

stratosphere the second layer of the atmosphere up to about 55 km above the Earth's surface, between the troposphere and the mesosphere

troposphere the layer of the atmosphere closest to the Earth's surface, containing closely packed particles of air

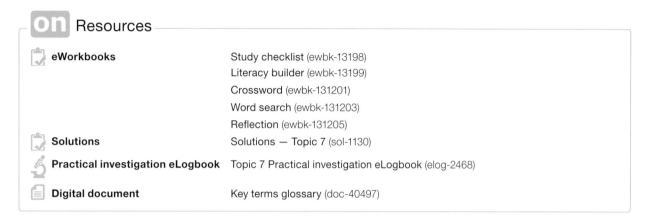

on Resources

📋 **eWorkbooks**	Study checklist (ewbk-13198)
	Literacy builder (ewbk-13199)
	Crossword (ewbk-131201)
	Word search (ewbk-131203)
	Reflection (ewbk-131205)
📋 **Solutions**	Solutions — Topic 7 (sol-1130)
🔬 **Practical investigation eLogbook**	Topic 7 Practical investigation eLogbook (elog-2468)
📄 **Digital document**	Key terms glossary (doc-40497)

7.9 Activities

learn on

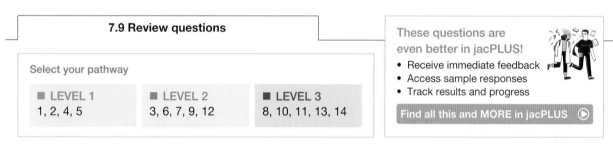

7.9 Review questions

Select your pathway

■ LEVEL 1	■ LEVEL 2	■ LEVEL 3
1, 2, 4, 5	3, 6, 7, 9, 12	8, 10, 11, 13, 14

These questions are even better in jacPLUS!
- Receive immediate feedback
- Access sample responses
- Track results and progress

Find all this and MORE in jacPLUS ⊙

Remember and understand

1. a. State whether the following statements are true or false.

Statements	True or false?
i. Palaeoclimates offer a unique perspective in that they can show the wide range of climates over various time scales, and transitions between them.	
ii. No-one can be certain about the actual consequences of global warming because so many variables influence climate that computer modelling cannot provide completely accurate predictions.	
iii. The biosphere includes all of the ecosystems on Earth.	
iv. Ocean currents are the result of the temperature difference between the tropics and poles, and Earth's rotation.	
v. Following the process of photosynthesis, high levels of carbon dioxide are released from the atmosphere.	
vi. CFCs are important molecules to help increase the amount of ozone molecules in the atmosphere.	

b. Justify any false responses. ▶

2. Match the following terms to the correct definition.

Term	Definition
a. Atmosphere	A. The rising temperature of Earth
b. Lithosphere	B. Earth's water from the surface, underground and air
c. The greenhouse effect	C. The maintenance of Earth's temperatures by the atmospheric gases
d. Global warming	D. The layer of gases around Earth
e. Biosphere	E. The outermost layer of Earth
f. Hydrosphere	F. The layer on Earth in which all life exists. It includes all living things and all ecosystems.
g. Troposphere	G. The layer of the atmosphere closest to Earth's surface
h. Biota	H. The living things within a region or geological period

3. Describe the relationship between the atmosphere, lithosphere, hydrosphere and biota. You may wish to construct a diagram to show this.
4. Identify each of the following greenhouse gases.
    a. The greenhouse gas associated with sources such as aerosols, refrigerators, plastic foam production and dry-cleaning
    b. The greenhouse gas associated with sources such as bacteria in fertilisers and burning fossil fuels
    c. The greenhouse gas associated with sources such as rice paddies, livestock (for example, cows), bacteria in bogs and landfill
    d. The greenhouse gas associated with sources such as deforestation, decomposition, burning fossil fuels and cellular respiration
5. Define the following terms.
    a. Climate sensitivity
    b. Enhanced greenhouse effect
    c. Species diversity
    d. Stratosphere
6. a. Outline four ways in which humans may contribute to the enhanced greenhouse effect.
    b. Provide suggestions for how we can reduce our impact on the enhanced greenhouse effect.

Apply and analyse

7. Construct a diagram showing the carbon cycle. Provide a brief statement outlining the importance of each of the components of the cycle.
8. Describe the relationships between each set of terms provided.
    a. Species, biodiversity, biodiversity loss, threatened, endangered, extinct, mass extinction
    b. Biosphere, lithosphere, hydrosphere, biota, atmosphere, troposphere, stratosphere
    c. Stratosphere, climate change, greenhouse gas, fossil fuels, global warming, carbon dioxide, methane, nitrous oxide, biodiversity loss, enhanced greenhouse effect, cellular respiration, lithosphere
    d. Carbon cycle, photosynthesis, cellular respiration, carbon dioxide
    e. Greenhouse effect, enhanced greenhouse effect, global warming
9. Using the knowledge you have gained from this topic, comment on the following statement:
    Constantly changing physical, chemical and biological cycles have contributed to the survival of various forms of life on Earth. Our life-support systems are not in good shape.
10. a. What is meant by biodiversity and why is loss of biodiversity a concern?
    b. Explain how invasive species introduced from overseas can affect the biodiversity of Australian flora and fauna.

Evaluate and create

11. **a.** The mountain pygmy possum is restricted to an area of 6 km^2 in the Australian Alps. Suggest how such a restricted habitat may influence its chances of survival.
    b. Suggest abiotic and biotic factors that may affect this possum.
    c. Suggest how warmer temperatures and reduced snow may affect its lifestyle. Be specific in your response by including examples of different scenarios.
    d. What is meant by the term *extinction*?
    e. If this species were to become extinct, suggest implications for other organisms within its ecosystem.

12. Rising sea levels and saltwater intrusion associated with climate change are threats that Kakadu National Park is experiencing.
    a. Suggest why these threats are associated with climate change.
    b. Suggest effects that these new threats may have on the **(i)** biotic and **(ii)** abiotic parts of this ecosystem.
    c. Suggest actions that could be taken to reduce the loss of biodiversity within Kakadu National Park.

13. Global warming and climate change is a current issue that is not going away.
    a. Outline the most accepted view within the scientific community of the cause of global warming.
    b. Describe examples of effects or consequences of global warming that have been suggested by scientists.
    c. State your opinion about the possible **(i)** cause, **(ii)** effects and **(iii)** solutions for global warming.
    d. View the top ten arguments about global warming that are used by sceptics. Rank these statements in order of most like your opinion to least like your opinion. Justify your ranking.
    e. State the difference between an opinion, a theory and a fact.
    f. Can scientists have opinions? If you agree, when, how and why should these be shared? If you do not agree, why not?
    g. Should science play a part in the making of climate policy? Justify your response.
    h. Suggest possible reasons for the climate debate.

14. Agriculture has had (and continues to have) a devastating effect on a number of marine ecosystems. Hypoxia in coastal zones from nitrogen and phosphorus outputs of agricultural and livestock industries is one such example.
    a. Using your knowledge of the nitrogen and phosphorus cycles, explain how these outputs may damage marine ecosystems.
    b. Suggest strategies that may reduce the negative impact that agriculture has on our ecosystems.

Fully worked solutions and sample responses are available in your digital formats.

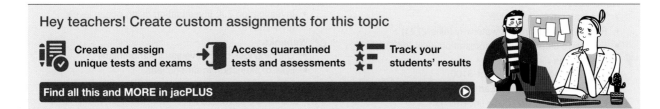

Hey teachers! Create custom assignments for this topic

Create and assign unique tests and exams

Access quarantined tests and assessments

Track your students' results

Find all this and MORE in jacPLUS

Online Resources

 Resources

Below is a full list of **rich resources** available online for this topic. These resources are designed to bring ideas to life, to promote deep and lasting learning and to support the different learning needs of each individual.

7.1 Overview

eWorkbooks
- Topic 7 eWorkbook (ewbk-13165)
- Starter activity (ewbk-13167)
- Student learning matrix (ewbk-13169)

Solutions
- Topic 7 Solutions (sol-1130)

Practical investigation eLogbook
- Topic 7 Practical investigation eLogbook (elog-2468)

Video eLessons
- Climate change protests (eles-4175)
- Asteroid impact (eles-2730)

Weblink
- Cool Antarctica

7.2 Global systems

eWorkbooks
- Recycling carbon (ewbk-13170)
- Cycles in nature (ewbk-13172)
- Labelling the water cycle (ewbk-13174)
- Labelling the carbon cycle (ewbk-13176)
- Labelling the nitrogen cycle (ewbk-13178)
- Labelling the phosphorus cycle (ewbk-13180)

Practical investigation eLogbooks
- Investigation 7.1: Effect of deforestation (elog-2470)
- Investigation 7.2: Effect of acid rain on plants (elog-2472)

Video eLessons
- The water cycle (eles-0062)
- Removing algae from a waterway (eles-2731)
- Excessive clearing and deforestation (eles-2905)

Interactivities
- Labelling the water cycle (int-8100)
- The different biomes on Earth (int-8228)
- Labelling the carbon cycle (int-8101)
- Labelling the nitrogen cycle (int-8102)
- Labelling the phosphorus cycle (int-8103)
- Lake Urmia (int-5605)

Weblinks
- Australian State of the Environment: Land
- WWF Footprint Calculator

7.3 Climate patterns

Practical investigation eLogbook
- Investigation 7.3: Exploring the axis tilt of Earth (elog-2474)

Video eLesson
- Weather station (eles-2733)

Interactivities
- Surface energy distribution (int-5882)
- Wind patterns (int-5883)

7.4 Climate change

eWorkbooks
- Secrets in the ice (ewbk-13182)
- Global warming and the greenhouse effect (ewbk-13184)

Video eLessons
- Analysing core samples (eles-4176)
- Global warming in Australia (eles-0057)
- Air temperature over Earth (eles-2732)

Interactivities
- Sources of greenhouse gases (int-5884)
- Secrets in the ice (int-5885)

Weblinks
- 2010 Sustainable Cities Index
- 2018 Sustainable Cities Index
- Australia thaws in mid-winter warmth – is winter over?

7.5 Biodiversity and climate change

eWorkbook
- How climate change impacts biodiversity (ewbk-13186)

Practical investigation eLogbook
- Investigation 7.4: Investigating rising ocean levels (elog-2476)

Video eLesson
- Polar bears (eles-2908)

Interactivities
- Changing climate indicators (int-5886)

Teacher-led video
- Investigation 7.4: Investigating rising ocean levels (tlvd-10819)

Weblinks
- Ongoing loss of biodiversity in amphibians
- Risk of extinction to amphibians

7.6 Cool solutions

eWorkbooks
- Slowing global warming (ewbk-13188)
- A dome away from home (ewbk-13190)

Practical investigation eLogbook
- Investigation 7.5: How much oxygen is produced in a closed system? (elog-2478)

Video eLesson
- Biosphere 2, southern Arizona (eles-2912)

Interactivies
- The survival game (int-0217)
- Water conservation within the Biosphere 2 wilderness environment (int-5890)

Weblinks
- United Nations climate change parties
- Biosphere 2

7.8 Project — Fifty years later

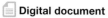 **ProjectsPLUS**
- Fifty years later (pro-0248)

7.9 Review

Digital document
- Key terms glossary (doc-40497)

eWorkbooks
- Topic review Level 1 (ewbk-13192)
- Topic review Level 2 (ewbk-13194)
- Topic review Level 3 (ewbk-13196)
- Study checklist (ewbk-13198)
- Literacy builder (ewbk-13199)
- Crossword (ewbk-131201)
- Word search (ewbk-131203)
- Reflection (ewbk-131205)

To access these online resources, log on to **www.jacplus.com.au**

8 Forces, energy and motion

CONTENT DESCRIPTION

Investigate Newton's laws of motion and quantitatively analyse the relationship between force, mass and acceleration of objects (AC9S10U05).

Source: F–10 Australian Curriculum 9.0 (2024–2029) extracts © Australian Curriculum, Assessment and Reporting Authority; reproduced by permission.

LESSON SEQUENCE

SCIENCE INQUIRY AND INVESTIGATIONS

Science inquiry is a central component of Science curriculum. Investigations, supported by a **Practical investigation eLogbook** and **teacher-led videos**, are included in this topic to provide opportunities to build Science inquiry skills through undertaking investigations and communicating findings.

LESSON
8.1 Overview

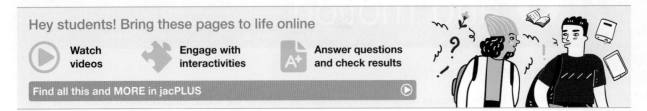
8.1.1 Introduction

The thrill of a rollercoaster ride allows you to experience sudden changes in motion. When the car suddenly falls, you seem to get left behind just for a moment. When you reach the bottom of the track and the car rises, your stomach seems to sink. And when you round a bend, your body seems to be flung sideways. Such a ride raises many questions about the way in which forces affect motion and energy.

The laws of motion discussed in this topic were established centuries ago, yet are still vital for people in many fields. They form the basis of calculations that allow for the construction of vehicles, precision sports measurements and safe theme park rides. Any further study of physics will require a thorough understanding of this topic.

FIGURE 8.1 The design of rollercoasters requires an understanding of forces, energy and motion.

---

on Resources

▶ **Video eLesson** Rollercoasters (eles-2734)

Watch this video to watch the motion of a rollercoaster. The rollercoaster shown is the fastest and most popular rollercoaster in the Netherlands.

8.1.2 Think about forces, energy and motion

1. Could a kangaroo win the Melbourne Cup?
2. How do radar guns measure the speed of cars?
3. Why do you feel pushed to the left when the bus you are in turns right?
4. Why do space rockets seem to take forever to get off the ground?
5. Why does it hurt when you catch a fast-moving ball with your bare hands?
6. What does doing work really mean?
7. Why can't a tennis ball bounce higher than the height from which it is dropped?
8. Why are cars deliberately designed to crumple in a crash?

8.1.3 Science inquiry

A world of forces, energy and motion

Find out what you already know about forces, energy and motion by answering the following questions.

1. Use a small toy car or a motorised car. Model each of the following motions and then describe the movement a passenger in a car would feel when it:
   a. accelerates suddenly
   b. stops suddenly
   c. takes a sharp left turn.
2. Copy and complete the following table to list the forces acting on each of the people shown in the photos in figure 8.2. The number of forces acting on each of them is provided in brackets.

TABLE Forces acting on people in different situations

Person	Forces acting on the person
a. Bungee jumper (3)	
b. Parachutist (3)	
c. Cyclist (5)	
d. Skier (2)	
e. Swimmer (4)	
f. Skater (5)	
g. Reader (2)	

3. Some of the people in the photographs in figure 8.2. are speeding up; others may be travelling at a steady speed or slowing down.

FIGURE 8.2 What forces are acting on the different people?

   a. Which three of the people are the most likely to be speeding up? How do you know?
   b. Which three people are most likely to be moving at a non-zero constant speed? How do you know?
   c. Which three people are clearly losing gravitational potential energy?
   According to the Law of Conservation of Energy, energy cannot be created or destroyed. It can only be transformed into another form of energy or transferred to another object.
   d. What happens to the lost gravitational potential energy of each of the three people referred to in part **c.**? What happens if they reach a steady speed?

4. Use a plastic or paper bag (or other appropriate material) and string to create a parachute. Attach a small mass at the end to represent the person. Carefully drop the parachute.
   a. Were you able to observe all the forces you listed in question **2**?
   b. Which forces would be classed as non-contact forces? How do you know?
   c. Compare dropping your parachute inside to dropping it outside. What differences do you notice in the movement of a parachute and the forces acting on it?
   d. Add more mass to the bottom of your parachute. What do you notice? Does this change the forces acting on the parachute?

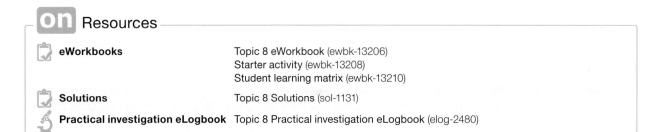

Resources

eWorkbooks	Topic 8 eWorkbook (ewbk-13206)
	Starter activity (ewbk-13208)
	Student learning matrix (ewbk-13210)
Solutions	Topic 8 Solutions (sol-1131)
Practical investigation eLogbook	Topic 8 Practical investigation eLogbook (elog-2480)

LESSON
8.2 Average speed, distance and time

LEARNING INTENTION

At the end of this lesson you will be able to explain the meaning of average speed and be able to use the average speed formula to calculate speed, distance or time. You will also be able to describe the vector quantities of displacement and velocity.

8.2.1 Calculating speed

Speed is a measure of the **rate** at which an object moves over a distance. In other words, it tells you how quickly distance is covered. We use v to represent speed. Speed is usually measured in kilometres per hour, km/h, or metres per second, m/s. The words speed and velocity are often used interchangeably in everyday life, however, they have very different meanings in science. This will be discussed further in section 8.2.4.

Imagine you are in a car travelling at 110 kilometres per hour (km/h). This means that if this speed were maintained, the car would travel 110 kilometres every hour.

In two hours, the car would be expected to travel 220 kilometres. However, a journey over that distance is unlikely to be completed at a constant speed.

DISCUSSION

List some reasons why the speed of the car might need to vary from the fixed speed of 110 km/h.

If your friend asked how fast you travelled on a journey where your actual speed was constantly changing, then without presenting them with a graph showing your speed at each instant in time, the best that you can do is to describe your **average speed**.

Average speed (represented by v_{av}) is based on how far you travelled in the total time of your journey.

speed a measure of the rate at which an object moves over a distance

rate how a physical quantity changes with respect to time

average speed total distance travelled divided by time taken

It does not take into account the fluctuations in speed that occurred during the journey. It can be calculated by dividing the total distance travelled by the total time taken.

Average speed

$$\text{average speed} = \frac{\text{total distance travelled}}{\text{total time taken}}$$

In symbols, this formula is usually expressed as:

$$v_{av} = \frac{d}{t}$$

Where v_{av} is the average speed
d is the total distance travelled
t is the total time taken.

Example 1: Calculating average speed in m/s

A cheetah sprints after a young impala, covering the 150 metre distance between them in 6 seconds. To calculate the average speed of the cheetah, use the average speed equation.

$$\begin{aligned} \text{average speed} &= \frac{\text{total distance travelled}}{\text{total time taken}} \\ &= \frac{150 \text{ m}}{6 \text{ s}} \\ &= 25 \text{ m/s (approximately 90 km/h)} \end{aligned}$$

Example 2: Calculating speed in cm/min

The average speed of a snail that takes 10 minutes to cross an 80 centimetre concrete paving stone in a straight line is:

$$\begin{aligned} v &= \frac{d}{t} \\ &= \frac{80 \text{ cm}}{10 \text{ min}} \\ &= 8 \text{ cm/min} \end{aligned}$$

FIGURE 8.3 The average speed of a snail is often calculated in cm/min.

8.2.2 Converting between units of speed

The speed of a vehicle is usually expressed in kilometres per hour (km/h). Sometimes it is more convenient to express speed in different units. Speed must, however, always be expressed as a unit of distance divided by a unit of time. The internationally accepted scientific units of speed are metres per second. These units should be used for more complex calculations.

The easy way to do this is to use the units in the equation.

If distance is in kilometres and time is in hours then:

$$\frac{\text{km}}{\text{h}} = \text{km/h (or km/h}^{-1})$$

For slower speeds, smaller units may be used. If distance is in metres and time is in seconds then:

$$\frac{\text{m}}{\text{s}} = \text{m/s (or m s}^{-1})$$

The speed at which grass grows could sensibly use units where distance is in millimetres and time is in weeks. In this case:

$$\frac{mm}{week} = mm/week \ (or \ mm \ week^{-1})$$

When converting into different units, you may:

1. Convert the distance and time into the correct units first and then calculate the average speed.
2. Use a conversion (refer to the following pink box) to convert between units of speed.

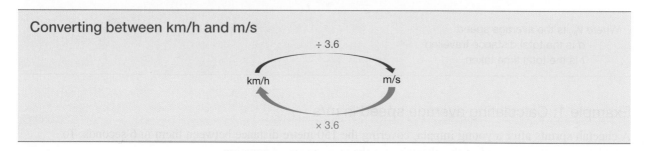

Converting between km/h and m/s

÷ 3.6

km/h   m/s

× 3.6

Example 3: Calculating average speed in km/h and m/s

The average speed of an aeroplane that travels from Perth to Melbourne, a distance of 2730 kilometres by air, in 3 hours is:

$$v_{av} = \frac{d}{t}$$
$$= \frac{2730 \ km}{3 \ hours}$$
$$= 910 \ km/h$$

FIGURE 8.4 Though aeroplanes change speed throughout a journey, the distance and time can be used to calculate their average speed.

The average speed formula can also be used to express the speed in metres per second (m/s) by converting kilometres to metres and hours to seconds. There are 1000 metres in 1 kilometre and 3600 seconds in 1 hour (60 s × 60 min).

$$v = \frac{d}{t}$$
$$= \frac{2730000 \ m}{3 \times 3600 \ s} \ (converting \ kilometres \ to \ metres \ and \ hours \ to \ seconds)$$
$$= 253 \ m/s$$

ACTIVITY: Converting from m/s to km/h

Look back at example 1, which calculated the speed of the cheetah as 25 m/s. Convert this speed into km/h.

8.2.3 Calculating distance and time

The formula used to calculate speed (and, in turn, average speed) can also be used to calculate the distance travelled or the time taken by rearranging the equation to make distance or time the subject.

$$For \ speed: \ v_{av} = \frac{d}{t}$$

(Remember, this formula calculates average speed; so we can only calculate the distance or time of an entire journey from this formula.)

If we know the average speed we are travelling at and want to calculate the distance we would travel in a given time, assuming we maintain a constant speed, distance must become the subject of the equation. Multiplying both sides of the equation by t we obtain:

$$v_{av} \times t = \frac{d}{t} \times t$$
$$v_{av} \times t = \frac{d}{\cancel{t}} \times \cancel{t}$$
$$d = v_{av}t$$

If we know the distance to be travelled and an estimate for our average speed, we can calculate how long a journey should take by making time the subject of the equation. It is important to do this carefully, because a common mistake is to simply move the distance to the left side of the equation. Using the formula to calculate distance and dividing both sides by velocity gives:

$$\frac{d}{v_{av}} = \frac{v_{av}t}{v_{av}}$$
$$\frac{d}{v_{av}} = \frac{\cancel{v_{av}}t}{\cancel{v}}$$
$$t = \frac{d}{v_{av}}$$

Using a formula triangle

Sometimes rearranging these equations correctly can be difficult. It can be helpful to use a formula triangle to find the right equation.

To use the triangle, simply cover the value that you want to calculate and if the remaining values are next to each other, you multiply them. If they are above each other, you divide the one on top by the one below.

Note: When using the speed formula to calculate distance or time, the units used for distance and time must be the same as those used in the unit for speed. If the units are not the same, you will need to convert them.

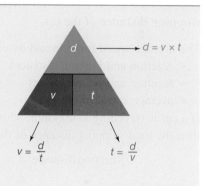

$d = v \times t$

$v = \dfrac{d}{t}$

$t = \dfrac{d}{v}$

Example 4: Calculating distance when speed and time are known

The distance covered in two and a half hours by a train travelling at an average speed of 70 km/h is:

$$d = v_{av}t$$
$$= 70\,\text{km/h} \times 2.5\,\text{h}$$
$$= 175\,\text{km}$$

Example 5: Calculating time when distance and speed are known

The time taken for a giant tortoise to cross a 6-metre-wide deserted highway at an average speed of 5.5 cm/s is:

$$t = \frac{d}{v_{av}}$$
$$= \frac{6 \times 100\ \text{cm}}{5.5\ \text{cm/s}}$$
$$= 109\ \text{s (to the nearest second)}$$
$$= 1\ \text{min}\ 49\ \text{s}$$

Note that a unit conversion was required for this question.

FIGURE 8.5 The time taken for a tortoise to cross a highway requires knowing both the distance and the speed.

8.2.3 Stopping and braking distances

During the school term, traffic around schools can be very busy and also hazardous as people try to cross the streets, sometimes without looking both ways. In school zones around Australia, most speed limits are set at 40 km/h during the school term.

So why has this value been set as the speed limit? It gives a driver enough time to apply the brakes and stop their car if there is a hazard ahead.

The **reaction distance** is the distance travelled by the car once the driver sees the hazard, before applying the brakes. Once the brakes have been applied, the car will still cover some distance (**braking distance**) until it stops. When these two distances are added together, it gives the total **stopping distance** of the car.

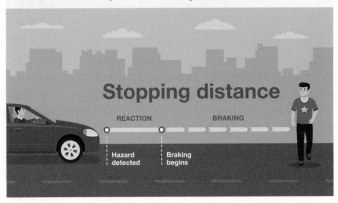

FIGURE 8.6 The stopping distance of the car is the reaction distance plus the braking distance.

The average speed formula and its variations can be used to calculate the:
- reaction and braking distances
- reaction and braking times
- average speed of the car.

If a car is travelling at 40 km/h, with a driver reaction time of 0.8 seconds, and a braking time of 0.9 seconds, then the total stopping distance of the car would be:

$$\text{Reaction distance:} \, d = v_{av} \times t$$
$$d = (40 \div 3.6) \times 0.8 \, (\text{convert 40 km/h into m/s by dividing by 3.6})$$
$$d = 8.9 \, \text{m}$$
$$\text{Braking distance:} \, d = v_{av} \times t$$
$$d = (40 \div 3.6) \times 0.9$$
$$d = 10 \, \text{m}$$
$$\text{Stopping distance} = 8.9 + 10$$
$$= 18.9 \, \text{m}$$

If the car was travelling at 50 km/h instead, with a driver reaction time of 0.8 seconds, and a braking time of 0.9 seconds, then the total stopping distance of the car would be:

$$\text{Reaction distance:} \, d = v_{av} \times t$$
$$d = (50 \div 3.6) \times 0.8 \, (\text{convert 50 km/h into m/s by dividing by 3.6})$$
$$d = 11.1 \, \text{m}$$
$$\text{Braking distance:} \, d = v_{av} \times t$$
$$d = (50 \div 3.6) \times 0.9$$
$$d = 12.5 \, \text{m}$$
$$\text{Stopping distance} = 11.1 + 12.5$$
$$= 23.6 \, \text{m}$$

While a stopping distance difference of 4.7 m might not be a lot, if a child was to run onto the road unexpectedly, that 4.7 m would make a difference. This doesn't take into account the condition of the car and tyres, its mass, road and weather conditions and individual reaction times.

If the roads were wet, would this affect the stopping distance of a car? Do you think speed limits should be lowered to 30 km/h in a school zone?

reaction distance the distance travelled once a hazard has been spotted, to just before applying the brakes

braking distance the distance travelled once the brakes have been applied and the vehicle stops

stopping distance the sum of the reaction and braking distances

8.2.4 When the direction matters

Imagine you are a pilot. You are flying through such thick fog that you can see nothing at all through the windows. You are relying entirely on your instruments and instructions from flight control. You are nearly out of fuel and must land. Over the radio you are told to approach the runway with a speed of 100 m/s. This information is useless to you because you don't know which way to fly at that speed. A more useful piece of information would be to fly at a **velocity** of 100 m/s in an easterly direction. This would at least point you at the right runway!

Velocity and speed can sometimes be used to mean the same thing. However, scientifically, speed and velocity have different meanings. They are calculated using different quantities and provide different information about situations. The distinction between speed and velocity can be summed up as follows:

- Speed is a measure of the rate at which distance is covered. It has a size or **magnitude** only and is described as a **scalar quantity** because it only has 'scale' or size information. The direction an object is travelling doesn't matter when speed is calculated; only the total distance travelled is important.
- Velocity is a measure of the rate of change in **position**. To describe a change in position, any change in direction of the object has to be considered and stated. Velocity has a direction as well as a magnitude (size) and is known as a **vector quantity**. The overall change in position of an object, including the direction, is known as **displacement**. In a straight line for instance, displacement is the change of position of the object from the start to the end position, and includes the direction of the line.

velocity a measure of rate of change in position, with respect to time, described by a magnitude and direction

magnitude size

scalar quantity any quantity that is described only by a magnitude

position the location of an object

vector quantity any quantity that is described by both a magnitude and direction

displacement the change in position of an object, including direction

Writing vectors

When writing vectors, we often write the symbol in bold, or with a right-facing arrow above the symbol (for example, $\boldsymbol{v}$ or $\vec{v}$). If the magnitude of the vector is only used (scalar quantity), it does not need to be bolded or include the right-facing arrow above the symbol.

This representation is used with all vector quantites, where the symbol is either bolded or has a right facing arrow. Scalar quantities are never in bold.

8.2.5 Displacement, distance, velocity and speed

Displacement and velocity are both vector quantities, and therefore are described by both a magnitude and direction.

Displacement is the change in position of an object. *Note*: The symbol s can also be used to represent displacement.

Displacement of an object

The displacement of an object that has moved from position x_1 to position x_2 can be expressed as:

$$\text{displacement} = \text{change in position}$$
$$= \text{final position} - \text{initial position}$$
$$\Delta \boldsymbol{x} = \boldsymbol{x}_2 - \boldsymbol{x}_1$$

Where $\Delta \boldsymbol{x}$ is the change in position
$\boldsymbol{x}_1$ is the initial position
$\boldsymbol{x}_2$ is the final position.

The triangle-shaped symbol, Δ, is used to represent the change in a quantity; this symbol is the uppercase Greek letter, delta.

Velocity is the rate of change of position, calculated by dividing the change in position by time.

Velocity of an object

$$\text{velocity} = \frac{\text{displacement}}{\text{time taken}}$$

$$= \frac{\Delta x}{t}$$

$$v = \frac{x_2 - x_1}{t}$$

Where v is the velocity
Δx is the displacement
t is the time taken

At a snail's pace

Imagine a race between two snails, Bo and Jo, between the points P and R, as shown in figure 8.7. Bo, being slower but smarter, takes the direct route. Jo, faster but not as clever, takes an indirect route via Q and travels a greater distance. The race is a dead heat — both snails finish in one minute.

Table 8.1 describes the motion of the two snails and shows the difference between their speed and velocity.

Notice that with no change in direction (as for Bo), the magnitude of the velocity is the same as the average speed. If direction changes, velocity and speed will be different (as for Jo).

TABLE 8.1 The race between Bo and Jo

	Bo	Jo
Average speed	$\dfrac{\text{Distance travelled}}{\text{time taken}}$ $= \dfrac{d}{t}$ $= \dfrac{5.7 \text{ cm}}{1 \text{ min}}$ $= 5.7 \text{ cm/min}$	$\dfrac{\text{Distance travelled}}{\text{time taken}}$ $= \dfrac{d}{t}$ $= \dfrac{8.0 \text{ cm}}{1 \text{ min}}$ $= 8.0 \text{ cm/min}$
Average velocity	$\dfrac{\text{Change in position}}{\text{time taken}}$ $= \dfrac{\Delta x}{t}$ $= \dfrac{5.7 \text{ cm NE}}{1 \text{ min}}$ $= 5.7 \text{ cm/min NE}$	$\dfrac{\text{Change in position}}{\text{time taken}}$ $= \dfrac{\Delta x}{t}$ $= \dfrac{5.7 \text{ cm NE}}{1 \text{ min}}$ $= 5.7 \text{ cm/min NE}$

FIGURE 8.7 Calculating displacement, distance, speed and velocity depends on both magnitude and direction.

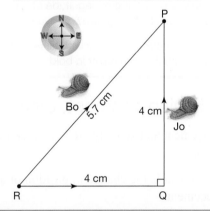

The sign and direction convention to determine the direction of a vector quantity

In order to perform calculations using vectors, the direction is indicated using positive and negative numbers. When dealing with motion in a straight line, it is important to define which direction will be represented by positive numbers and which direction will be represented by negative numbers. Common conventions are:

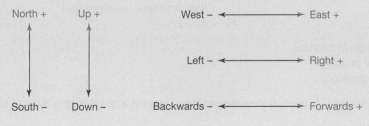

8.2.6 Using position-time and velocity-time graphs

Position-time graphs

The movement of an object over a time period can be observed using a position-time graph (or a displacement-time graph). An example can be seen in figure 8.8. These graphs also allow for the speed and velocity to be determined.

For example, if figure 8.8 shows the journey of a drone, we can determine the following.

- It travelled 4 metres in the first four seconds.
- It did not move (had a speed of 0 m/s) between 4 and 8 seconds.
- At 8 seconds, it turned around and went back to the same position it started at.
- It travelled at a faster speed between 8 and 10 seconds than it did between 0 and 4 seconds: between 0 and 4 seconds, they travelled 4 metres in 4 seconds (a speed of 1 m/s) whereas between 8 and 10 seconds, it travelled 4 metres in 2 seconds (a speed of 2 m/s). A steeper slope represents a faster speed.
- It had a negative velocity between 8 and 10 seconds. This is because the displacement is negative (−4 m), therefore its velocity must be −2 m/s (or 2 m/s in the opposite direction). Here the negative sign indicates the direction of the velocity.

FIGURE 8.8 A position-time graph

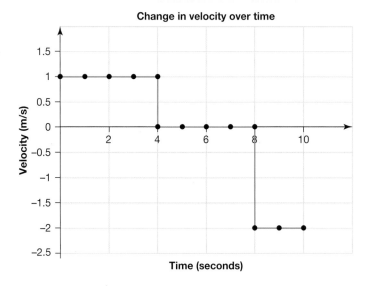

Velocity-time graphs

The velocity-time graph is a similar type of graph to the position-time graph. Velocity-time graphs also provide information about the movement of an object over a period of time. These can be created from position-time graphs, or through using other data on velocity.

The graph shown in figure 8.9 was created using the data in figure 8.8.

- The velocity between 0 and 4 seconds was 1 m/s.
- The velocity between 4 and 8 seconds was 0 m/s (the drone was stationary
- The velocity between 8 and 10 seconds was −2 m/s (the drone was moving in the opposite direction to the original motion).

FIGURE 8.9 A velocity-time graph

Resources

eWorkbook Speed and velocity (ewbk-13211)

Weblink   Adding vectors

8.2 Activities

These questions are even better in jacPLUS!
- Receive immediate feedback
- Access sample responses
- Track results and progress

Find all this and MORE in jacPLUS

Remember and understand

1. Explain the difference between scalar and vector quantities and give one example of each.
2. Classify each of the following quantities as either a scalar or a vector.
   a. Mass
   b. Force
   c. Speed
   d. Distance
   e. Displacement
   f. Temperature
   g. Acceleration
   h. Density
3. Explain the difference between speed and velocity. Use an example to support your explanation.

Apply and analyse

4. Determine the average speed of each of the following. Give your answer in the units specified in the bracket.
   a. A racehorse that wins the 3200 m Melbourne Cup in a time of 3 min 20 s (in m/s)
   b. A kangaroo that flees from a dingo, bounding a distance of 2.5 km in 3 min (in m/s)
   c. A dolphin that just manages to keep up with a speeding boat for a distance of 2 km over a period of 3 min (in km/h)
   d. A sea turtle that is able to maintain its maximum speed for 0.5 h and in that time swims a distance of 16 km (in km/h)
   e. An Olympic swimmer who completes a 1500 m training swim in 16 min (in km/h)
   f. A mosquito that flies a distance of 2 m in 4 s (in cm/s)
5. John travels 400 km north but then backtracks south for 150 km to pick up a friend. What is John's total displacement?
6. Calculate how long it would take you to walk from Melbourne to Sydney, a distance of 900 km, if you walked at the following average speeds.
   a. 5 km/h without stopping
   b. 5 km/h for 10 h each day
   c. 1.5 m/s without stopping
7. Calculate how far a snail can crawl if it moves at an average speed of 8.0 cm/min for the following durations.
   a. 3 minutes
   b. 3 hours
8. In a heat of a swimming trial, a swimmer swims the 100 m breaststroke event in 68 s. The event is completed in a pool that is 50 m long. She finishes the event at the same end of the pool from which she started. If she begins the event by swimming due north, and takes 35 s to swim the first 50 m, calculate the following.
   a. Her average speed for the whole swim
   b. Her average velocity for the first 50 m
   c. Her average velocity for the whole swim
9. A swimmer completes a 1500 m race in 870 s. Calculate the swimmer's average speed in the following units.
   a. m/s
   b. km/h
10. A 200 m race is held on a circular track, 400 m in circumference. The race starts on the east side of the track and finishes on the west side. The winner finishes in 23 seconds. Calculate the following.
    a. The average speed of the winner
    b. The average velocity of the winner
    c. The displacement and velocity of a runner who completed the whole 400 m in 50 seconds.

Evaluate and create

11. **SIS** **a.** Design a chart with pictures that compares the speeds of a range of animals.

b. Calculate the distance each animal can run in 8 minutes (assuming they can maintain the same speed) and show this data on a bar graph.

12. **SIS** A film crew is making a documentary about Australian wildlife. They want to carry out a measurement of the typical speeds that the creatures travel at. Considering the equipment likely to be at their disposal, design an experiment to gather these measurements. State the equipment used, measurements to be made and calculations that would need to be carried out.

Fully worked solutions and sample responses are available in your digital formats.

LESSON
8.3 Measuring speed

LEARNING INTENTION

At the end of this lesson you will be able to describe how speed is measured with the help of different devices and understand the importance of controlling speed on the roads for safety.

8.3.1 Keeping track of speed

When British Formula One racing driver Lewis Hamilton broke the Australian Grand Prix qualifying lap record in 2019, he completed a 5.303 km lap in 80.486 seconds.

His average speed was:

$$v_{av} = \frac{d}{t}$$

$$= \frac{5303\,m}{80.486\,s}$$

$$= 65.89\,m/s \times 3.6$$

$$\approx 237\,km/h$$

FIGURE 8.10 The average speed and instantaneous speed of a car racing in the Grand Prix varies.

However, he was able to travel down the straight at speeds of up to 320 km/h. Clearly, the average speed does not provide much information about the speed at any particular instant during the race.

The full story of Lewis Hamilton's qualifying lap could be more accurately told if his average speed was measured over many short intervals throughout the event. For example, if stopwatches were placed at every 100 metre point along the track, his average speed for each 100 metre section of the circuit could then be calculated. If stopwatches were placed every metre along the track, his average speed for each 1 metre section could be calculated. By using more stopwatches and placing them closer together, a more accurate estimate of his **instantaneous speed** can be obtained. The instantaneous speed is the speed at any particular instant of time.

> **instantaneous speed** speed at any particular instant of time

8.3.2 Measuring speed: Ticker timers

If a car were to drip oil on the road, with one drop landing every second, this information could be used to find the speed of the car. If a big puddle formed under the car, then the car must be stationary. If the drops are 10 metres apart, the car must be travelling at 10 m/s. If the drops get further apart, the car must be speeding up.

Rather than dripping oil, a ticker timer provides a simple way of recording motion in a laboratory. When the ticker timer is connected to an AC power supply, its vibrating arm strikes its base 50 times every second. Paper ticker tape attached to a moving object is pulled through the timer. A disc of carbon paper between the paper tape and the vibrating arm ensures that a dot is left on the paper 50 times every second; that is, a dot is made every fiftieth of a second.

The average speed between each pair of dots is determined by dividing the distance between them by the time interval. To make calculating the speed easier, every fifth dot can be marked, as shown in figure 8.12. Each marked interval on the tape represents five-fiftieths of a second — that is, 0.1 seconds.

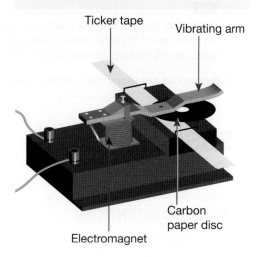

FIGURE 8.11 Motion can be recorded with a ticker timer.

Ticker tape · Vibrating arm

Electromagnet

Carbon paper disc

FIGURE 8.12 Each marked interval represents a time of 0.1 s.

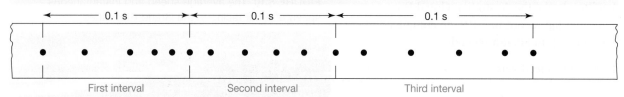

First interval · Second interval · Third interval

The distance between the first and last dot within the first interval is 4.2 cm and the dots are getting closer together, indicating the speed is getting smaller. The average speed during the first interval on the tape shown is:

$$v_{av} = \frac{4.2\ \text{cm}}{0.1\ \text{s}}$$
$$= 42\ \text{cm/s}.$$

During the second interval, the dots are quite evenly spaced, indicating the speed was constant. The dots are furthest apart during the third interval, indicating this is when the speed was the largest.

EXTENSION: Measuring nautical speed

A common nautical measurement of speed is the knot. This dates from the 17th century, when sailors measured the speed of their ship using something called a 'common log'. This was a coil of rope with regularly spaced knots, attached to a piece of wood. By using a sand timer and counting the number of knots that passed through their fingers, a sailor could get an estimate of the speed of the ship. One knot is approximately equivalent to 1.85 km/h.

8.3.3 Measuring speed: Motion detectors

In many classrooms, ticker timers have been replaced with **sonic motion detectors**. These devices send out pulses of ultrasound at a frequency of about 40 kHz (kilohertz is used to measure frequency and is equal to 1000 cycles per second) and detect the reflected pulses from the moving object. The device uses the time taken for the pulses to return to calculate the distance between itself and the object. A small computer in the motion detector calculates the speed of the object.

sonic motion detector
device that sends out pulses of ultrasound at a frequency of about 40 kHz and then detects the reflected pulses from the moving object

A light beam can be used as a switch. When a light detector stops detecting light, the beam must have been cut. This triggers an alarm. The same technology can be used to measure speed.

At the PyeongChang Winter Olympics in 2018, sensor systems were used for the first time to provide continuous measurements from the start to the finish of events, meaning that athletes could gain an immediate understanding of where they gained or lost time, or where they won or lost points. Additionally, this information meant that people in the venues, as well as those at home watching on television, could have a far greater understanding of the athlete's motion as it happened.

FIGURE 8.13 Sonic motion detectors are used on the bumpers of cars to help the driver detect the distance between the car and another object.

This technology can also be found in **light gates**. When an object passes through the 'gate' where the beam is, a detector in the gate can no longer sense the light, so triggers a timer. When the object leaves the gate, the beam is now detected so the timer stops. By knowing the length of the object (usually a piece of card attached to a trolley) and the length of time as recorded by the gate, the speed of the object can be calculated.

8.3.4 Measuring speed: Speedometers

The speedometer inside a vehicle has a pointer that rotates further to the right as the wheels of the vehicle turn faster. It provides a measure of the instantaneous speed.

Older speedometers use a rotating magnet that rotates at the same rate as the car's wheels. This is shown in figure 8.14. The rotating magnet creates an electric current in a device connected to the base of the pointer. As the car's speed increases, the magnet rotates faster, the electric current increases and the pointer rotates further to the right. You may have fitted a speedometer to your bicycle that also works in this way.

Newer electronic speedometers use a rotating toothed wheel that interrupts a stationary **magnetic field**. An electronic sensor detects the interruptions and sends a series of pulses to a computer, which calculates the speed using the **frequency** of the pulses.

Car speedometers are not 100 per cent accurate. In Australia, an error of up to 10 per cent is common. Speedometers are manufactured according to the diameter of the tyres on the vehicle. Any change in that diameter will make the reading on the speedometer inaccurate. This is why many people are caught speeding when they may think that their speed is legal.

8.3.5 Measuring speed: GPS

Many people rely on their satellite navigation system to give a more accurate measure of their speed. The **global positioning system (GPS)** uses radio signals from at least four of up to 32 satellites orbiting Earth to accurately map your position (refer to figure 8.15), whether you are in a vehicle or on foot. GPS navigation devices usually calculate speed about once every second. To do this they measure the small perceived change in frequency of the received radio waves due to your motion relative to them. This is called the **Doppler effect**.

light gates gates used to measure the speed of an object using a timer and a detector

magnetic field area where a magnetic force is experienced

frequency the number of waves or pulses passing a single location in one second

global positioning system (GPS) device that uses radio signals from satellites orbiting the Earth to accurately map the position of a vehicle or individual

Doppler effect observed change in frequency of a wave when the wave source and observer are moving in relation to each other

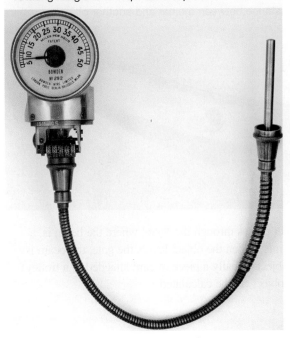

FIGURE 8.14 An older speedometer with a rotating magnet to help detect speed

FIGURE 8.15 GPS works using numerous satellites orbiting Earth

SCIENCE AS A HUMAN ENDEAVOUR: The use of GPS locators in the AFL

In the AFL, coaches and sports scientists use GPS locators to track the movement of players around the field. The locators are strapped to the player's upper back. A computer is used to analyse the data to provide information about distance covered, speed, time spent moving at different speeds, maximum speed and acceleration. A built-in sensor also monitors heart rate.

FIGURE 8.16 GPS used on AFL players

SCIENCE AS A HUMAN ENDEAVOUR: Speed and road safety

One of the major causes of road accidents and subsequent fatalities and injuries is excessive speed or driving at speeds that are unsafe for the road or weather conditions. Speed limits and speed advisories are set in an effort to minimise such accidents. The police use three different methods to monitor driving speeds as accurately as possible to ensure that speeding drivers are penalised.

- Radar guns (shown in figure 8.17) and mobile radar units in police cars send out radio waves. The radio waves are reflected from the moving vehicle. However, the frequency of the waves is changed owing to the movement of the vehicle. The change in the frequency, called the Doppler effect, depends on the speed of the moving vehicle. The

FIGURE 8.17 Radar guns detect the speed of passing motorists.

altered waves are detected by the radar gun or mobile unit, which calculates the instantaneous speed. One type of radar unit is linked to cameras that automatically photograph any vehicle that the radar detects as travelling above the speed limit.

- Police also use laser guns that send pulses of light that are reflected by the target moving vehicle. The time taken for each pulse to return is recorded and compared with that of previous pulses. This allows the average speed over a very small time interval to be calculated. Laser guns are useful when traffic is heavy because they can target single vehicles with the narrow light beams. Radio waves (as used in radar guns) spread out and, in heavy traffic, it is difficult to tell which car reflected the waves.
- Digitectors consist of two cables laid across the road at a measured distance from each other. Each cable contains a small microphone that detects the sound of a moving vehicle as it crosses the cable. The measured time interval between the sounds is used to calculate the average speed of the vehicle between the cables. Although digitectors were phased out after the 1980s, they are regaining popularity as an alternative to radar and laser guns.

DISCUSSION

Why do you think digitectors are being used again? What advantages do you think they might have over radar guns and laser guns?

elog-2482

INVESTIGATION 8.1

Ticker timer tapes

Aim

To record and analyse motion using a ticker timer

Materials

- ticker timer
- power supply
- scissors
- G-clamp
- ticker tape (in 60 cm lengths)

Method

1. Clamp the ticker timer firmly to the edge of a table or bench so that you will be able to pull 60 cm of ticker tape through it.
2. Connect the ticker timer to the AC terminals of the power supply and set the voltage as instructed by your teacher (start with the lowest voltage first if unsure).
3. Thread one end of the ticker tape through the ticker timer so that it goes under the carbon paper disc.
4. Turn on the power supply and check that the ticker timer leaves a mark on the ticker tape.
5. Hold the end of the ticker tape and walk away from the ticker timer so that the ticker tape moves through at a steady speed.
6. Remove the ticker tape and mark off the first clear dot made and every fifth dot after the first. (There should be four dots between each of the marked-off dots on the ticker tape.)

Results

1. Measure the distance travelled during each 0.1 s interval and write it on your tape. Label the intervals as interval 1, interval 2, interval 3, and so on.
2. Cut your ticker tape into 0.1 s intervals.

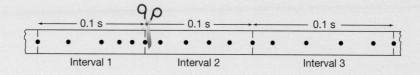

3. Glue the strips in order onto a sheet of paper as shown. Each strip shows the distance travelled during a 0.1 s time interval.

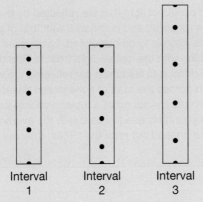

Interval 1   Interval 2   Interval 3

4. Use this to create a graph, showing how the speed changed with time.

Discussion

1. How much time elapsed between the printing of the first clear dot and the last dot marked off?
2. Calculate the average speed for the motion that took place between the printing of the first clear dot and the last marked dot.
3. Calculate the average speed during each 0.1 s interval.
4. Did you succeed in keeping your speed steady?
5. Suggest improvements in your investigations that would enable you to better keep a steady speed.

Conclusion

Summarise the findings for this investigation.

on Resources

eWorkbook   Ticker tapes (ewbk-13213)

Weblink   How do radar guns work?

8.3 Activities

8.3 Quick quiz on	8.3 Exercise

Select your pathway

■ LEVEL 1 1, 2	■ LEVEL 2 3, 5	■ LEVEL 3 4, 6, 7

These questions are even better in jacPLUS!
- Receive immediate feedback
- Access sample responses
- Track results and progress

Find all this and MORE in jacPLUS ⊙

Remember and understand

1. Explain the difference between instantaneous speed and average speed. Use an example to support your explanation.
2. Which methods are used by the police to measure:
   a. average speed?
   b. instantaneous speed?
3. Explain the difference between digitectors, radar and laser guns.

Apply and analyse

4. Explain why a speedometer reading might not be accurate. Include anything that could change the diameter of the vehicle's tyres.
5. An experiment investigating motion was conducted. The first three intervals on the ticker tape measured 5.4 cm, 8.3 cm and 4.8 cm. Calculate the average speed during each of these 0.1 s intervals.
6. **SIS** Use data-logging equipment with a motion detector or light gates to record the motion of a toy car or cart down a slope. Use the software to produce a graph of distance versus time and a graph of speed versus time. Comment on the shape of your graphs.

Evaluate and create

7. **SIS** The following table shows data collected by students in a ticker-tape experiment where a trolley rolls down a gently sloping ramp. The total distance travelled from the starting position to each dot is recorded. (At 50 Hz there is a 20 ms time interval between dots.)

TABLE Data collected from a ticker tape of a trolley moving down a ramp

Time (ms)	Distance (mm)
20	1
40	4
60	9
80	16
100	25
120	36
140	50
160	67

a. Plot the distance travelled against time on a graph.
b. What does the shape of your graph tell you?
c. Calculate the average speed of the trolley between each pair of adjacent dots on the ticker tape.

Fully worked solutions and sample responses are available in your digital formats.

TOPIC 8 Forces, energy and motion **547**

LESSON
8.4 Acceleration and changes in velocity

LEARNING INTENTION

At the end of this lesson you will be able to describe what acceleration is and calculate acceleration of an object that is speeding up or slowing down.

8.4.1 Getting faster

The accelerator pedal of a car is given that name because pushing on it usually makes the car accelerate. When an object moves in a straight line, its **acceleration** is a measure of the rate at which it changes velocity. The difference between speed and velocity is important when calculating acceleration as the velocity must be used rather than the speed.

acceleration rate of change in velocity

Average acceleration

The average acceleration can be calculated by dividing the change in velocity by the time taken for the change.

$$\text{average acceleration} = \frac{\text{change in velocity}}{\text{time taken}}$$

$$a_{av} = \frac{v - u}{t}$$

$$a_{av} = \frac{\Delta v}{t}$$

Where: a_{av} is the acceleration (in m/s^2).
   Δv is the change in velocity (in m/s). where $\Delta v = v - u$
   v = final velocity (in m/s)
   u = initial velocity (in m/s)
   t is the time taken (in s).

The internationally accepted, or SI units of acceleration are m/s/s (metres per second per second) or m/s^2 (metres per second squared). So, 10 m/s/s or 10 m/s^2 means 10 metres per second faster every second. However, sometimes other appropriate units may be used instead. When calculating acceleration, it is often useful to convert the units for velocity and time before placing them in the average acceleration fomula.

Example 6: Calculating average acceleration

A car travelling at 60 km/h north (or +60 km/h) increases its velocity to 100 km/h north (or +100 km/h) in 5.0 seconds. This is a positive change of +40 km/h. The car has an average acceleration of:

$$\text{average acceleration} = \frac{\text{change in velocity}}{\text{time taken}}$$

$$= \frac{\text{final velocity} - \text{initial velocity}}{\text{time taken}}$$

$$= \frac{100 \text{ km/h} - 60 \text{ km/h}/3.6}{5.0 \text{ s}}$$

$$= \frac{40 \text{ km/h}/3.6}{5.0\ 4 \text{ s}} \text{ (divide by 3.6 to convert km/h into m/s)}$$

$$= \frac{11.1 \text{ m/s}}{5.0 \text{ s}}$$

$$= 2.2 \text{ m/s}^2 \text{ North}$$

That is, on average, the car increases its velocity by 2.2 m/s each second (2.2 m/s^2 north).

8.4.2 Slowing down

In straight line motion, if the acceleration of an object is in the opposite direction to its velocity, the object slows down. This is commonly referred to as deceleration. However, scientifically it is still described as acceleration. Acceleration is simply the rate of change of velocity — whether this motion is increasing or decreasing doesn't matter, as the velocity is changing.

Example 7: Calculating negative acceleration

A truck is travelling at 35 m/s forwards (or +35 m/s) at the bottom of a hill but 10 seconds later at the top of the hill it has a velocity of 15 m/s forwards (or +15 m/s). The truck has an acceleration of:

$$
\begin{aligned}
\text{average acceleration} &= \frac{\text{change in velocity}}{\text{time taken}} \\
&= \frac{\text{final velocity} - \text{initial velocty}}{\text{time taken}} \\
&= \frac{15 - 35}{10} \\
&= \frac{-20}{10} \\
&= -2.0 \, \text{m/s}^2
\end{aligned}
$$

That is, on average, the truck decreases its velocity by 2 m/s each second, because its acceleration is in the opposite direction to its velocity.

8.4.3 Comparing positive and negative acceleration

The sport of drag racing is a test of acceleration. From a standing start, cars need to cover a distance of 400 metres in the fastest possible time. To do this, they need to reach high velocities very quickly. The fastest drag-racing cars can reach speeds of more than 500 km/h in less than 5.0 seconds. As drag races occur along a straight track with no changes in direction, the magnitude of a car's velocity will be the same as the speed of the car.

FIGURE 8.18 A drag race is a race between two vehicles on a short, flat, straight course.

The average acceleration of a drag-racing car that reaches a velocity of 506 km/h in 4.6 seconds is:

$$
\begin{aligned}
\text{average acceleration} &= \frac{\text{change in velocity}}{\text{time taken}} \\
&= \frac{\text{final velocity} - \text{initial velocity}}{\text{time taken}} \\
&= \frac{506 \, \text{km/h} - 0 \, \text{km/h}}{4.6 \, \text{s}} \\
&= \frac{140 \, \text{m/s}}{4.6 \, \text{s}} \quad \text{(divide by 3.6 to convert km/h into m/s)} \\
&= 31 \, \text{m/s}^2
\end{aligned}
$$

It means that, on average, the car increases its speed by 31 m/s each second (31 m/s^2). If this is examined in km/h, the car's velocity increases 110 km/h every second!

Once the drag-racing car has completed the required distance of 400 metres, it needs to stop before it reaches the end of the track. The fastest cars release parachutes so that they can stop in time. The acceleration of a car that comes to rest in 5.4 seconds from a speed of 506 km/h is:

$$
\begin{aligned}
\text{average acceleration} &= \frac{\text{change in velocity}}{\text{time taken}} \\
&= \frac{\text{final velocity} - \text{initial velocity}}{\text{time taken}} \\
&= \frac{0 \text{ km/h} - 506 \text{ km/h}}{5.4 \text{ s}} \quad \text{(divide by 3.6 to convert km/h into m/s)} \\
&= \frac{-140.5 \text{ m/s}}{5.4 \text{ s}} \\
&= -2.6 \text{ m/s}^2
\end{aligned}
$$

This means, on average, that the car decreases its velocity by 26 m/s each second (or by 94 km/h each second).

DISCUSSION

A student wrote a statement that 'A positive value acceleration always means an object must be speeding up'. Do you agree with this statement? Discuss reasons you might agree or disagree with this.

INVESTIGATION 8.2

elog-2484

Drag strips

Aim

To analyse the motion of an accelerating object using ticker tape

Materials

- ticker timer
- G-clamp
- power supply
- ticker tape (in 60 cm lengths)
- toy car (or dynamics trolley)
- ticky tape or masking tape
- clear length of bench at least 60 cm long

Method

1. Clamp the ticker timer firmly to the edge of a table or bench so that you will be able to pull 60 cm of ticker tape through it. Connect the ticker timer to the AC terminals of the power supply and set the voltage as instructed by your teacher.
2. Thread one end of the ticker tape through the ticker timer so that it goes under the carbon paper disc.
3. Attach the end of the ticker tape to the toy car or trolley.
4. Turn on the power supply and check that the ticker timer leaves a black mark on the ticker tape.
5. Model a drag-racing car by pushing the toy car or trolley forward, starting from rest, so that it reaches a maximum speed near the halfway mark. Make it come to a gradual stop near the end of the 'track'.
6. Remove the ticker tape and mark off the first clear dot and every fifth dot after the first. Each interval between the marks represents a time of $\frac{5}{50}$ of a second; that is, 0.1 seconds. Measure the distance travelled during each 0.1 second interval and write it on your tape. Label the intervals as interval 1, interval 2, interval 3 and so on.

Results

1. Construct a table like the following in which to record your data. Calculate the average speed during each interval and record it in the table.

TABLE Distances and speeds determined from ticker tape

Interval	Distance travelled (cm)	Average speed (cm/s)
Example	3.6	$\dfrac{3.6 \text{ cm}}{0.1 \text{ s}} = 36$
1		
2		
3		

2. Now cut your ticker tape into 0.1 second intervals and use the strips as described in Investigation 8.1 to construct a graph of speed versus time.

Discussion

1. Describe the motion of the toy car or trolley during the period over which it was recorded. Ensure that the words *speed, accelerated* and *decelerated* are used in your description.
2. Between which intervals was the acceleration:
   a. positive?
   b. negative?
3. During which interval did the greatest average speed occur?
4. When did the greatest positive acceleration take place?
5. Identify the dependent, independent and controlled variables in your investigation.
6. Comment on the accuracy and precision of your results.

Conclusion

Summarise the findings for this investigation about the motion of an accelerating object.

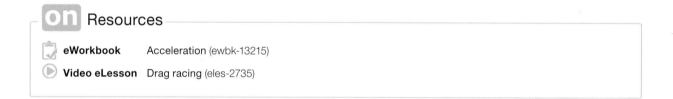

on Resources

📋 **eWorkbook** Acceleration (ewbk-13215)

▶ **Video eLesson** Drag racing (eles-2735)

8.4 Activities

learn on

8.4 Quick quiz on	**8.4 Exercise**

Select your pathway

■ LEVEL 1	■ LEVEL 2	■ LEVEL 3
1, 2	3, 5, 7	4, 6

These questions are even better in jacPLUS!
- Receive immediate feedback
- Access sample responses
- Track results and progress

Find all this and MORE in jacPLUS ⊙

Remember and understand

1. Consider the accelerator in a car. With reference to its use, why is it called an accelerator?
2. Explain the difference between positive and negative acceleration with the help of a numerical example or a well-labelled diagram.

▶

Apply and analyse

3. A car that has stopped at a set of traffic lights sets off when the lights turn green. It increases its forwards speed by 5 m/s during each of the first 3 seconds after it sets off, and by 3 m/s during the following 2 seconds.
    a. Calculate the forwards speed at the end of the following time periods.
        i. 1 s
        ii. 2 s
        iii. 5 s
    b. Determine the average acceleration of the car during the first 5 seconds after it sets off.
4. Two drag cars are racing next to each other. The first car increases its forwards speed to 122 m/s in 7.5 seconds. The second car increases its forwards speed to 137 m/s in 9.2 seconds.
    a. Calculate the acceleration of each car.
    b. Which car has the greater acceleration?
    c. If both cars reach a top speed of 512 km/h and it takes them 8.3 seconds to come to a stop at 0 km/h, what is the magnitude of this acceleration or negative acceleration in m/s^2?

Evaluate and create

5. **SIS** The following table shows data collected by students in a ticker-tape experiment where a trolley rolls down a gently sloping ramp. The total distance travelled from the start to each dot is recorded.

TABLE Distance and velocity over different times for trolley on a sloping ramp

Time (ms)	Distance (mm)	Velocity (mm/ms)
20	1	0.05
40	4	0.15
60	9	0.25
80	16	0.35
100	25	0.45
120	36	0.55
140	50	0.65
160	67	0.75

    a. Plot a graph of average velocity against time.
    b. Comment on the shape of your graph. What does it show?
    c. Calculate the average acceleration of the trolley.
    d. Suggest how the students could improve the precision of their experiment.
6. **SIS** Repeat investigation 8.2 using data-logging equipment and a motion detector (or research expected results). Print a hard copy of your graph.
    a. On a copy of your graph, use coloured pencils or a highlighter pen to indicate the part of the graph that shows the car or trolley speeding up. Use a different colour to indicate the part of the graph that shows the car or trolley slowing down.
    b. What is the maximum forwards speed of the car or trolley?
    c. At what time was the maximum forwards speed reached?
    d. How would your graph be different if the trolley sped forwards more quickly, yet still reached the same maximum forwards speed?
    e. Calculate the average acceleration of the car or trolley used in your own experiment.
7. **SIS** A student running a race wishes to find their acceleration in the first 5 seconds after the starter gun goes off.
    a. Design an investigation that would allow for this to be tested, including any equipment that may be used.
    b. Comment on the accuracy and precision of the data you would collect through your method.

Fully worked solutions and sample responses are available in your digital formats.

LESSON
8.5 Forces and Newton's First Law of Motion

LEARNING INTENTION

At the end of this lesson you will be able to explain what forces are and how an object will stay in its state of motion or rest unless an external unbalanced force is applied.

8.5.1 What is a force?

A **force** is an interaction between two objects. Commonly this is a a push, pull or twist applied to one object by another. It is a vector quantity, which means it requires a magnitude and a direction to describe it. The unit of force is the **newton** (N), named after the famous scientist Sir Isaac Newton, who discovered the three Laws of Motion.

The four fundamental forces in nature are:
- *Gravitational force:* This is the force of attraction between two objects with **mass**. It is the weakest of the four fundamental forces but it acts over very large distances.
- *Electromagnetic forces:* These are the forces associated with electric and magnetic fields, and describe interactions between charged particles. They can be split into two types of forces: magnetic forces and electrostatic forces.
- *Strong nuclear forces:* These forces are only observed inside the nucleus of an atom. They are the strongest of all the forces, but act over extremely small distances. They are beyond the scope of this topic.
- *Weak nuclear forces:* These forces are associated with particle transformations within the nucleus, and are beyond the scope of this topic.

> **force** an interaction between two objects that can cause a change; commonly a push, pull or twist applied to one object by another, measured in newtons (N)
>
> **newton** unit of force defined as the force required to provide a mass of 1 kg with an acceleration of 1 m/s²
>
> **mass** the amount of matter in an object, measured in kilograms (kg)

FIGURE 8.19 Some different types of forces

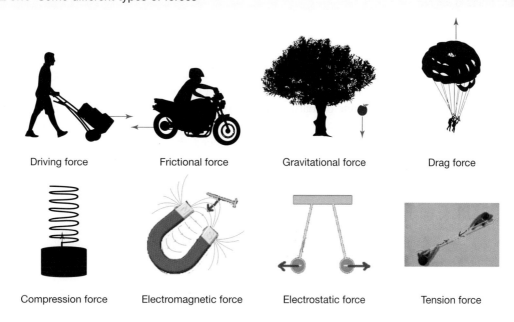

Driving force   Frictional force   Gravitational force   Drag force

Compression force   Electromagnetic force   Electrostatic force   Tension force

There are other common ways two objects interact and lead to other common forces. These include:
- *Normal force:* The force applied by a surface at right angles to the surface in response to a push on it by an object.
- *Resistive forces:* The forces applied to an object to prevent or oppose its motion. Examples of resistive forces are drag forces and frictional forces, including air resistance.

- *Thrust or driving forces:* The forces causing an object to move or accelerate forward. These forces commonly exist due to the engines of cars or planes, but can also include forces created by oars moving a boat or a person pushing on the pedals of a bicycle.
- *Tension or compression forces:* The forces caused by pulling on an object to stretch it or pushing on an object to compress it. The tension force commonly exists in ropes and wires used to move objects or hold them stationary.

Calculating the force due to gravity at Earth's surface

The force of gravitational attraction on an object towards a huge body such as a planet is commonly called **weight**. Weight is a force and is, therefore, measured in newtons (N). While the word weight is used in everyday life, scientists refer to this force as the force of gravity, force due to gravity or the gravitational force, represented by F_g.

> **weight** a commonly used word to describe the force due to gravity on an object

Close to Earth's surface, this force is calculated as $F_g = mg$, where g is Earth's gravitational field strength (acceleration due to Earth's gravitational field), which is equal to approximately 9.8 m/s. This is a special case of Newton's second law applied to an object in free fall; that is, when the only force acting on an object is the force due to gravity.

Force due to gravity

The weight or the force due to gravity (in newtons) of any object can be calculated using the formula:

$$F_g = mg$$

Where m is the mass (in kg)

g is Earth's gravitational field strength (in N / kg)

F_g is the force due to gravity (in N).

At Earth's surface, the gravitational field strength is 9.8 N/kg. That is, the gravitational force acting on each kilogram of mass is 9.8 N.

Mass or force due to gravity

Force due to gravity is different from mass. Mass is the amount of matter in an object and is measured in kilograms (kg). It is a scalar quantity while force due to gravity is a vector quantity. The mass of an object stays the same everywhere (that is, it is the same on Earth or the Moon); however, the force due to gravity changes according to the gravitational field strength where the object is located. For example, consider an object with a mass of 50 kg on Earth. As the mass of the object remains unchanged, the object will have same the mass on the Moon or Mars. The force due to gravity of the object can be calculated using the following:

Force due to gravity on Earth:

$$m = 50 \text{ kg}, g_{Earth} = 9.8 \text{ N/kg}$$
$$F_g = m \times g_{Earth}$$
$$= 50 \times 9.8$$
$$F_g = 490 \text{ N}$$

Force due to gravity on the Moon:

$$m = 50 \text{ kg}, g_{Moon} = 1.6 \text{ N/kg}$$
$$F_g = m \times g_{Moon}$$
$$= 50 \times 1.6$$
$$F_g = 80 \text{ N}$$

The 50 kg object will have a force due to gravity of 490 N near Earth's surface but only 80 N on the surface of the Moon.

DISCUSSION

Near Earth's surface the value of g is 9.8 N/kg (or m/s^2). On the Moon, however, this value is around 1.6 N/kg. Discuss why those on the Moon describe themselves as feeling lighter. Why do you think astronauts on the International Space Station describe themselves as feeling 'weightless'? Do they still have a gravitational force acting on them?

8.5.2 Adding forces

Forces are vector quantities. That means that direction is important. When two forces are applied in the same direction, you can simply add the magnitudes. If forces are acting in opposite directions, they oppose each other and their magnitudes must be subtracted. (Remember, the direction of a vector can be shown using positive and negative numbers. Positive numbers indicate one direction while negative numbers indicate the opposite direction.)

If the **net force** (the result of combining all of the forces) is zero, the forces are balanced. If the force in one direction is greater than the force in the opposite direction, the forces are unbalanced.

int-5894

FIGURE 8.20 Net forces can be calculated when forces are unbalanced using sign and direction convention.

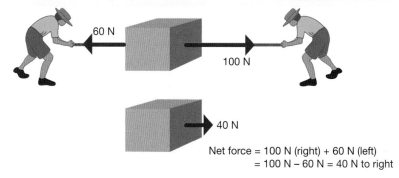

Net force = 100 N (right) + 60 N (left)
= 100 N − 60 N = 40 N to right

For example, when two people pull in opposite directions on the box shown in figure 8.20 the net force is 40 N to the right.

In figure 8.20, the force due to gravity and the normal forces have not been considered. Because the surface is horizontal, the force of gravity (weight) acting on the box and the upward normal force are equal in size and opposite in direction. In other words, they are balanced. The vertical forces add up to zero.

8.5.3 Newton's laws

Sir Isaac Newton's laws of motion were first published in 1687, many years before buses, cars, trains and aeroplanes were invented. He realised that it wasn't enough to say that an object moved because of forces acting on it. A book sitting on a perfectly level table has forces on it, a force due to gravity and a normal reaction force.

net force the net (total or resultant) force of two or more forces acting on an object

The reason the book doesn't move has nothing to do with an absence of forces on the book. It doesn't move because those forces are balanced and no horizontal forces are moving the book across the table.

Newton's three laws of motion explain the effect of all of the forces acting on the motion of objects.

Newton's laws of motion

1. **Newton's First Law of Motion** states that an object will remain at rest, or will not change its speed or direction, unless it is acted upon by an outside, unbalanced force.
2. **Newton's Second Law of Motion** states that the acceleration of an object depends on the size of the net (total or resultant) force and the mass of the object.
3. **Newton's Third Law of Motion** states that for every force that exists, a second force of equal size and opposite direction also exists. That is, when an object applies a force to a second object, the second object applies an equal and opposite force to the first object.

8.5.4 Newton's First Law

It isn't just stationary objects that have balanced forces on them. Imagine an aeroplane flying through the air at a constant height and with a constant velocity (see figure 8.21). The magnitude of the **thrust** on the plane is equal to the magnitude of the drag produced as the plane pushes through the air. These balanced horizontal forces mean the plane travels with a steady velocity forwards. The magnitude of the lift is also equal to the magnitude of the force due to gravity. These balanced vertical forces mean the plane doesn't change its vertical velocity and therefore maintains a constant height. Forces are vectors, so because these forces are the same size but in opposite directions, the net force on the plane is equal to zero both horizontally and vertically. This means that the plane travels at a steady velocity because it cannot accelerate without an unbalanced net force.

thrust forward push

ewbk-13219

int-8151

FIGURE 8.21 The plane flies at a steady velocity because no unbalanced net force is acting on it.

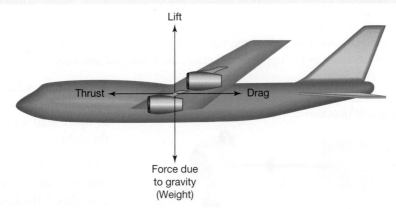

In many manoeuvres that you may experience as a passenger on a bus, an unbalanced force must act on the vehicle to make it change its speed or direction.

For example, the bus stops suddenly when the brakes are applied. The resistance forces are large and there is no thrust. Your seat is rigidly attached to the bus, so it also stops suddenly. However, the resistance forces are not acting on you. You continue to move forward at the speed that you were travelling at before the brakes were applied until there is a force to stop you. That force could be provided by a seatbelt, the back of any seat in front of you, a passenger in front of you or you desperately holding onto a handrail! It may feel as if you are thrown forwards in this situation, but in reality you continue to travel forwards as there was no unbalanced force acting on you to bring you to a stop.

When the bus makes a sudden right turn, the unbalanced net force acting on the bus to change its direction is not acting directly on you. You continue to move in the original direction. The inside wall of the bus moves to the right but you don't. So it seems like you've been flung to the left.

Figure 8.22 shows the forces acting on a bus moving along a straight, horizontal road at a constant speed. The upward push of the road must be the same as the force due to gravity. If that were not the case, the bus would fall through the road or be pushed up into the air. If the thrust is greater than the resistance forces, the bus speeds up. If the resistance forces are greater than the thrust, the bus slows down.

FIGURE 8.22 The forces acting on a moving bus. The forces are in balance when the bus is not changing speed or direction.

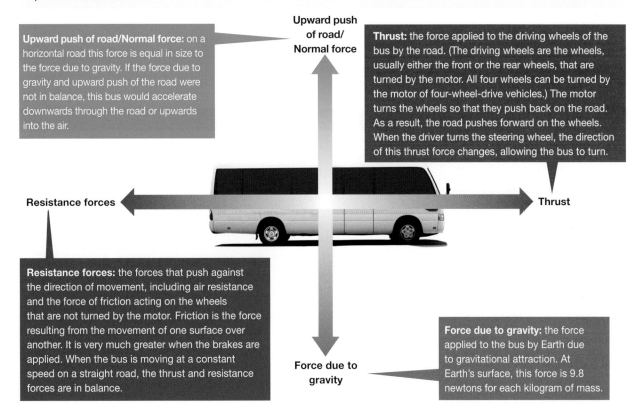

Upward push of road/Normal force: on a horizontal road this force is equal in size to the force due to gravity. If the force due to gravity and upward push of the road were not in balance, this bus would accelerate downwards through the road or upwards into the air.

Upward push of road/ Normal force

Thrust: the force applied to the driving wheels of the bus by the road. (The driving wheels are the wheels, usually either the front or the rear wheels, that are turned by the motor. All four wheels can be turned by the motor of four-wheel-drive vehicles.) The motor turns the wheels so that they push back on the road. As a result, the road pushes forward on the wheels. When the driver turns the steering wheel, the direction of this thrust force changes, allowing the bus to turn.

Resistance forces

Thrust

Resistance forces: the forces that push against the direction of movement, including air resistance and the force of friction acting on the wheels that are not turned by the motor. Friction is the force resulting from the movement of one surface over another. It is very much greater when the brakes are applied. When the bus is moving at a constant speed on a straight road, the thrust and resistance forces are in balance.

Force due to gravity: the force applied to the bus by Earth due to gravitational attraction. At Earth's surface, this force is 9.8 newtons for each kilogram of mass.

Force due to gravity

Only two significant forces are acting on you as a passenger of the bus when its speed is steady:

- the force applied on you by Earth's gravitational attraction (your weight)
- the push of the bus seat, which pushes upwards and forwards. The upward part of this force is just enough to balance your weight. The forwards part of this force is what makes you speed up with the bus. If the bus is moving at a constant speed forwards, there is no forwards part of this force as you do not need a force to keep moving at the same velocity.

Newton's First Law of Motion
law that states that an object remains at rest or continues to move with the same speed in the same direction unless acted on by an outside, unbalanced force

EXTENSION: Newton's First Law and inertia

How does the magician pull the tablecloth without moving everything else in figure 8.23? The expensive crockery on the table is quite safe if the magician is fast enough. **Newton's First Law of Motion** provides an explanation. The magician is pulling on the tablecloth — not on the expensive crockery. A small unbalanced force is acting on the crockery due to friction. However, if the tablecloth is pulled away quickly enough, this force does not act for long enough to make the crockery move. If the tablecloth is pulled too slowly, the force of friction on the crockery will pull it off the table as well.

FIGURE 8.23 The magician pulls the tablecloth without moving the expensive crockery.

Inertia is the property of objects that makes them resist changes in their motion. The inertia of the crockery keeps it on the table. The inertia of the tablecloth is not large enough to allow it to resist the change in motion. The greater the mass of an object, the more inertia it has. For example, it takes a much larger force to change the motion of a cricket ball than it does for a table tennis ball travelling at the same speed.

ACTIVITY: Observing inertia

Try pulling an A4 sheet of paper out from under a 20-cent coin. Then try it with a 5-cent coin. Do this a few times for each coin and explain your observations using Newton's First Law.

---

elog-2486

INVESTIGATION 8.3

Forces on cars

Aim

To investigate the forces acting on a toy car

Materials

- toy car

Method

1. Rest a toy car on a smooth, level surface.
2. Push the car quickly forwards and then let it go.

Results

Note down your observations.

Discussion

1. Would you expect individuals to get the same results with the same method? Suggest ways to improve the method that may allow for this.
2. What forces are acting on the car while it is at rest?
3. How do you know that more than one force is acting on the car while it is at rest?
4. Are the forces on the car in balance after you stop pushing? How do you know?
5. Which force or forces cause the car to slow down after you stop pushing it forwards?
6. How would the car's motion be different if you pushed it forwards and let it go on the following?
   a. A much smoother surface
   b. A rough surface

Conclusion

Summarise the findings for this investigation about forces acting on a toy car.

8.5.5 Driverless cars

A driverless car uses computers to make decisions about the size of the thrust and resistive forces on the car. The computer needs a sensor to detect the frictional forces with the road and then needs to adjust the thrust from the car engine to ensure the car speeds up, maintains a constant speed or slows down as required. The computer also uses information about the distance to the cars in front and on either side to help it make decisions about the size of the thrust force.

When a driverless car needs to brake, the computer needs to perform more complicatedcalculations than when the car is travelling at a constant speed. If the computer applies a large braking force to the car the passengers could be injured, however, if the braking force is too small the car could crash. Manufacturers of driverless cars need to ensure the computers inside each car are receiving enough information from the car's sensors and have information about appropriately sized forces to ensure a smooth ride for passengers.

8.5 Activities

learn on

8.5 Quick quiz on	8.5 Exercise

Select your pathway

■ LEVEL 1	■ LEVEL 2	■ LEVEL 3
1, 2, 8	4, 6, 7, 10	3, 5, 9, 11

These questions are
even better in jacPLUS!
• Receive immediate feedback
• Access sample responses
• Track results and progress

Find all this and MORE in jacPLUS ▶

Remember and understand

1. Name the force that prevents a bus from falling through the surface of a road.
2. Name two forces that resist the forward motion of a bus.
3. Explain what inertia is.
4. Explain the difference between mass and weight.

Apply and analyse

5. The gravitational field strength on Mars is only 3.7 N/kg. Calculate your weight on Mars.
6. **MC** A small boat is resting on a beach. The owner tries to drag it down to the sea. She pulls with a force of 2000 N but the boat does not move. What was the value of friction?
   A. Greater than 2000 N
   B. Equal to 2000 N
   C. Less than 2000 N
   D. There is no friction.
7. Use your memories of bus trips to help you answer the following questions. Assume that you are comfortably seated, not wearing a seatbelt and that the bus seats have no head restraints.
   a. Describe your immediate resulting motion (as a passenger on the bus) if the bus performed the following manoeuvres.
      i. A very quick start from rest
      ii. A forward motion at constant speed
      iii. A very sharp right turn
      iv. A slow left turn
      v. An emergency stop from a speed of 60 km/h
      vi. A forward jerk as the bus is struck from behind by another vehicle
   b. During which of the manoeuvres listed in part a. does the bus move with the same speed and direction as the passenger?
   c. Explain how properly fitted seatbelts would change the resulting motion of the passenger.
   d. Explain how a head restraint would change the resulting motion of the passenger in a bus that is struck from behind.
8. In each scenario, identify if the thrust or the resistance force acting on the bus is larger. Explain each of your responses.
   a. Increasing speed
   b. Decreasing speed
   c. Constant speed

▶

Evaluate and create

9. Explain in terms of Newton's First Law of Motion why you should never step off a bus, tram or train before it has completely stopped. Present this as a poster, report, brochure or campaign.
10. The car shown in the figure is travelling to the left at a constant velocity. The four major forces acting on the car are represented by arrows labelled A, B, C and D.
    a. State which two forces combine to provide the force represented by arrow C.
    b. Describe how the size of the force represented by arrow A compares with that of arrow C.
    c. Describe two different changes in the forces acting on the car that could cause it to slow down.
    d. Use a toy car to model different magnitudes of C by pushing the car along surfaces with different amounts of friction. Summarise your investigation and findings.

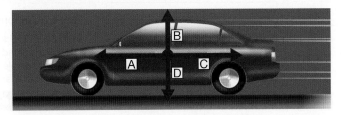

11. **SIS** Read the following newspaper article and answer the questions.
    a. Explain why the passengers and crew were injured during the flight.
    b. What was really being thrown around — the passengers or the aircraft? Explain your answer.
    c. How would a seatbelt protect a passenger in an incident like the one described?

SEVERAL HURT WHEN UNITED FLIGHT HITS TURBULENCE

WELLINGTON, New Zealand (AP) — A United Airlines flight from Sydney to San Francisco detoured to Auckland late Wednesday local time after several people on board were injured when the plane hit severe turbulence over the Pacific Ocean, an airline spokesman said.

A female cabin attendant broke a leg and a male cabin crew member had back and shoulder injuries from being thrown around in the turbulence.

Three passengers were taken to hospital with 'bruising and muscular discomfort'. Two other passengers with minor injuries were treated by ambulance staff at the airport.

Passenger Julie Greenwood told *The New Zealand Herald* newspaper the turbulence lasted about 30 seconds. 'It was like an earthquake in the air — I was lifted out of my chair twice,' she said.

Airline spokesman Jonathan Tudor in Auckland told *The Associated Press* that United Airlines flight 862, carrying 269 passengers and 21 crew, had taken off from Sydney at 3:35 pm New Zealand time and flew into 'clear air turbulence' after about four hours.

Those who were uninjured would be housed overnight in Auckland hotels, he added. It was not immediately clear when the flight would continue on to San Francisco.

Mary Brander, 77, from Sydney, said the bottom seemed to be falling out of the plane when it struck the turbulence.

'One minute we were in clear blue sky and it hit,' she told *The New Zealand Herald*.

Fully worked solutions and sample responses are available in your digital formats.

LESSON
8.6 Newton's Second Law of Motion

First Nations Australian readers are advised that this lesson and relevant resources may contain images of and references to people who have died.

LEARNING INTENTION

At the end of this lesson you will be able to apply Newton's Second Law to determine the acceleration of an object.

8.6.1 Newton's Second Law of Motion

Newton's Second Law of Motion describes how the mass of an object affects the way that it moves when acted upon by one or more forces.

Newton's Second Law of Motion

Newton's second law can be expressed as:

$$a = \frac{F_{net}}{m}$$

Where: **a** is the acceleration (in m/s^2)

F_{net} is the net force on the object (in N)

m is the mass of the object (in kg).

This formula describes the observation that larger masses accelerate less rapidly than smaller masses acted on by the same total force. For example, a car will accelerate faster with only the driver inside than when it is fully loaded down with passengers. The forward thurst acting on the car is the same in both cases; however, the greater mass means that the same force has produced less acceleration.

> **Newton's Second Law of Motion** law that states that the acceleration of an object is equal to the total force on the object divided by its mass

The formula also describes how a particular object accelerates more rapidly when a larger net force is applied. When using Newton's Second Law of Motion it is important to remember that an object's acceleration will always be in the same direction as the net force acting on the object. A key part of this formula is identifying that when the total force on an object is zero, the acceleration of the object will also be zero. This confirms and adds to Newton's First Law which states that an object only accelerates when the forces on it are unbalanced.

8.6.2 Newton's Second Law in action

The launching of any rocket at Cape Canaveral in Florida is a spectacular sight. At launch, a rocket like the one pictured in figure 8.24 has a mass of about 580 000 kilograms. Most of this mass is fuel, which is burned during the launch.

Two forces are acting on such a spacecraft, including the main rocket and four or five smaller boosters, as it blasts off:
- the downward pull of gravity. The downward pull of gravity of the spacecraft at launch is about 5.8 million newtons.
- the upward thrust provided by the burning of fuel, which is typically about 12 million newtons.

The thrust (upwards) is about 6.2 million newtons greater than the weight (downwards). That is, the net force on the spacecraft is about 6.2 million newtons upwards.

Newton's second law can be used to estimate the acceleration of the spacecraft at blast-off:

$$a = \frac{F_{net}}{m}$$
$$= \frac{6\,200\,000}{580\,000}$$
$$= 10.7 \text{ m/s}^2 \text{ upwards}$$

In other words, the spacecraft gains speed at the rate of only 10.7 m/s (or 39 km/h) each second. Its acceleration is upwards because the net force on the spacecraft acts in an upwards direction.

Newton's second law also explains why the small acceleration at blast-off is not a problem. As the fuel is rapidly burned, the mass of the space shuttle decreases. As this happens, the acceleration gradually increases, and the rocket gains speed more quickly.

8.6.3 Newton's Second Law in sport

Newton's Second Law is important in many sports as it helps athletes prevent injury and can suggest ways to improve their performance.

In sports such as gymnastics and dance, athletes land on soft mats or bouncy floors, which can help to prevent injuries. The soft mats and bouncy floors mean it takes longer for an athlete to come to a stop after jumping or leaping or falling from a height. In turn, this reduces the size of the negative acceleration experienced by the athlete and therefore reduces the force felt by the athlete. A smaller force is less likely to cause injuries.

A similar idea is behind the padding wrapped around goal posts during football and netball games. If an athlete accidentally runs into the goal post, the padding will extend the time it takes the athlete (or their head) to come to a stop, reducing the acceleration and therefore the force they experience.

Using the same idea in reverse can help athletes improve their sporting performance. If an athlete is using the largest force possible, the acceleration can be further increased by reducing the time needed to perform an action. For example, an athlete who can release a javelin more quickly will give the javelin a larger acceleration and cause it to travel more quickly, and therefore farther than an athlete who releases a javelin more slowly.

This idea of performing actions in a shorter time to increase acceleration was used by First Nations Australians in their spear throwers, also known as a woomera, and bows. The action of throwing or releasing spears, arrows and other hunting devices was sped up, leading to a larger acceleration and a faster maximum speed. In turn, the spear or arrow would travel farther and cause more damage when it landed.

In figure 8.25, you can see the spear thrower attached to the left-hand end of the spear. When the spear is thrown, the spear thrower quickly accelerates the left-hand end of the spear so the spear leaves with a higher speed than if it was thrown by hand. This high speed means the spear can travel farther and will be more deadly upon impact, thus making hunting more efficient.

FIGURE 8.24 A spacecraft is launched by powerful rockets — yet it seems to take a long time to get off the ground. Newton's Second Law provides an explanation.

F_{thrust} = 12 million N

F_{net} = 6.2 million N

F_{weight} = 5.8 million N

FIGURE 8.25 A spear-thrower is a tool that enables a user to throw a spear farther and faster.

elog-2488

tlvd-10820

INVESTIGATION 8.4

Force, mass and acceleration

Aim

To investigate Newton's Second Law of Motion

Materials

- dynamics trolley
- string
- one 2-kg mass
- masking tape

- pulley
- four 500 g masses
- stopwatch
- metre ruler

Method

1. On your benchtop, use the masking tape and the metre ruler to mark the starting line and finishing line for a 1 metre course for your dynamics trolley.
2. Tie one end of the string to your trolley. Place the trolley at the starting line and bring the string forward so that it passes over the finish line and then hangs over a pulley at the end of the bench. Tie the 2-kg mass to this end of the string. Hold the trolley in place while you do this (see the figure provided).
3. Start the stopwatch at the same moment the trolley is released, and time how long the unladen trolley takes to cross the finish line. Enter the time in the data table. Repeat this step twice more and then determine the average race time.
4. Place a 500-g mass on the trolley and repeat the previous step.
5. Repeat the experiment for increasing masses of 1000 g and 1500 g.

Results

Copy and complete the following table.

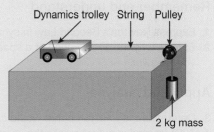

Dynamics trolley String Pulley

2 kg mass

TABLE The effect of a changing mass on the acceleration of a moving trolley

Mass on the trolley (g)	Time (s)			
	Trial 1	Trial 2	Trial 3	Average
0				
500				
1000				
1500				

Discussion

1. Identify the dependent, independent and controlled variables in this investigation.
2. Which of the trolleys had the fastest race time? How can you tell if it had the greatest acceleration?
3. The equation $d = ut + \frac{1}{2}at^2$ allows you to calculate the size of the acceleration that was acting on the trolley each time, where d = distance, u = starting speed, and t is the average race time to cover the distance. Because d = 1 m and u = 0 m/s, this equation simplifies so that we get $a = \frac{2}{t^2}$. Use this equation and the average race times in the table to determine the average acceleration of each trolley.
4. The force due to gravity of the 2-kg mass provided the force to move the trolley. Calculate the size of this force.
5. Was the force due to gravity of the 2-kg mass the only force acting on the trolley? What other forces can you identify that would have affected the trolley's acceleration?
6. Were the forces acting on the trolley each time balanced or unbalanced? How can you tell?
7. Give a general statement about the relationship between the mass of the trolley and its acceleration when a constant force is applied.

Conclusion

Summarise the findings of this investigation about Newton's Second Law of Motion.

8.6 Activities

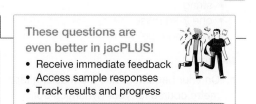

8.6 Quick quiz on	8.6 Exercise

Select your pathway

■ LEVEL 1 1, 4	■ LEVEL 2 2, 3, 6	■ LEVEL 3 5, 7

These questions are even better in jacPLUS!
- Receive immediate feedback
- Access sample responses
- Track results and progress

Find all this and MORE in jacPLUS ▶

Remember and understand

1. Express Newton's Second Law in symbols.
2. Explain why the acceleration of a spacecraft being launched increases as it rises. Use an equation to justify your answer.

Apply and analyse

3. Determine the total force that would cause a 1.5-kg glass salad bowl to accelerate across a table at 0.30 m/s^2.
4. A 10-kg sled is pulled across the snow so that the total force acting on it is 12 N. Calculate the average acceleration of the sled.
5. Two identical toy carts, A and B, each with a mass of 1.0 kg, are pushed across a smooth, level tabletop with the same force. One of them contains a heavy brick. Cart A accelerates more rapidly than cart B.
   a. Which toy cart contains the brick? Justify your answer.
   b. If the acceleration of cart A is 2.0 m/s^2, calculate the total force acting on each cart.
   c. If the acceleration of cart B is 0.5 m/s^2, calculate the mass of the brick.

Evaluate and create

6. A person pulls a cart with a force of 500 N. The friction opposing the motion is 200 N. If the cart has a mass of 60 kg, calculate the acceleration of the cart. Draw a diagram to show this scenario.
7. **SIS** A parachutist jumps out of a plane. Using a device to measure drag, the parachutist records the air resistance that she experiences as she falls. The data is in the following table.

TABLE The change in friction force over time acting on a parachutist

Time (s)	Friction force (N)
0	0
2	400
4	725
6	950
8	1000
10	1000
12	2500
14	1650

16	1000
18	1000
20	1000

a. Plot a graph of the friction force against time.
b. From your data, estimate the weight of the parachutist. Explain, using the idea of the net force, how you arrived at your answer.
c. Calculate the acceleration of the parachutist at 12 seconds into the fall.

Fully worked solutions and sample responses are available in your digital formats.

LESSON
8.7 Newton's Third Law of Motion

LEARNING INTENTION

At the end of this lesson you will be able to apply Newton's Third Law to two interacting objects.

8.7.1 It takes two

Newton's Third Law of Motion states that forces are created in pairs as they are an interaction between two objects.

That is, when an object applies a force to a second object, the second object applies an equal and opposite force of the same type to the first object.

The first two Laws of Motion look at the forces on an object — singular. They describe the effect of those forces on that one object. Newton's Third Law is different. It is the only law where you have to identify which two objects are interacting with each other. The two objects will experience the same size force but in opposite directions (figure 8.26)

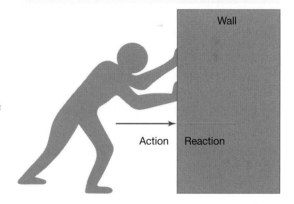

FIGURE 8.26 The person and the wall are experiencing the same size force but in opposite directions.

8.7.2 Pairs of forces

Forces always occur in pairs. Sometimes this is painfully obvious. An easy way to always spot the paired force is to identify what object is pushing and on what.

For example, when you catch a fast-moving softball or cricket ball with your bare hands, your hands are interacting with the ball.
- One force in the pair is the force of your hands on the ball.
- The second force in the pair is the force of the ball on the hands (which is what hurts you).

Newton's Third Law of Motion
law that states that forces are created in pairs as they are an interaction between two objects

On the move with Newton's Third Law of Motion

Whether you are getting around on the ground, in the water, in the air or even in outer space, pairs of forces are needed.

- When an athlete pushes back and down on the starting block, the starting block pushes the athlete forwards and upwards (figure 8.27).
- The forward push of the road on the driving wheels of a car occurs only because the wheels push back on the road.
- When you swim through water, you push back on the water with your arms and legs so the water pushes you forward.

FIGURE 8.27 In order to move forward quickly, the athletes need to push back. Why?

INVESTIGATION 8.5

Just a lot of hot air

Aim

To observe Newton's Third Law of Motion in action

Materials

- a balloon

Method

1. Write a clear hypothesis for the investigation.
2. Inflate a balloon and hold the opening closed.
3. Release the balloon and observe its motion through the air.

Results

Describe your observations for each of the following questions.
1. What happens to the air inside the balloon when you release the balloon?
2. Which way does the balloon move as the air is pushed out?

Discussion

1. What provides the force that pushes the balloon through the air?
2. What is similar about the way in which the balloon is propelled and the way a jet engine works?
3. How is the motion of the balloon different from the motion of a jet engine?

Conclusion

Summarise the findings for this investigation about Newton's Third Law of Motion in action.

8.7.3 Up and away

The force that pushes a jet aeroplane forward is provided by the exhaust gases that stream from its engines. As the jet engines push the exhaust gases backwards, the gases push forwards with an equal and opposite force. This forward push is called the thrust. In order to equal or exceed the resistance to the jet's motion, the thrust needs to be very large. The huge blades inside a jet engine compress the air flowing into the engine and push it into the combustion chamber behind the blades. In the combustion chamber, fuel is added and burns rapidly in the compressed air. The exhaust gases are forced out of the engine at very high speed.

INVESTIGATION 8.6

Balloon rocket

Aim

To model an application of Newton's Third Law of Motion

Materials

- drinking straw (plastic)
- masking tape
- balloon (sausage-shaped if possible)
- scissors
- fishing line (about 20 m)

Method

1. Write a clear hypothesis for your investigation.
2. Cut two short pieces from the drinking straw and thread a length of fishing line through them. Attach the ends of the fishing line to two fixed points so that the line is taut.
3. Inflate the balloon and hold it closed while your partner attaches it to the pieces of drinking straw with masking tape, as shown in the figure.
4. Release the balloon and observe its motion along the string.

Results

Record the total distance travelled by your balloon rocket and note down any observations made.

Discussion

1. Compare your balloon rocket with those of others in your class. What features of the balloon rocket seemed to determine its range?
2. Suggest how your balloon rocket could be improved.
3. What would be expected if your balloon was more inflated? Justify your response.

Conclusion

Summarise the findings of this investigation.

SCIENCE AS A HUMAN ENDEAVOUR: Blast off

The rockets used to launch spacecraft use a pair of forces to propel themselves upwards. Like jet engines, they push exhaust gases rapidly out behind them. As the rocket pushes the gases out, the gases push in the opposite direction on the rocket.

Unlike jet engines, rockets used to launch spacecraft do not use air to burn fuel. They carry their own supply of oxygen so that the fuel can burn quickly enough to lift huge loads into space. The oxygen is usually carried as a liquid or a solid. Once a spacecraft is in orbit, smaller rockets can be used to make the craft speed up, slow down or change direction.

FIGURE 8.28 The rocket pushes away the exhaust gases, and the exhaust gases push the rocket forwards.

Resources

 eWorkbook   Newton's third law (ewbk-13225)

 Video eLessons   Newton's third law (eles-2892)
                    Acceleration (eles-2893)

8.7 Activities

| 8.7 Quick quiz on | 8.7 Exercise |

Select your pathway

| ■ LEVEL 1 | ■ LEVEL 2 | ■ LEVEL 3 |
| 1, 2, 4 | 3, 5, 7 | 6, 8, 9, 10 |

These questions are even better in jacPLUS!
- Receive immediate feedback
- Access sample responses
- Track results and progress

Find all this and MORE in jacPLUS ⊙

Remember and understand

1. Describe Newton's Third Law of Motion with the help of a well-labelled diagram.
2. List three pairs of action and reaction forces. Be sure to state what each force is acting on.
3. **MC** When you walk forwards, which of the following provides the forward push?
   A. The normal reaction force
   B. Your leg muscles pushing you forwards
   C. The reaction from the surface that you are walking on
   D. Your weight
4. **MC** A yacht uses the push of the air on its sails to propel it forwards. If the push of the air on the sails is an action, what is the corresponding reaction?
   A. The normal reaction force
   B. The weight of the yacht
   C. The upthrust of the water on the boat
   D. The push of the sails on the air

Apply and analyse

5. Explain two similarities and two differences between rockets and jet engines.
6. Describe the reaction force that propels a Murray River paddleboat forwards. Use a well-labelled diagram to show clearly the action force that makes up the other part of the action–reaction pair.
7. Research and explain how a hovercraft works. State what action–reaction pairs are involved in its operation.
8. A skydiver is falling towards Earth. Ignoring air resistance, the only force on the skydiver is their force due to gravity. State the Newton's Third Law paired force in this situation.

Evaluate and create

9. **SIS** A student has designed an investigation using a spring balance of 10 N and a spring balance of 20 N. They wish to explore Newton's Third Law of Motion by connecting the spring balances together, with one spring balance attached to a retort stand and the other spring balance being pulled by their hand, to see if the force of the retort stand on their hand is the same as the force of their hand on the retort stand.
   a. State the dependent variable in this investigation.
   b. Suggest some improvements for this investigation.
   c. Outline a clearly reproducible scientific method.
10. Explain the two main differences between Newton's First Law and Third Law with the help of well-labelled diagrams.

Fully worked solutions and sample responses are available in your digital formats.

LESSON
8.8 Work and energy

LEARNING INTENTION

At the end of this lesson you will be able to explain that energy can be stored, transformed or transferred, and that transferring energy from one type to another is known as work, and you will be able to calculate the work done on an object as its change in energy.

8.8.1 What is work?

When a force is applied to an object causing it to move in the direction of the force, **work** is done. A person lifting weights (figure 8.29) does work when lifting — the person applies an upward force and the weights move up. However, when they hold the weights still, no work is being done on the weights. The person is applying an upward force equal to the downward pull of gravity on each weight, but there is no movement in the direction of that force. It is easy to get confused by this because it obviously takes effort to hold a heavy object, but the work that you do goes on in your muscles. No work is being done on each of the weights.

work transfer in energy that occurs when a force moves an object in the direction of the force

FIGURE 8.29 Different examples of work being done

Out of contact

In the examples in figure 8.30, the object on which the work is done is in contact with the object doing the work, and moves in the direction of the force applied to it. However, contact is not always necessary for work to be done. The planet Earth does work on you when you fall because of the gravitational force between it and you. When you pick up a small piece of paper with a charged plastic pen, work is done on the paper because of the force created by the electric charge in the pen and the paper. When you charge a balloon and place it near a stream of water, work is done on the water, causing it to bend.

FIGURE 8.30 A non-contact force, causing work to be done on the water

8.8.2 What is energy?

In physics, **energy** is described as the capacity to do work. That means that energy has the ability to make things change. Some forms of energy are obvious because you can see movement. Other forms, such as the energy stored in the muscles of your body or a car battery, are not as obvious, but they have the capacity to make things change. The unit of energy is joule (J), named after the scientist James Prescott Joule.

FIGURE 8.31 Forms of energy

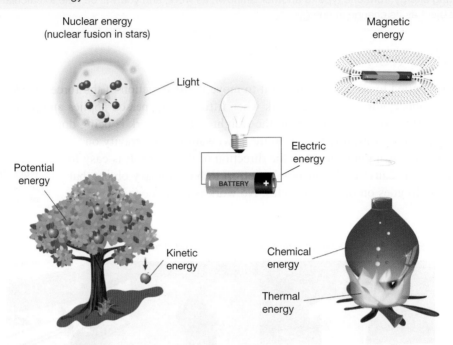

8.8.3 The relationship between work and energy

Energy can be transferred to an object by doing work on it. Doing work on an object can also convert the energy an object possesses from one form to another.

In fact, work is a measure of change in energy. The amount of energy transferred or converted when 1 newton of force moves an object 1 metre is 1 joule.

Work

$$W = \Delta E$$

Where: W = is the work done (in J)

ΔE = is the change in the energy (in J).

8.8.4 Potential energy

All stored energy is called **potential energy**. Energy can be stored in several different ways.

- **Elastic potential energy** (also called strain energy) is present in objects when they are stretched or compressed. Stretched rubber bands and springs have elastic potential energy, as do compressed springs. When the spring is released, the elastic potential energy in the compressed spring is mostly converted into kinetic energy.

energy the capacity of an object or a substance to do work

potential energy energy that is stored due to the position or state of an object

elastic potential energy the potential energy stored in a stretched or compressed object or substance

- **Gravitational potential energy** is present in objects in a position from which they could fall as a result of the force of gravity. The water in a hydro-electric dam has gravitational potential energy. When the water is released, the force of gravity pulls down on it, doing work and converting the gravitational potential energy mostly into kinetic energy.
- **Electrical potential energy** is present in objects or groups of objects in which positively and negatively charged particles are separated. It is also present when like electric charges are brought close together. The most obvious evidence of electrical potential energy is in clouds during thunderstorms. When enough electrical potential energy builds up, electrons move as lightning between clouds or to the ground.
- **Chemical potential energy** is present in all substances as a result of the electrical forces that hold atoms together. When chemical reactions take place, the stored energy can be converted to other forms of energy or it can be transferred to other atoms. Chemical potential energy is a form of electrical potential energy.
- **Nuclear energy** is the potential energy stored within the nucleus of all atoms. In radioactive substances, nuclear energy is naturally converted to other forms of energy. In nuclear reactions, such as those in nuclear power stations, in nuclear weapons and in the Sun and other stars, nuclei are split or combined. As a result, some of the energy stored in the reacting nuclei is converted into other forms of energy.

Note that when work is done some energy is generally dissipated ('lost' to the environment), in the form of heat, light or sound energy (think about the noise a waterfall makes for instance). This will be discussed further in lesson 8.9.

Gravitational potential energy

Gravitational potential energy is the amount of energy available to an object by virtue of its position and can be calculated as:

$$E_g = mg\Delta h$$

Where: E_g is the gravitational potential energy (in J)
m is the mass (in kg)
g is the magnitude of the gravitational field strength (in N/kg)
Δh is the change in height or position (in m).

Example 8: Calculating gravitational potential energy of an object

A 1.5 kg book falls from the top of a 2.3 m bookshelf. Its gravitational potential energy before falling to the ground can be calculated as follows.

$$E_g = mg\Delta h,$$
$$m = 1.5\,\text{kg}, \Delta h = 2.3\,\text{m}, g = 9.8\,\text{m/s}^2.$$
$$= 1.5 \times 9.8 \times 2.3$$
$$E_g = 33.8\,\text{J}$$

8.8.5 Kinetic energy

Kinetic energy is the energy possessed by an object due to its motion. Potential energy can be converted into kinetic energy. For example, when a ball is dropped from a height (and possesses gravitational potential energy) the ball is in motion and has kinetic energy.

Kinetic energy

Kinetic energy is energy due to motion of an object and can calculated as

$$E_k = \frac{1}{2}mv^2$$

Where: E_k is the kinetic energy (in J)
m is the mass (in kg)
v is the magnitude of the velocity of an object (in m/s).

---

gravitational potential energy
energy stored due to the height of an object above a base level

electrical potential energy
energy present in objects or groups of objects in which positively and negatively charged particles are separated

chemical potential energy
energy present in all substances as a result of the electrical forces that hold atoms together

nuclear energy the energy stored at the centre of atoms, the tiny particles that make up all substances

kinetic energy energy due to the motion of an object

Example 9: Calculating the kinetic energy of an object

An athlete has a mass of 60 kg and is running at a speed of 15 m/s. Their kinetic energy can be calculated as follows.

$$E_k = \frac{1}{2}mv^2$$
$$m = 60\,\text{kg},\ v = 15\,\text{m/s}$$
$$= \frac{1}{2} \times 60 \times (15)^2$$
$$= 6750\,\text{J}$$

EXTENSION: Work, force and distance

work done = force × distance travelled in the direction of the force

$$W = Fd$$

If the magnitude of the force is measured in newtons (N) and the distance is measured in metres, work done is measured in units of newton metres (N m). This unit is more commonly referred to as joules (J).

If you lift a 5-kilogram bowling ball with a force of 49 newtons (just enough to overcome its weight) to a height of 40 centimetres, the amount of work done is given by:

work done = force × distance
= 49 N × 0.40 m
= 19.6 Nm
= 19.6 J

By doing work on the bowling ball, you have transferred 19.6 joules of energy to it. The additional energy is stored in the ball as gravitational potential energy. This stored energy has the potential to be converted into other forms of energy or transferred to other objects. For example, if you drop the ball, the force of gravity can do work on the ball, increasing its kinetic energy at the expense of its gravitational potential energy. Kinetic energy is the energy associated with movement. If your toe happens to be in the way when the bowling ball reaches floor level, the kinetic energy is transferred to your toe — ouch!

Now you can see where the formula for gravitational potential energy came from. When you do work to lift the ball, the amount of work done to lift it at constant speed is:

work done = force × distance
= weight × change in height
= mg × Δh

---

on Resources

8.8 Activities

learn on

| 8.8 Quick quiz **on** | 8.8 Exercise |

These questions are even better in jacPLUS!
- Receive immediate feedback
- Access sample responses
- Track results and progress

Find all this and MORE in jacPLUS ▶

Select your pathway

■ LEVEL 1	■ LEVEL 2	■ LEVEL 3
2, 3, 5, 11	1, 4, 6, 7, 9	8, 10, 12

Remember and understand

1. Explain the similarities between work and energy.
2. Suggest a type of force, other than electrical and gravitational forces, that can do work on an object without being in contact with it.
3. Describe three types of energy.
4. Do you agree or disagree with the statement 'Energy is a vector quantity'? Justify your answer.

Apply and analyse

5. Name an example of a child's toy that converts the following.
   a. Gravitational potential energy into kinetic energy
   b. Elastic potential energy into kinetic energy
6. Classify each of the following situations as work done or no work done.
   a. An ice skater glides across the ice at constant speed.
   b. A roller skater gradually comes to a stop.
   c. An ice dancer is lifted into the air.
   d. An ice dancer is held in the air and is moving at a steady speed.
   e. An ice dancer is held in the air and is speeding up.
7. Calculate how much work is done on a car when three people unsuccessfully try to push it out of the mud. Each of the three people applies a forward force of 400 N to the car (but the car does not move).
8. An Olympic diver with a weight of 540 N dives into a pool from a height of 10 m.
   a. Calculate the diver's gravitational potential energy when she stands on the platform.
   b. How much of her gravitational potential energy would you expect to be converted into kinetic energy during her dive?
9. A car with a mass of 1800 kg, travelling at a constant speed, has 250 kJ of energy (250 kJ = 250 000 J). What is the speed of the car?
10. A ball with a mass of 400 g is dropped from rest from a height of 8 m. What is its speed as it reaches the ground? (g= 9.8 N/kg)

Evaluate and create

11. Research and report on the life of James Joule. What did Joule achieve to deserve the honour of having the unit of energy named after him?
12. **SIS** Investigate and write a report on the energy changes that take place when you cut the lawn with a lawnmower. Identify the forces that do work while you are mowing the lawn and state how you would find the work done.

Fully worked solutions and sample responses are available in your digital formats.

LESSON
8.9 Conservation of energy

LEARNING INTENTION

At the end of this lesson you will be able to explain how energy can neither be created nor destroyed, but is simply stored or changed to another form. You will also be able to determine the efficiency of a system.

8.9.1 Never created, never destroyed

Whenever work is done, energy is transferred to another object or converted into another form of energy. Energy is never created or destroyed.

In fact, the total amount of energy in the Universe remains constant. This observation is known as the **Law of Conservation of Energy**. The energy of a bouncing basketball is converted from kinetic energy to stored energy and back again. However, because it 'loses' energy during these conversions, it never bounces back to its original height. Nevertheless, its energy is not really lost. It is transferred to other objects — the air and the ground. The atoms in these are given more energy due to the collisions with the ball. The greater the kinetic energy of the atoms in a substance, the higher the temperature. This means that when energy is 'lost', we usually mean that it is converted to heat energy.

Law of Conservation of Energy law that states that energy cannot be created or destroyed

---

SCIENCE AS A HUMAN ENDEAVOUR: How much energy is in the Universe?

One of humanity's greatest ambitions is to understand our Universe, and we have developed telescopes that have studied the most distant reaches of the observable Universe.

One of the most bizarre discoveries is the total amount of energy in the Universe. If the theory is correct, for every particle with energy, there is an equivalent warping of space producing 'negative energy'. This means that for all of the amazing things that we can see in the Universe, the total energy is ... zero!

FIGURE 8.32 An illustration of the Milky Way

8.9.2 Ups and downs

The energy changes that take place while bouncing on a trampoline occur when work is done in turn by the force of gravity, the force of the trampoline on the person or the force of the person on the trampoline.

Consider Sam in figure 8.33 while bouncing on the trampoline.

FIGURE 8.33 Changes in energy and work being done while bouncing on a trampoline

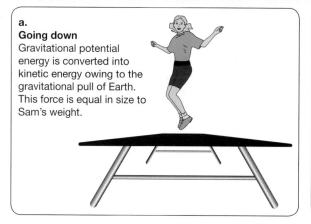

a.
Going down
Gravitational potential energy is converted into kinetic energy owing to the gravitational pull of Earth. This force is equal in size to Sam's weight.

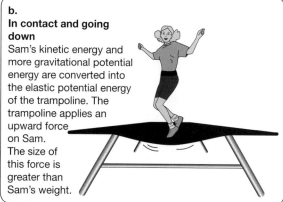

b.
In contact and going down
Sam's kinetic energy and more gravitational potential energy are converted into the elastic potential energy of the trampoline. The trampoline applies an upward force on Sam. The size of this force is greater than Sam's weight.

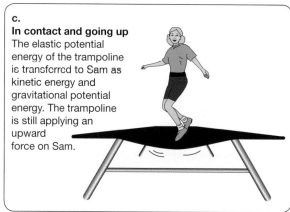

c.
In contact and going up
The elastic potential energy of the trampoline is transferred to Sam as kinetic energy and gravitational potential energy. The trampoline is still applying an upward force on Sam.

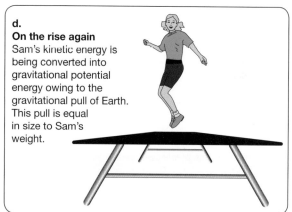

d.
On the rise again
Sam's kinetic energy is being converted into gravitational potential energy owing to the gravitational pull of Earth. This pull is equal in size to Sam's weight.

The energy changes involved:

1. At the top of the bounce, work is being done by the force of gravity to increase her kinetic energy (going down) at the expense of gravitational potential energy.
2. When she lands on the trampoline, she does work on the trampoline to convert her kinetic energy into elastic potential energy in the trampoline.
3. As she is pushed back up into the air, the trampoline is doing work on her. This converts the stored elastic potential energy into her kinetic energy.
4. Some energy will be converted into heat in the material of the trampoline (and the girl).
5. As she then rises, her kinetic energy is converted back into gravitational potential energy. She will not go as high as her starting point due to the loss of energy to heat during the bounce.

In order to jump higher on the trampoline, she would have to jump at the right moment, converting the energy stored in her body to extra kinetic energy and in turn to greater gravitational potential energy.

8.9.3 Systems and efficiency

Consider a basketball bounced against a concrete surface. The basketball and the concrete surface are a system. The **efficiency** of a system is the fraction of energy supplied to a device that is useful energy. It is usually expressed as a percentage.

efficiency the fraction of energy supplied to a device as useful energy

---

Efficiency of a system

$$\text{efficiency} = \frac{\text{useful energy output}}{\text{energy input}} \times 100\%$$

The 'useful' energy within the system is the sum of the kinetic energy, elastic potential energy and gravitational potential energy (E_g) of the basketball. The energy input for the basketball is its initial E_g. If this system were 100 per cent efficient, the basketball would bounce back to its original height and keep bouncing forever. The useful energy output would be the same as the energy input.

Another example of this may be in a light globe. A set amount of potential energy (electrical potential energy) is transferred to the light globe. The aim of the light globe is to convert this into light energy. However, some heat energy is also produced. The light energy is the useful energy. The more efficient a light globe, the better it is at converting electrical energy to light energy.

Example 10: Calculating efficiency

A basketball is dropped from a height of 120 centimetres and bounces back to a height of 72 centimetres. The gravitational potential energy ($mg\Delta h$) is the energy being explored that is useful.

$$\begin{aligned} \text{Efficiency} &= \frac{mg\,h_2}{mg\,h_1} \times 100\% \\ &= \frac{h_2}{h_1} \times 100\% \\ &= \frac{72}{120} \times 100\% \\ &= 60\%. \end{aligned}$$

elog-2494

INVESTIGATION 8.7

Follow the bouncing ball

Aim

To investigate the energy transfers and transformations of a bouncing ball

Materials

- tennis ball
- metre ruler

Method

1. Write a clear hypothesis for your investigation.
2. Drop the tennis ball from a height of 1 metre onto a hard surface. Take care that the ball is dropped from rest and not assisted on its way down.
3. Watch the top of the ball closely as it hits the ground and rebounds.
4. Measure the height of the rebound from the top of the ball to the ground.
5. Drop the ball again from the same height and measure the height to which it rebounds after a single bounce.
6. Repeat your measurements at least five times and find the average bounce height.

Results

Record your results in a table similar to the following.

TABLE The rebound height of a tennis ball

	Trial 1	Trial 2	Trial 3	Trial 4	Trial 5
Height of rebound (m)					

1. Construct a flow chart using text descriptions similar to those used in the illustrations in this lesson of the girl on the trampoline to show the energy changes that take place as the ball is:
   a. falling through the air
   b. slowing down while in contact with the ground
   c. speeding up just before leaving the ground
   d. rising through the air.
2. Identify the largest force that is acting on the ball as it:
   a. falls through the air
   b. is in contact with the ground
   c. rises through the air.
3. Do you agree with the statement that 'there is a loss of energy'? Justify your answer.
4. If you agree with the preceding statement, explain where this energy is lost.
5. Comment on the accuracy and precision of your results. How might you improve each of these?

Conclusion

Summarise the findings for this investigation about the energy transfers and transformations of a bouncing ball.

8.9.4 Swinging pendulums

A pendulum is a suspended object that is free to swing to and fro. The most well-known use of pendulums is in clocks, mostly very old ones. A playground swing is simply a large pendulum and provides an excellent model of a system in which energy is alternately transformed from gravitational potential energy to kinetic energy. Without a push, the swing slows to a stop. The potential and kinetic energy of the system have dropped to zero. But that energy is not lost; it has been gradually transferred to the surroundings by resistive forces which have heated the air around the swing and the bar at the top of the swing.

FIGURE 8.34 A pendulum is a good demonstration of conversion of energy between kinetic and gravitational potential energy.

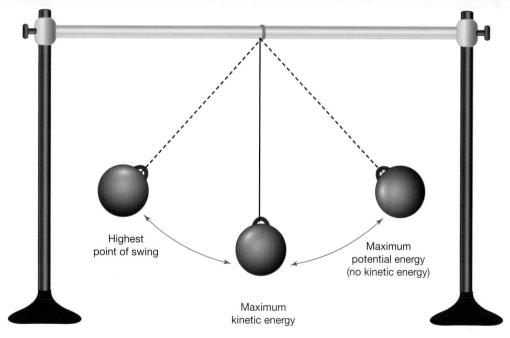

Highest point of swing

Maximum potential energy (no kinetic energy)

Maximum kinetic energy

DISCUSSION

A student collected a Newton's cradle, as shown in figure 8.35.

When one ball from one end is released, an equal and opposite reaction occurs and the other ball on the other end extends out. This is a highly efficient system, other than the small sound of balls hitting.

Would a Newton's cradle be able to continue this motion indefinitely? Discuss your response.

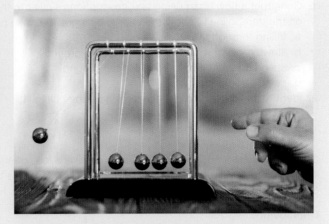

FIGURE 8.35 A Newton's cradle

INVESTIGATION 8.8

Swing high, swing low

Aim

To investigate energy changes of a pendulum (a swing)

Materials

- swing
- partner

Method

1. Push a friend on a swing.
2. Carefully observe their motion, considering their speed of movement and their height at different periods of time.

Results

Construct a flow chart, including text descriptions like those shown in the illustration in figure 8.34 to show how the energy of your friend changes through one complete backwards and forwards swing.

Discussion

1. The initial kinetic energy of the person on the swing is zero. Where does the person's initial increase in kinetic energy come from?
2. In order to keep the person swinging up to the same height over and over again, you must continue to push. If energy is conserved, why do you need to continue to provide additional energy by pushing?
3. Design an investigation using a pendulum to show the change in kinetic energy and gravitational energy and how this can be affected by the length of the pendulum swing.

Conclusion

Summarise the findings for this investigation about energy changes of a pendulum.

 Resources

 eWorkbook   Conservation of energy (ewbk-13229)

 Weblink   Exploring the conservation of energy – energy skate park

These questions are even better in jacPLUS!
- Receive immediate feedback
- Access sample responses
- Track results and progress

Find all this and MORE in jacPLUS

Remember and understand

1. When you travel down a playground slide, the amount of kinetic energy that you gain on the way down is less than the amount of gravitational potential energy that you lose.
   a. Explain where the 'missing' energy goes.
   b. Describe what can be done to maximise the amount of your initial gravitational potential energy that is converted into your kinetic energy.
2. A girl on a trampoline is able to return to the same height after each bounce. Explain how this is achieved when the system of the girl and the trampoline is not really 100 per cent efficient.
3. Consider the kinetic energy of a tennis ball just before it bounces on a concrete surface.
   a. Explain what happens to the kinetic energy during the period of time when the ball is in contact with the ground.
   b. Explain what happens to the remaining energy in the ball as the ball bounces upwards.

Apply and analyse

4. If energy is conserved, meaning that the sum of the gravitational potential energy and the kinetic energy is the same everywhere on a rollercoaster ride, determine the gravitational potential and kinetic energy in terms of 'mg' at the following points of the rollercoaster shown in the figure.
   a. Point A
   b. Point B
   c. Point C

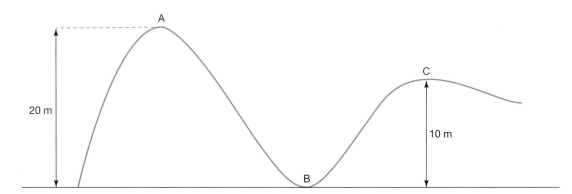

5. A 50-kg girl drops from a height of 1.0 m onto a trampoline.
   a. Calculate the weight of the girl. Assume that $g = 9.8$ N/kg.
   b. Calculate the amount of gravitational potential energy that has been lost by the girl at the instant she makes contact with the trampoline.
   c. Determine the girl's kinetic energy when she makes contact with the trampoline.
6. Calculate the efficiency of the bounce of a cricket ball dropped vertically onto a concrete floor from a height of 1.40 m if it rebounds to a height of 35 cm.

Evaluate and create

7. Construct a clear diagram or poster that shows what happens to your gravitational potential energy during either a bungee jump or a rollercoaster ride.

8. **SIS** Design and build a device that uses the gravitational potential energy stored in a tennis ball (or another type of ball) to act as a simple timer. Explain how your system works even though it is not 100 per cent efficient. Evaluate the usefulness of your device.

9. **SIS** Fuel efficiency of vehicles is an important factor to consider when designing or buying a vehicle. Shown in the following table are some estimated data on the average number of kilometres a typical US car could travel on 1L of petrol over the years 1975 to 2020.

Year	km/L
1975	5
1980	8
1985	9
1990	8
1995	8
2000	8
2005	9
2010	10
2015	11
2020	13

a. Present the data in an appropriate graphical form.
b. Describe what the shape of your graph tells you about the changing fuel efficiency of vehicles over the last few decades.
c. Research the ways engineers will have brought about the observed changes from part **b**.
d. Suggest why the data suggests that pressure has been placed on manufacturers to change their designs over the years.

Fully worked solutions and sample responses are available in your digital formats.

LESSON
8.10 Designing safety devices

LEARNING INTENTION

At the end of this lesson you will be able to explain how forces and energy considerations are used to design safety devices. You will be able to describe how a series of safety devices in a car work to prevent harm.

8.10.1 Forces shaped our evolution

This topic has discussed how forces govern the way an object accelerates.

For instance, human bones are incredibly strong but also slightly flexible. This allows them to withstand the forces of impact from running or combat, both being important factors for our survival in the ancient past. Our bones have not evolved to withstand the forces involved in a typical high-speed collision, however, which is why these collisions can be so deadly to us. When astronauts spend a long time in space, the reduced stress on their skeletons results in the bones becoming weaker. Regular physical exercise can go some way towards slowing the problem down, but if we are ever to go further into the Solar System, on journeys that may take years, we will have to find a way to overcome our biological limitations.

8.10.2 Forces shaped our technology

You may have heard of periods in human history such as the Stone Age, Bronze Age and Iron Age. In these periods, our technology was limited by the forces that the materials could withstand. In the past, the strength of stone blocks and later bricks limited the height of buildings. In the past, many buildings and structures were built with stone blocks (and later bricks), with many structures still standing today including the pyramids, aqueducts and cathedrals. If a structure was too tall, the lower levels could be under such massive forces that the blocks could crumble. However, an understanding of engineering saw arches incorporated to lower the mass but keep them strong or ensuring the joints between stones were tight so no water could get through. Modern structures are enormous because they are built with materials including steel, to make them strong and light.

As for our vehicles, we have gone from simple carts to trains, aeroplanes, cars and rockets. We are now capable of producing enormous forces, propelling us to high velocities. It is no surprise that we must also design safety equipment into these vehicles, as the forces acting on the occupants can be deadly.

In 1970, about 6 million vehicles were being driven on Australia's roads. During that year, 3798 people lost their lives in road accidents. Now there are about 20 million registered vehicles on Australia's roads, and during the ten years prior to the end of 2022, the average number of lives lost in road accidents was just under 1200 people.

One of the key reasons for the reduction in the road toll is that the cars we drive today are safer than ever before. Cars are designed by engineers who use scientific knowledge and experimentation to make cars lighter, stronger and, most importantly, safer.

eles-2895

SCIENCE AS A HUMAN ENDEAVOUR: Safety devices

Car safety

Safety features such as seatbelts, collapsible steering wheels, padded dashboards, head restraints, airbags and crumple zones have to be tested by engineers before being introduced. The testing continues after introduction as car manufacturers strive to make cars even safer.

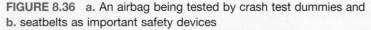

FIGURE 8.36 **a.** An airbag being tested by crash test dummies and **b.** seatbelts as important safety devices

Testing of safety features involves deliberately crashing cars with crash test dummies as occupants. The dummies are constructed to resemble the human body and numerous sensors are used to detect and measure the effects of a collision. Up until recently, most crash test dummies were the 'average' male and did not take into account the differences between male and female bodies, which may have contributed to higher hospitalisation rates for females involved in car accidents. Understanding these differences means we now see female crash test dummies being used to investigate the effects of collisions. These tests are also taking into account the differences between adults, children and babies and will allow for the continual safety improvements of cars.

Before real crash testing takes place, engineers use computer modelling to simulate crashes with virtual cars.

The effect of inertia

Most deaths and serious injuries in road accidents are caused when the occupants collide with the interior of the vehicle or are thrown from the vehicle. In a head-on collision the vehicle stops suddenly. However, unrestrained occupants continue to move at the pre-collision velocity of the vehicle until they collide with the steering wheel, ▶

dashboard or windscreen. Seatbelts provide an immediate force on the occupants so that they don't continue moving forwards. Front airbags reduce injuries caused by collisions between the upper body (which is still moving) and the steering wheel, dashboard or windscreen.

FIGURE 8.37 This inertia reel seatbelt is shown in the locked position. If rapid acceleration occurs, the small pendulum is able to swing forward with the inertia of the vehicle and lock the reel holding the seatbelt, preventing the passenger from moving forward.

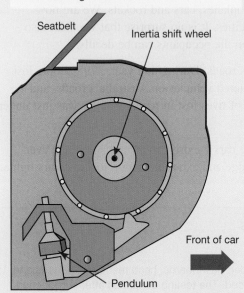

Seatbelt

Inertia shift wheel

Front of car

Pendulum

FIGURE 8.38 How airbags inflate

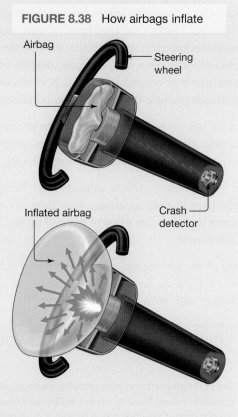

Airbag

Steering wheel

Inflated airbag

Crash detector

FIGURE 8.39 A stationary car is struck from behind by another vehicle. Without a head restraint, your head remains at rest until pulled forwards by your neck.

Car and seat pushed forward

Head remains at rest

Side airbags are becoming more common. They protect occupants in the event of a side-on collision. More recent airbag technology measures the size of the impact and delays inflation until just the right moment. These improvements are the direct result of computer modelling and real crash testing.

Head restraints on seats reduce neck and spinal injuries, especially in a vehicle that is struck from behind by another vehicle. An impact from the rear pushes a vehicle forwards suddenly. Your body is pushed forwards by your seat. However, without a head restraint, your head remains where it was. The sudden strain on your neck can cause serious spinal injuries. Neck injuries caused this way are often referred to as whiplash injuries. Some new cars have head restraints that automatically move forward and up when a collision occurs.

The zone defence

The occupants of a car sit in a very strong and rigid protection zone designed to prevent outside objects (including the car's engine, other cars and tree trunks) from entering the passenger 'cell' and causing injuries during a collision. The roof panel is supported by strong pillars to make it less likely to be crushed.

The rigid passenger cell is flanked by **crumple zones** at the front and rear of the vehicle (see figure 8.40). These zones are deliberately designed to crumple, absorbing and spreading much of the energy transferred to the vehicle during a collision. As a result, less energy is transferred to the protection zone carrying the occupants, reducing the chance of injuries. The crumple zone also allows the vehicle to stop more gradually. Without a crumple zone, the vehicle would stop more suddenly and perhaps even rebound. Occupants would make contact with the interior at a greater speed, and the chance of serious injury or death would be greater.

crumple zones zones in motor vehicles that are deliberately designed to crumple, absorbing and spreading much of the energy transferred to the vehicle during collision

FIGURE 8.40 Car safety features employed in a frontal collision

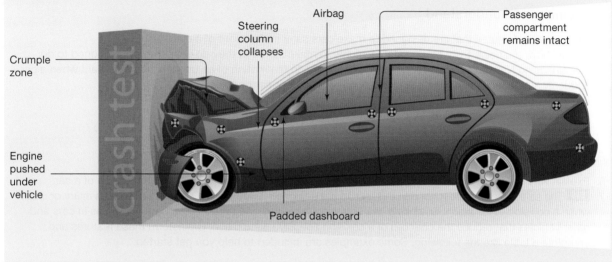

Airbag

Steering column collapses

Passenger compartment remains intact

Crumple zone

Engine pushed under vehicle

Padded dashboard

DISCUSSION

Discuss the safety features in cars. What other features do you think should be incorporated for safety? How do these compare to other transport devices such as buses, trains and motorcycles?

ACTIVITY

Research why females are at greater risk of injury in a car accident and how the different car safety features can pose a risk to them.

on Resources

Interactivities How an airbag works (int-5896)
Car safety features employed in a frontal collision (int-5895)

Weblink Car safety — RACV

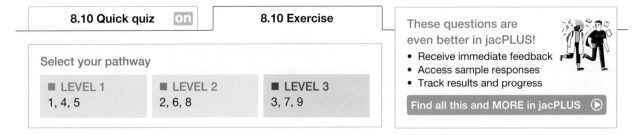

Remember and understand

1. List six safety features that are designed to make cars safer in the event of a collision.
2. Explain how engineers test vehicle safety features to make sure that they do what they are designed to do.
3. Explain, using Newton's Laws of Motion, what happens to the motion of an unrestrained occupant when a car suddenly stops because it has collided head-on with another car or object.
4. Describe how each of the following features protect occupants during a collision.
   a. Seatbelts   b. Airbags   c. Head restraints
5. What are crumple zones, and how do they protect the occupants of a vehicle during a collision?

Apply and analyse

6. Explain why it is important that there is a strong and rigid zone between the two crumple zones of a car.
7. **SIS** The safety features described in this lesson are designed to reduce the chances of serious injury or death when a collision takes place. Scientists and engineers have designed many other safety features in cars and other vehicles that reduce the chances of a collision actually taking place. Brainstorm these features and complete a table like the following. Some examples are included to help you get started.

TABLE Safety features designed to prevent collisions

Feature	How the feature works
Tyre tread	Increases friction and makes steering and braking more reliable, especially in wet weather. The tread even pushes water out from beneath the tyre when the road is wet.
Windscreen wipers	Keep the windscreen clear to ensure good visibility for the driver.
Speed alarm	The driver selects a maximum speed. If that speed is exceeded an alarm sounds, warning the driver to slow down.

Evaluate and create

8. **SIS** Use the internet to research and report on the following.
   a. How anti-lock brake systems (ABS) make braking in an emergency situation safer
   b. The benefits of electronic stability control (ESC)
   c. How driverless cars detect objects around them and avoid crashing
9. **SIS** Collect two or more advertising brochures for recently manufactured passenger vehicles, then complete the following tasks.
   a. Outline the features of the cars that relate to the safety of the occupants of the vehicle.
   b. List the claims made about the safety features of the cars that do not make reference to scientific evidence.
   c. Comment on missing scientific evidence that you would want to see included to support claims made in the brochures.

Fully worked solutions and sample responses are available in your digital formats.

LESSON
8.11 Thinking tools — Cycle maps

8.11.1 Tell me

What is a cycle map?

A cycle map is also known as a cycle chart or cyclical map. It is a type of graphic organiser and flow chart that shows events in a repeating cycle, showing how different events are related to each other.

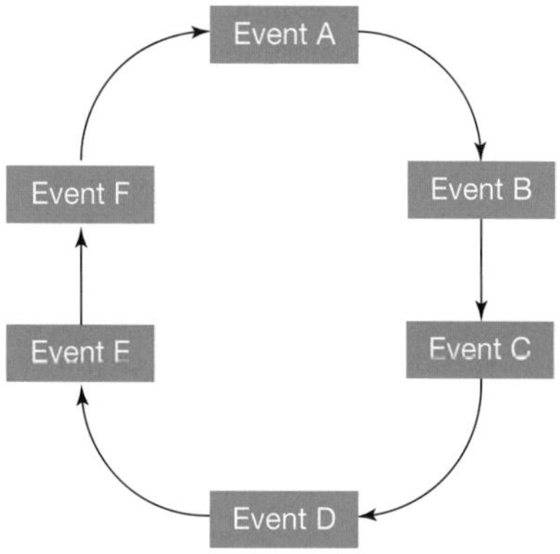

Why use cycle maps?

Cycle diagrams have no beginning and no end because they illustrate continually repeating processes, and so help show a sequence of events.

The ability to show a repeating process is helpful. Consider, for example, the Law of Conservation of Energy. This has no beginning and end because the energy is transferred from one form to another.

This makes cycle maps preferable in many situation to diagrams such as storyboards, which can show similar concepts, but do not tend to show them as a repeating process.

FIGURE 8.41 A typical cycle map

Event A → Event B → Event C → Event D → Event E → Event F → Event A

8.11.2 Show me

In order to create a cycle map you should:
1. List actions or steps that are relevant to a particular cycle on small pieces of paper.
2. Order your pieces of paper and then position the steps in a circle.
3. Review your cycle — are steps in the wrong order, missing or irrelevant? If so, make changes.
4. Write your cycle with each step placed in a box and join the boxes using arrows.

Figure 8.42 shows an example cycle map illustrating the motion of a rollercoaster.

FIGURE 8.42 A cycle map illustrating the motion of a rollercoaster

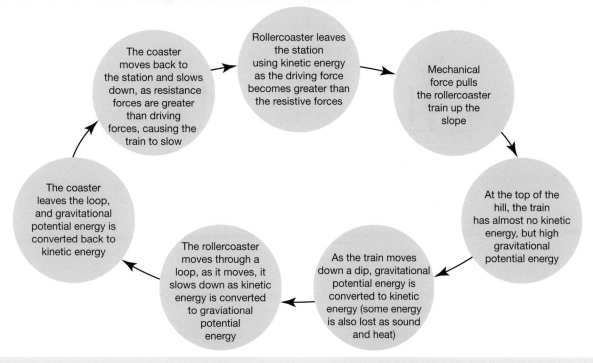

The coaster moves back to the station and slows down, as resistance forces are greater than driving forces, causing the train to slow

Rollercoaster leaves the station using kinetic energy as the driving force becomes greater than the resistive forces

Mechanical force pulls the rollercoaster train up the slope

At the top of the hill, the train has almost no kinetic energy, but high gravitational potential energy

As the train moves down a dip, gravitational potential energy is converted to kinetic energy (some energy is also lost as sound and heat)

The rollercoaster moves through a loop, as it moves, it slows down as kinetic energy is converted to graviational potential energy

The coaster leaves the loop, and gravitational potential energy is converted back to kinetic energy

8.11.3 Let me do it

8.11 Activity

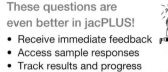
1. The cycle map provided represents the motion of a girl on a trampoline. The vertical arrows represent the direction of motion of the girl.
   a. Some of the boxes describing the energy changes that are taking place are empty. Copy the cycle map and complete it by describing the missing energy changes.
   b. During which stage (or stages) of the cycle is the girl doing the following?
      i. Accelerating upwards
      ii. Accelerating downwards
      iii. Slowing down
      iv. Speeding up
      v. Moving with zero velocity
   c. During which stage (or stages) of the trampoline cycle is the net force acting on the girl the following?
      i. Up
      ii. Down
   d. During which stages of the trampoline cycle is the force of gravity acting on the girl?
   e. What is happening to the energy lost by the girl after coming into contact with the trampoline?

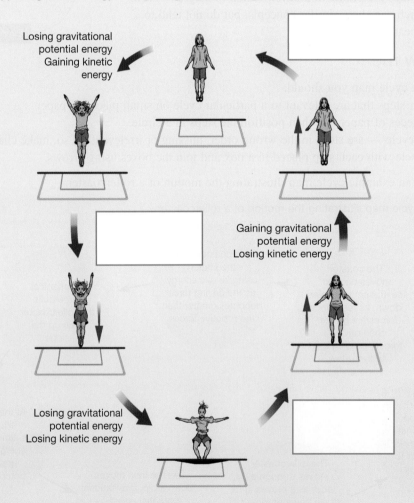

2. Construct a cycle map that describes the energy changes that take place when a tennis ball is dropped from a height and is allowed to bounce several times.

Fully worked solutions and sample responses are available in your digital formats.

LESSON
8.12 Project — Rock and rollercoaster

Scenario

Many psychologists think that the reason rollercoasters are so popular is tied up with the 'rush' that follows stimulation of the fear response. When exposed to the combination of speed, noise, high hills, twists, loops and steep descents of the ride, our brains tell us that some element of threat or danger is present. This triggers our 'fight or flight' instinct, sending adrenaline coursing through our bodies in a way that many people find very stimulating. Of course, some of us just throw up rather than finding it fun!

FIGURE 8.43 Designing rollercoasters relies on an understanding of forces, energy and motion.

Thrill-ride engineers say that the aim of a good ride is to provide a simulation of danger without actually putting people at risk. By manipulating the characteristics of gravitational acceleration, periodic motion and velocity, these engineers use physics to trick the body into thinking that it is in a lot more trouble than it really is. But the line between a ride that thrills and a ride that kills is a very narrow one, and the structural and mechanical engineers who design and build these rides must test their designs rigorously by using computer models or even physical models before the first steel rail leaves the factory! So how would you, as part of a team of rollercoaster engineers working at the new theme park, design a rollercoaster that was high on the thrill but zilch on the kill?

Your task

- Use your knowledge of physics and forces to design and build a model rollercoaster that has a set length and includes a minimum number of loops, hills and turns. Your design will also include a rollercoaster car that will travel along the length of the track. In order for the design to be considered successfully tested, your car must be able to travel the length of the track and then be brought to rest within the last 50 cm without derailing.
- Draw a plan or diagram of your rollercoaster identifying the positions and types of components used — hill, turn, twist or loop — and the points at which the car has the highest and lowest values of kinetic energy and gravitational potential energy.
- Finally, set up a blog that includes:
  (a) a summary of how the kinetic energy and gravitational potential energy values change over the course length
  (b) a log describing the development, building and testing of your rollercoaster and its different sections, including the method used to bring the car to a safe stop at the end
  (c) your drawing or plan of your completed rollercoaster
  (d) video footage of your rollercoaster in action from start to finish.

 Resources

ProjectsPLUS Rock and rollercoaster (pro-0116)

LESSON
8.13 Review

Access your topic review eWorkbooks

 Resources

■ Topic review Level 1
ewbk-13231

■ Topic review Level 2
ewbk-13233

■ Topic review Level 3
ewbk-13235

8.13.1 Summary

Average speed, distance and time

- Speed is a measure of the rate at which an object moves over a distance.
- Speed can be calculated using the formula $v_{av} = \frac{d}{t}$, where v is the speed, d is the distance travelled and t is the time.
- Speed is a scalar quantity and only has a size or magnitude.
- Displacement is the change in position of an object. It is a vector quantity, and so has both magnitude and direction. It can be calculated using Δx = final position − initial position = $x_2 - x_1$.
- Velocity is a measure of the change in position (displacement) and can be calculated using $v = \frac{\Delta x}{t}$, where v is the velocity, Δx is the displacement (change in position) and t is the time. Velocity is a vector quantity and has both magnitude and direction.

Measuring speed

- Various devices to measure speed include ticker timers, motion detectors, speedometers, GPS, radar guns, laser guns and digitector.

Acceleration and change in velocity

- Acceleration is a measure of the rate at which an object changes velocity.
- Acceleration is a vector quantity and has both magnitude and direction.
- This can be calculated as $a = \frac{\Delta v}{t}$, where a is the acceleration, Δv is the change in velocity (final velocity − initial velocity) and t is the time.

Forces and Newton's First Law of Motion

- A force is simply an interaction between two objects. It is a vector quantity and is measured in newtons (N).
- Newton's First Law of Motion states that an object will remain at rest, or will not change its speed or direction, unless it is acted upon by an outside, unbalanced force.
- Inertia is the property of objects that makes them resist changes in their motion.
- Weight is the force of gravitational attraction on an object towards a huge body such as a planet. It is a vector quality and is measured in newtons (N).
- The weight or force due to gravity of any object can be calculated using $F_g = mg$, where F_g is the force due to gravity, m is the mass of the object and g is the gravitational field strength.

Newton's Second Law of Motion

- Newton's Second Law of Motion states that the acceleration of an object depends on the size of the net force and the mass of the object.
- This can be shown through the formula $a = \dfrac{F_{net}}{m}$, where F_{net}, is the net force, m is the mass of an object and a is the acceleration.

Newton's Third Law of Motion

- Newton's Third Law of Motion states that for every action there is an equal and opposite reaction. That is, when an object applies a force to a second object, the second object applies an equal and opposite force to the first object.
- The individual forces act on two different objects to cause different accelerations according to Newton's second law.

Work and energy

- Energy is the capacity to do work. That means that energy has the ability to make things move.
- Energy can be potential energy (stored energy) or kinetic energy (energy associated with movement).
- Energy can be transferred to an object by doing work on it. Doing work on an object can also convert the energy an object possesses from one form to another. This can be calculated as $W = \Delta E$
- Energy and work both are scalar quantities and are measured in joules (J).
- Kinetic energy is energy due to motion of an object and can be calculated using $E_k = \dfrac{1}{2}mv^2$, where E_k is the kinetic energy (in J), m is the mass (in kg) and v is the velocity of an object (in m/s).
- Gravitational potential energy is the amount of energy available to an object by the virtue of its position and can be calculated using $E_g = mg\Delta h$, where E_g is the gravitational potential energy (in J), m is the mass (in kg), g is the magnitude of the gravitational field strength (in N/kg) and Δh is the change in height or position (in m).

Conservation of energy

- Whenever work is done, energy is transferred to another object or converted into another form of energy. Energy cannot be created or destroyed.
- The efficiency of a system is the fraction of energy supplied to a device as useful energy. It can be calculated as $\text{efficiency} = \dfrac{\text{useful energy output}}{\text{energy input}} \times 100\%$.

Designing safety devices

- It is vital to consider forces and energy when designing various technologies particularly around safety.
- The use of seatbelts, crumple zones and airbags in cars are all vital, preventing us from colliding with objects such as the windshield.
- Crumple zones at the front and back of cars are deliberately designed to crumple, absorbing and spreading much of the energy transferred to the vehicle during a collision.

8.13.2 Key terms

acceleration   rate of change in velocity
average speed   total distance travelled divided by time taken
braking distance   the distance travelled once the brakes have been applied and the vehicle stops
chemical potential energy   energy present in all substances as a result of the electrical forces that hold atoms together
crumple zones   zones in motor vehicles that are deliberately designed to crumple, absorbing and spreading much of the energy transferred to the vehicle during collision
displacement   the change in position of an object, including direction
Doppler effect   observed change in frequency of a wave when the wave source and observer are moving in relation to each other
efficiency   the fraction of energy supplied to a device as useful energy
elastic potential energy   the potential energy stored in a stretched or compressed object or substance
electrical potential energy   energy present in objects or groups of objects in which positively and negatively charged particles are separated

energy   the capacity of an object or a substance to do work

force   an interaction between two objects that can cause a change; commonly a push, pull or twist applied to one object by another, measured in newtons (N)

frequency   the number of waves or pulses passing a single location in one second

global positioning system (GPS)   device that uses radio signals from satellites orbiting the Earth to accurately map the position of a vehicle or individual

gravitational potential energy   energy stored due to the height of an object above a base level

inertia   property of objects that makes them resist changes in their motion

instantaneous speed   speed at any particular instant of time

kinetic energy   energy due to the motion of an object

Law of Conservation of Energy   law that states that energy cannot be created or destroyed

light gates   gates used to measure the speed of an object using a timer and a detector

magnetic field   area where a magnetic force is experienced

magnitude   size

mass   the amount of matter in an object, measured in kilograms (kg)

net force   the net (total or resultant) force of two or more forces acting on an object

newton   unit of force defined as the force required to provide a mass of 1 kg with an acceleration of 1 m/s^2

Newton's First Law of Motion   law that states that an object remains at rest or continues to move with the same speed in the same direction unless acted on by an outside, unbalanced force

Newton's Second Law of Motion   law that states that the acceleration of an object is equal to the total force on the object divided by its mass

Newton's Third Law of Motion   law that states that forces are created in pairs as they are an interaction between two objects

nuclear energy   the energy stored at the centre of atoms, the tiny particles that make up all substances

position   the location of an object

potential energy   energy that is stored due to the position or state of an object

rate   how a physical quantity changes with respect to time

reaction distance   the distance travelled once a hazard has been spotted, to just before applying the brakes

scalar quantity   any quantity that is described only by a magnitude

sonic motion detector   device that sends out pulses of ultrasound at a frequency of about 40 kHz and then detects the reflected pulses from the moving object

speed   a measure of the rate at which an object moves over a distance

stopping distance   the sum of the reaction and braking distances

thrust   forward push

vector quantity   any quantity that is described by both a magnitude and direction

velocity   a measure of rate of change in position, with respect to time, described by a magnitude and direction

weight   a commonly used word to describe the force due to gravity on an object

work   transfer in energy that occurs when a force moves an object in the direction of the force

---

8.13 Activities

learn on

Select your pathway

■ LEVEL 1	■ LEVEL 2	■ LEVEL 3
1, 2, 5, 6, 10, 11, 17	3, 7, 8, 13, 14, 15, 18	4, 9, 12, 16, 19, 20

Remember and understand

1. The following table provides information about four laps completed by one of the drivers in an Australian Formula One Grand Prix. The distance covered during one complete circuit of the course is 5.3 km.
   a. Copy and complete the table.

TABLE The speeds in various laps of a Formula One race

Lap no.	Time (s)	Average speed (m/s)	Average speed (km/h)
5	90		
15		60	216
25	110		
35	92	57.6	

   b. Suggest two likely reasons for the lower average speed during lap 25.
2. Complete these statements about Newton's laws of motion.
   a. An object remains at rest, or will not change its speed or direction, unless …
   b. When an unbalanced force acts on an object, the mass of an object affects …
   c. For every action, there is …
3. Explain why the following statement is false: *A car travelling along a straight road at constant speed has no forces acting on it.*
4. Explain in terms of Newton's First Law of Motion why it is dangerous to have loose objects inside a moving car.
5. Explain the difference between the following terms.
   a. Velocity and speed
   b. Distance and displacement
   c. Gravitational potential energy and potential energy
   d. Energy and work
   e. Reaction and action
6. a. State whether the following statements are true or false.

Statements	True or false?
i. Energy can be created but never destroyed.	
ii. Energy can be transferred from one object to another.	
iii. Energy can be transformed from one type to another.	
iv. Energy cannot be stored.	
v. Energy is measured in joules.	

   b. Justify any false responses.

Apply and analyse

7. List the energy changes that take place as a tennis ball dropped from a height of 2 m does the following.
   a. Falls towards the ground
   b. Is in contact with the ground
   c. Rebounds upwards through the air
8. List the forces acting on the tennis ball described in question **7** while it is doing the following.
   a. Moving through the air
   b. In contact with the ground
9. **SIS** The following figure shows part of the ticker tape record of the motion of a toy car as it is pushed along a table. As the tape moves through the ticker timer, a new black dot is produced every fiftieth of a second (0.02 s). The ticker tape has been divided into four equal time intervals labelled A, B, C and D. The distance from the start of interval A to the end of interval D is 12.8 cm.

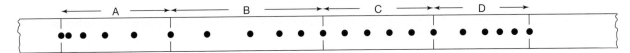

   a. During which time intervals is the speed of the toy car:
      i. increasing?
      ii. decreasing?
   b. During which of the four time intervals is the following occurring?
      i. The speed of the toy car is constant.
      ii. The acceleration of the toy car is constant.
   c. During which of the four intervals was the total force acting on the toy car zero?
   d. During which of the four intervals was the unbalanced force acting on the toy car in the same direction as the car was moving?
   e. What period of time is represented by each of the four time intervals? Express your answer in decimal form.
   f. What is the average speed during the entire time interval represented by the ticker tape?
10. Two people push on a table. The first person exerts a force of 60 N down on the table. The second person exerts a 30 N force upwards on the table. What is the net force acting on the table? What is the direction of this force?
11. Where is more work being done — the work a body-builder does on a 60-kg barbell while holding it above her head or the work a student does on a 50-g pen while writing the answer to this question? Explain your answer.
12. An object is launched into space. It has a mass of 60 kg on the surface of Earth. Calculate the following.
    Note: Jupiter's acceleration of gravity is 24.8 m/s^2.
    a. The mass of the object if the object lands on Jupiter
    b. The weight of the object on Jupiter
13. A person is at a tenpin bowling alley and preparing to bowl. She is holding the ball 1.2 metres high. The bowling ball has a mass of 5.5 kg. Determine the gravitational potential energy of this ball.
14. Use Newton's Second Law to answer the following.
    a. A 1400-kg car accelerates at 3 m/s^2. Calculate the size of the net force needed to cause this acceleration.
    b. A net force of 160 N causes an object to accelerate at 2 m/s^2. Determine the object's mass.
    c. A net force of 210 N acts on a mass of 70 kg. Calculate its acceleration.
15. A hare and a tortoise decide to have a race along a straight 100-m stretch of highway. They both head due north. However, at the 80-m mark, the hare realises that he dropped his phone at the 20-m mark. He dashes back, grabs his phone and resumes the race, arriving at the finishing line at the same time as the tortoise. Calculate the following.
    a. The displacement of the hare during the entire race
    b. The distance travelled by the hare during the race
    c. The distance travelled by the tortoise during the race
    d. The displacement of the hare during his return to collect his phone

Evaluate and create

16. Many older people who drove cars more than 50 years ago make the comment, 'They don't make them like they used to' in discussions about crumple zones. They describe how older cars were stronger and tougher, and therefore safer. Write a letter to a person who has made such a statement explaining why it is better that 'they don't make them like they used to'. Include three main arguments in your response.

17. By increasing the time over which a collision occurs, the force of the impact on a car and its passengers can be decreased. List and describe three safety features of a car that decrease the force of impact in this way. Draw a diagram to show these features.

18. Use the information given in the graph provided to answer the following questions.

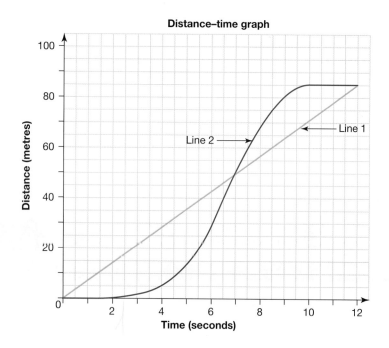

Distance–time graph

a. Describe the journey of an individual whose distance and time is shown by line 1.
b. Calculate the average speed for each of the journeys the lines represent.
c. Explain the difference between lines 1 and 2.

19. **SIS** Carefully examine the following position–time graph, which shows the position of an individual. Note that the positive values represent points to the east of a person's house, and negative values represent points west of the house.

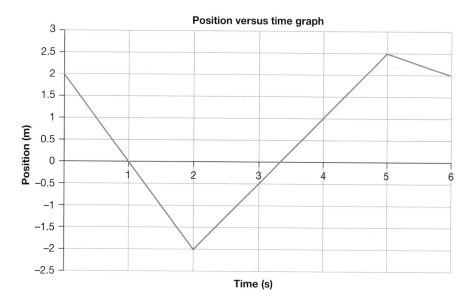

Position versus time graph

a. Describe the journey of the individual in this graph.
b. Determine the total distance the individual travelled.
c. Calculate the average speed of the individual over the journey.

d. Calculate the average velocity of the individual over the journey.
e. Calculate the velocity of the individual between the following times.
i. 0 and 1 seconds
ii. 2 and 4 seconds
iii. 5 and 6 seconds
f. At what point of the journey was the magnitude of the speed the greatest?
g. Is the individual accelerating at all between 2 and 5 seconds? Justify your response.

20. **SIS** A student wishes to conduct an experiment showing the acceleration of a ball as it moves down a slope.
a. State a hypothesis for this investigation.
b. Write a clear reproducible method that allows for this hypothesis to be tested.
c. Draw an outline of a graph that shows expected results for the following.
i. Position versus time
ii. Velocity versus time
iii. Acceleration versus time

Fully worked solutions and sample responses are available in your digital formats.

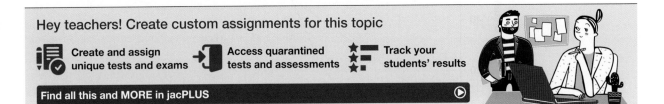

Hey teachers! Create custom assignments for this topic

Create and assign unique tests and exams

Access quarantined tests and assessments

Track your students' results

Find all this and MORE in jacPLUS

Online Resources

 **Resources**

Below is a full list of **rich resources** available online for this this topic. These resources are designed to bring ideas to life, to promote deep and lasting learning and to support the different learning needs of each individual.

8.1 Overview

eWorkbooks
- Topic 8 eWorkbook (ewbk-13206)
- Starter activity (ewbk-13208)
- Student learning matrix (ewbk-13210)

Solutions
- Topic 8 Solutions (sol-1131)

Practical investigation eLogbook
- Topic 8 Practical investigation eLogbook (elog-2480)

Video eLesson
- Rollercoasters (eles-2734)

8.2 Average speed, distance and time

eWorkbook
- Speed and velocity (ewbk-13211)

Weblink
- Adding vectors

8.3 Measuring speed

eWorkbook
- Ticker tapes (ewbk-13213)

Practical investigation eLogbook
- Investigation 8.1: Ticker timer tapes (elog-2482)

Weblink
- How do radar guns work?

8.4 Acceleration and changes in velocity

eWorkbook
- Acceleration (ewbk-13215)

Practical investigation eLogbook
- Investigation 8.2: Drag strips (elog-2484)

Video eLesson
- Drag racing (eles-2735)

8.5 Forces and Newton's First Law of Motion

eWorkbooks
- Force and gravity (ewbk-13217)
- Labelling the forces acting on a plane (ewbk-13219)
- Inertia and motion (ewbk-13221)

Practical investigation eLogbook
- Investigation 8.3: Forces on cars (elog-2486)

Video eLessons
- Science demonstrations (eles-1076)
- Newton's laws (eles-0036)

Interactivities
- Labelling the forces acting on a plane (int-8151)
- Net force (int-5894)

8.6 Newton's Second Law of Motion

eWorkbook
- Newton's second law (ewbk-13223)

Practical investigation eLogbook
- Investigation 8.4: Force, mass and acceleration (elog-2488)

Teacher-led video
- Investigation 8.4: Force, mass and acceleration (tlvd-10820)

Video eLesson
- The Space Shuttle (eles-2737)

8.7 Newton's Third Law of Motion

eWorkbook
- Newton's third law (ewbk-13225)

Practical investigation eLogbooks
- Investigation 8.5: Just a lot of hot air (elog-2490)
- Investigation 8.6: Balloon rocket (elog-2492)

Video eLessons
- Newton's third law (eles-2892)
- Acceleration (eles-2893)

8.8 Work and energy

Video eLessons
- Elastic potential energy (eles-2739)
- Applying force to an object (eles-2738)

8.9 Conservation of energy

eWorkbook
- Conservation of energy (ewbk-13229)

Practical investigation eLogbooks
- Investigation 8.7: Follow the bouncing ball (elog-2494)
- Investigation 8.8: Swing high, swing low (elog-2496)

Video eLesson
- A Newton's cradle (eles-2740)

Weblink
- Exploring the conservation of energy — energy skate park

8.10 Designing safety devices

eWorkbook
- Labelling the safety equipment of a car (ewbk-13245)

Video eLesson
- Virtual crash test dummies (eles-2895)

Interactivities
- How an airbag works (int-5896)
- Car safety features employed in a frontal collision (int-5895)
- Labelling the safety equipment of a car (int-8152)

Weblink
- Car safety — RACV

8.12 Project — Rock and rollercoaster

ProjectsPLUS
- Rock and rollercoaster (pro-0116)

8.13 Review

eWorkbooks
- Topic review Level 1 (ewbk-13231)
- Topic review Level 2 (ewbk-13233)
- Topic review Level 3 (ewbk-13235)
- Study checklist (ewbk-13237)
- Literacy builder (ewbk-13238)
- Crossword (ewbk-13240)
- Word search (ewbk-13242)
- Reflection (ewbk-13244)

Digital document
- Key terms glossary (doc-40498)

To access these online resources, log on to **www.jacplus.com.au**

9 Science quests

online only

SCIENCE INQUIRY AND INVESTIGATIONS

Science inquiry is a central component of Science curriculum. Investigations, supported by a **Practical investigation eLogbook**, are included in this topic to provide opportunities to build Science inquiry skills through undertaking investigations and communicating findings.

Online Resources

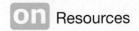
Below is a full list of **rich resources** available online for this topic. These resources are designed to bring ideas to life, to promote deep and lasting learning and to support the different learning needs of each individual.

9.1 Overview

 eWorkbooks
- Topic 9 eWorkbook (ewbk-13247)
- Student learning matrix (ewbk-13249)
- Starter activity (ewbk-13251)

Solutions
- Topic 9 Solutions (sol-1132)

Practical investigation eLogbooks
- Topic 9 Practical investigation eLogbook (elog-2498)
- Investigation 9.1: Life in 2050 (elog-2500)

 Video eLesson
- Nanocoating (eles-4185)

9.2 Endless possibilities

 eWorkbook
- Inventions and innovations (ewbk-13252)

Video eLesson
- Faster computing (eles-1077)

Weblinks
- Is AI humankind's last invention?
- Facts and statistics on transplants in Australia
- Organ donation in the United States

9.3 From superheroes to super science

 eWorkbook
- Science under scrutiny (ewbk-13254)

Practical investigation eLogbook
- Investigation 9.2: Using biomimicry to design a car (elog-2502)

Video eLessons
- Diatoms (eles-4283)
- The lotus effect (eles-4282)
- Velcro (eles-4281)

Weblink
- Elon Musk's plans to colonise Mars

9.4 The nanoworld

 eWorkbook
- Killer-bot (ewbk-13256)

Practical investigation eLogbook
- Investigation 9.3: Making a spider-bot (elog-2504)

Video eLessons
- Nanobot (eles-4284)
- Nanofabrication centre (eles-4285)

 Weblinks
- Nanobots propel through the eye
- The sounds of nanotechnology
- Translating proteins to music
- Making music from proteins
- Vibranium in the real world

9.5 Stem cells

 eWorkbook
- The uses of stem cells (ewbk-13258)

Weblinks
- Scientist mass produces minihearts
- Stem cells

9.6 Moving to Mars

 eWorkbook
- Moving to Mars (ewbk-13260)

Practical investigation eLogbook
- Investigation 9.4: Building a cardboard prototype for life on Mars (elog-0623)

Video eLesson
- Curiosity — a year of discovery (eles-2752)

Weblinks
- Space station science
- Mars 2020 mission overview

9.7 The science quest continues

 eWorkbook
- Great ideas and lucky breaks (ewbk-13262)

9.8 Review

 eWorkbooks
- Topic review Level 1 (ewbk-13264)
- Topic review Level 2 (ewbk-13266)
- Topic review Level 3 (ewbk-13268)
- Study checklist (ewbk-13270)
- Literacy builder (ewbk-13271)
- Crossword (ewbk-13273)
- Word search (ewbk-13275)
- Reflection (ewbk-13277)

 Digital document
- Key terms glossary (doc-40499)

To access these online resources, log on to **www.jacplus.com.au**

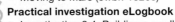
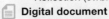

10 Psychology

LESSON SEQUENCE

online only

SCIENCE INQUIRY AND INVESTIGATIONS

Science inquiry is a central component of Science curriculum. Investigations, supported by a **Practical investigation eLogbook**, are included in this topic to provide opportunities to build Science inquiry skills through undertaking investigations and communicating findings.

Online Resources

Below is a full list of **rich resources** available online for this topic. These resources are designed to bring ideas to life, to promote deep and lasting learning and to support the different learning needs of each individual.

10.1 Overview

 eWorkbooks
- Topic 10 eWorkbook (ewbk-12682)
- Student learning matrix (ewbk-13278)
- Starter activity (ewbk-12685)

 Solutions
- Topic 10 Solutions (sol-1133)

10.2 Introducing psychology

Interactivity
- Gallery: Specialist areas in psychology (int-7051)

10.3 The brain

 eWorkbooks
- Labelling parts of the brain (ewbk-12687)
- Investigating neurons (ewbk-12689)

Interactivity
- Labelling parts of the brain (int-8133)

10.4 Intelligence

 eWorkbooks
- Multiple intelligences (ewbk-12691)
- Types of intelligence (ewbk-12693)

10.5 Emotions and communication

 eWorkbook
- Emotions (ewbk-12695)

10.6 Memory

 Video eLessons
- Neural activity in the brain (eles-2565)

Weblink
- Memory

10.7 Sleep and sleep disorders

Weblink
- Sleep Health Foundation

10.9 Treatment of mental health disorders

Weblink
- Beyond Blue

10.12 Review

 eWorkbooks
- Topic review Level 1 (ewbk-12697)
- Topic review Level 2 (ewbk-12699)
- Topic review Level 3 (ewbk-12701)
- Study checklist (ewbk-12703)
- Literacy builder (ewbk-12704)
- Crossword (ewbk-12706)
- Word search (ewbk-12708)
- Reflection (ewbk-12710)

 Digital document
- Key terms glossary (doc-40500)

To access these online resources, log on to **www.jacplus.com.au**.

11 Forensics

SCIENCE INQUIRY AND INVESTIGATIONS

Science inquiry is a central component of Science curriculum. Investigations, supported by a **Practical investigation eLogbook**, are included in this topic to provide opportunities to build Science inquiry skills through undertaking investigations and communicating findings.

Online Resources

Below is a full list of **rich resources** available online for this topic. These resources are designed to bring ideas to life, to promote deep and lasting learning and to support the different learning needs of each individual.

11.1 Overview

 eWorkbooks
- Topic 11 eWorkbook (ewbk-12477)
- Student learning matrix (ewbk-13279)
- Starter activity (ewbk-12480)

 Solutions Topic 11 Solutions (sol-1134)

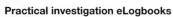 **Practical investigation eLogbooks**
- Topic 11 Practical investigation eLogbook (elog-2508)
- Investigation 11.1: Recreating a crime scene (elog-2510)

 Video eLesson
- How a gun fires bullets in slow motion (eles-4218)

11.2 *The Forensic Herald*

 eWorkbook
- How observant are you? (ewbk-12482)

11.3 Who knows who dunnit?

 eWorkbook
- Forensic scientists (ewbk-12484)

11.4 Forensic toolkit

 eWorkbook
- Forensic toolkits (ewbk-12486)

 Practical investigation eLogbooks
- Investigation 11.2: Are you a loop, arch, whorl or composite? (elog-2512)
- Investigation 11.3: Exploring UV light (elog-2514)

 Video eLesson
- Creating a rolling fingerprint (eles-4219)

 Weblinks
- Snapshot facial reconstruction
- Osceola County uses Laser Scanner for crime scene analysis and forensics

11.5 Discovering the truth through forensics

 eWorkbook
- Using DNA evidence to solve crimes (ewbk-12488)

 Video eLesson
- Comparing and matching DNA profiles (eles-4220)

 Weblink
- The story of Lindy Chamberlain — 40 years on

11.6 Clues from blood

 eWorkbook
- Evidence from blood (ewbk-12490)

 Practical investigation eLogbook
- Investigation 11.4: Exploring blood splatters (elog-2516)

 Weblink
- Australian Police — Bloodstain pattern analysis

11.7 Clues in hair, fibres and tracks

 eWorkbooks
- Exploring fibres, hair and tracks (ewbk-12492)
- Read my lips (ewbk-12494)

 Practical investigation eLogbooks
- Investigation 11.5: Comparing animal and human hair (elog-2518)
- Investigation 11.6: Analysing shoe prints (elog-2520)

 Weblink
- Different evidence used in forensics

11.8 Life as a forensic scientist

 eWorkbook
- Mystery in the house (ewbk-12496)

 Weblink
- How accurate are crime shows on TV?

11.9 Forensics and the future

 Weblinks
- Nanotechnology and forensics
- The Australian Synchrotron

11.10 Review

 eWorkbooks
- Topic review Level 1 (ewbk-12498)
- Topic review Level 2 (ewbk-12500)
- Topic review Level 3 (ewbk-12502)
- Study checklist (ewbk-12504)
- Literacy builder (ewbk-12505)
- Crossword (ewbk-12507)
- Word search (ewbk-12509)
- Reflection (ewbk-12511)

 Digital document
- Topic 11 Key terms glossary (doc-40501)

To access these online resources, log on to **www.jacplus.com.au.**

PERIODIC TABLE OF THE ELEMENTS

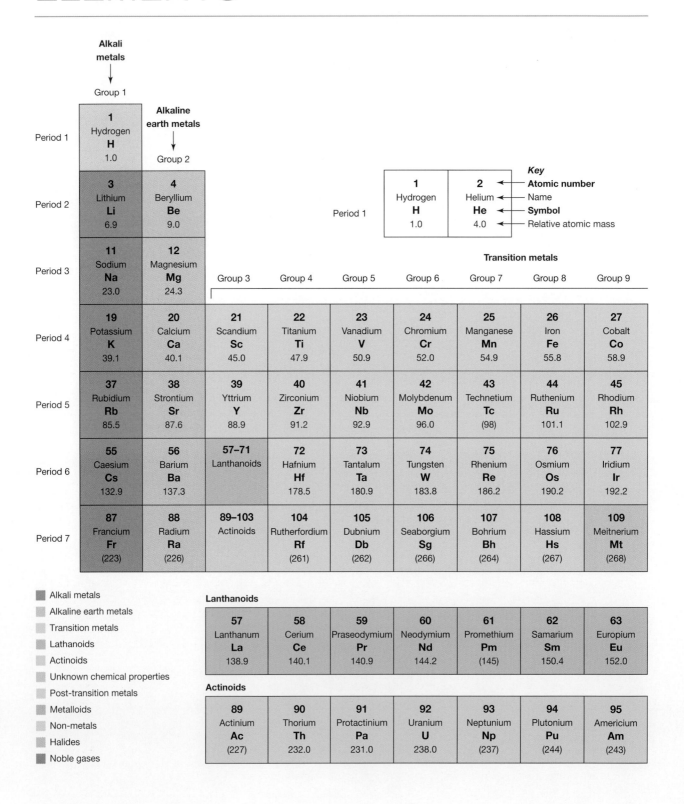

Alkali metals → Group 1

Alkaline earth metals ← Group 2

Transition metals

Group 1	Group 2									Group 3	Group 4	Group 5	Group 6	Group 7	Group 8	Group 9

Key
1	2
Hydrogen	Helium
H	**He**
1.0	4.0

← Atomic number
← Name
← Symbol
← Relative atomic mass

Period 1

Period 1
1
Hydrogen
H
1.0

Period 2
3	4
Lithium	Beryllium
Li	**Be**
6.9	9.0

Period 3
11	12
Sodium	Magnesium
Na	**Mg**
23.0	24.3

Period 4
19	20	21	22	23	24	25	26	27
Potassium	Calcium	Scandium	Titanium	Vanadium	Chromium	Manganese	Iron	Cobalt
K	**Ca**	**Sc**	**Ti**	**V**	**Cr**	**Mn**	**Fe**	**Co**
39.1	40.1	45.0	47.9	50.9	52.0	54.9	55.8	58.9

Period 5
37	38	39	40	41	42	43	44	45
Rubidium	Strontium	Yttrium	Zirconium	Niobium	Molybdenum	Technetium	Ruthenium	Rhodium
Rb	**Sr**	**Y**	**Zr**	**Nb**	**Mo**	**Tc**	**Ru**	**Rh**
85.5	87.6	88.9	91.2	92.9	96.0	(98)	101.1	102.9

Period 6
55	56	57–71	72	73	74	75	76	77
Caesium	Barium	Lanthanoids	Hafnium	Tantalum	Tungsten	Rhenium	Osmium	Iridium
Cs	**Ba**		**Hf**	**Ta**	**W**	**Re**	**Os**	**Ir**
132.9	137.3		178.5	180.9	183.8	186.2	190.2	192.2

Period 7
87	88	89–103	104	105	106	107	108	109
Francium	Radium	Actinoids	Rutherfordium	Dubnium	Seaborgium	Bohrium	Hassium	Meitnerium
Fr	**Ra**		**Rf**	**Db**	**Sg**	**Bh**	**Hs**	**Mt**
(223)	(226)		(261)	(262)	(266)	(264)	(267)	(268)

Legend:
- ■ Alkali metals
- ■ Alkaline earth metals
- ■ Transition metals
- ■ Lathanoids
- ■ Actinoids
- ■ Unknown chemical properties
- ■ Post-transition metals
- ■ Metalloids
- ■ Non-metals
- ■ Halides
- ■ Noble gases

Lanthanoids
57	58	59	60	61	62	63
Lanthanum	Cerium	Praseodymium	Neodymium	Promethium	Samarium	Europium
La	**Ce**	**Pr**	**Nd**	**Pm**	**Sm**	**Eu**
138.9	140.1	140.9	144.2	(145)	150.4	152.0

Actinoids
89	90	91	92	93	94	95
Actinium	Thorium	Protactinium	Uranium	Neptunium	Plutonium	Americium
Ac	**Th**	**Pa**	**U**	**Np**	**Pu**	**Am**
(227)	232.0	231.0	238.0	(237)	(244)	(243)

Non-metals →

Metals ↓

			Group 13	Group 14	Group 15	Group 16	Group 17	Group 18
								2 Helium **He** 4.0
			5 Boron **B** 10.8	6 Carbon **C** 12.0	7 Nitrogen **N** 14.0	8 Oxygen **O** 16.0	9 Fluorine **F** 19.0	10 Neon **Ne** 20.2
Group 10	Group 11	Group 12	13 Aluminium **Al** 27.0	14 Silicon **Si** 28.1	15 Phosphorus **P** 31.0	16 Sulfur **S** 32.1	17 Chlorine **Cl** 35.5	18 Argon **Ar** 39.9
28 Nickel **Ni** 58.7	29 Copper **Cu** 63.5	30 Zinc **Zn** 65.4	31 Gallium **Ga** 69.7	32 Germanium **Ge** 72.6	33 Arsenic **As** 74.9	34 Selenium **Se** 79.0	35 Bromine **Br** 79.9	36 Krypton **Kr** 83.8
46 Palladium **Pd** 106.4	47 Silver **Ag** 107.9	48 Cadmium **Cd** 112.4	49 Indium **In** 114.8	50 Tin **Sn** 118.7	51 Antimony **Sb** 121.8	52 Tellurium **Te** 127.6	53 Iodine **I** 126.9	54 Xenon **Xe** 131.3
78 Platinum **Pt** 195.1	79 Gold **Au** 197.0	80 Mercury **Hg** 200.6	81 Thallium **Tl** 204.4	82 Lead **Pb** 207.2	83 Bismuth **Bi** 209.0	84 Polonium **Po** (210)	85 Astatine **At** (210)	86 Radon **Rn** (222)
110 Darmstadtium **Ds** (271)	111 Roentgenium **Rg** (272)	112 Copernicium **Cn** (285)	113 Nihonium **Nh** (280)	114 Flerovium **Fl** (289)	115 Moscovium **Mc** (289)	116 Livermorium **Lv** (292)	117 Tennessine **Ts** (294)	118 Oganesson **Og** (294)

64 Gadolinium **Gd** 157.3	65 Terbium **Tb** 158.9	66 Dysprosium **Dy** 162.5	67 Holmium **Ho** 164.9	68 Erbium **Er** 167.3	69 Thulium **Tm** 168.9	70 Ytterbium **Yb** 173.1	71 Lutetium **Lu** 175.0

96 Curium **Cm** (247)	97 Berkelium **Bk** (247)	98 Californium **Cf** (251)	99 Einsteinium **Es** (252)	100 Fermium **Fm** (257)	101 Mendelevium **Md** (258)	102 Nobelium **No** (259)	103 Lawrencium **Lr** (262)

GLOSSARY

abiotic factors the non-living things in an ecosystem

absolute age number of years since the formation of a rock or fossil

absolute dating determining the age of a fossil and the rock in which it is found using the remaining amount of unchanged radioactive carbon

absolute magnitude actual brightness of a star

absolute referencing used in a spreadsheet when a cell address in the formula remains constant, no matter where it is copied to

absolute zero temperature at which the particles that make up an object or substance have no kinetic energy, approximately −273.15 °C

absorption spectrum the pattern of dark lines against a rainbow background obtained when white light is passed through a gas

accelerant a substance that spreads fire easily

acceleration rate of change in velocity

accuracy how close a measurement is to the true value

accurate an experimental measurement that is close to a known value

acid chemical that reacts with a base to produce a salt and water; edible acids taste sour

acid rain rainwater, snow or fog that contains dissolved chemicals that make it acidic

activation energy minimum energy required to start the reaction

activity series a classification of metals in decreasing order of reactivity

adaptation a feature that aids in the survival of an organism

adaptive radiation when divergent evolution of one species results in the formation of many species that are adapted to a variety of environments

affiliation the human need for involvement and belonging to a social group

afforestation planting trees in areas that had no recent cover

aim a statement outlining the purpose of an investigation

alkali base that dissolves in water

alkali metals very reactive metals in group 1 of the periodic table

alkaline earth metals reactive metals in group 2 of the periodic table

allele alternate form of a gene for a particular characteristic

allopatric speciation a type of speciation where populations are separated by a geographical barrier

allotrope different forms of a chemical element, for instance different arrangement of atoms in crystalline solids

alloy a mixture of several metals or a mixture of a metal and a non-metal

amino acid an organic compound that forms the building blocks of proteins

amniocentesis a type of prenatal screening in which a sample is collected from the amniotic fluid around a fetus

amygdala the emotional centre of the brain that processes primal feelings, such as fear and rage

amylase enzyme in saliva that breaks down starch into sugar

analyse examine methodically and in detail to answer a question or solve a problem

anions atoms that have gained electrons to become negatively charged ions

anthropologists people who are experts in humans and human remains

anti-anxiety drugs drugs that aim to inhibit anxiety by restoring the balance of certain chemicals in the brain

antipsychotic drugs drugs that treat mental health disorders such as depression, anxiety and schizophrenia

anxiety a natural and usually short-lived reaction to a stressful situation; a disorder whereby anxious thoughts, feelings and physical symptoms occur frequently and persistently, disrupting daily life

apparent magnitude brightness of a star as seen from Earth

aqueous solutions solutions in which water is the solvent

argument value that a function in a spreadsheet will operate on

arousal the state of being physiologically or psychologically activated

arson the criminal act of deliberately setting fire to property

artificial intelligence technologies that are able to perform tasks that usually require human intelligence

artificial selection the process in which humans breed animals or plants in such a way to increase the proportion of desired traits

asexual reproduction reproduction that does not involve fusion of sex cells (gametes)

assailant someone who physically attacks another person

astronomical unit unit of length equal to the mean distance between Earth and the Sun, defined as approximately 150 million kilometres

atmosphere the layer of gases around the Earth

atomic number the number of protons in the nucleus of an atom

atoms very small particles that make up all things

autocratic leadership a leadership style whereby the leader makes all decisions and has complete control

autosomes non-sex chromosomes

average speed total distance travelled divided by time taken

base chemical substance that will react with an acid to produce a salt and water; edible bases taste bitter

beamline a part of a synchrotron that directs radiation through a monochromator and into an experimental station

behaviour therapy a type of therapy that aims to identify and help fix self-destructive or unhealthy behaviours

beliefs feelings or mental acceptance that things are true or real

biased inclined towards a preference or prejudice; not objective; taking personal experiences and ideologies into account

Big Bang theory a theory that states that the Universe began about 15 billion years ago with the explosive expansion of a singularity

Big Chill theory a theory that proposes that the expansion of the Universe will continue indefinitely until stars use up their fuel and burn out

Big Crunch theory a theory that proposes that the Universe will snap back on itself resulting in another singularity

Big Rip theory a theory that proposes that the Universe will rip itself apart due to accelerating expansion

binomial system of nomenclature system devised by Carolus Linnaeus giving organisms two names, the genus and another specific name

biodiversity total variety of living things on Earth

bioinformatics the science of analysing biological data through computers, particularly around genomics and molecular genetics

biologists scientists who study the science of life

biomedical therapy a type of therapy that uses medications to treat mental health disorders

biomes regions of the Earth divided according to dominant vegetation type

biomimicry the practice of developing new products and technologies based on replicating or imitating design in nature

biosilification the ability to manipulate silicon at the nanoscale that involves the use of living cells

biosphere the layer on Earth in which all life exists; includes all living things (biota) and all ecosystems

biota the living things within a region or geological period

bipolar disorder a disorder whereby a person has extreme mood swings between mania and depression

black hole the remains of a star, which forms when the force of gravity of a large neutron star is so great that not even light can escape

blog a personal website or web page where an individual can upload documents, diagrams, photos and short videos

blue shift shift of lines of a spectral pattern towards the blue end when a light source is moving rapidly towards the observer

bonding electrons shared electrons holding two atoms together

booster ring a part of a synchrotron that increases the energy of the electrons

botanists scientists who study the life of plants

braking distance the distance travelled once the brakes have been applied and the vehicle stops

brittle breaks easily into many pieces

calibrate to check or adjust a measuring instrument to ensure accurate measurements

carbon dating a radiometric dating technique that uses an isotope of carbon-14 to determine the absolute age of fossils

carrier an individual heterozygous for a characteristic who does not display the recessive trait

catalase enzyme in the liver involved in the breakdown of hydrogen peroxide, a toxic waste product from cells in the body

catalyst chemical that speeds up reactions but is not consumed in the reaction

cations atoms that have lost electrons to become positively charged ions

cell the smallest unit of life and the building blocks of living things

cellular respiration the chemical reaction involving oxygen that moves the energy in glucose into the compound ATP

centromere a section of a chromosome that links sister chromatids

cerebral cortex the outer layer of the brain, which processes information and is linked to memory and problem solving

cerebral hemispheres the two halves of the cerebral cortex, both responsible for different ways of thinking

Chargaff's rule a rule that states the pairing of adenine with thymine and cytosine with guanine

chemical potential energy energy present in all substances as a result of the electrical forces that hold atoms together

chemical reaction the breaking of chemical bonds between reactant molecules and the forming of new bonds in product molecules

chemists scientists who study the atoms and molecules that make up all substances

chorionic villus sampling (CVS) a type of prenatal screening in which tissue from the chorion is tested during pregnancy

chromatid one identical half of a replicated chromosome

chromosomes tiny thread-like structures inside the nucleus of a cell that contain the DNA that carries genetic information

civil court a court that handles legal disputes, not crimes

climate sensitivity the measure of temperature change in the climate, dependent on the amount of carbon dioxide released into the atmosphere

clones genetically identical copies

coarseness roughness

codon sequence of three bases in mRNA that codes for a particular amino acid

coevolution the process in which two species evolve in partnership so that they depend on each other

cognitive therapies types of therapy that aim to teach people more constructive ways of thinking

collective a collection of two or more people who do not interact with all other members of the collection

combination reaction chemical reaction in which two substances, usually elements, combine to form a compound; also synthesis reaction

combustion reaction chemical reaction in which a substance reacts with oxygen and heat is released

comparative anatomy exploring similarities and differences in the anatomical structures of various species

comparative embryology exploring similarities and differences between embryos of various species

comparison microscope a microscope that allows two samples to be viewed simultaneously

compensation money that is awarded to someone in recognition of loss or injury

complementary base pairs in DNA, specific base pairs will form between the nitrogenous bases adenine (A) and thymine (T) and between the bases cytosine (C) and guanine (G)

complete dominance a type of inheritance where traits are either dominant or recessive

conductors materials that allow electric charge to flow through them

constellations groups of stars that were given a particular name because of the shape the stars seem to form in the sky when viewed from Earth

control an experimental set-up in which the independent variable is not applied

control group a group of participants within an experiment who are not exposed to the variable being tested, and are used to compare to the experimental group

controlled variables   the conditions that must be kept constant throughout an experiment

convection currents   process of heat transfer in gases and liquids in which lighter and hotter materials rise, and cooler and denser materials sink

convergent evolution   tendency of unrelated organisms to acquire similar structures due to similar environmental pressures

corpus callosum   the part of the brain that connects the two hemispheres, allowing them to function interactively

corrosion   chemical reaction that wears away a metal

corrosive   chemical that wears away the surface of substances, especially metals

cosmetology   the study and application of beauty treatment

cosmology   the study of the beginning and end of the Universe

counterfeit   an imitation or fake replica of money or a document

covalent bond   a shared pair of electrons holding two atoms together

covalent compounds   compounds in which the atoms are held together by covalent bonds

criminal profiling   a profile detailing the physical and behavioural traits of a criminal that is used by detectives and police

crossing over   exchange of alleles between maternal and paternal chromosomes

crumple zones   zones in motor vehicles that are deliberately designed to crumple, absorbing and spreading much of the energy transferred to the vehicle during collision

cryotherapy   a form of cold therapy which exposes the body to extremely low temperatures for up to three minutes to promote healing at the cellular level

cult   a social group defined by intense belief and devotion to unusual religious or spiritual beliefs

cuticle   the outer layer of hair

cybernetics   the science of communications and automatic control systems in both machines and living things

cytogenetics   the study of heredity at a cellular level, focusing on cellular components such as chromosomes

dangerous good   chemical that could be dangerous to people, property or the environment

dangerousness   the likelihood of an offender to commit a serious act of violence with little provocation

dark energy   a theoretical form of energy that may exist throughout the Universe

dark matter   a component of the universe that can be detected from its gravitational attraction

decomposition reaction   chemical reaction in which one single compound breaks down into two or more simpler chemicals

deforestation   the process of clearing trees to convert the land for other uses

democratic leadership   a balanced approach to leadership that allows group members to participate in decision making

dendrimer   branched molecules that make up certain nanoparticles

deoxyribonucleic acid (DNA)   a substance found in all living things that contains its genetic information

deoxyribose   the sugar in the nucleotides that make up DNA

dependent variable   the variable that is being affected by the independent variable; that is, the variable you are measuring

depression   lasting and continuous, deeply sad mood or loss of pleasure

desertification   the process in which fertile regions become more dry and arid

diatomic   molecule containing two atoms

diatoms   tiny, single-celled algae

diploid   the possession of two copies of each chromosome in a cell

displacement   the change in position of an object, including a direction

displacement reactions   reactions in which an atom or a set of atoms is displaced by another atom in a molecule; also chemical reactions involving the transfer of electrons from the atoms of a more reactive metal to the ions of a less reactive metal

distress   negative or harmful stress that overwhelms an individual's ability to cope and adapt

divergent evolution   when a population is divided into two or more new populations that are prevented from interbreeding

DNA   a substance found in all living things that contains genetic information

DNA hybridisation a technique that can be used to compare the DNA in different species to determine how closely related they are

DNA ligase an enzyme that joins DNA fragments together

DNA replication process that results in DNA making a precise copy of itself

Doha Amendment change made to the Kyoto Protocol that added new carbon emission targets

dominant a trait (phenotype) that requires only one allele to be present for its expression in a heterozygote

Doppler effect observed change in frequency of a wave when the wave source and observer are moving in relation to each other

double covalent bond a covalent bond involving the sharing of two pairs of electrons

ductile capable of being drawn into wires or threads; a property of most metals

duet rule the tendency of elements that only contain electrons in shell 1 to gain electrons to have two

duty moral obligation or responsibility

ecliptic the path that the Sun traces in the sky during the year

ecosystem community of living things that interact with each other and with the environment in which they live

efficiency the fraction of energy supplied to a device as useful energy

elastic potential energy the potential energy stored in a stretched or compressed object or substance

electrical potential energy energy present in objects or groups of objects in which positively and negatively charged particles are separated

electrolysis the decomposition of a chemical substance by the application of electrical energy

electromagnetic radiation waves produced by the acceleration of an electric charge and that have an electric field and a magnetic field at right angles to each other

electron very light, negatively charged particle inside an atom that moves around the central nucleus

electron configuration an ordered list of the number of electrons in each electron shell, from inner (low energy) to outer (higher energy) shells

electron dot diagrams diagrams using dots to represent the electrons in the outer shell of atoms and to show the bonds between atoms in molecules

electron gun a part of a synchrotron that allows electrons to be fired from a highly-charged tungsten filament

electron shell diagram a diagram showing electrons in their shells around the nucleus of an atom

electrostatic attraction the attraction between two oppositely charged bodies

electrovalency the number of positive or negative charges on an ion

elements pure substances made up of only one type of atom

emigration a type of gene flow in which an individual leaves a population

emission spectra the pattern of coloured lines obtained when emitted light from a sample is observed through a spectroscope

emotion a complex pattern of bodily and mental changes

endogenous stem cells adult stem cells that are tissue-specific

endothermic reaction that absorbs energy

energy the capacity of an object or a substance to do work

engineering the field of science and technology that applies scientific principles to design and build structures

entomologist a person who studies insects

enzyme biological catalyst that speeds up reactions

epigenetics the study of the effect of the environment on the expression of genes

equation statement describing a chemical reaction, with the reactants on the left and the products on the right separated by an arrow

eras divisions of geological time defined by specific events in the Earth's history, divided into periods

ethics the system of moral principles on the basis of which people, communities and nations make decisions about what is right or wrong

eustress positive or beneficial stress that motivates and energises an individual

eutrophication a form of water pollution involving an excess of nutrients leaching from soils

evidence facts or information that can be examined to determine whether a proposition is true

evolution the process in which traits in species gradually change over successive generations

excited state the highest energy state of an atom, in which electrons jump to a higher shell

exothermic reaction that releases energy

experimental group a group of participants within an experiment who are exposed to the variable being tested

experiment station areas positioned around a synchrotron, in which different samples can be analysed by different teams of scientists

extinction complete loss of a species when the last organism of the species dies

fair testing a test that changes only one variable and controls all other variables when attempting to answer a scientific question

falsifiable can be proven false

falsification a credible hypothesis or theory should be able to be tested to be potentially disproved or contradicted by evidence

fault a break in a rock structure causing a sliding movement of the rocks along the break

fermentation a chemical process in which the glucose is broken down to ethanol by the action of yeast

fertilisation penetration of the ovum by a sperm

fibres a thread that makes up a material such as a piece of clothing

flame test a type of test in which an element is heated so the atom moves from a ground state to an excited state, and releases coloured light as the electrons fall back down

fold a layer of rock bent into a curved shape, which occurs when rocks are under pressure from both sides

force an interaction between two objects that can cause a change; commonly a push, pull or twist applied to one object by another, measured in newtons (N)

forensic psychology the application of psychological knowledge and principles to legal issues

forged something that has been faked

formulae set of chemical symbols showing the elements present in a compound and their relative proportions

frequency the number of waves that pass a particular point each second

functions common type of calculation built into spreadsheets

galaxies very large groups of stars and dust held together by gravity

galvanised metal coated with a more reactive metal that will corrode first, preventing metal from corroding

gametes reproductive or sex cells such as sperm or ova

gel electrophoresis a technique used to separate molecules based on their size through an agarose gel

gene segment of a DNA molecule with a coded set of instructions in its base sequence for a specific protein product; when expressed, may determine the characteristics of an organism

gene flow the movement of individuals and their alleles between populations

gene pool the total genetic information of a population, usually expressed in terms of allele frequency

gene therapy technique that uses the modification of genes to treat or prevent disease

genetic diversity variation in genes between members of the same species

genetic drift changes in allele frequency due to chance events such as floods and fires

genetic engineering one type of biotechnology that involves working with DNA

genetic engineers scientists who use special tools to cut, join, copy and separate DNA

genetic genealogy the use of DNA along with other genealogical tests to infer relatedness between individuals

genetic modification the technique of modifying the genome of an organism

genetically modified organism any organism that has had their genome altered in some way

genetics study of genes and inheritance

genome the complete set of genes present in a cell or organism

genome maps maps that describe the order and spacing of genes on each chromosome

genomics the study of genomes

genotype genetic instructions (contained in DNA) inherited from parents at a particular gene locus

geosequestration the process that involves separating carbon dioxide from other flue gases, compressing it and piping it to a suitable site

global positioning system (GPS) device that uses radio signals from satellites orbiting the Earth to accurately map the position of a vehicle or individual

global warming the observed rise in the average near-surface temperature of the Earth

goal something that you want to achieve

graphene   an allotrope of carbon consisting of a single layer of atoms arranged in a two-dimensional honeycomb lattice nanostructure

graphology   the study of handwriting

gravitational potential energy   energy stored due to the height of an object above a base level

greenhouse effect   a natural effect of the Earth's atmosphere trapping heat, which keeps the Earth's temperature stable

ground state   the lowest energy state of an atom

group   collection of two or more people who interact with and influence one another and who share a common purpose

groups   columns of the periodic table containing elements with similar properties

half-life   time taken for half the radioactive atoms in a sample to decay; that is, change into atoms of a different element

halogens   non-metal elements in group 17 of the periodic table

haploid   the possession of one copy of each chromosome in a cell

hazardous substance   chemical that has an effect on human health

heterozygous   a genotype in which the two alleles are different

hippocampus   the area of the brain stem able to transfer information between short- and long-term memory

holobiont   a host and their associated microbiota

hologenome   the sum of genetic information of a host and its microbiota

homologous   chromosomes with matching centromeres, gene locations, sizes and banding patterns

homologous structures   body structures that can perform a different function but have a similar structure due to their evolutionary origin

homology   similar characteristics that result from common ancestry

homozygous   a genotype in which the two alleles are identical

homozygous dominant   a genotype where both alleles for the dominant trait are present

homozygous recessive   a genotype where both alleles for the recessive trait are present

human error   mistakes made by the person performing the investigation

hydrosphere   Earth's water from the surface, underground and air

hypothesis   a suggested, testable explanation for observations or experimental results; it acts as a prediction for the investigation

hypoxic   a condition in which oxygen levels are reduced

ice cores   samples of ice extracted from ice sheets containing a build-up of dust, gases and other substances trapped over time

immigration   a type of gene flow in which an individual enters a population

independent variable   the variable that you deliberately change to during an experiment observe its effect on another variable

indicator   substance that changes colour when it reacts with acids or bases; the colour shows how acidic or basic a substance is

induced mutation   a mutation of DNA in which the cause can be identified

inductionism   a theory stating that with enough evidence, scientific theories can become universal laws

inertia   property of objects that makes them resist changes in their motion

infra-red radiation   invisible radiation emitted by all warm objects

inheritance   genetic transmission of characteristics from parents to offspring

insomnia   a sleep disorder whereby a person has difficulty either falling asleep or staying asleep

instantaneous speed   speed at any particular instant of time

intelligence   has no agreed definition — some definitions include: 'the ability to learn and solve problems' and 'the capacity for logic, understanding, self awareness and creativity'

ionic bond   an attractive force between ions with opposite electrical charge

ionic compounds   compounds containing positive and negative ions held together by the electrostatic force

ions   atoms that have lost or gained electrons and carry an electrical charge

isotopes   atoms of the same element that differ in the number of neutrons in the nucleus

judgements   opinions formed after considering available information

karyotype   an image that orders chromosomes based on their size

kinesics the study of body language as a type of non-verbal communication

kinetic energy energy due to the motion of an object

kinetochore a region on a chromosome associated with cell division

kleptoplasts captured plastids

kleptoplasty a process in which one organism captures the plastids from another organism and integrates them into its cellular structure

Kyoto Protocol an international agreement with the goal of reducing the amount of greenhouse gases produced by industrialised nations

laissez-faire leadership a person-oriented approach to leadership style whereby the leaders are friendly and helpful but do not have direct control over the group

landfills areas set aside for the dumping of rubbish

latent fingerprints fingerprints that are invisible to the human eye

lattice a structure in which particles are bonded to each other to form crystal structures

Law of Conservation of Energy law that states that energy cannot be created or destroyed

leaching process of using water or another liquid to remove soluble matter from a solid

liability an obligation, responsibility, hindrance or something that causes a disadvantage

light gates gates used to measure the speed of an object using a timer and a detector

light-year the distance travelled by light in one year

linac the linear accelerator in a synchrotron that accelerates the electrons to 99.9987% of the speed of light

linkage analysis use of markers to scan the genome and map genes on chromosomes

lithosphere the outermost layer of the Earth; includes the crust and uppermost part of the mantle

locus position occupied by a gene on a chromosome

logbook a complete record of an investigation from the time a search for a topic is started

long-term storage a function of the brain that enables memories to be stored, sorted and retrieved

lotus effect the repellence of water on the surface of lotus leaves; also known as superhydrophobicity

lustre the shiny appearance of many metals

magnetic field area where a magnetic force is experienced

magnitude size

main sequence area on the Hertzsprung–Russell where the majority of stars are plotted that produce energy by fusing hydrogen to form helium

malleable able to be beaten, bent or flattened into shape

mania an elevated mood involving intense elation or irritability

mass the amount of matter in an object, measured in kilograms (kg)

mass number the number of protons and neutrons in the nucleus of an atom

maternal chromosomes chromosomes from the ovum

meiosis cell division process that results in new cells with half the number of chromosomes of the original cell

Mendelian inheritance an inheritance pattern in which a gene inherited from either parent segregates into gametes at an equal frequency

meniscus the curve seen at the top of a liquid in response to its container

messenger RNA (mRNA) single-stranded RNA transcribed from a DNA template that then carries the genetic to a ribosome to be translated into a protein

metagenomics technology that combines DNA sequencing with molecular and computational biology

metalloids elements that have features of both metals and non-metals

metals elements that conduct heat and electricity and are shiny solids that can be made into thin wires and sheets

mineral ores rocks mined to obtain a metal or other chemical within them

mitosis cell division process that results in new genetically identical cells with the same number of chromosomes as the original cell

mnemonic a strategy to help you to remember things

molecular biology the study of the structure and composition of molecules within a cell

molecular clock a technique that uses known mutation rates of biomolecules to determine the time two species diverged

molecular formula   a shorthand statement of the elements in a molecule showing the relative number of atoms of each kind of element

molecular genetics   study of genetics at a molecular level

molecular imprinting   a process by which molecules that match a target molecule can be made

molecular ions   groups of atoms that have an overall charge and are treated as an entity; e.g. OH^-, SO_4^{2-}, NH_4^+

molecules   small particles of a chemical substance consisting of one or more types of atoms

monohybrid ratio   the 3:1 ratio of a particular characteristic for offspring produced by heterozygous parents, controlled by autosomal complete dominant inheritance

monomers   molecules that are the building blocks of larger molecules known as polymers, usually containing carbon and hydrogen, and sometimes other elements

monosomy   a condition where there is only one copy of a particular chromosome instead of two

Montreal Protocol   an international agreement with the goal of reducing and eliminating the use of substances such as CFCs that contribute to the depletion of the ozone layer

multipotent stem cells   stem cells that can give rise to only certain cell types

mutagen   agent or factor that can induce or increase the rate of mutations

mutated gene   the gene with an altered genetic sequence

mutations   changes to DNA sequence, at the gene or chromosomal level

nanobots   miniscule sized robots on a nanoscale that can be used in medicine

nanometer   one billionth of a metre

nanotechnology   a rapidly developing field that includes studying and investigating our environment at a molecular level (on the nanoscale) and then applying what we learn to a variety of technologies

nanotubes   cylindrical molecules usually made of a single layer of carbon atoms

nanowires   extremely thin wires on the nanoscale

natural selection   process in which organisms better adapted for an environment are more likely to pass on their genes to the next generation

nebulae   clouds of dust and gas that may be pulled together by gravity and heat up to form a star

need   something that you require

net force   the net (total or resultant) force of two or more forces acting on an object

network lattice   a lattice in which each atom is covalently bonded to adjacent atoms

neutralisation reaction   between an acid and a base

neutron   uncharged particle found in the nucleus of an atom

neutron star   extremely dense remnants of a supernova in which protons and electrons in atoms are fused to form neutrons

newton   unit of force defined as the force required to provide a mass of 1 kg with an acceleration of 1 m/s^2

Newton's First Law of Motion   law that states that an object remains at rest or continues to move with the same speed in the same direction unless acted on by an outside, unbalanced force

Newton's Second Law of Motion   law that states that the acceleration of an object is equal to the total force on the object divided by its mass

Newton's Third Law of Motion   law that states that forces are created in pairs as they are an interaction between two objects

nitrogenous base   a component of nucleotides that may be one of adenine, thymine, guanine, cytosine or uracil

noble gases   elements in the last column of the periodic table that are unreactive

non-bonding pairs   electron pairs not involved in covalent bonding (also known as lone pairs)

non-metals   elements that do not conduct electricity or heat and usually have a lower boiling and melting point than metals

non-rapid eye movement sleep (NREM sleep)   stages of sleep characterised by the absence of rapid eye movements

nuclear energy   the energy stored at the centre of atoms, the tiny particles that make up all substances

nuclear fusion   joining together of the nuclei of lighter elements to form another element, with the release of energy

nuclear reactions reactions involving the breaking of bonds between the particles (protons and neutrons) inside the nuclei of atoms

nucleic acids molecules composed of building blocks called nucleotides, which are linked together in a chain

nucleotides compounds (DNA building blocks) containing a sugar part (deoxyribose or ribose), a phosphate part and a nitrogen-containing base that varies

nucleus the central part of an atom, made up of protons and neutrons; contains DNA and acts as the control centre for the cell

nylon a group of polymers called polyamides used in clothing

octet rule the tendency of atoms to gain or lose electrons to have eight in their outer shell

odontologists scientists who study the structure and diseases of teeth

opinions personal views or judgements about something

ova female reproductive cells or eggs

oxidation chemical reaction involving the loss of electrons by a substance

palaeoclimate the climate at a specific point in geological history

pandemic a widespread outbreak of a disease that affects multiple countries in different regions globally

paradigms generally accepted perspectives, ideas or theories at a particular time

paralanguage non-verbal parts of communication; for example, the way something is said, rather than the words that are used

parallax apparent movement of close stars against the background of distant stars when viewed from different positions around Earth's orbit

parsec unit for expressing distances to stars and galaxies, it represents the distance at which the radius of Earth's orbit subtends an angle of one second of arc

partial prints a small and incomplete section of a full fingerprint

patent fingerprints fingerprints that leave a visible mark on a surface

paternal chromosomes chromosomes carried in the sperm

pathologists people who study the causes and effects of diseases, examining samples for diagnostic or forensic purposes

pedigree chart diagram showing the family tree and a particular inherited characteristic for family members

peer-reviewed evaluated scientifically, academically, or professionally worked by others working in the same field

periodic law elements with similar properties occur at regular intervals when all elements are listed in order of atomic weight

periodic table the table listing all known elements, grouped according to their properties and in order of atomic number (number of protons in their nucleus)

periods subdivisions of geological time, divided into epochs; also rows of elements on the periodic table

permafrost ground that is frozen for at least two continuous years

personal space the space within a small distance of a person

personality the combination of characteristics or qualities that form an individual's distinctive character

pH scale the scale from 1 (acidic) to 14 (basic) that measures how acidic or basic a substance is

phenotype characteristics or traits expressed by an organism that result from the expression of genes, which may also be influenced by the environment

phobia an intense or irrational fear of something that poses little or no threat

photochromic describes lenses made from glass that darkens in bright light

photosynthesis food-making process in plants that takes place in chloroplasts, and uses carbon dioxide, water and energy from the Sun

phyletic evolution when a population of a species progressively changes over time to become a new species

placebo a medicine or procedure that has no therapeutic effect, used as a control in testing new drugs

planetary nebula ring of expanding gas caused by the outer layers of a star less than eight times the mass of our sun being thrown off into space

plastic synthetic substance capable of being moulded

platelet rich plasma blood plasma which is rich in a patient's own platelets that can be used to treat injuries and in cosmetology

pluripotent stem cells stem cells that can give rise to most cell types

point mutation a mutation at one particular point in the DNA sequence, such as a substitution or single base deletion or insertion

polyatomic ions ions containing more than one type of atom; another term for a molecular ions, in which a group of atoms have an overall charge

polyethylene material formed from the monomer ethylene

polygraph a machine that attempts to detect deceptive behaviour by measuring arousal

polymerase chain reaction (PCR) a process which amplifies small amounts of DNA

polymerisation chemical reaction joining monomers in long chains to form a polymer

polymers molecules made of repeating subunits of monomers joined together in long chains

polyvinyl chloride polymer formed from the monomer vinyl chloride; also known as PVC

population members of one species living together in a particular place at the same time; the entire group of people who are being studied from which a sample is selected

position the location of an object

positrons positively charged electrons

potassium–argon dating a radiometric dating technique based on the measure ratio of potassium-40 to argon-40

potential energy energy that is stored due to the position or state of an object

precipitate solid product of a chemical reaction that is insoluble in water

precipitation reaction chemical reaction in which there is a water insoluble product

precise multiple measurements of the same investigation being close to each other

prenatal screening testing a fetus during pregnancy to detect any abnormalities

prevailing winds winds most frequently observed in each region of the Earth

prevalence the percentage or proportion of the population that have a certain illness or disorder

proteins molecules, such as enzymes, haemoglobin and antibodies made up of amino acids

proton a positively charged particle found in the nucleus of an atom

proximate rules rules that govern the physical distance that is comfortable between people

pseudosciences fields that are not sciences, despite having scientific sounding names or claiming to be scientific

psychiatrists medical doctors that specialise in the diagnosis and treatment of medical illness

psychological therapies structured interaction between a professional and a client to help overcome psychological problems

psychologists an expert or specialist in psychology

psychology the systematic study of thoughts, feelings and behaviours

psychopathology the scientific study of mental health disorders

pulsating star star that periodically expands due to an increase in core temperature and then cools and contracts under gravity

Punnett square a diagram used to predict the outcome of a genetic cross

radiant heat heat transferred by radiation, such as from the Sun to the Earth

radio telescope a telescope that can detect radio waves from distant objects

radiometric dating a technique in which radioactive substances are used to calculate the age of rocks or dead plants and animals

random error an error that affects the precision of results

rapid eye movement sleep (REM sleep) a stage of sleep characterised by the repetitive, brief and erratic movements of the eyeballs under eyelids

rate how a physical quantity changes with respect to time

reaction distance the distance travelled once a hazard has been spotted, to just before applying the brakes

recessive a trait (phenotype) that will only be expressed in the absence of the allele for the dominant trait

recombinant DNA a molecule of DNA that contains fragments from more than one source

recombinant DNA technology technology that can form DNA that does not exist naturally, by combining DNA sequences that would not normally occur together

record the first step in forming a memory

red giant star in the late stage of its life in which helium in its core is fused to form carbon and other heavier elements

red shift shift of lines of a spectral pattern towards the red end when the source is moving away from the observer

redox oxidation and reduction considered together

reduction a chemical reaction involving the gain of electrons by a substance

reflectiveness being self-aware, open-minded and sometimes standing back and looking at the big picture

reforestation planting trees in forest areas that have been destroyed or damaged

relative age age of a rock compared with the age of another rock

relative atomic mass mass of the naturally occurring mixture of the isotopes of an element

relative dating method of dating that determines the age of a rock layer by relating it to another layer using superposition and the fossils contained

relative referencing used in a spreadsheet when the cell address in the formula is changed

reliable data consistent data that is achieved when an investigation is replicated

remote sensing data collection about Earth's biosphere completed from space by devices such as satellites

replication repeating of an experiment to make sure you have collected reliable data

resilience the ability to tolerate feeling a little uncomfortable sometimes

resourcefulness taking responsible risks and using a range of appropriate learning tools and strategies

responsibility being able to manage yourself and your learning

restriction enzymes enzymes that cut DNA at specific base sequences (recognition sites)

restriction fragment length polymorphisms (RFLPs) variations in the lengths of DNA fragments in individuals with different alleles of a gene

reticular activation system a part of the brain that filters incoming information on the basis of importance

retrieve the ability to recall memories

retroposons segments of DNA that can break off a chromosome and paste themselves elsewhere in the genome

ribonucleic acid (RNA) a type of nucleic acid that contains ribose sugar

ribose the sugar found in nucleotides of RNA

ribosomal RNA (rRNA) a special type of RNA that forms the structure of ribosomes

ribosome organelle found in the cells of all organisms in which translation occurs

right something you feel that you are entitled to

risk assessment a procedure that identifies the potential hazards of an experiment and gives protective measures to minimise the risk; a document outlining these potential hazards and protective measures

safety data sheet (SDS) document containing important information about hazardous chemicals

salt one of the products of the reaction between an acid and a base

sample a smaller group of individuals selected from a larger group that has been chosen to be studied

scalar quantity any quantity that is described only by a magnitude

schizophrenia a disorder that affects a person's ability to think, feel and behave clearly

scientific method a systematic and logical process of investigation to test hypotheses and answer questions based on data or observations

scientific notation a way of writing very large or small numbers, using power notation; e.g. 1.5×10^8 km

selective agents the different living (biotic) and non-living (abiotic) agents that influence the survival of organisms

selective pressures factors that contribute to selecting which variations will provide the individual with an increased chance of surviving over others

self-identity the personal awareness of one's self and where they fit into society

sensory register a system in the brain that filters incoming information on the basis of importance

sex chromosomes chromosomes that determine the sex of an organism

sex-linked inheritance an inherited trait coded for by genes located on sex chromosomes

sexual reproduction reproduction that involves the joining together of male and female gametes

shells energy levels surrounding the nucleus of an atom into which electrons are arranged

silica a hard colourless mineral; also known as silicon dioxide

single covalent bond a covalent bond involving the sharing of one pair of electrons

single displacement reaction reaction in which one element is substituted for another element in a compound

single nucleotide polymorphisms (SNPs) genetic differences between individuals that can result from single base changes in their DNA sequences

singularity a single point of immense energy present at the time of the Big Bang

sister chromatids identical chromatids on a replicated chromosome

sleep disorders problems that disrupt the normal sleep cycle

sleep phenomenon behaviours that occur at night including sleepwalking, sleep talking, nightmares and night terrors

smelting process to obtain a metal from its ore; the production of a metal in its molten state

somatic cells cells of the body that are not sex cells

somnambulism another word for sleepwalking

sonic motion detector device that sends out pulses of ultrasound at a frequency of about 40 kHz and then detects the reflected pulses from the moving object

speciation formation of new species

species taxonomic unit consisting of organisms capable of mating and producing viable and fertile offspring

species diversity the number of different species within an ecosystem

spectroscope an instrument that breaks up the light it receives into distinct frequencies, colours or wavelengths

spectrum light from a source separated out into the sequence of colours, showing the different frequencies

speed a measure of the rate at which an object moves over a distance

speedometer the instrument that displays the speed of a vehicle such as a car or truck

sperm male reproductive cell

spontaneous mutation a mutation of DNA that cannot be explained or identified

Steady State theory a theory that states that there was no beginning to the Universe and that the Universe does not change in appearance

stem cells special cells that are able to develop into many different cell types

stopping distance the sum of the reaction and braking distances

storage ring a part of a synchrotron where electrons are trapped in circular orbits by large magnets that steer them around

store storage of a memory in the appropriate part of the brain

stratosphere the second layer of the atmosphere up to about 55 km above the Earth's surface, between the troposphere and the mesosphere

structural formula a diagram showing the arrangement of atoms in a substance with covalent bonds drawn as dashes

substrate a molecule acted upon by an enzyme

supercomputers computers with the fastest processing capacity available

supergiant very large star that is expanding while running out of fuel, and will eventually explode

supernova huge explosion that happens at the end of the life cycle of supergiant stars

suspects people who are thought to be guilty of a crime

symbiotic a very close relationship between two organisms of different species

symbols one- or two-letter code(s) used for the elements

sympatric speciation a type of speciation where populations are separated by a barrier that is not geographical

symptoms physical or psychological features of a disease or disorder

synchrotron a building-sized device that uses electrons accelerated to near the speed of light to produce intense electromagnetic radiation. This is used to produce data that can describe objects as small as a single molecule.

synthetic manufactured by people

systematic desensitisation a type of therapy that aims to desensitise people to their phobias

systematic error an error (often due to equipment) that affects the accuracy of results

telomerase an enzyme involved in maintaining and repairing a telomere

telomere a cap of DNA on the tip of a chromosome that enables DNA to be replicated safely without losing valuable information

terraforming the modification of an environment, including its atmosphere and temperature, to make a location habitable for Earth-like organisms

testable able to be supported or proven false through the use of observations and investigation

thalamus the part of the brain at the top of the brain stem that passes sensory information to key areas in the brain for processing

theory a well-supported explanation of a phenomenon based on facts that have been obtained through investigations, research and observations

thrust forward push

totipotent stem cells the most powerful stem cells that can rise to all cell types

transcription the process by which the genetic message in DNA is copied into a mRNA molecule

transfer RNA (tRNA) molecules located in the cytosol that transport specific amino acids to complementary mRNA codons in the ribosome

transgenic organism an organism with genetic information from another species in its genome

transition metal block a block of metallic elements in the middle of the periodic table

translation the process in which amino acids are joined in a ribosome to form a protein

transposition the ability of a gene to change position on the chromosome

transposons a section of chromosome that moves about the chromosome within a cell through the method of transposition

triple covalent bond a covalent bond involving the sharing of three pairs of electrons

triplet a sequence of three nucleotides in DNA that can code for an amino acid

trisomy a condition where there are three copies of a particular chromosome instead of two

troposphere the layer of the atmosphere closest to the Earth's surface, containing closely packed particles of air

ultraviolet light light that is invisible to the human eye but can be used to make certain substances glow

ultraviolet radiation invisible radiation similar to light but with a slightly higher frequency and more energy

universal indicator mixture of indicators that changes colour as the strength of an acid or base changes, indicating the pH of the substance

valence electrons outer shell electrons involved in chemical reactions

valency the number of electrons that each atom needs to gain, lose or share to 'fill' its outer shell

valid sound or true experiment that can be supported by other scientific investigations

values a deep commitment to a particular issue and serve as standards for decision making

Van der Waals forces very weak molecular interactions

variation differences between cells or organisms

vasoconstriction narrowing or constriction of the blood vessels

vector quantity any quantity that is described by both a magnitude and direction

velocity a measure of rate of change in position, with respect to time, described by a magnitude and direction

vestigial structures structures with no apparent function that are remnants from a past ancestor

wavelength the distance between the crest of one wave and the next; also invisible radiation similar to light but with a slightly higher frequency and more energy

weight a commonly used word to describe the force due to gravity on an object

weighted mean the average mass of an element that is calculated from the percentage of each isotope in nature

white dwarf the core remaining after a red giant has shed layers of gases

work transfer in energy that occurs when a force moves an object in the direction of the force

X-rays high energy electromagnetic waves that can be transmitted through solids and provide information about their structure

xenotransfusion a process in which other animals are used for blood transfusions for humans

xenotransplant a process in which organs for transplantation are taken from animals other than humans

zeolite crystalline substance consisting of aluminium, silicon and oxygen that is used as a catalyst to break up large molecules in crude oil

zygote a cell formed by the fusion of male and female reproductive cells

INDEX